AF720508

Biocomposite Materials for Wastewater Remediation

Biocomposite Materials for Wastewater Remediation

Preparation, Properties and Applications

Edited by

Rachid Bouhfid
International Water Research Institute (IWRI)
Mohammed VI Polytechnic University
Ben Guerir, Morocco

Nadia Zari
Moroccan Foundation for Advanced Science
Innovation and Research (MAScIR)
Mohammed VI Polytechnic University
Ben Guerir, Morocco

Abou el kacem Qaiss
College of Chemical Sciences and Engineering (CCSE)
Department of Materials Science
Energy and Nano-engineering (MSN)
Mohammed VI Polytechnic University (UM6P)
Ben Guerir, Morocco

WILEY

Copyright © 2026 by John Wiley & Sons, Inc. All rights reserved, including rights for text and data mining and training of artificial intelligence technologies or similar technologies.

Published by John Wiley & Sons, Inc., Hoboken, New Jersey.

No part of this publication may be reproduced, stored in a retrieval system, or transmitted in any form or by any means, electronic, mechanical, photocopying, recording, scanning, or otherwise, except as permitted under Section 107 or 108 of the 1976 United States Copyright Act, without either the prior written permission of the Publisher, or authorization through payment of the appropriate per-copy fee to the Copyright Clearance Center, Inc., 222 Rosewood Drive, Danvers, MA 01923, (978) 750-8400, fax (978) 750-4470, or on the web at www.copyright.com. Requests to the Publisher for permission should be addressed to the Permissions Department, John Wiley & Sons, Inc., 111 River Street, Hoboken, NJ 07030, (201) 748-6011, fax (201) 748-6008, or online at http://www.wiley.com/go/permission.

The manufacturer's authorized representative according to the EU General Product Safety Regulation is Wiley-VCH GmbH, Boschstr. 12, 69469 Weinheim, Germany, e-mail: Product_Safety@wiley.com.

Trademarks: Wiley and the Wiley logo are trademarks or registered trademarks of John Wiley & Sons, Inc. and/or its affiliates in the United States and other countries and may not be used without written permission. All other trademarks are the property of their respective owners. John Wiley & Sons, Inc. is not associated with any product or vendor mentioned in this book.

Limit of Liability/Disclaimer of Warranty: While the publisher and the authors have used their best efforts in preparing this work, including a review of the content of the work, neither the publisher nor the authors make any representations or warranties with respect to the accuracy or completeness of the contents of this work and specifically disclaim all warranties, including without limitation any implied warranties of merchantability or fitness for a particular purpose. No warranty may be created or extended by sales representatives, written sales materials or promotional statements for this work. The fact that an organization, website, or product is referred to in this work as a citation and/or potential source of further information does not mean that the publisher and authors endorse the information or services the organization, website, or product may provide or recommendations it may make. This work is sold with the understanding that the publisher is not engaged in rendering professional services. The advice and strategies contained herein may not be suitable for your situation. You should consult with a specialist where appropriate. Further, readers should be aware that websites listed in this work may have changed or disappeared between when this work was written and when it is read. Neither the publisher nor authors shall be liable for any loss of profit or any other commercial damages, including but not limited to special, incidental, consequential, or other damages.

For general information on our other products and services or for technical support, please contact our Customer Care Department within the United States at (800) 762-2974, outside the United States at (317) 572-3993 or fax (317) 572-4002.

Wiley also publishes its books in a variety of electronic formats. Some content that appears in print may not be available in electronic formats. For more information about Wiley products, visit our web site at www.wiley.com.

Library of Congress Cataloging-in-Publication Data Applied for:

Hardback ISBN: 9781394246458

Cover Design: Wiley
Cover Image: Courtesy of Rachid Bouhfid

Set in 9.5/12pt STIXTwoText by Straive, Chennai, India

This book is dedicated to the soul of my beloved son, Adam Bouhfid.
Brilliant in mind, strong in spirit, boundless in kindness.
You illuminated my path in life, and your light lives within my heart forever.

Adam Bouhfid

July 9, 2009–May 17, 2023

إلى روح ابني الحبيب، آدم بوحفيظ

Contents

List of Contributors

Yaser Açıkbaş
Department of Materials Science and Nanotechnology Engineering
Usak University, Usak, Turkey

Tania Ali
Nanosciences & Technology Department (NS&TD), National Centre for Physics (NCP)
Quaid-i-Azam University, Islamabad
Pakistan

and

Department of Chemistry
Hazara University, Mansehra, Pakistan

Abdelhay El Amri
Mohammed VI Polytechnic University
Ben Guerir, Morocco

Kenza Bahsaine
Moroccan Foundation for Advanced Science, Innovation and Research (MAScIR)
Mohammed VI Polytechnic University
Ben Guerir
Morocco

and

Laboratory of Analytical Chemistry
Faculty of Medicine and Pharmacy
Mohammed V University
Rabat
Morocco

Hanane Benzeid
Laboratory of Analytical Chemistry
Faculty of Medicine and Pharmacy
Mohammed V University
Rabat
Morocco

Rachid Bouhfid
International Water Research Institute (IWRI)
Moroccan Foundation of Advanced Science Innovation and Research
MAScIR
Mohammed VI Polytechnic University
Ben Guerir, Morocco

Chunguang Chen
Department of Chemistry, School of Materials and Chemistry
University of Shanghai for Science and Technology, Shanghai, China

Xiaxia Chen
College of Chemical Engineering
Qingdao University of Science and Technology, Qingdao, China

Soulaima Chkirida
College of Chemical Sciences and Engineering (CCSE)
Department of Materials Science Energy and Nano-engineering (MSN)
Mohammed VI Polytechnic University (UM6P)
Ben Guerir
Morocco

Zeynep Ciğeroğlu
Department of Chemical Engineering
Faculty of Engineering
Usak University, Usak, Turkey

Yilei Ding
Department of Chemistry, School of Materials and Chemistry
University of Shanghai for Science and Technology, Shanghai, China

Subhrata Dutta
Department of Chemistry, Faculty of Paramedical Science
Assam Down Town University
Guwahati, Assam, India

André Luís Rossoni Ferreira
Physical Organic Chemistry Group
Department of Chemistry
Institute of Exact and Biological Sciences, Ouro Preto, Brazil

and

Graduate Program in Environmental Engineering
School of Mines, Ouro Preto, Brazil

Brener Felipe Melo Lima Gomes
Physical Organic Chemistry Group
Department of Chemistry
Institute of Exact and Biological Sciences, Ouro Preto, Brazil

and

Graduate Program in Environmental Engineering
School of Mines, Ouro Preto, Brazil

Asiya Gul
Nanosciences & Technology Department (NS&TD), National Centre for Physics (NCP)
Quaid-i-Azam University, Islamabad
Pakistan

and

Department of Chemistry
Hazara University, Mansehra, Pakistan

Leandro Vinícius Alves Gurgel
Physical Organic Chemistry Group
Department of Chemistry
Institute of Exact and Biological Sciences, Ouro Preto, Brazil

Md Enamul Hoque
Department of Biomedical Engineering
Military Institute of Science and Technology, Dhaka, Bangladesh

Aziz Ihammi
International Water Research Institute (IWRI)
Mohammed VI Polytechnic University
Ben Guerir, Morocco

Bushra Iqbal
Institute of Chemistry
University of Okara, Okara, Pakistan

Dipika Jaspal
Symbiosis Institute of Technology (SIT)
Symbiosis International (Deemed University), (SIU), Pune, Maharashtra India

Abdullah Al Jobair
Department of Civil and Environmental Engineering
Islamic University of Technology
Gazipur, Bangladesh

Sadia Khalid
Nanoscience & Technology Department, National Centre for Physics
Quaid-i-Azam University, Islamabad Pakistan

Abdellah Lacherai
Laboratory of Applied Chemistry and Environment, Faculty of sciences
Ibn Zohr University, Agadir, Morocco

Feihu Li
Collaborative Innovation Center of Atmospheric Environment and Equipment Technology, Jiangsu Key Laboratory of Atmospheric Environment Monitoring and Pollution Control
School of Environmental Science and Engineering, Nanjing, China

Chao Liu
College of Chemical Engineering
Qingdao University of Science and Technology, Qingdao, China

Mengyu Ma
Collaborative Innovation Center of Atmospheric Environment and Equipment Technology, Jiangsu Key Laboratory of Atmospheric Environment Monitoring and Pollution Control
School of Environmental Science and Engineering, Nanjing, China

Prashant Mahajan
Symbiosis Institute of Technology (SIT)
Symbiosis International (Deemed University), (SIU), Pune, Maharashtra India

Anca Mazare
Department of Materials Science and Engineering
Friedrich Alexander University of Erlangen Nürnberg, Erlangen Germany

Fanqing Meng
College of Chemical Engineering
Qingdao University of Science and Technology, Qingdao, China

Noureddine El Messaoudi
Laboratory of Applied Chemistry and Environment, Faculty of sciences
Ibn Zohr University, Agadir, Morocco

Youssef Miyah
Laboratory of Materials, Processes Catalysis, and Environment, Higher School of Technology
University Sidi Mohamed Ben Abdellah, Fez, Morocco

and

Ministry of Health and Social Protection
Higher Institute of Nursing Professions and Health Techniques, Fez, Morocco

Abdelaziz El Mouden
Laboratory of Applied Chemistry and Environment, Faculty of sciences
Ibn Zohr University, Agadir, Morocco

Nadeem Muhammad
School of Life Science
Wuchang University of Technology
Wuhan, China

and

Laboratory for Chemical Technology
Ghent University, Ghent, Belgium

Amara Nasir
Department of Physics
The University of Lahore
Pakistan

Laila Noureen
MTA-SZTE Lendület Biocolloids Research Group, Department of Physical Chemistry and Materials Science, Interdisciplinary Excellence Centre
University of Szeged, Szeged, Hungary

Abdeladim Oulguidoum
Moroccan Foundation for Advanced Science, Innovation and Research (MAScIR)
Mohammed VI Polytechnic University
Ben Guerir, Morocco

Sathish Kumar Palaniappan
Natural Composites Research Group Lab, Department of Materials and Production Engineering, The Sirindhorn International Thai-German Graduate School of Engineering
King Mongkut's University of Technology North Bangkok, Bangsue
Thailand

Vandana Patyal
Symbiosis Institute of Technology (SIT)
Symbiosis International (Deemed University), (SIU), Pune, Maharashtra
India

P.K. Praseetha
Department of Nanotechnology, Noorul Islam Centre for Higher Education, Kumaracoil, Tamil Nadu
India

Nusrat Zahan Ramesha
Department of Mechanical Engineering, Military Institute of Science and Technology, Dhaka
Bangladesh

K. Ravichandran
PG & Research Department of Physics
AVVM Sri Pushpam College (Affiliated to Bharathidasan University Tiruchirappalli), Thanjavur
Tamil Nadu, India

Ismaila Olalekan Saheed
Green Analytical Chemistry Laboratory, School of Chemical Sciences, Minden, Malaysia
Department of Chemistry and Industrial Chemistry, Kwara State University, Ilorin, Nigeria

Zeynep Mine Şenol
Sivas Cumhuriyet University, Faculty of Health Sciences, Department of Nutrition and Diet, Sivas, Turkey

R. Shalini
PG & Research Department of Physics
AVVM Sri Pushpam College (Affiliated to Bharathidasan University Tiruchirappalli), Thanjavur
Tamil Nadu, India

Faiz Bukhari Mohd Suah
Green Analytical Chemistry Laboratory, School of Chemical Sciences, Minden, Malaysia

A.R. Umayal Sundari
Department of Physics, Periyar Maniammai Institute of Science and Technology (Deemed to be University)
Thanjavur, Tamil Nadu, India

Muhammad Tayyab
Tsinghua University, Shenzhen, China

M. Varshini
PG & Research Department of Physics
AVVM Sri Pushpam College (Affiliated to Bharathidasan University Tiruchirappalli), Thanjavur
Tamil Nadu, India

A. Viji
Department of Physics, Kongunadu College of Engineering and Technology (Affiliated to Anna University), Thottiyam, Tamil Nadu, India

Yinglong Wang
College of Chemical Engineering
Qingdao University of Science and Technology, Qingdao, China

Ranran Wei
College of Chemical Engineering
Qingdao University of Science and Technology, Qingdao, China

Xuewen Wei
College of Chemical Engineering
Qingdao University of Science and Technology, Qingdao, China

Hongwei Xu
College of Chemical Engineering
Qingdao University of Science and Technology, Qingdao, China

Congfei Yao
Department of Chemistry, School of Materials and Chemistry, University of Shanghai for Science and Technology
Shanghai, China

Junkai Zhang
Department of Chemistry, School of Materials and Chemistry, University of Shanghai for Science and Technology
Shanghai, China

Xiaoying Zhang
College of Chemical Engineering
Qingdao University of Science and Technology, Qingdao, China

Nadia Zari
Moroccan Foundation of Advanced Science Innovation and Research
MAScIR
Mohammed VI Polytechnic University
Ben Guerir, Morocco

Preface

Water scarcity and water pollution are no longer abstract global concerns; they are concrete challenges affecting societies, industries, and ecosystems worldwide. Accelerated urban growth, industrial activity, agricultural practices, and climate-related pressures have significantly altered the quality and availability of freshwater resources. While conventional wastewater treatment technologies have played a crucial role in protecting public health and the environment, their reliance on energy-intensive processes and non-renewable materials raises legitimate concerns regarding long-term sustainability.

Against this backdrop, biocomposite materials have gained increasing attention as credible and forward-looking alternatives for wastewater remediation. By combining natural polymers, bio-derived carbons, renewable fibers, and functional nanomaterials, biocomposites offer a unique balance between environmental responsibility and technical performance. Their structural versatility allows them to be tailored for adsorption, membrane filtration, photocatalysis, antimicrobial control, and hybrid treatment systems addressing complex wastewater matrices.

This edited book, *Biocomposite Materials for Wastewater Treatment: Principles, Technologies, and Future Perspectives*, aims to present a coherent and critical overview of recent scientific advances in this rapidly evolving field. The opening chapters establish the foundation by examining natural fiber-based biocomposites, chitosan-based materials, lignin valorization, and lignocellulosic systems, with particular emphasis on sustainability, material availability, and adsorption-driven treatment mechanisms relevant to municipal and industrial wastewater.

Subsequent contributions focus on cellulose-derived materials, including cellulose nanomaterials, nanocrystalline cellulose biocomposites, and advanced cellulose-based membranes. These chapters highlight how surface modification, nanostructuring, and functionalization strategies directly influence separation efficiency, selectivity, fouling resistance, and long-term material stability.

Alginate-based systems form another central theme of this book. Dedicated chapters address alginate biocomposites, alginate–titania (TiO_2) composites, and alginate nanocomposite membranes, illustrating their adaptability in adsorption- and photocatalysis-based treatment processes for heavy metals, dyes, pharmaceuticals, and microbial contaminants.

The volume further explores multifunctional and emerging biocomposite systems, including graphene bio-nanocomposites, magnetic biocomposites, and biochar-based materials. These chapters demonstrate how the integration of carbon nanostructures, magnetic components, and biomass-derived carbons can enhance pollutant removal, enable material recovery, and support the development of hybrid and process-integrated wastewater treatment technologies.

Membrane-based approaches are examined in detail through chapters dedicated to chitosan–clay composite membranes, nanocellulose biocomposite membranes, and photocatalytic–photothermal membranes for microbial remediation. Particular attention is given to antibiofouling strategies, multifunctional membrane design, and practical challenges associated with membrane durability and scale-up.

Finally, the book extends beyond material development to system-level considerations. Chapters addressing wastewater treatment plants based on biocomposite materials discuss the feasibility of integrating these materials into existing and future treatment infrastructures, while the concluding chapter provides a critical perspective on current limitations, scalability challenges, environmental impacts, and future research directions.

By bringing together contributions from researchers working across disciplines and geographical regions, this book is intended to serve as a reliable reference for graduate students, researchers, engineers, and professionals in environmental engineering, materials science, chemistry, and water resource management. It is our hope that this volume will not only document current progress, but also stimulate further innovation toward sustainable, circular, and resilient wastewater treatment solutions.

1

Natural-Fiber-Based Biocomposites for Wastewater Treatment

Soulaima Chkirida[1], Rachid Bouhfid[2], and Nadia Zari[3]

[1] *College of Chemical Sciences and Engineering (CCSE), Department of Materials Science, Energy and Nano-engineering (MSN), Mohammed VI Polytechnic University (UM6P), 43150, Ben Guerir, Morocco*
[2] *International Water Research Institute (IWRI), Mohammed VI Polytechnic University, Lot 660 – Hay Moulay Rachid, 43150, Ben Guerir, Morocco*
[3] *Moroccan Foundation of Advanced Science Innovation and Research MAScIR, Mohammed VI Polytechnic University, Ben Guerir, Morocco*

1.1 Introduction

In line with Sustainable Development Goal SDG6: Clean Water and Sanitation, the need for sustainable water treatment technology has never been more critical and pressing. With freshwater supplies depleting and water pollution reaching worrisome levels, there is an urgent need for solutions that not only properly treat wastewater but also reduce their environmental footprint [1]. Conventional water treatment technologies frequently rely on chemical additives and energy-intensive procedures that are hazardous to ecosystems and human health. The need for safe and environmentally friendly water treatment methods has become paramount in this regard [2].

In recent years, the notion of "waste-to-wealth" has gained tremendous backing in response to rising environmental concerns and the urgent need for sustainable solutions [3]. This paradigm shift challenges traditional ideas about waste by reframing it as a precious resource ready for transformation. The use of discarded natural fibers to create novel biocomposite materials is at the forefront of this movement, offering a synergy of scientific innovation and environmental care. In fact, untapped potential exists for natural fibers as raw materials for biocomposite synthesis since they are plentiful in nature and sometimes regarded as waste products of industrial and agricultural activities. By using these discarded fibers, not only trash buildup and accumulation can be managed but also value-added materials with diverse applications can be created. This chapter delves into the significant potential of waste natural fibers for developing biocomposite materials for wastewater treatment. Through a multidisciplinary perspective, it explores the natural fibers' sustainable characteristics, structural properties, and the numerous applications of biocomposites in wastewater treatment. Furthermore, it includes a critical assessment of these materials' effects on the environment, highlighting the need for safe and sustainable solutions to safeguarding water resources.

1.2 Overview of Natural Fibers

Natural fibers, versatile and ecologically sound, have emerged as pivotal contributors across a spectrum of industries, offering eco-friendly substitutes that align with the principles of sustainability practices. Originating from animal, plant, and mineral sources [4], these fibers have been used for ages and have a wide range of qualities and uses, showcasing a myriad of properties and applications that have evolved over time. The trend toward sustainable development and increased environmental consciousness has led to a significant resurgence in the interest in and use of natural fibers, which have spread from traditional industries into more modern ones like wastewater treatment, construction, and textiles [5].

Natural fibers can be classified based on many considerations such as their origin, structure, and features. They are frequently classified into three distinct categories: plant-based, animal-based, and mineral-based, depending on the source they come from. Plants like cotton, flax, and bamboo trees are the sources of plant-based fibers. These fibers are predominantly composed of cellulose and encompass bast fiber, leaf or hard fiber, seed, fruit, wood, cereal straw, and other grass fibers. This type provides an attractive option due to their lower density. They frequently have great tensile strength and stiffness and are biodegradable and renewable [6]. On the other hand, animal-based fibers are derived from living organisms and are primarily composed of proteins such as hair, wool, and silk. Lastly, mineral-based fibers such as asbestos and fiberglass are formed from minerals and rocks, as shown in Figure 1.1. However, because mineral fibers, especially asbestos, are nonrenewable and often hazardous, they are not recommended for use in biocomposite applications [7].

Another categorization scheme classifies natural fibers into four categories based on their structural composition: fruit, seed, leaf, and bast. Bast fibers are known for their resilience; they are widely used in textiles and composites and are derived from the inner bark of plants such as hemp and flax [7]. Extracted from the leaves of plants like abaca and sisal, leaf fibers are known for their exceptional strength and stiffness, which makes them perfect for use in paper and textile products. Cotton is an example of a seed fiber, which is used in the textile industry for clothing and bedding because of its softness and absorbency [8].

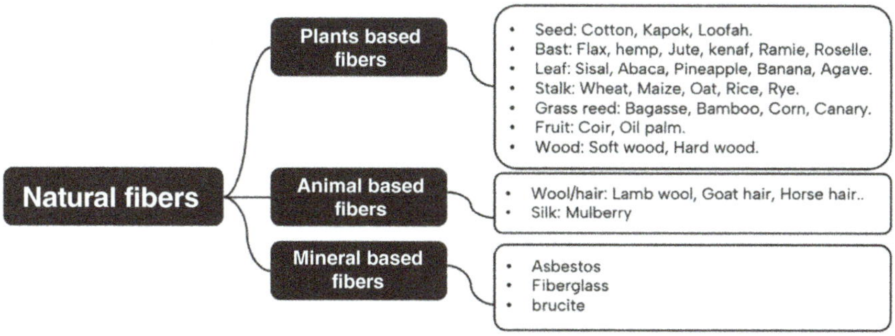

Figure 1.1 Classification of some of the most common natural fibers.

1.2.1 Sustainable Characteristics of Natural Fibers

Natural fibers boast an array of sustainable characteristics that position them as environmentally friendly alternatives in various applications, we cite:

- Renewable resources – Unlike synthetic fibers that have a limited supply, natural fibers may be regenerated by continuous farming and extraction procedures. As a result of its renewable nature, the supply chain is continuous and environmentally benign, according to the principles of sustainable practices.
- Biodegradability – Natural fibers have the inherent tendency to biodegrade, setting them distinct from their manufactured counterparts. They spontaneously decompose after their life cycle without leaving persistent environmental residues. This quality greatly minimizes the environmental effect of waste disposal, supporting an eco-friendly and circular methodology [9].
- Carbon sequestration (in certain cases) – Carbon is actively sequestered by some natural fibers, particularly those originating from plants like bamboo. It also suggests that these plants help to mitigate the impacts of climate change by absorbing and storing carbon dioxide from the environment during their development [10]. This unique feature offers certain kinds of natural fibers a greater level of sustainability.
- Cultural and economic sustainability – Natural fiber materials promote cultural and economic sustainability in areas where natural fiber farming is common. By promoting regional corporations that manufacture natural fibers, communities' standards of living conditions improve and a sustainable economic ecosystem is established [11].
- Diversity in application – The versatility of natural fibers extends to a wide range of applications, including textiles, construction, and, as explored in this chapter, wastewater treatment. This diversity in application ensures that the benefits of natural fibers are leveraged across various industries, further enhancing their overall sustainability [9].

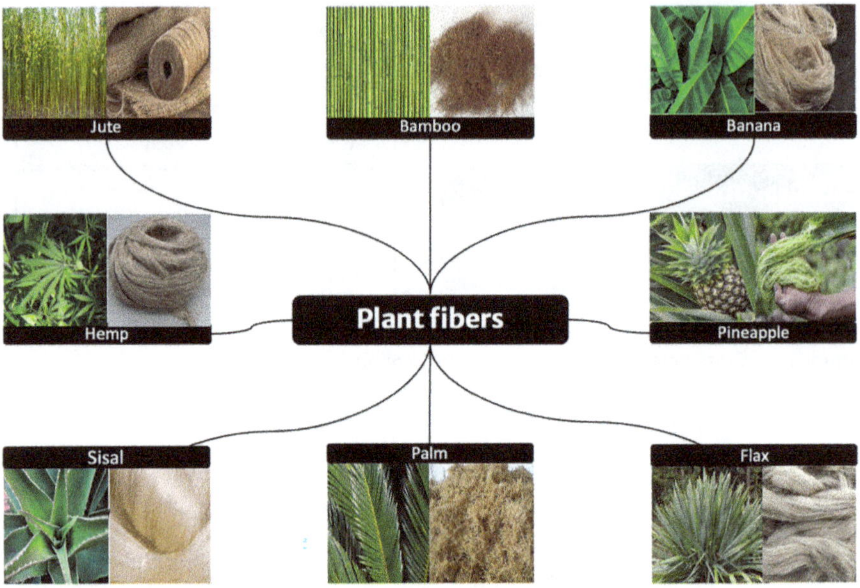

Figure 1.2 Variety of natural plant fibers. Reproduced from Pulikkalparambil et al., (2023) / MDPI / CC BY 4.0.

1.2.2 Structure of Natural Fibers

Prevalent in biocomposite research are animal and plant fibers. Plant cells are defined by their rigid cell wall, a feature that distinguishes them from animal cells. Figure 1.2 illustrates the variety of natural plant fibers found in nature. Within these fibers, their lignocellulosic aspect is derived from the helical arrangement of cellulose microfibrils inside a lignin and hemicellulose mixture that constitutes most of these fibers except for cotton [12]. Cellulose, a long linear macromolecule made up of repetitive D-anhydro-glucose ($C_6H_{11}O_5$) units joined by β-1,4-glycosidic bonds with a degree of polymerization (DP) usually about 10,000, serves as the primary building block, imparting strength and stability to both plant cell walls and fibers [9]. Roughly 40–45% of plant fibers are made up of cellulose, which is composed of 49.4% oxygen, 44.4% carbon, and 6.2% hydrogen [13]. Unlike cellulose, hemicellulose, which is a complex substance made up of different sugar monomers, acts as a bonding agent between cellulose microfibrils forming hydrogen bonds with cellulose to create the cellulose–hemicellulose network. Lignin is a biochemical polymer that acts as a hydrophobic network and is distinguished by its amorphous and varied composition of aromatic polymers and phenyl propane monomers. By acting as a coupling agent and a structural support (Figure 1.3a), this lignin network boosts the cellulose/hemicellulose composite's stiffness and influences the characteristics of other structural elements [9].

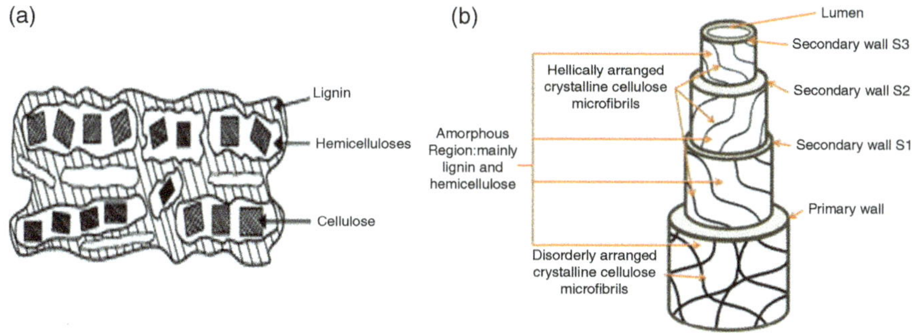

Figure 1.3 (a) Arrangement of the three major components within the cell wall of natural fibers. (b) Structure of natural fibers. Adapted from [12, 14] / with permission of Elsevier.

The arrangement of these components (cellulose, hemicellulose, and lignin) within a basic fiber or the fiber cell wall is nonhomogeneously structured, consisting of several layers, as can be seen in Figure 1.3b.

Drawing parallels, cellulose, lignin, and hemicelluloses in the cell wall mirror construction materials in reinforced concrete as the iron core, cement, and reinforcement, respectively, fortifying cohesion and integrity. Although in lesser amounts, additional components such as pectin, oils, waxes, starches, and soluble compounds are also frequently found [12].

This nonuniform structure is characteristic of natural fibers, which consist of microscopic tubes featuring a core lumen enveloped by cell walls. The primary (outer) wall, formed during initial cell growth, is composed of a framework made of randomly organized cellulose microfibrils connected to an unidentified phase of hemicellulose and lignin, and it acts as a support structure for the cellulose bundles. The secondary wall comprises three other layers (S1, S2, and S3), with the middle layer crucial for mechanical strength. Within this secondary wall, as mentioned before, cellulose microfibrils are arranged helically, influencing fiber properties through the microfibrillar angle [14]. This latter represents the angle between the microfibrils and the fiber axis, and usually varies from one fiber to another. For instance, jute fibers have an angle of 8°, while flax fibers range from 6° to 10°. Hemp fibers typically exhibit an angle of 6°, whereas sisal fibers show a wider range from 10° to 25° [6].

1.2.3 Chemical, Physical, and Mechanical Properties of Natural Fibers

Diverse characteristics like microfibrillar angles, ratios of cellulose, hemicellulose, lignin, and moisture content varied greatly among natural fibers. Moreover, natural fiber plants exhibit varied traits impacted by their cultivation and growing environments. This variety plays a role in the wide range of mechanical qualities seen in natural fibers as well as the unique chemical, physical, and mechanical properties displayed by various plant fiber types. These distinctions are detailed in Table 1.1, which compiles data on the most frequently studied and utilized fibers.

1.2.3.1 Chemical Properties

As previously demonstrated, the predominant biopolymers forming plant fibers are cellulose, hemicelluloses, lignin, and pectin, with each contributing unique property. Among these, cellulose stands out as the main structural component, dictating the mechanical strength of the fibers. Fiber behavior is also impacted by the cellulose type, which is determined by variables such as cell geometry. Pectin, which is made up of complex polysaccharides cross-linked with calcium ions and arabinose sugars, also has a major influence on the properties of fiber. Plant fiber composition also contains inorganic components that contribute to fiber abrasiveness and organic extractives that influence characteristics like color and decay resistance [12]. Table 1.1 provides an extensive summary of the amounts and composition of components within various natural fibers.

1.2.3.2 Physical and Mechanical Properties

Many natural fibers exhibit noteworthy physical and mechanical properties, though they display a greater degree of variability compared to their synthetic counterparts. Table 1.1 illustrates the many benefits that natural fibers have with regard to these particular properties.

- Microfibrillar Angle (MFA)

A key factor influencing the characteristics of fiber strength is the microfibrillar angle (MFA) in the secondary wall layers. Lower MFAs and higher cellulose contents are associated with stronger fibers [4]. Notably, lower angles increase the fiber's stiffness and strength while greater MFAs, on the other hand, with a drop in cellulose content often impart fiber ductility, such as in the case of coir fibers [17]. That being said, not all natural plant fibers exhibit this phenomenon, as evidenced by variations observed in cotton and bamboo fibers [15].

- Cellulose Crystallinity

Cellulose content and crystallinity are important elements in determining fiber mechanical qualities, especially in structural applications where greater levels of both are preferable. Fiber strength is greatly enhanced by crystalline cellulose, which is renowned for its exceptional rigidity [18]. Reduced crystallinity in terms of increased hemicellulose amount, on the other hand, increases the absorption of water due to its hydrophilic properties, thereby impacting mechanical traits and

Table 1.1 Chemical, physical, and mechanical properties of natural plant fibers.

Fiber	Cellulose (wt.%)	Hemicellulose (wt.%)	Lignin (wt.%)	Pectin (wt.%)	MFA	Moisture content (wt.%)	Cellulose crystallinity	Tensile strength (MPa)	Young's modulus (GPa)	Specific strength (GPa g^{-1} cm^{-3})	Specific modulus (GPa g^{-1} cm^{-3})	Elongation at break (%)
Jute	61–71.5	12–35	12–13	0.2	8	12.6	50–80	187–773	20–55	140–320	14–39	1.5–3.1
Kenaf	45–57	3–33	8–21.5	2	2–6.2	9	—	295–930	22–60	246–993	18–50	2.7–6.9
Hemp	70–74	18–22	3–5.7	—	2–6.2	—	50–90	580–1110	30–70	210–510	20–41	1.6
Banana	63–64	12–25	5–10	4	11	—	—	529–914	27–32	392–677	20–24	3–10
Pineapple	80–83	17.5	8.3–12.7	4	10–22	—	44–60	170–1627	60–82	287–1130	42–57	1–3
Bamboo	26–55	20.5–30	15–32	—	2–10	—	40–60	391–1000	11–36	383	18	2
Sisal	57–73	11.5–16	8–12	1.2	10–22	17	50–70	507–885	9.4–22	55–580	6–15	1.9–3
Coir	36–43	0.15–25	41–45	4	30–49	—	27–33	175	6	92–152	5.2	30
Ramie	68–77	13–16.7	0.7	2	7.5	—	64	400–938	61.4–128	590	29	1.2–3.8
Flax	62–81	4–20.6	22.5	0.9	5–10	10	50–90	343–1035	27.6	345–620	34–48	1.2–3
Abaca	56–63	21.7	12–13	0.8	8–15	—	52	430–813	31–34	—	—	3–10
Cotton	85–90	5.7	0.7–1.6	6	20–30	—	—	287–397	6–10	194–452	4–6.5	7–8

Source: [4, 15, 16].

dimensional stability. Table 1.1 presents typical values for the cellulose content and its crystallinity of various natural plant fibers [15].

- Moisture Content

While high moisture content can diminish mechanical properties and dimensional stability, it may also enhance the biodegradability of biocomposites [19]. This inherent property arises from the presence of free hydroxyl and polar groups in the fibers. In fact, water absorption in lignocellulosic fibers is predominantly driven by hemicellulose, which has extremely hydrophilic polysaccharide chains that are more accessible than cellulose chains in crystalline areas [4].

- Young's Modulus and Tensile Strength

As the cellulose content of natural fibers increases, so does their tensile strength and Young's modulus, which makes them stiff and resistant to stretching when the fibrils line up parallel to the fiber axis [15].

When subjected to stress, microfibrils align along the fiber axis. Fiber failure occurs when the hydrogen bonds in cellulose microfibrils break due to a weakening of the bond between matrix components and reinforcing fibrils [7]. This results in decreased tensile strength for fibers with a lower cellulose amount. Given that hemicelluloses are amorphous and nonuniform, it is obvious from the chemical composition (Table 1.1) that an increase in hemicellulose proportions diminishes tensile strength [20].

- Elongation at Break

Expressed as the ratio of the altered length to the initial length post-breakage, elongation at break, also known as fracture strain, measures the amount of deformation a natural plant fiber can withstand prior to fracturing. It illustrates how resilient the fiber is to changes in shape without cracking [21]. Tensile testing is commonly used to evaluate elongation at break.

While synthetic fibers are typically greater in terms of physical and mechanical features compared to natural fibers, natural fibers play an essential role in polymer engineering composites owing to their greatest specific modulus and elongation at break. There is a trend suggesting an inverse relationship between elongation at break and both Young's modulus and strength values among different types of fibers [8]. Table 1.1 amply illustrates this statement; fibers with higher elongation values, showing more capacity for deformation before breaking, have lower strength and Young's modulus, which represent their ability to bear applied force and resist deformation, respectively [22].

1.2.4 Downsides of Natural Fibers

While natural fibers show promise in a variety of applications, their use in wastewater treatment is especially remarkable. However, it is critical to acknowledge and

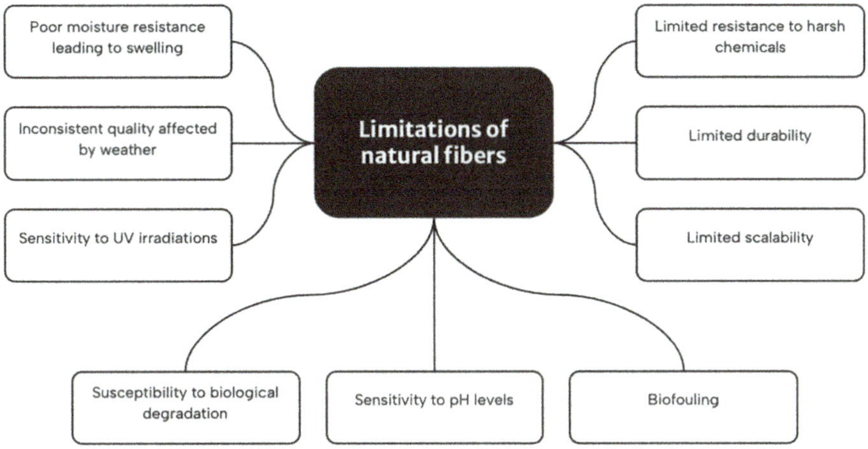

Figure 1.4 Limitations of natural fibers in wastewater treatment applications.

address the limitations of their application in this realm. This section digs into the unique issues that occur when using natural fibers in wastewater treatment applications [4]. Understanding these constraints allows better strategizing and development to optimize the efficacy of natural-fiber-based solutions in dealing with wastewater treatment difficulties. Figure 1.4 summarizes these downsides, which are as follows:

- Poor moisture resistance leading to swelling – Natural fibers are well known for their high tendency to absorb moisture from their surroundings, affecting their dimensional stability and mechanical features [19].
- Inconsistent quality affected by weather conditions – Weather conditions can cause many fluctuations in fiber qualities such as strength, composition, and efficacy in water treatment procedures [14]. As a consequence, it becomes challenging to maintain a consistent quality of natural fibers for wastewater treatment applications.
- Sensitivity to ultraviolet (UV) – Natural fibers can be vulnerable to UV light, causing structural degradation and reducing durability over time. Exposure to sunlight, especially over long periods of time, can promote photodegradation of natural fibers, resulting in changes to their mechanical characteristics and appearance [23].
- Limited scalability – Such challenges include getting consistent and dependable natural fibers, as well as upgrading infrastructure to suit their use.
- Limited durability – Continuous exposure to wastewater conditions can cause natural fibers to deteriorate over time, resulting in a reduction in their treatment effectiveness and potential leaching.

- Limited resistance to harsh chemicals – The natural fibers deterioration may be exacerbated when they are exposed to specific chemicals often present in wastewater, resulting in increased degradation and loss of mechanical properties and function [24].
- Susceptibility to biological degradation – Microorganisms present in wastewater can degrade natural fibers over time, reducing their effectiveness and longevity [14].
- Sensitivity to pH levels – When exposed to varying pH levels in wastewater, natural fibers may exhibit fluctuations in effectiveness and durability, limiting their performances in particular treatment methods [25].

1.3 Natural-Fiber-Based Biocomposites

Previously, we discovered that natural fibers had a number of outstanding features, including stiffness and strength. Notwithstanding these benefits, a major obstacle to the widespread application and processing of natural fibers is their unpredictable endurance and their fibrous structure that make them difficult to use on their own. As a result, including these natural fibers as reinforcements in combination with biopolymers as matrix appears as a very promising strategy to develop innovative composite materials with distinct properties, popularly known as "biocomposites" [16]. Incorporating natural fibers into a biodegradable matrix has a significant impact on the physical properties of the resulting biocomposite, including mechanical, thermal, decomposition, and crystallinity characteristics. Ongoing advances in understanding component materials allow for more control over various design factors, hence improving the qualities of biocomposite materials.

Biocomposites are the result of combining various sciences such as biology, materials science, chemistry, and engineering to create green and eco-friendly materials from renewable resources that can be used in a variety of applications without harming the environment [2]. They seem intriguing as they combine material features in ways that are not common in nature. The green composite materials frequently provide lightweight structures with high stiffness and features specifically designed for a certain purpose, saving weight and reducing energy consumption [26]. In fact, the concept of green composites has been around for an extended time, yet barely any are biocomposites. The word "biocomposite" is commonly used, but its connotation varies. Materials made from biologically produced fibers and polymers are explicitly referred to as "true biocomposites"; furthermore, both components must be biodegradable [27]. One of the prime earliest examples of biocomposite development may be found in the pioneering work done in the late 1990s by Herrmann and colleagues [28], where they created biocomposites by combining various natural fibers—such as ramie, hemp, and flax—with biodegradable polymers, including modified starch and modified cellulose.

Biofiber-reinforced biocomposites also provide an alternative, value-added source of revenue for the agricultural community. Natural fibers such as jute, coir, sisal, kenaf, flax, hemp, and ramie are grown and manufactured in different parts of the world, each with its own set of qualities and benefits [9]. Sisal, a tropical plant, is widely grown in African countries, the West Indies, and the Far East, notably Tanzania and Brazil. Originating in Bangladesh and India, jute is extensively grown in these areas. Coir production is mostly concentrated in tropical nations, with India accounting for a sizable portion of global output. Kenaf is cultivated commercially in the United States. Flax, a versatile commodity crop, is grown in a variety of agricultural systems worldwide, including the European Union, Canada, Argentina, and India. Hemp, which originated from Central Asia, has migrated to China and is now grown in a variety of temperate zone nations. Ramie fibers are the longest and most durable fine fibers, and they are mostly accessible and utilized in China, Japan, and Malaysia [9]. The cost of natural fibers fluctuates based on the economic situation of the countries where they are cultivated and harvested.

1.4 Application of Natural-Fiber-Based Biocomposites in Wastewater Treatment

In light of the fundamental aspects previously discussed, we now shift our focus to the practical applications of natural-fiber-based biocomposites in the realm of wastewater treatment. This section will delve into the cutting-edge developments and potential breakthroughs associated with the emerging application of biocomposites in this crucial process, highlighting the scientific evidence supporting their efficacy in removing a panoply of emerging wastewater contaminants. A comprehensive Table 1.2 classifying these natural-fiber-based biocomposites according to the pollutants removed will be provided for reference.

1.4.1 Natural-Fiber-Based Biocomposites as Adsorbent Materials

Cellulose-fiber-based biocomposites have a wide range of uses in wastewater treatment. Several investigations have confirmed their efficiency in removing emerging contaminants from wastewater streams. Zhao et al. [29] developed an effective adsorbent for removing Pb(II) from aqueous solutions. Sodium alginate and cellulose nanofibers were utilized in a simple cross-linking process to create porous hydrogel beads. Adsorption studies demonstrated the adsorbent's efficacy,

Table 1.2 Studies conducted on wastewater treatment by means of natural-fiber-based biocomposite.

Fiber used	Matrix	Biocomposite	Natural fiber %	Form	Pollutant	Technology	Capacity ($mg\ g^{-1}$) or efficiency (%)	Reference
Cellulose NF	Alginate	P-CNF-SA		Hydrogel beads	Pb^{2+}	Adsorption/ complexation	82% 318.47 $mg\ g^{-1}$	[29]
Water hyacinth	Chitosan	WHF/CS	10%	Biofilm	Cu^{2+}	Adsorption/ complexation	$34.1 \pm 2.2\ mg\ g^{-1}$	[30]
Cellulose nanofibers	Chitosan and Alginate		15%	Biofoam	EBT dye	Adsorption	2297 $mg\ g^{-1}$	[31]
Cellulose microfibers	Chitosan/ ZnO	ZnO/Chi-MCM	—	Cellulose mat	MO dye	Adsorption/ photo-degradation	42.8 $mg\ g^{-1}$	[32]
Cotton fibers	Cotton cloth	Cotton fiber/ CS	—	Cotton cloth	RR dye	Adsorption/ complexation	169.33 $mg\ g^{-1}$	[33]
Micro-fibrillated cellulose	Chitosan	Dialdehyde MF cellulose/ CS	>6 wt.%	Membrane	CR dye	Adsorption	99.5% 152.5 $mg\ g^{-1}$	[34]
Carboxy-lated cellulose MF	Polylactide/ poly(butylene adipate-co-terephthalate) polymer blend	PLA/PBAT-MEFB	30 wt.%	Membrane	MB dye	Adsorption	97.2%	[35]

Banana trunk fibers (BTF)	Chitosan	BTF-i-CTS	0.2 wt.%	Biofilm	Cr^{6+}	Adsorption/ complexation	97.79%	[36]
Cellulose fibers	Chitosan	Cu/CH-FP	—	Cellulose mat	MO dye CR dye	Catalytic reduction	100% 100%	[37]
Cellulose acetate fibers	Chitosan	(C-T/CA)	10 wt.%	Membrane	MO dye	Adsorption/ photocatalysis	98%	[38]
Cellulose microfibrils	Chitosan	Cellulose/ $(TiO_2/CHI)_n$/ Ag-NP	—	Filter paper	*S. aureus* *E. coli*	Antibacterial activity	100% 100%	[39]

while acid reclamation tests confirmed its suitability for reuse. In an attempt to develop three-dimensional porous biofoams for dye removal from wastewater, Mokhtari et al. [31] have used cellulose nanofibers (NFs) in conjunction with alginate and chitosan to achieve polyelectrolyte complex biocomposites. Leveraging their numerous surface functional groups and high surface area, cellulose NFs were investigated not only for the improvement of the adsorption efficiency of chitosan but also for the improvement of the hydrolyzability of resultant hydrogels before freeze-drying. The addition of cellulose NFs also introduces a physical cross-linking mechanism through hydrogen bonding interactions with the matrix. With 15 wt.% cellulose nanofibers ($CSAC_{15}$), biocomposite exhibited exceptional adsorption efficiency, notably for the elimination of the Eriochrome black-T (EBT) anionic dye. With a maximum adsorption capacity qm of 2297 $mg\,g^{-1}$, these findings outperform earlier research on EBT adsorbents. Notably, $CSAC_{15}$ retained its stability and strong adsorption capability across a variety of pH solutions and demonstrated high durability, preserving its dye removal capacity even after five successive adsorption cycles.

Cotton fibers have also found application in the context of wastewater treatment. Using a simple, yet inventive technique, cotton cloth acting as the matrix and primarily consisting of 92% cellulose was treated by depositing chitosan biopolymer on its surface to give rise to chitosan-treated cotton biocomposite [33]. With a hierarchical structure, a functional morphology, and numerous adsorptive sites, the developed biocomposite exhibited a maximum adsorption capacity for removing the reactive dye (Remazol Brilliant Red F3B [RR]) reaching 169.33 $mg\,g^{-1}$, which is comparable to other materials. Furthermore, through adsorption-elution-regeneration cycles, the biocomposite revealed its reusability and stability, suggesting its viability for treating real-world wastewater samples. Overall, these findings highlight the chitosan–cotton biocomposite's potential as an effective, cost-efficient, and ecologically friendly dye removal material in water, helping to safeguard the environment and human health.

1.4.2 Natural-Fiber-Based Biocomposites as Membranes

Cellulose fiber's varied features make it perfect for developing multifunctional wastewater treatment membranes. Its hardness and flexibility at the nano/micro scale, along with its nontoxic, biocompatible and biodegradable properties, result in environmentally friendly membranes [40].

By combining dialdehyde microfibrillated cellulose (DAMFC) with chitosan to create a composite material, a high-efficiency membrane was created [34]. DAMFC was developed specifically to remove the anionic dye Congo red (CR) through the sorption process. At first, microfibrillated cellulose was derived from microcrystalline cellulose via high-pressure homogenization. The DAMFC was subsequently mixed with chitosan using solvent casting to create the composite membrane. It is noteworthy that the mechanical strength and dye solution stability of the chitosan composites were significantly enhanced by the addition of DAMFC. This novel DAMFC/chitosan membrane, in only 10 minutes, obtained an astounding 99.5% removal efficiency for the CR dye and demonstrated a high adsorption capacity of 152.5 $mg\,g^{-1}$.

As a result of the insertion of carboxyl groups, carboxylated cellulose microfibers (CCMFs) possess a large surface area and enhanced hydrophilicity, enhancing their dispersion in aqueous solutions. These fibers have higher tensile strength and flexibility, making them excellent for reinforcing composites and biocomposites. The carboxyl groups further increase chemical reactivity, allowing for further functionalization, and considerably enhance the adsorption capacity for contaminants, making CCMFs ideal for membranes for water treatment. Ampawan [35] has leveraged these advantages to craft a bio-based membrane for cationic dyes removal via phase inversion method, which is straightforward and inexpensive. Based on CCMF extracted from raw empty fruit bunch and polylactide blend, which is a biodegradable polymer, well suitable for membrane preparation, the biocomposite membrane has shown great potential for selective filtration of cationic pollutants and wastewater treatment due to its economical production and remarkable performance. In fact, filtration studies with a mixed dye solution including anionic methyl orange (MO) and cationic methylene blue (MB) revealed that the biocomposite membrane adsorbs MB rapidly while allowing MO to pass through. The incorporation of carboxylated cellulose fibers significantly boosted the pure water flux, as evidenced by the biocomposite membrane's substantially higher flux rate of 1214 $Lm^{-2}\,h^{-1}$ compared to the membrane's rate of 371 $Lm^{-2}\,h^{-1}$ during dynamic adsorption processes. Notably, this membrane was straightforward to regenerate and reliably maintained good separation performance across numerous cycles.

1.4.3 Natural-Fiber-Based Biocomposites as Photocatalyst Materials

Plenty of research on photocatalytic materials derived from cellulose fibers has been conducted, revealing cellulose fiber's considerable contribution to integrated adsorption/photocatalysis processes for wastewater treatment. Kamal et al. [32] have shown that the use of cellulose microfibers uniformly coated with a biocomposite layer of ZnO and chitosan offers a potential method for successfully removing MO azo dye from wastewater.

The biocomposite zinc oxide/chitosan-coated adsorbent (ZnO/Chi-MCM) was found significantly accelerating the adsorption process of MO dye as compared to both untreated microfibrillar cellulose mats and those coated simply with chitosan (Chi-MCM). Simultaneous adsorption and photocatalytic degradation processes can be enhanced through the use of the synergistic properties of chitosan and ZnO within the ZnO/Chi-MCM structure. Therefore, the effectiveness of MO dye removal from wastewater is significantly increased in the presence of UV light.

Shi [38] has successfully produced an integrated adsorbent/photocatalyst membrane based on cellulose acetate, chitosan, and titanium dioxide (TiO_2) C-T/CA fiber membranes for adsorption/photodegradation of MO. The approach was two-pronged: electrospinning of cellulose acetate (CA) and electro spraying of chitosan/TiO_2 concave hemispheres onto the prepared cellulose acetate electrospun fibers. The concept was based on choosing CA fibers as a substrate in order to enhance water flux, allowing aqueous solutions of the organic pollutant methyl orange to rapidly penetrate the membranes. Meanwhile, chitosan (CS) hemispheres enabled the uniform dispersion of TiO_2 nanoparticles, which absorbed MO and transported it to TiO_2-active sites for photocatalytic degradation. For five consecutive cycles, the fiber membranes consistently achieved a 98% MO removal ratio, demonstrating maintained adsorption and photocatalytic degradation abilities. This stability was ascribed to the uniform distribution of TiO_2 through interactions with CS and the enhanced aqueous stability of CS through hydrogen bonding with CA.

1.4.4 Natural-Fiber-Based Biocomposites as Reduction Catalyst

For immobilizing metallic nanoparticles in catalytic reduction processes, cellulose fibers serve as an excellent support system. Their great surface area, biocompatibility, and availability make them ideal substrates. In addition to ensuring homogeneous dispersion and effective mass transfer to catalytic sites, cellulose fibers also inhibit nanoparticle agglomeration [41]. This eco-friendly method reduces environmental impact while allowing for efficient pollutant removal. Kamal [37] has developed a streamlined approach for simultaneously synthesizing and immobilizing zero-valent copper (Cu) nanoparticles over chitosan (CH) coating layers on cellulose microfibers of filter paper. Excellent catalytic performance was demonstrated by the resultant Cu/CH-FP catalyst in water treatment applications. Cu/CH-FP was easily separated by simply drawing the strip out of the reaction fluid. Hence, the development of Cu/CH-FP represents a substantial improvement in nanoparticle production and separation procedures. The creation of metallic nanoparticles within a thin layer of chitosan put on top of filter paper substrate is made possible by this innovative technique that allows the production of nanoparticles with great catalytic efficacy while also providing unprecedented separation ease. Overall, Cu/CH-FP's catalytic proficiency was very excellent, demonstrating its ability to rapidly and efficiently reduce both MO and CR dyes using $NaBH_4$ as the reducing agent. With a total reduction of MO in 13 minutes and CR in 17 minutes, the biocomposite film demonstrated remarkable efficiency. Furthermore, its recyclability was notable, with Cu/CH-FP maintaining its catalytic activity over multiple cycles, showcasing sustained efficacy and practical utility in catalytic applications.

1.4.5 Natural-Fiber-Based Biocomposite as Antibacterial Agents

Water pollution takes many forms, including organic, inorganic, and microbiological contamination caused by harmful microorganisms such as viruses, bacteria, and parasites. These microorganisms represent substantial health hazards, causing a wide range of infectious illnesses in humans [2]. In contrast to conventional methods, natural-fiber-based biocomposites provide a viable avenue for the introduction of new, eco-friendly, and highly effective materials for antibacterial applications.

When it comes to antibacterial activity, chitosan's qualities are well-established. Chitosan's antibacterial activities are mostly explained by the interactions of glucosamine's positively charged amino groups with the negatively charged microbial cell membranes; this interaction disrupts the integrity of the membranes, leading to the efflux of intracellular constituents of microorganisms [32]. Through layer-by-layer self-assembly, Xiao et al. [39] created a nanometer-thick titania/chitosan/Ag-NP bionanocomposite film and coated it on cellulose microfibrils. Using both Gram-positive (*Staphylococcus aureus*) and Gram-negative (*Escherichia coli*) bacteria, the antibacterial activity of this biocomposite was assessed. With increasing antibacterial activities in the order of cellulose/$(TiO_2)_n$, cellulose/$(TiO_2/CHI)_n$, and cellulose/$(TiO_2/CHI)_n$/Ag-NP, the synergistic combination of cellulose, chitosan, titania, and silver nanoparticles significantly disinfected the inoculation bacteria.

1.4.6 Natural-Fiber-Based Biocomposites as Biofilms

Seeking to mitigate concerns related to lower surface area for interactions with target contaminants, researchers often opt to engineer biofilms. These biofilms are designed to enhance adsorption efficiency by maximizing the available interaction surface, thereby mitigating concerns regarding efficacy. In this vein, fibers from water hyacinth leaves were extracted and used to make chitosan-based biocomposite biofilms for the adsorption of Cu^{2+} ions from aqueous solutions [30]. Through a series of steps including shredding, alkaline treatment, neutralization, and filtration methods, water hyacinth fibers (WHFs) were extracted. Cellulosic fibers measuring $11 \pm 2\,\mu m$ in diameter on average were obtained as a result, and solution casting was used to prepare the WHF/chitosan composite biofilms. With 5% fiber loading in the biocomposite, tensile tests demonstrated a significant increase in Young's modulus. At 10% loading, the biocomposite biofilm demonstrated a remarkable fivefold increase in Cu^{2+} ion uptake from water when compared to pure chitosan biofilm: Cu^{2+} adsorption increased from 6.4 ± 2.4 to $34.1 \pm 2.2\, mg\, g^{-1}$.

In the same vein, a biocomposite film incorporating banana trunk fibers (BTFs) into chitosan (CTS), termed BTF-i-CTS, was developed [36], revealing a remarkable Cr^{6+} removal efficiency (97.79%) and showcasing its superior affinity for metal ion removal. SEM images displayed a smooth biofilm surface, indicating an effective dispersion and entrapment of BTF within the CTS matrix. The biocomposite demonstrated fast absorption of water, thanks to the entrapped natural fiber (BTF) network, acting as a conduit for water infiltration. Notably, while CTS alone has a high affinity for Cr^{6+} removal (97.78%), its brittleness and difficulties in adsorption make the biocomposite film a more practical alternative.

1.5 Sustainability Assessment and Life Cycle Assessment

Environmental conservation and sustainability are frequently considered as being closely connected. This relationship stems from a shared commitment to preserving equitable well-being throughout generations, given that nature's resources and services are seen as the cornerstone of society [42]. In the current era of increased environmental consciousness and sustainability standards, evaluating the environmental implications connected with various materials and processes has become crucial. Life cycle assessment (LCA) is a reliable process for systematically evaluating the environmental burdens and benefits of a product or service across its full life cycle, from raw material extraction to disposal [43].

As much as employing natural fiber waste appears to be a sustainable strategy that fits with the circular economy, it is critical to ensure that the entire upcycling process is sustainable. Examples include fiber extraction, manufacture, and application. However, the valorization process's overall eco-friendliness may be threatened in certain cases of the upcycling process since it might use potentially hazardous substances or need high energy inputs [44].

While various studies have investigated LCA in the context of wastewater treatment, the focus has generally been on procedures already established within the industry and conventional materials, rather than specific materials such as natural-fiber-based biocomposites [45].

Since different natural fiber types are utilized to create distinct biocomposites, each LCA methodology needs to be tailored specifically to the research in question. Although LCA of natural-fiber-based biocomposites has been investigated in various contexts [46], it is conspicuously absent in the field of wastewater treatment. This has left a gap in the literature regarding an in-depth assessment of the environmental impacts induced by the use of natural-fiber-based biocomposites for wastewater treatment.

It is deemed useful to take LCA into account when treating wastewater with natural-fiber-based biocomposites for a number of reasons [43]:

- Sustainability assessment – Based on their environmental performances, LCA makes it possible to compare the new, suggested natural-fiber-based technologies with traditional wastewater treatment methods. This allows for the development of more sustainable alternatives that decrease environmental impact while reaching wastewater treatment goals.
- Resource optimization – LCA enables resource optimization and waste reduction by assessing natural resource reliance, energy consumption, and emissions throughout the life cycle of natural-fiber-based materials and associated treatment technologies. This might lead to lower ecological impacts, more effectiveness, and financial savings.
- Stakeholder involvement and transparency – LCA promotes transparency and stakeholder involvement by providing objective and strictly scientific evaluations of environmental impacts. This stimulates meaningful debate among interested parties, including the general public, corporate leaders, politicians, and researchers, therefore furthering collective initiatives for more environmentally friendly wastewater treatment methods.
- Policy and regulation compliance – Implementing environmentally friendly wastewater treatment materials and procedures is demanded by increasingly rigorous ecological standards as well as sustainability criteria. LCA insights are especially valuable for ensuring regulatory compliance and guiding policy development in the wastewater treatment industry.

1.6 Research Gaps and Areas for Improvement

Despite several factors promoting their usage, biocomposites are less widespread in the wastewater treatment industry than conventional materials. The majority of developed biocomposite materials seek to replace materials without altering the production infrastructure or product design. As will be explained in the following and depicted in Figure 1.4, an array of technical limitations and perceptual challenges are responsible for this hesitation toward biocomposite adoption [47].

- Technical challenges – Natural-fiber-based biocomposites have inherent characteristics that make them difficult to process. These characteristics include their hydrophilicity, variability in properties, unexpected durability, and low resistance to moisture and heat impacts. As a result, their use is limited compared to conventional composites made of synthetic polymers, which are more widely used and have the benefit of an established production infrastructure. The scarcity of information among stakeholders and designers further limits the practical implementation of biocomposite materials. Yet, it's crucial to recall that these biocomposites can provide unique benefits in terms of material quality and sustainability if additional funding is invested in the development of technologies for the synthesis of biopolymers and the processing of biocomposites [7].

- Biodegradability profile – Researchers tend to interpret biodegradability as an absolute rather than a gradient property, therefore the concept is typically unambiguous. Although different studies may declare their fiber-based biocomposites to be "biodegradable," the actual mechanisms, by-products, and times needed for biodegradation may vary significantly between fibers. This variety emphasizes the need for a more thorough understanding of biodegradability and its link with ecosystem sustainability [48].
- Assessment of long-term performance – Research on natural-fiber-based biocomposites and their short-term efficacy in wastewater treatment applications is numerous. However, to evaluate these materials' durability, flexibility, and efficacy over extended periods of usage, long-term performance evaluations are required. Longitudinal research conducted in actual wastewater treatment conditions can provide considerable insights into the durability and performance of biocomposites made from natural fibers [47].
- Scale-up and commercialization – Natural-fiber-based biocomposites have been shown to be effective in treating wastewater through laboratory-scale experiments; however, information on the technology's potential for scaling up and commercialization is lacking. Major hurdles must be overcome in order to scale up biocomposite production processes and incorporate them into the existing wastewater treatment infrastructure. These challenges necessitate cross-disciplinary collaboration as well as innovative technical solutions [7].
- Lack of standardized LCA studies – LCA especially made for natural-fiber-based biocomposites in wastewater treatment is still uncommon. Environmental performance evaluations would be more credible if standard LCA methodologies were adopted, ensuring uniformity and comparability between studies. The following section will take an in-depth look at LCA, providing a comprehensive analysis of its use in analyzing the environmental implications of natural-fiber-based biocomposites for wastewater treatment [48].

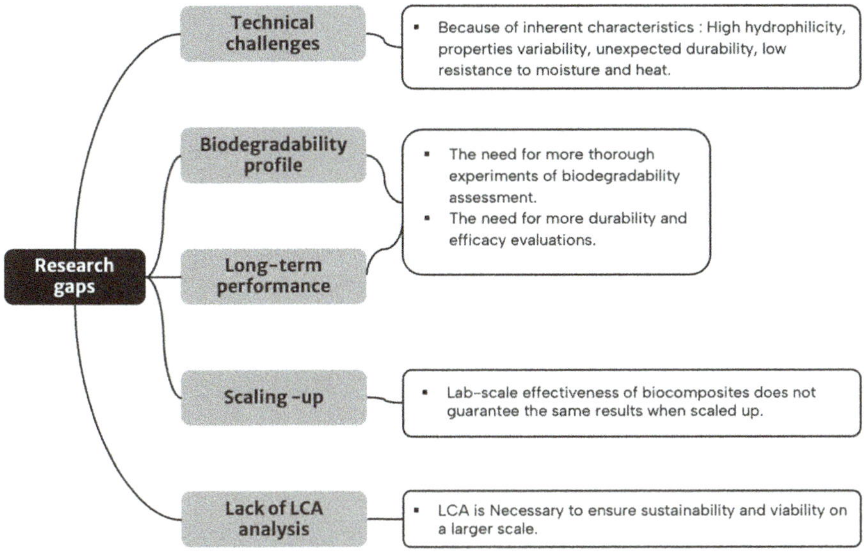

1.7 Conclusion

This chapter looks at natural-fiber-based biocomposites as a potential solution for wastewater treatment. Their various characteristics are presented, including their durability, structural properties, and intrinsic capacities, and their drawbacks are identified. The focus then shifts to understanding the various applications of natural-fiber-based biocomposites in wastewater treatment. These are shown to be versatile, including adsorbents, membranes, photocatalysts, reduction catalysts, antibacterial agents, and biofilms, demonstrating their ability to effectively address water pollution problems while being environmentally friendly. In addition, the need for sustainability assessment and LCA has been taken into account to evaluate the environmental impact and sustainability of these biocomposites throughout their life cycle. This approach highlights the importance of considering the potential environmental impacts and not just the positive aspects when evaluating these materials. This chapter also highlights the need for further research and innovation to advance the development of natural-fiber-based biocomposites for more effective wastewater treatment. In summary, natural-fiber-based biocomposites offer a sustainable and effective solution for wastewater treatment. However, their full potential can only be realized through in-depth LCA evaluations and by filling current research gaps.

References

1 Obaideen, K., Shehata, N., Sayed, E.T. et al. (2022). The role of wastewater treatment in achieving sustainable development goals (SDGs) and sustainability guideline. *Energy Nexus* 7 (January): 100112. https://doi.org/10.1016/j.nexus.2022.100112.

2 Chkirida, S., Zari, N., Bouhfid, R., and Qaiss, A.e.k. (2021). Insight into the bionanocomposite applications on wastewater decontamination: review. *Journal of Water Process Engineering* 43 (August): 102198. https://doi.org/10.1016/j.jwpe.2021.102198.

3 Yu, J., Tang, T., Cheng, F. et al. (2021). Waste-to-wealth application of wastewater treatment algae-derived hydrochar for Pb(II) adsorption. *MethodsX* 8: 101263. https://doi.org/10.1016/j.mex.2021.101263.

4 Bavan, D.S. and Kumar, G.C.M. (2014). A view on cellulosic nanocomposites for treatment of wastewater. In: *Nanocomposites in wastewater treatment*, vol. 6, 47.

5 John, M.J. and Thomas, S. (2008). Biofibres and biocomposites. *Carbohydrate Polymers* 71 (3): 343–364. https://doi.org/10.1016/j.carbpol.2007.05.040.

6 Mohanty, A.K., Misra, M., and Hinrichsen, G. (2000). Biofibres, biodegradable polymers and biocomposites: an overview. *Macromolecular Materials and Engineering* 276–277: 1–24. https://doi.org/10.1002/(SICI)1439-2054(20000301)276:1<1::AID-MAME1>3.0.CO;2-W.

7 Mohanty, A.K., Misra, M., and Drzal, L.T. (2018). Sustainable bio-composites from renewable resources: opportunities and challenges in the green materials world. *Renewable Energy: Four Volume Set* 3–4 (April) 2002: 396–409. https://doi.org/10.4324/9781315793245-107.

8 A. K. Mishra, *Nanocomposites in Wastewater Treatment*. 2014.

9 Elfaleh, I., Abbassi, F., Habibi, M. et al. (2023). A comprehensive review of natural fibers and their composites: an eco-friendly alternative to conventional materials. *Results in Engineering* 19 (April): 101271. https://doi.org/10.1016/j.rineng.2023.101271.

10 Gu, L., Zhou, Y., Mei, T. et al. (2019). Carbon footprint analysis of bamboo scrimber flooring-implications for carbon sequestration of bamboo forests and its products. *Forests* 10 (1): 1–14. https://doi.org/10.3390/f10010051.

11 Karimah, A., Ridho, M.R., Munawar, S.S. et al. (2021). A comprehensive review on natural fibers: technological and socio-economical aspects. *Polymers (Basel)* 13 (24): https://doi.org/10.3390/polym13244280.

12 Kabir, M.M., Wang, H., Lau, K.T., and Cardona, F. (2012). Chemical treatments on plant-based natural fibre reinforced polymer composites: an overview. *Composites. Part B, Engineering* 43 (7): 2883–2892. https://doi.org/10.1016/j.compositesb.2012.04.053.

13 Poletto, M., Pistor, V., and Zattera, A.J. (2013). Structural characteristics and thermal properties of native cellulose. In: *Cellulose—Fundamental Aspects*, 1–13. Vienna, Austria: InTech.

14 Azwa, Z.N., Yousif, B.F., Manalo, A.C., and Karunasena, W. (2013). A review on the degradability of polymeric composites based on natural fibres. *Materials and Design* 47: 424–442. https://doi.org/10.1016/j.matdes.2012.11.025.

15 Djafari Petroudy, S.R. (2017). *Physical and Mechanical Properties of Natural Fibers*. Elsevier Ltd.

16 Mochane, M.J., Magagula, S.I., Sefadi, J.S., and Mokhena, T.C. (2021). A review on green composites based on natural fiber-reinforced polybutylene succinate (PBS). *Polymers (Basel)* 13 (8): 1–38. https://doi.org/10.3390/polym13081200.

17 Satyanarayana, K.G., Ravikumar, K.K., Sukumaran, K. et al. (1986). Structure and properties of some vegetable fibres. *Journal of Materials Science* 21 (1): 57–63. https://doi.org/10.1007/BF01144699.

18 Blaker, J.J., Lee, K.Y., and Bismarck, A. (2011). Hierarchical composites made entirely from renewable resources. *Journal of Biobased Materials and Bioenergy* 5 (1): 1–16. https://doi.org/10.1166/jbmb.2011.1113.

19 Mukhopadhyay, S. and Fangueiro, R. (2009). Physical modification of natural fibers and thermoplastic films for composites—a review. *Journal of Thermoplastic Composite Materials* 22 (2): 135–162. https://doi.org/10.1177/0892705708091860.

20 Mohanty, A.K., Misra, M., and Drzal, L.T. (2002). Sustainable bio-composites from renewable resources in green materials world. *Journal of Polymers and the Environment* 10 (April): 19–26.

21 Tewari, M., Singh, V.K., Gope, P.C., and Chaudhary, A.K. (2012). Evaluation of mechanical properties of bagasse-glass fiber reinforced composite. *Journal of Materials and Environmental Science* 3 (1): 171–184.

22 Komuraiah, A., Kumar, N.S., and Prasad, B.D. (2014). Chemical composition of natural fibers and its influence on their mechanical properties. *Mechanics of Composite Materials* 50 (3): 359–376. https://doi.org/10.1007/s11029-014-9422-2.

23 Sørensen, L., Groven, A.S., Hovsbakken, I.A. et al. (2021). UV degradation of natural and synthetic microfibers causes fragmentation and release of polymer degradation products and chemical additives. *Science of the Total Environment* 755: 143170. https://doi.org/10.1016/j.scitotenv.2020.143170.

24 Koohestani, B., Darban, A.K., Mokhtari, P. et al. (2019). Comparison of different natural fiber treatments: a literature review. *International journal of Environmental Science and Technology* 16 (1): 629–642. https://doi.org/10.1007/s13762-018-1890-9.

25 Bin Bakri, M.K., Jayamani, E., Hamdan, S. et al. (2016). Fundamental study on the effect of alkaline treatment on natural fibers structures and behaviors.

Asian Research Publishing Network (ARPN) Journal of Engineering and Applied Sciences 11 (14): 8759–8763.

26 Chkirida, S., Zari, N., Qaiss, A.E.K., and Bouhfid, R. (2019). Nanocomposite materials based on TiO_2/clay for wastewater treatment. In: *Advanced Research in Nanosciences for Water Technology*, 363–380. Springer International Publishing.

27 Chkirida, S., Zari, N., el Kacem Qaiss, A., and Bouhfid, R. (2022). Future challenges and perspectives in water purification by hybrid materials. In: *Water Nanohybrid Materials for Water Purification*, 217–314. Springer Nature Singapore.

28 Herrmann, A.S., Nickel, J., and Riedel, U. (1998). Construction materials based upon biologically renewable resources—from components to finished parts. *Polymer Degradation and Stability* 59 (1–3): 251–261. https://doi.org/10.1016/s0141-3910(97)00169-9.

29 Zhao, H., Ouyang, X.K., and Yang, L.Y. (2021). Adsorption of lead ions from aqueous solutions by porous cellulose nanofiber–sodium alginate hydrogel beads. *Journal of Molecular Liquids* 324: 115122. https://doi.org/10.1016/j.molliq.2020.115122.

30 Pisitsak, P., Phamonpon, W., Soontornchatchavet, P. et al. (2019). The use of water hyacinth fibers to develop chitosan-based biocomposites with improved Cu2+ removal efficiency. *Composites Communications* 16 (August): 1–4. https://doi.org/10.1016/j.coco.2019.08.003.

31 Mokhtari, A., Sabzi, M., and Azimi, H. (2021). 3D porous bioadsorbents based on chitosan/alginate/cellulose nanofibers as efficient and recyclable adsorbents of anionic dye. *Carbohydrate Polymers* 265 (December 2020): 118075. https://doi.org/10.1016/j.carbpol.2021.118075.

32 Kamal, T., Ul-Islam, M., Khan, S.B., and Asiri, A.M. (2015). Adsorption and photocatalyst assisted dye removal and bactericidal performance of ZnO/chitosan coating layer. *International Journal of Biological Macromolecules* 81: 584–590. https://doi.org/10.1016/j.ijbiomac.2015.08.060.

33 Salman, M.S., Sheikh, M.C., Hasan, M.M. et al. (2023). Chitosan-coated cotton fiber composite for efficient toxic dye encapsulation from aqueous media. *Applied Surface Science* 622 (February): 157008. https://doi.org/10.1016/j.apsusc.2023.157008.

34 Zheng, X., Li, X., Li, J. et al. (2018). Efficient removal of anionic dye (Congo red) by dialdehyde microfibrillated cellulose/chitosan composite film with significantly improved stability in dye solution. *International Journal of Biological Macromolecules* 107 (Part A): 283–289. https://doi.org/10.1016/j.ijbiomac.2017.08.169.

35 Ampawan, S., Phreecha, N., Chantarak, S., and Chinpa, W. (2022). Selective separation of dyes by green composite membrane based on polylactide with carboxylated cellulose microfiber from empty fruit bunch. *International Journal of Biological Macromolecules* 225 (November 2022): 1607–1619. https://doi.org/10.1016/j.ijbiomac.2022.11.218.

36 Rahim, M. and Mas Haris, M.R.H. (2019). Chromium (VI) removal from neutral aqueous media using banana trunk fibers (BTF)-reinforced chitosan-based film, in comparison with BTF, chitosan, chitin and activated carbon. *SN Applied Sciences* 1 (10): 1–11. https://doi.org/10.1007/s42452-019-1206-9.

37 Kamal, T., Khan, S.B., and Asiri, A.M. (2016). Synthesis of zero-valent Cu nanoparticles in the chitosan coating layer on cellulose microfibers: evaluation of azo dyes catalytic reduction. *Cellulose* 23 (3): 1911–1923. https://doi.org/10.1007/s10570-016-0919-9.

38 Shi, X., Zhang, X., Ma, L., and Xiang, C. (2019). TiO_2-doped chitosan microspheres supported on cellulose acetate fibers for adsorption and photocatalytic degradation of methyl orange. *Polymers (Basel)* 11: 1293.

39 Xiao, W., Xu, J., Liu, X. et al. (2013). Antibacterial hybrid materials fabricated by nanocoating of microfibril bundles of cellulose substance with titania/chitosan/silver-nanoparticle composite films. *Journal of Materials Chemistry B* 1 (28): 3477–3485. https://doi.org/10.1039/c3tb20303d.

40 Abdelhamid, H.N. and Mathew, A.P. (2021). Cellulose-based materials for water remediation: adsorption, catalysis, and antifouling. *Frontiers in Chemical Engineering* 3 (December 2021): 1–23. https://doi.org/10.3389/fceng.2021.790314.

41 Chen, M., Kang, H., Gong, Y. et al. (2015). Bacterial cellulose supported gold nanoparticles with excellent catalytic properties. *ACS Applied Materials & Interfaces* 7 (39): 21717–21726. https://doi.org/10.1021/acsami.5b07150.

42 Hauschild, M.Z., Rosenbaum, R.K., and Olsen, S.I. (2017). Life cycle assessment: theory and practice. *Life Cycle Assessment (LCA): Theory and Practice* 1–1216. https://doi.org/10.1007/978-3-319-56475-3.

43 Johnson, E. (2003). Handbook on life cycle assessment operational guide to the ISO standards. *Environmental Impact Assessment Review* 23 (1): 129–130. https://doi.org/10.1016/s0195-9255(02)00101-4.

44 Peydayesh, M., Bagnani, M., Soon, W.L., and Mezzenga, R. (2023). Turning food protein waste into sustainable technologies. *Chemical Reviews* 123 (5): 2112–2154. https://doi.org/10.1021/acs.chemrev.2c00236.

45 Corominas, L., Byrne, D.M., Guest, J.S. et al. (2020). The application of life cycle assessment (LCA) to wastewater treatment: a best practice guide and critical review. *Water Research* 184: 116058. https://doi.org/10.1016/j.watres.2020.116058.

46 Ita-Nagy, D., Vázquez-Rowe, I., Kahhat, R. et al. (2020). Life cycle assessment of bagasse fiber reinforced biocomposites. *Science of the Total Environment* 720: https://doi.org/10.1016/j.scitotenv.2020.137586.

47 Manu, T., Nazmi, A.R., Shahri, B. et al. (2022). Biocomposites: a review of materials and perception. *Materials Today Communications* 31 (February): https://doi.org/10.1016/j.mtcomm.2022.103308.

48 Faruk, O., Bledzki, A.K., Fink, H.P., and Sain, M. (2014). Progress report on natural fiber reinforced composites. *Macromolecular Materials and Engineering* 299 (1): 9–26. https://doi.org/10.1002/mame.201300008.

2

Synthesis and Characterization of Chitosan Biopolymer Material for Application in Wastewater Treatment via Adsorption

Zeynep Ciğeroğlu[1], Noureddine El Messaoudi[2,3], Youssef Miyah[4,5], Zeynep Mine Şenol[6], Yaser Açıkbaş[7], Abdelaziz El Mouden[2], and Abdellah Lacherai[2]

[1]*Department of Chemical Engineering, Faculty of Engineering, Usak University, Usak, Turkey*
[2]*Laboratory of Applied Chemistry and Environment, Faculty of sciences, Ibn Zohr University, Agadir, Morocco*
[3]*Department of Chemistry, College of Science, Imam Mohammad Ibn Saud Islamic University (IMSIU), Riyadh 11623, Saudi Arabia*
[4]*Laboratory of Materials, Processes, Catalysis, and Environment, Higher School of Technology, University Sidi Mohamed Ben Abdellah, Fez, Morocco*
[5]*Ministry of Health and Social Protection, Higher Institute of Nursing Professions and Health Techniques, Fez, Morocco*
[6]*Sivas Cumhuriyet University, Faculty of Health Sciences, Department of Nutrition and Diet, Sivas, Turkey*
[7]*Department of Materials Science and Nanotechnology Engineering, Usak University, Usak, Turkey*

2.1 Introduction

Major impacts of water pollution relate to a few problems: interference in the ecological balance or loss of biodiversity, risk to human health, and so on. Degradation in aquatic habitat owing to industrial effluent, agricultural runoff, sewage from urban areas, etc., has caused loss of biodiversity or disrupted ecological balance. It may include heavy metals, organic pollutants, and nutrients that could be used in potential effects on the health or habitat of an organism [1, 2]. Eutrophication in water bodies is a result of excessive nutrient inputs, especially from nitrogen and phosphorus sources emanating from agricultural fertilizers, sewage, and runoff [3, 4]. It promotes algal blooms, oxygen depletion, and dead zone creation; such conditions have frequently been linked to fish kills, loss of biodiversity, and decline in water quality [5]. Bioaccumulation and Biomagnification: The toxicants, mainly the persistent organic pollutants (POPs) such as Polychlorinated Biphenyls (PCBs), dioxins, and certain pesticides that have been accumulated by the aquatic organisms, resulted in biomagnification via the food chain [6].

This threatens the lives of top predators, including human beings, through adverse health impacts resulting from their use of contaminated prey or altered ecosystems [7]. Water pollution decreases the fundamental ecosystem services such as cleaning of water, nutrient cycling, and regulation of floods [8, 9]. Aquatic ecosystems are degraded in such a way that minimizes their ability to provide humankind with a clean supply of water, fisheries, and other recreational opportunities [10]. Further alterations in hydrological regimes result in increased runoff, sedimentation, and flash flooding commonly associated with urbanization and land use conversion [11]. This disrupts the hydrological cycle, consequently increasing erosion and sedimentation and transporting more pollutants into the water body. The worst part is the further deterioration in water quality and ecosystem health [12]. Other methods of entry into and propagation within the aquatic body in invasive species take place as ballast water discharge, aquaculture, and the accidental release [13, 14]. Invasive species outcompete native species for food and space, disturb nutrient cycling and food webs, and alter ecosystem dynamics, thus causing cascading ecological effects and biodiversity loss [15]. For such environmental challenges, a holistic approach to water management is required, from pollution prevention to the sustainable use of the resource and even policy intervention and ecosystem restoration. This is the reason why, in the effort of reducing water pollution and maintaining ecosystem health, this involves maintaining the integrity of all aquatic environments for their continued uses from the present to the future.

Treatment of wastewater is, therefore, an important feature of environmental stewardship trying to reduce the negative impact that pollutants have on ecosystems and human health as a whole [16, 17]. For a long period, wastewater treatment has been considered a major means of ensuring public health, environmental protection, and sustainable development [18]. Wastewater treatment removes or reduces the contaminants, minimizing the spread of waterborne diseases and thus safeguarding public health. Discharging untreated wastewater into natural water bodies may lead to eutrophication, habitat degradation, and loss of biodiversity [16]. Wastewater treatment avoids accumulation of pollutants in the environment, maintains aquatic ecosystems, and reduces the negative impacts of pollution on flora and fauna [19]. Treated wastewater can, therefore, be directly recycled and reused in many purposes such as irrigation, industrial purposes, and nonpotable uses, thus preventing over-extraction of freshwater resources and alleviating pressures on water supplies [20, 21]. The treatment of wastewater enables the sustainable management of water, hence enhancing resilience to water scarcity and climate change [16]. Many governments worldwide have set regulations and standards that control wastewater effluent into the environment [16]. Wastewater treatment brings economic development and an improved lifestyle to many communities worldwide. Finally, wastewater treatment protects public health, preserves the environment and resources, maintains regulatory requirements, and promotes economic growth. Effective wastewater treatment will provide the basis for resilient, sustainable communities and protect the health and well-being of current and future generations.

Chitin and chitosan have many interesting physicochemical and biological properties (biocompatibility, bioactivity, biodegradability), which point to a wide range of potential applications. Chitosan has attracted more interest than chitin and has been used in economically buoyant sectors such as the food, cosmetics, pharmaceutical, and agricultural industries (Figure 2.1).

The high adsorption capability, cationic nature, and versatile chemical characteristics of chitosan, a natural polysaccharide derived from chitin, have made it a highly promising candidate for wastewater treatment [22, 23]. This chapter, therefore, dwells on the synthesis and characterization of chitosan biopolymer material and its applications in wastewater treatment. In the last couple of years, much effort has been devoted to the fabrication of chitosan-based materials tailored for effective withdrawal of contaminants from aqueous solutions [24, 25]. Because of its unique physicochemical properties, such as large surface area, rich functional groups, and excellent biocompatibility, chitosan is a highly suitable candidate in various applications of wastewater treatment mechanisms involving adsorption, flocculation, and ion exchange [26]. These authors tried to optimize the performance of chitosan by capitalizing on its inherent properties and incorporating novel synthesis and characterization techniques, intending to extend its applicability to various scenarios of wastewater treatment [27].

This chapter highlights the synthesis methodologies that have been developed to improve adsorption properties and stability of chitosan biopolymer materials.

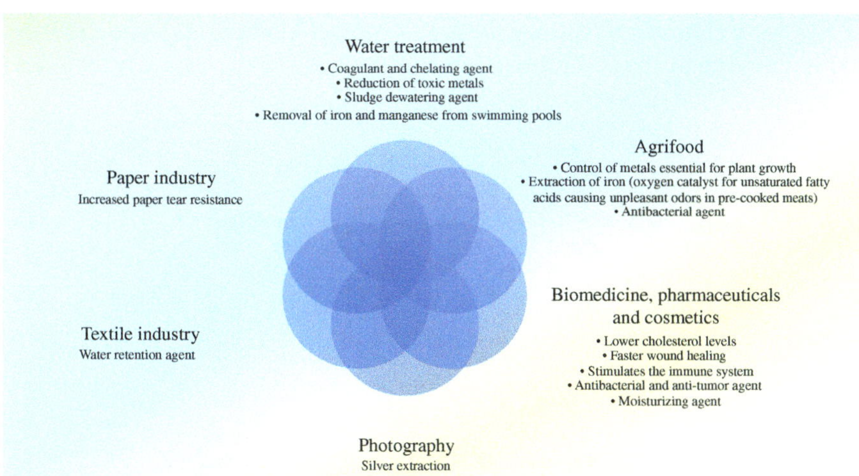

Figure 2.1 Some areas of application for chitosan.

Furthermore, it also highlights the different characterization techniques employed for determining the structural, morphological, and physicochemical properties of chitosan-based adsorbents, important for giving deeper insight into the adsorption mechanisms and performances of such materials under different environmental conditions. It will also discuss current progress on the application of chitosan biopolymer materials for the removal of such pollutants, heavy metals, dyes, pharmaceuticals, and organic contaminants in wastewater streams. Generally, synthesis and characterization of chitosan biopolymer materials are promising routes to achieve sustainable and cost-effective methods for the treatment of wastewater. This chapter outlines the basic principles that govern the design and fabrication of chitosan-based adsorbents and describes their performance in their applications to develop viable alternatives for contributing to the advancement of the address of global challenges resulting from water pollution.

2.2 Synthesis of Chitosan Biopolymer

2.2.1 Natural Sources of Chitosan and Structure

It is, in fact, the case that Felix Hoppe-Seyler coined the term "chitosan" in 1894 [28]. It was not until 1950 that the chemical structure of chitosan was identified [29] (Figure 2.2). Improvements in methods for extraction, purification, and modification made it possible to produce chitin and chitosan on an industrial scale, which led to a sea change in the commercialization of these substances in the 1970s [30]. The yearly output of chitin in the biosphere is predicted to be approximately 1000 billion tons [31]. The creation of efficient and economical processes for removing chitin from shrimp shells [32], fungal biomass [33], and mollusks [34] and turning them into chitosan was a major factor in the successful commercialization of these important biopolymers.

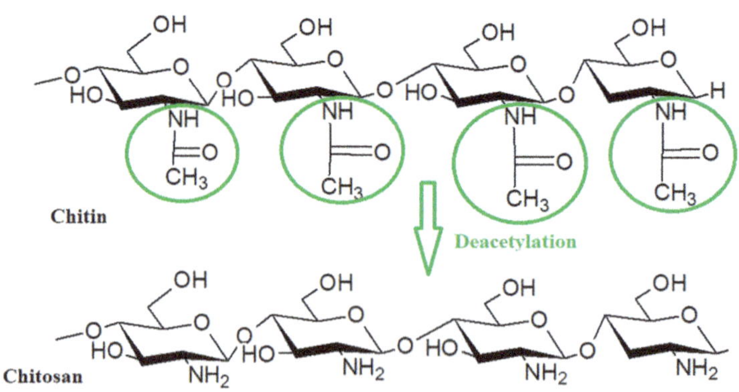

Figure 2.2 Chitosan chemical composition structure.

Depending on how the deacetylation and extraction processes were carried out, chitosan's degree of deacetylation (DDA), or proportion of free NH_2 groups, can change [35]. Chitosan, the deacetylated version of chitin, is created through this procedure, which entails removing acetyl groups from the chitin molecule [36]. Chitosan's deacetylation levels have been shown to vary, usually ranging from 50% to 90%, according to studies [37]. It is composed of N-acetylglucosamine and glucosamine units connected by β (1-4)-glycosidic linkages (Figure 2.1). The chitosan production parameters and preparations are greatly influenced by the supply of natural chitin. Several factors, including the origin of chitin, the technique of extraction, and the level of deacetylation, affect the yield of chitosan derived from chitin. Numerous techniques, including Fourier transform infrared (FTIR) spectroscopy [38], nuclear magnetic resonance (NMR) spectroscopy [39], potentiometric titration [40], and elemental analysis [41], can be used to compute the DDA.

The insolubility of chitosan in neutral or alkaline pH solutions is primarily caused by the presence of acetamide or primary amino group residues, which are necessary for the formation of strong intra- and/or intermolecular hydrogen bonds. Chitosan is typically not soluble in water due to the presence of robust hydrogen bonds, which contribute to the stability of its crystalline structure. Also, the reduction of chitosan's molecular weight changes its properties and restricts its ability to dissolve, which therefore hinders its extensive applications. Several studies were able to investigate the solubility of chitosan at various acidic conditions, and they were able to illustrate the effect of protonation on the solubility characteristics of materials. Rinaudo et al. [42] did research to test the solubility of chitosan in solutions of acetic acid [42]. The study explained the importance of protonation in enhancing solubility. When the pH drops below 6.5, the amine groups in chitosan become protonated to form positively charged ammonium groups. This enhances the electrostatic repulsion between the molecules [43]. This action has the result of making chitosan more soluble in the above-mentioned acids when they are diluted, including hydrochloric acid, malic acid, lactic acid, acetic acid, and succinic acid [44].

Chitosan can be categorized into three primary groups according to their molecular weight ranges: low molecular weight chitosan (LMWC) with a weight below 150 kDa, medium molecular weight chitosan (MMWC) with a weight between 150 and 700 kDa, and high molecular weight chitosan (HMWC) with a weight above 700 kDa [45]. Chitosan's structural and chemical properties, such as its solubility, viscosity, and biological activity, are significantly impacted by its molecular weight, which is an essential element in determining these properties. It is primarily the degree of polymerization and deacetylation that chitosan possesses that is responsible for determining its molecular weight [46]. The pharmaceutical industry mostly favors low molecular weight (MW) chitosan over high MW chitosan because of its distinct function in drug delivery applications [47].

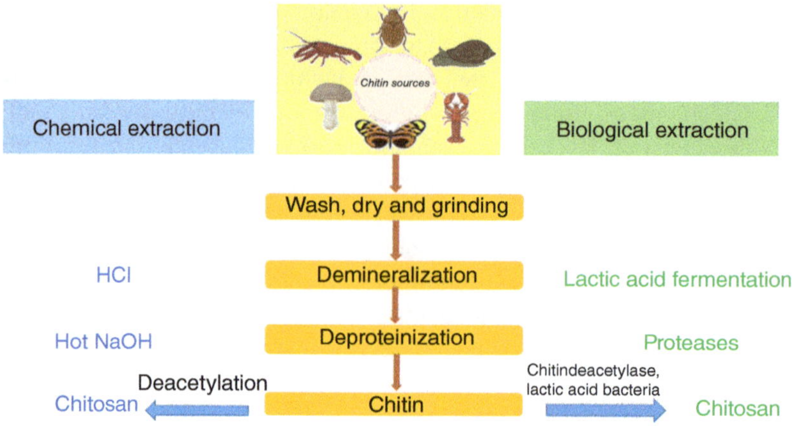

Figure 2.3 Extraction of chitin using chemical and biological techniques.

2.2.2 Chemical and Biotechnological Methods of Chitin Extraction and Conversion to Chitosan

Chitosan can be extracted from chitin sources using chemical, biochemical, or biological methods. Chemical and biological techniques are utilized in the extraction of chitin, as shown in Figure 2.3.

2.2.2.1 Chemical Extraction

The chemical approach comprises four consecutive processes: demineralization, deproteinization, discoloration, and deacetylation. The acquisition of chitosan from chitin entails a series of fundamental stages. At first, chitin is obtained from chitin sources by employing inorganic acids (HCl, HNO_3, H_2SO_4, CH_3COOH, and HCOOH) to remove minerals like calcium carbonate ($CaCO_3$) and calcium phosphate ($Ca_3(PO_4)_2$) and high alkali (>40 wt.% NaOH) to remove proteins [48]. The following alkaline solutions for the deproteinization step are included: NaOH, Na_2CO_3, KOH, K_2CO_3, $Ca(OH)_2$, Na_2SO_3, $NaHSO_3$, $CaHSO_3$, Na_3PO_4, and Na_2S [49].

The next step in making chitosan is deacetylating the chitin. Sodium hydroxide (NaOH) is a typical solvent for this deacetylation reaction; however, it requires careful attention to detail. For example, chitosan may be synthesized by dissolving chitin in an alkaline solution at a specific concentration and temperature for a specified amount of time [50, 51]. Factors influencing deproteinization include temperature, alkali concentration, and the mass-to-alkali ratio (w/v) of the shells. An important consideration is the solid-solvent ratio and the temperature of the solution [52]. After the deproteinization and demineralization procedures, the decolorization stage in the synthesis of chitosan is an important step that comes after their completion. Pigments like melanin and carotenoids can be removed by using an organic solvent [49]. A wide range of techniques are employed for the purpose of decolorization, which encompasses the utilization of chemicals such as chloroform, hydrogen peroxide, and acetone and the exposure of the substance to sunlight [53]. The discoloration procedure significantly reduced the MW of the chitosan polymer chain by breaking it down [54].

The deacetylation stage is crucial since it has a substantial influence on the characteristics and excellence of the ultimate chitosan product. Research has demonstrated that the DDA of chitosan, which indicates the proportion of acetyl groups eliminated, is crucial in defining the functional properties of chitosan. Increased levels of deacetylation are frequently linked to the enhanced solubility, biocompatibility, and antibacterial efficacy of chitosan [55].

2.2.2.2 Biological Extraction

Biological treatments provide an alternate method for extracting chitin from crustacean shells. This is done in order to avoid the use of acidic and alkaline treatments, which have the potential to cause environmental issues. Fermentation has been considered to be among the alternative technologies that are the most benign to the environment, risk-free, technologically adaptable, and economically viable [56]. In order to demineralize crustacean shells, microorganisms that produce lactic acid have been used. The lactic acid that is created by bacteria is able to react with the calcium carbonate that is present in the waste from biomass, which results in the development of calcium lactate. This calcium lactate can then be precipitated and removed by washing [57]. The advantages of solid-state fermentation over the conventional methods are many, and the bacterial strains used most frequently are *Lactobacillus* sp. and *Clostridium* sp. [58]. Pacheco et al. [59] produced chitin from shrimp waste using lactic acid fermentation and obtained high rates of demineralization and deproteinization [59]. The method of extraction of chitin from biological sources in their case involved the use of bacteria producing protease enzymes, which degrade proteins, and bacteria producing lactic acid for the removal of minerals [60]. This study indicated that lactic acid fermentation can be an effective source for the extraction of chitin and chitosan from wastes, a feasible ecological method in the manufacture of chitosan.

Lee et al. [61] investigated the production of proteases and chitin from chitin-containing fish by-products by using *Paenibacillus elgii* [61]. The study revealed several proteases produced by *P. elgii*, which showed that microbial strains could be utilized for the effective extraction of chitin and proteases from various wastes [61]. Jabeur et al. [61] conducted a study that specifically examined the process of extracting and characterizing chitin, chitosan, and protein hydrolysate from the invasive Pacific blue crab [61]. The study placed particular emphasis on investigating the biological activity of these recovered chemicals. The research used a central composite design to enhance the synthesis of protease, underscoring the significance of effective enzyme production for chitin extraction procedures.

In that perspective, deproteinization is necessary in order for chitin to be isolated while in the process of extracting it through the production of protease enzymes. Deproteinization by enzymes refers to a process where the enzymes catalyze the proteins contained within the chitin source into smaller fragments and eventually obtain efficient chitin. Some literature reveals that enzymatic deproteinization helps to obtain chitin. Manni et al. [62] used enzymatic deproteinization of shrimp shell wastes using a protease from Bacillus cereus SV1 to assist in synthesis of chitin [62]. This approach highlights enzyme-based methods for protein removal in chitin extraction processes. The enzymatic technique offers an environmentally benign alternative to conventional chemical procedures and has shown potential for effectively extracting chitin from crustacean shells [63]. The method normally consists of three phases: deproteinization, demineralization, and deacetylation. These processes are necessary in order to create chitosan with the needed characteristics (Figure 2.3) [64].

This means that selection between biological or chemical techniques of chitosan extraction is to be decided on various bases such as required nature, environmental problems, cost-effectiveness, and other relevant grounds. The biological processes promise a greener path where purer grades of chitosan may also be achieved, but chemical ones stand tall in efficiency and are amenable to scale-up.

2.2.3 Modifications of Chitosan-Based Materials

Chitosan, in its raw form, is still unable to fulfill the requirements of each and every application. As a result, techniques of modification that include physical, chemical, and biological processes have been established (Figure 2.4).

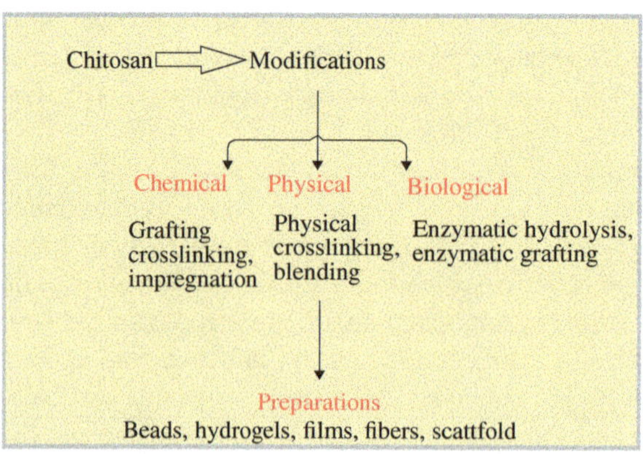

Figure 2.4 Diverse techniques for modifying chitosan.

2.2.3.1 Physical Modification

Modifying the physical characteristics of chitosan is essential for customizing its features for different uses, while keeping its chemical makeup the same. Chitosan may be physically cross-linked using many techniques, including temperature variations, pH adjustments, or ionic interactions. This method is essential for improving the mechanical characteristics and stability of materials made from chitosan.

Chitosan has the ability to be blended with other substances to create chitosan-based composites that possess a diverse array of characteristics. The often used additives for blending with chitosan encompass polymers, enzymes, microorganisms, medicines, nutrition, carbon-based chemicals, magnetic materials, and metal nanoparticles [65].

2.2.3.2 Chemical Modification

Chitosan molecules are amenable to simple chemical modification due to the polymer chain's reactive amino (—NH_2) and hydroxyl (—OH) groups. Chemical modification is recognized for its ability to change the physicochemical characteristics of polymeric surfaces, including electrostatic charge and penetration [66]. It is possible to change the molecular structure of chitosan using a number of different chemical modifications, including acylation, carboxylation, and etherification, in particular graft copolymerization and blending modification. This, in turn, results in the production of a variety of chitosan derivatives, which, among other things, have the potential to improve the fluidity of chitosan, its biological activity, and its mechanical qualities [67].

- Acetylation – Acetylation is the treatment of imparting acetyl groups to chitosan, thereby reversing its deacetylation process, hence rendang it closer to chitin. Lamarque et al. [68] studied the physicochemical properties of a set of acetylated chitosan in water, focusing on the effects of several structural factors on acetylated chitosan [68].
- Carboxylation – During carboxylation of chitosan, the carboxyl groups get attached to the structure of chitosan. This is done in order to attribute some special properties and uses to the material. In this regard, Shen et al. [69], while studying carboxylation effects on the antibacterial properties and cytocompatibility of the carboxylated chitosan/silver-hydroxyapatite hybrid microspheres [69], showed that the carboxylation resulted in improving the material properties with respect to its antibacterial property as well as its cytocompatibility and thus reflected merits of the carboxylation method for material property improvement [69].
- Etherification – The hydroxyl group of chitosan reacts with methyl ether, benzyl ether, and ether, among others, to form the ether molecule as a product [67].

- Cross-linking – Chitosan cross-linking includes the formation of covalent linkages either between chains of chitosan or between chains and other molecules to form a network structure. Chen et al. [70] studied adsorption of Cu(II), Zn(II), and Pb(II) ions using chitosan that was cross-linked with epichlorohydrin [70]. This study has identified that chitosan cross-linked with epichlorohydrin enhances its resistive nature against corrosive action of acids and chemicals [70].
- Quaternization – The quaternization of chitosan is done with the addition of the quaternary ammonium group to the chitosan molecule, hence leading to the development of a new form of chitosan that is water-soluble and has different properties. In the study, Wang et al. [71] considered the quaternization process for improving the aqueous solubility, pH sensitivity, and targeting potential of chitosan derivatives for pharmaceutical applications [71].
- Chitosan phosphorylation – This involves the attachment of phosphate groups to the chitosan molecule. In this, various kinds of phosphorylated chitosan derivatives having variable nature are obtained. Suchyta and Soto, in 2017, worked on the specific monophosphorylation of chitosan through the use of phosphorus oxychloride [72]. They intended to increase hydrosolubility without the generation of polyphosphates.
- Grafting – This involves the chemical attachment of polymer chains or monomers onto the chitosan backbone. In this case, chitosan-based copolymers with value-added characteristics are produced.
- Acylation – Acylation of chitosan involves the interaction of the chitosan with a great number of organic acids and their derivatives, leading at the end to the attachment of acyl groups to chitosan.
- Sulfonation – The sulfonation of chitosan involves the attachment of sulfonic acid groups on the chitosan molecule, thus forming sulfonated chitosan derivatives that exhibit different properties.

2.2.3.3 Biological Modifications

Chitosan may be modified biologically by adding bioactive molecules, such as peptides, proteins, growth factors, or other biological agents, to give it specialized biological activities.

- Enzymatic deacetylation of chitosan is the process of using enzymes to eliminate acetyl groups from chitosan, which leads to the creation of chitosan with a reduced degree of acetylation.
- Enzymatic degradation of chitosan: The enzymatic degradation of chitosan includes the degradation of chitosan molecules using enzymes, a process which results in either chitooligosaccharides or low numerical weight products.

- Attaching enzymes to chitosan-based carriers is the process of enzyme immobilization for chitosan. This process enables better catalytic characteristics, increased stability, and reusability of the enzymes.
- Biological grafting of chitosan involves the attachment of physiologically important molecules to chitosan, thus changing the original biopolymer and producing value-added polysaccharides. This approach aims at the optimization of the properties and potentials of chitosan for a wide range of applications. Several research have concentrated on grafting phenolic compounds and other biological molecules onto chitosan in order to enhance the bioactivities of chitosan. The grafting procedure involves the bonding of active groups to the chitosan structure by exploiting the reactive hydroxyl and amino groups that are naturally present in chitosan. This technique enables the augmentation of the biological activities and characteristics of materials based on chitosan.

Hydrogels are three-dimensional networks capable of retaining considerable amounts of water and, therefore, suitable for drug delivery. Chitosan-based hydrogels have been extensively studied in terms of their ability to encapsulate pharmaceuticals and provide controlled release [73]. Xerogels and aerogels, obtained from hydrogels through drying processes, present low density and high porosity [74]. These attributes render them appropriate for applications requiring lightweight materials with high surface area, such as drug delivery and tissue engineering [75]. Chitosan xerogels offer favorable results in drug loading and release, thanks to the intrinsic characteristics of chitosan that enhance drug solubility and bioavailability. Aerogels can be engineered to provide precise release profiles and targeted delivery, which makes them appropriate for sophisticated therapeutic applications [75, 76]. Chitosan-derived films are one of the important techniques in drug delivery. They may act as protective barriers and also as matrices for controlled release. Chitosan films have excellent mechanical properties and biodegradability, which make them useful in packaging applications and for direct drug delivery [77]. The addition of various additives can enhance chitosan film functionality by way of improvement in barrier properties or modification of release kinetics [78].

2.3 Characterization and Mechanical Proprieties of Chitosan Biopolymer

2.3.1 Characterization

Physicochemical, structural, and functional properties of chitosan biopolymer have been characterized with various techniques that provide information into its structure and possible applications. These characterization methods, when singly or combination employed, provide comprehensive knowledge about physicochemical and structural as well as functional properties of chitosan. FTIR spectroscopy is among the most generally used analysis technique for the recognition of some functional groups that are contained in chitosan upon absorption of radiation in infrared region [9]. It provides information about the chemical structure and composition of chitosan. When the FTIR spectrum of chitosan presented in Figure 2.5a is examined, it is seen that the biopolymer has a wide characteristic band at 3320 cm^{-1} due to NH_2 and OH stretching. The peaks at 2925 and 2860 cm^{-1} are attributed to CH_3 and CH_2 stretching vibrations, respectively [79]. The peaks observed at 1650 and 1587 cm^{-1} are attributed to stretching vibrations that can be attributed to N—H deformations. The peak at 1420 cm^{-1} is attributed to C—O stretching vibrations in amide-I and the peak at 1327 cm^{-1} is attributed to C—N stretching vibrations in amide-II. The peaks at 1150 and 1040 cm^{-1} are attributed to —C—O—C— glycosyl bond vibrations [17]. The peak at 1020 cm^{-1} is attributed to C—O skeleton vibration [80].

Crystallinity and crystal structure of chitosan could be determined by X-ray diffraction (XRD) analysis. It gives details about the arrangement of the chains inside the polymer. In general, the XRD of chitosan indicates an amorphous structure of the polymer. An example is shown in Figure 2.5b. Since the chitosan

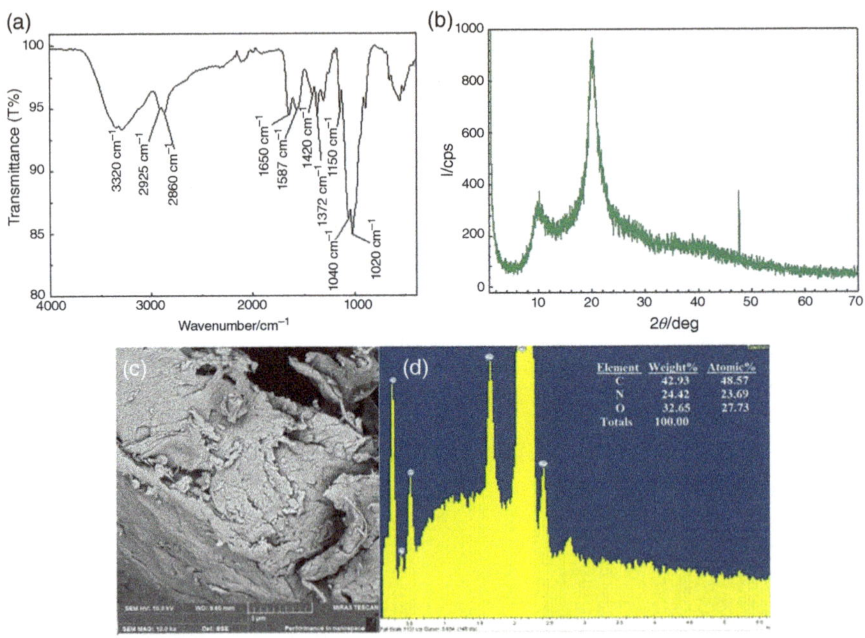

Figure 2.5 (a) FTIR spectra, (b) XRD patterns, (c) SEM images, and (d) EDX spectra of chitosan biopolymer.

has semi-crystalline structure polymer, the unit cell arrangement is orthorhombic type; hence, two characteristic peaks related to (020) and (040) plane reflections corresponding to hydrated and amorphous crystal structure are seen in the XRD spectrum [81]. Even when completely dried, the $2\theta = 9.60$ reflection observed in chitosan containing 5% water is considered to indicate the hydrated chitosan crystal structure called "crystal form I," and the $2\theta = 20.10$ reflection is considered to be an indicator of the relatively regular crystal lattices of chitosan called "crystal form II" [82]. Two strong peaks observed at $2\theta = 9.60$ and $2\theta = 20.10$ in the XRD spectrum of chitosan indicate the high crystallinity degree of chitosan. The average size (D_{hkl}) and interlayer distance (d_{hkl}) values of chitosan crystals are presented in Table 2.1.

Scanning electron microscopy and energy dispersive X-ray spectroscopy (SEM-EDX) are commonly used to visualize the surface morphology and microstructure of chitosan. SEM morphology helps understand chitosan, surface roughness, porosity, shape, particle size, and particle size distribution. EDX analysis is a technique used to analyze the elemental composition of chitosan biopolymers. The SEM view of chitosan describes its typical amorphous morphology (Figure 2.5c). The SEM morphology of chitosan exhibits a dense morphology that appears nonporous, indented, irregular, microfibrillar, and visibly rough [83]. When the EDX spectrum of chitosan (Figure 2.5d) is examined, it is seen that it contains C, O, and N elements [84].

The thermogravimetric analysis (TGA) analysis helps in the determination of the temperature at which chitosan degrades, together with the accompanying weight loss as a function of temperature. It has been used for the analysis of the thermal stability and decomposition behavior of chitosan. Differential scanning calorimetry (DSC) measures heat flux associated with phase transitions and chemical reactions as a function of temperature. It gives information about the thermal properties of chitosan, including glass transition temperature, melting point, and crystallization behavior. When the TGA and DSC graphs of chitosan are evaluated together, they reveal three temperature ranges where maximum mass losses are seen (Figure 2.6a). These are: 30–150 °C (highest rate at 76 °C), where 5% mass loss is observed; 200–350 °C (highest rate at 273 °C), where 54% mass loss is observed; and 350–600 °C (highest rate at 460 °C), where 40.4% mass

Table 2.1 Characteristic peaks observed in the XRD spectrum of Ch and D_{hkl} and d_{hkl} values from these peaks.

Material	$2\theta_{hkl}/^0$	FWHM/^{0}	D_{hkl}/nm	d_{hkl}/nm
Chitosan	9.6	4.11	2.30	0.92
	20.1	2.16	3.89	0.44
ORT ± SS			3.10 ± 1.12	0.52 ± 0.37

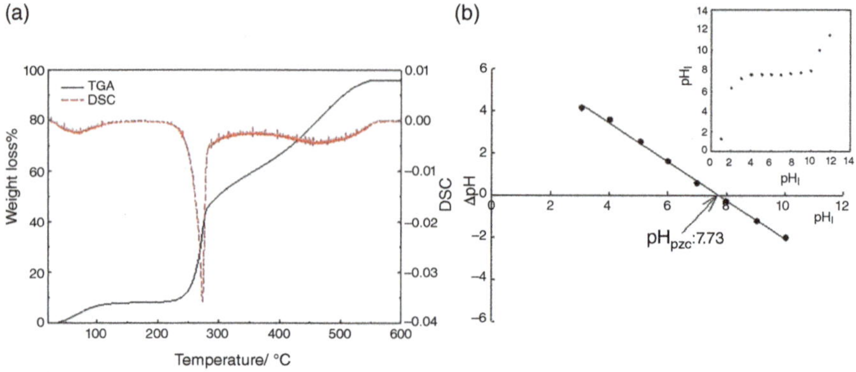

Figure 2.6 (a) TGA (solid black line) and the DSC (dotted red line) curves, and (b) the point of zero charge of chitosan biopolymer.

loss is observed. The results found are as explained in previous research. In the 25–200 °C range, water bound to chitosan chains with weak hydrogen bonds or physically adsorbed evaporates. The peak observed in the 200–350 °C range is the main degradation peak of chitosan, and the ring opening reactions and degradation of pyranose rings follow the depolymerization, dehydration, and amine structures of chitosan chains. Finally, it is explained as the formation of volatile decomposition products between 360 °C and 600 °C [85].

Zeta potential measurements are used to analyze the surface charge of chitosan particles in solution. It provides information about the stability and colloidal behavior of chitosan dispersions. The solution pH value where the surface charge is zero is defined the point of zero charge (PZC). One of the methods used to determine as the PZC is the solid addition method [86]. In this method, chitosan is kept in solutions containing KNO_3 with different initial pH values (pH_i) for 24 hours, and then the equilibrium pH values (pH_f) are measured. ΔpH values are calculated with $\Delta pH = pH_f - pH_i$ and $\Delta pH_f - pH_i$ graphs are drawn. The point where the resulting lines intersect the x-axis is determined as the PZCs. The PZC graph of chitosan drawn with this method is presented in Figure 2.6b. The $pH_f - pH_i$ graph presented in the figure explains the pH ranges in which the material has buffering properties. The pH_{PZC} value found for chitosan with this method is 7.73, and the range in which chitosan has buffering properties is pH: 3.5–10 [87].

2.3.2 Mechanical Properties

Chitosan has an important place among many polymers due to its biocompatibility, biodegradability, nontoxicity, high water permeability and mechanical strength, excellent film-forming ability, and sensitivity to chemical modifications

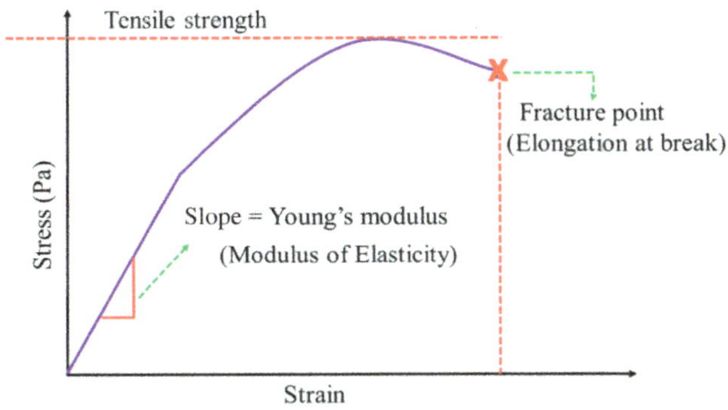

Figure 2.7 A symbolic presentation of the stress–strain curve.

[88]. Because of these properties, chitosan, a natural polymer, has numerous applications in various industries, such as pharmaceuticals, cosmetics, medicine, and agriculture. While this natural polymer enables outstanding applications, it can suffer from serious limitations in some applications. This disadvantage is due to the weak mechanical stability of CS's films, scaffolds, and beads. In this context, it is important to improve the mechanical properties of CS. In recent years, different approaches have been adopted to increase this property.

Generally, the investigation of the mechanical properties of different materials is performed by tensile and tension tests. In these tests, the sample to be analyzed is clamped to the measuring system, and a load (F) is applied longitudinally to the sample. The load F is gradually increased to record the elongation of the sample. Then, the stress–strain curve is generated, which is symbolically represented in Figure 2.7.

Parameters such as modifying agent, form of CS, and modification type play an important role in determining suitable approaches to enhance the mechanical properties of materials. Sen et al. modified the chitosan matrix by chemical cross-linking using poly(ethylene glycol) diglycidyl ether (PEGDGE) and thus developed novel hybrid hydrogel sponges with tailor-made elasticity. Using this approach, the elastic modulus values obtained for these sponges varied in the range of 0.1–2 MPa [89]. The effect of glutaraldehyde (GA) concentration on the mechanical properties of CS hollow fibers was investigated by Tasselli and co-workers using chemical cross-linking modification types. The results showed that the elongation at break decreased as the GA concentration increased, while the tensile strength and modulus obtained their maximum values at the lowest GA concentration value [90]. In the other study, whose modified type was selected as chemical cross-linking, chitosan/poly(vinyl alcohol)/genipin or glutaraldehyde (CS/PVA/GEN and CS/PVA/GA) hydrogels were formed by Palafox and Arévalo to improve their mechanical properties. The results revealed that GEN-based hybrid hydrogels showed similar elongation rates and structural parameters as GA-based hydrogels. Furthermore, the elastic moduli of GEN and GA-based hydrogels were determined to be 2.08 ± 0.11 and 2.82 ± 0.33 MPa, respectively [91]. Zang et al. investigated the effects of vanillin cross-linking agents on chitosan films as well as their mechanical properties, such as tensile strength and elongation at break, using different characterization techniques. The characterization results show that the vanillin-added chitosan film caused a significant decrease in % E, elongation at break values, and a significant increase in tensile strength values compared to the pure chitosan film [92].

In addition to these studies, unlike the chemical cross-linking modification type, incorporation of plasticizers was preferred as the modification type and the mechanical properties of CS-based films prepared by using glycerol as modifying agent were investigated by Prateepchanachai et al. [93]. In the other study, whose modified type was selected as the incorporation of plasticizers by Ma et al., glycerol was added as modifying agent to conjugate chitosan as plasticizer. Results show that with the increase of glycerol, the tensile strength decrease about 78% while the tensile elongation increased about 186% [94].

Besides chemical cross-linking and plasticizer modification types, blending with polymers or distinct compounds is also preferred as a modification type to enhance the mechanical properties of CS. While using this modification type, gallic acid [95], gelatin [96], sporopollenin [97], and anthocyanin extract [98] were selected as modifying agents to form CS-glycerol films.

2.4 Wastewater Treatment Application via Adsorption

The polyelectrolyte, chelating, and complexing properties of chitosan make it a good candidate for environmental applications in wastewater treatment. These applications are based on chitosan's polycationic character in acidic media and its solubility, which opens the way to numerous applications in coagulation-flocculation, chelation, adsorption, or ultrafiltration. Indeed, its protonated amine groups in acidic media are capable of binding all transition metals and radionuclides. Chitosan can be used as a coagulant, chelating agent, or adsorbent to complex a wide range of pollutants, such as aromatic and phenolic molecules and organic dyes (food, paper, and textile effluents). Another recent application is the immobilization of microorganisms or sludge in chitosan-based polymer matrices for wastewater treatment, again based on the same encapsulation principle. Chitosan is a biopolymer obtained by partial deacetylation of chitin, one of nature's most abundant polysaccharides. Chitin is present in the shells of crustaceans, the walls of certain filamentous fungi, and the cuticle of insects. This biopolymer is the acetylated precursor (N-acetyl-D-glucosamine units associated with β (1 → 4) bonds) of chitosan, a heteropolymer made up of D-glucosamine units and N-acetyl-D-glucosamine units. Chitosan is rarely found in nature, is identified only in a few crustacean organelles or in the walls of Mucorales, and is obtained commercially by deacetylation of chitin. As such, it would be more appropriate to speak of chitosans, as the properties of the biopolymer can vary according to its characteristics: molecular weight affects the viscosity of the biopolymer solubilized in acid medium, its DDA, the solubility of the macromolecule, and its degree of crystallinity and hydration capacity. Thus, changing its DDA and/or its molecular weight will considerably modify its mode of use and performance. The presence of numerous hydroxyl groups confers significant hydration properties, while amino functions influence the polymer's ability to be modified both physically (solubilization and shaping) and chemically (grafting of reactive groups, cross-linking, etc.). These amino functions give chitosan unique cationic properties among polysaccharides. Protonation of these amino functions leads to solubilization of the biopolymer, which can be used in dissolved form for water treatment operations, notably coagulation-flocculation, complexation-assisted ultrafiltration, or to prepare original forms (hydrogel beads, membranes, sponges/foams, fibers, hollow fibers, etc.). This protonation also ensures the material's ion-exchange properties, vis-à-vis anionic solutes, for example. In this way, the material can be used directly in solid form for adsorption applications, taking advantage of the presence of amino functions (complexation of cations in near-neutral media, electrostatic attraction/ion exchange in acidic media for the attachment of anions to protonated amino functions). It can also be used in the liquid state (dissolved in acid medium) for charge neutralization and flocculation, coagulation of anionic compounds (dissolved solids, aloids, and suspended particles), and metal ion complexation combined with retention by semi-permeable membranes in the case of complexation-assisted ultrafiltration. Chitosan has remarkable potential in wastewater treatment, in particular, due to its adsorption capacity, which enables it to effectively capture various contaminants present in wastewater. This section explores in detail the various applications of chitosan in wastewater treatment, highlighting its adsorption mechanisms, its effectiveness in removing heavy metals and organic pollutants, and the prospects and challenges associated with its use.

2.4.1 Adsorption Mechanism

Chitosan is effective as a biopolymer for the purification of wastewater owing to its high affinity for many contaminants, including metal ions, dyes, pesticides, and volatile organic compounds (VOCs) [23, 99–103]. The chitosan adsorption mechanism includes physicochemical interactions, namely, electrostatic attraction, coordination bonds, and hydrophobic interactions. Figure 2.8 illustrates a schematic model of these mechanisms.

The specific interactions between chitosan and the compounds to be adsorbed are highly dependent on the chemical structure of chitosan [99]. Chitosan's

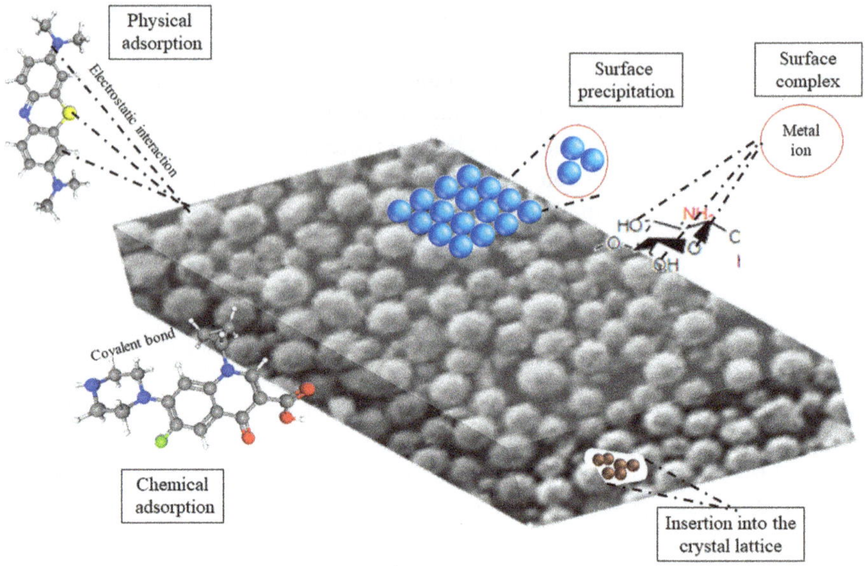

Figure 2.8 Main adsorption mechanisms.

hydroxyl and amine groups play a crucial role in these interactions. While amine groups can be ionized as a function of pH, enabling electrostatic interactions with anionic compounds, nonionized amine groups can interact with cations thanks to their free electron doublet [104]. This, among other characteristics, makes the degree of acetylation an important structural parameter. Higher degrees of deacetylation lead to greater complexation capability with metal ions [105]. The presence of acetylated groups in chitosan provides an amphiphilic character to it; thus, it can also interact with compounds that are slightly hydrophobic. However, the exact mechanisms involved in the different types of adsorptions by chitosan require further investigation [106]. At present, most research has focused on elucidating the mechanisms of metal ion adsorption. Although hydroxyl groups help stabilize interactions between the amine group and metal ions, it is generally accepted that adsorption occurs primarily via amine groups [107].

Hydroxyl group contribution to metal adsorption occurs through several mechanisms, including the complexation of nitrogen atoms or the formation of ternary complexes, and/or electrostatic attraction by ion exchange [108]. The protonation status of amine groups provides a competition between the metal cation adsorption-ternary complex formation processes. pH is an essential factor in determining the speciation of metal species and, hence, the ionic or neutral form of an amine group. Various models have been advanced to describe the coordination mechanism involved in forming metal-chitosan complexes [109].

According to the bridge model, the metal ion is bonded to several nitrogen atoms in either the same or different chains, and according to the pendant model, it is bonded with an individual amine group [110].

2.4.2 Heavy Metal Removal

Heavy metals emission into the environment, stimulated by industrialization and urbanization, is one of the major challenges nowadays all over the world [14, 111, 112]. The pollution of the environment, including greenhouse gases and water-soluble toxic pollutants, is a serious problem all over the world. Among them, water cleaning and health are the most important, considering the importance of water for all living organisms. The methods of treatment include desalination, purification, osmosis, disinfection, and contaminant deposition, but one considered advantageous is adsorption. Chitosan is considered a pollutant adsorbent due to its low cost and very high adsorption potential as a result of hydroxyl and amino groups that are in high proportion and capable of binding to the heavy metal ions [113–115]. Heavy metal ions, the main water pollutants, affect human health via the food chain. They are widely present in the environment due to industrial activities. Chitosan and its derivatives are widely used in heavy metal ion adsorption for their high hydrophilicity due to the large number of hydroxyl and amine groups and the high chemical reactivity of these functional groups [116]. Such efficiency in heavy metal removal from wastewater by chitosan is based mainly on amine ($—NH_2$) and hydroxyl (—OH) functional groups. In chitosan, active sites can be established because of such groups, offering opportunities for selective adsorption in metal species such as Pb(II), Cd(II), Ni(II), and Cr(III). The present adsorption is based mainly on the electrostatic and complexation interactions between metallic ions and the chitosan functional group. If chitosan is in an aqueous solution, it could protonate the amine groups into positively charged protonated amine groups. The produced protonated amine groups would further interact with the negatively charged metal ions in water. For instance, for Pb, Cd, Ni, and Cr, the adsorption reaction can be described by Eq. (2.1), where M is heavy metal:

$$CH_2NH_3^+ + M^{2+} \rightarrow CH_2NH_3M^{2+} \tag{2.1}$$

The pH value of the solution has an influential role in this adsorption, since it determines the protonation status of chitosan amine groups. Increasing the pH may cause deprotonation of the amine groups and reduce their tendency to adsorb heavy metals through the loss of favorable electrostatic interactions. Chitosan, with active site-rich functional groups that act based on solution pH, selectively adsorbs heavy metals and presents an effective solution toward the mitigation of environmental risks due to heavy metal contamination. Indeed, chitosan proves to be a very effective heavy metal adsorbent via its chelation sites, which attract metal ions either by ion exchange or coordination bonds. Chitosan modified with amine groups, carboxymethylation, quaternization, metal oxides, or conductive polymers enhances its metal ion adsorption efficiency [117]. In addition, cross-linking chitosan improves its stability under acidic conditions and its mechanical properties, although this may reduce its adsorption capacity [117]. It is already realized that chitosan and its derivatives can serve as very active adsorbents of metal ions from solutions. Such complexation could be facilitated by several mechanisms, such as electrostatic interactions and complexation. An illustrative example of an effective adsorbent was represented by a sponge-type adsorbent comprising chitosan-lignin G and polyethyleneimine, showing fast and effective removal of mercury (II) ions within 6 hours, reaching a maximal adsorption capacity of 663.5 mg g^{-1}. Spectroscopic analyses revealed that mercury (II) ions were complexed by the functional groups of the nitrogen-containing adsorbent [118]. In another related work, ethylenediaminetetraacetic acid and graphene oxide added to chitosan increased the adsorption of lead (II) ions by increasing the chelating functional groups [119]. Similarly, the modification of chitosan with graphene oxide and Fe_3O_4 nanoparticles has created a magnetic composite effective for the adsorption of chromium (VI) and copper (II) ions [120]. Finally, a biosorbent complex of chitosan and alginate proved effective in removing cadmium (II) and lead (II) ions from real water samples, thanks to the combination of various functional groups present in both materials [121]. Table 2.2 summarizes several examples of heavy metal ion removal using chitosan-based materials and their removal characteristics. For Pb^{2+}, a chitosan-iron oxide-amine-thiol nanocomposite recorded a capacity of 282.4 mg g^{-1} [122], chitosan-coated magnetic silica nanoparticles showed a capacity of 268.01 mg g^{-1} [123], while magnetic cross-linked chitosan nanoparticles modified with ethylenediamine recorded 138 mg g^{-1} [124], chitosan ceramic beads recorded 38.89 mg g^{-1} [125], and cross-linked chitosan/poly(vinyl alcohol) nanofibers functionalized with an ionic liquid showed an adsorption capacity of the order of 166.34 mg g^{-1} [126]. For Co^{2+}, a chitosan-magnetite nanocomposite showed a capacity of 53.19 mg g^{-1} [128], while a chitosan/cellulose fiber composite recorded 23.6 mg g^{-1} [129]. A chitosan/cellulose/carboxymethylcellulose graphene oxide composite showed a capacity of 43.55 mg g^{-1} [130]. For Ni^{2+}, a chitosan-magnetite nanocomposite could record an adsorption capacity of between 30.03 mg g^{-1} [128] and 108.9 mg g^{-1} [131]. For Cd^{2+}, a chitosan/sodium alginate/calcium ion double-lattice hydrogel showed a capacity of 110.7 mg g^{-1} [132], while a graphene oxide/chitosan/carboxymethylcellulose composite recorded a capacity of 91.38 mg g^{-1} [130]. For Zn^{2+}, a carboxymethylated chitosan resin showed a capacity of 238.1 mg g^{-1} [127], while a chitosan-coated coconut shell recorded 60.41 mg g^{-1} [133] and chitosan ceramic beads showed a capacity of 29.87 mg g^{-1} [125].

Table 2.2 Adsorption capacities of heavy metal ions on chitosan-based materials surfaces.

Heavy metal ion	Chitosan-based materials	Adsorption capacity (mg g^{-1})	Reference
Pb^{2+}	Chitosan-iron oxide amine-thiol nanocomposite	282.4	[122]
Pb^{2+}	Chitosan-coated magnetic silica nanoparticles	268.01	[123]
Pb^{2+}	Magnetic cross-linked chitosan nanoparticles modified with ethylenediamine	138	[124]
Pb^{2+}	Chitosan ceramic beads	38.89	[125]
Pb^{2+}	Cross-linked chitosan/poly(vinyl alcohol) nanofibers	166.34	[126]
Pb^{2+}	Carboxymethyl-chitosan resin	4.78	[127]
Co^{2+}	Chitosan-magnetite nanocomposite	53.19	[128]
Co^{2+}	Chitosan/cellulose fiber composite	23.6	[129]
Co^{2+}	Graphene oxide/chitosan/carboxymethyl cellulose composite	43.55	[130]
Ni^{2+}	Chitosan-magnetite nanocomposite	30.03	[128]
Ni^{2+}	Magnetic chitosan composite	108.9	[131]
Cd^{2+}	Physically cross-linked chitosan/sodium alginate/calcium ion double-network hydrogel	110.7	[132]
Cd^{2+}	Graphene oxide/chitosan/carboxymethyl cellulose composite	91.38	[130]
Zn^{2+}	Carboxymethyl-chitosan resin	238.1	[127]
Zn^{2+}	Chitosan-coated coconut shell	60.41	[133]
Zn^{2+}	Chitosan ceramic beads	29.87	[125]
Mn^{2+}	Graphene oxide/chitosan/carboxymethyl cellulose composite	57.78	[130]
Hg^{2+}	Chitosan-nanocopper oxide composite	672	[134]
Hg^{2+}	Lanthanum-metal organic framework/citrate cross-linked chitosan composite sponge	765.22	[135]
CrO_4^{2-}	Graphene oxide/chitosan/carboxymethyl cellulose composite	77.7	[130]
Cr^{3+}	Chitosan-based adsorptive membrane modified by carboxymethyl cellulose	230.12	[136]
As^{3+}	Chitosan-based adsorptive membrane modified by carboxymethyl cellulose	234.78	[136]

Despite the wealth of scientific information in the literature on the adsorption of heavy metals by chitosan and its derivatives, little research has been carried out into the feasibility of these processes, leading to the common practice of discarding adsorbents after use, posing a serious problem for solid waste treatment. Consequently, it is inevitable to infer the viability of the application of developed chitosan-based adsorbents by increasing their adsorption performance and turnover capacity, as well as decreasing their usage time. Different desorption agents such as acids, alkalis, salts, and chelating agents were explored and found that acids are the most effective desorption agents, achieving a desorption rate of 49.3%; chelating agents were the second most effective agents, with a desorption rate of 26.9%, followed by alkalis at 14.9% and salts at 8.9%, all within an optimum desorption time estimated at between 0.84 and 1.37 hours [133]. In addition, research needs to be carried out into the leaching probability of chitosan-based adsorbents to prevent the oxidation of this material in moist air, which can then transform into other materials that are likely to release less effective organic compounds. Furthermore, real wastewater contains a wide range of contaminants, so it's crucial to study the selectivity of chitosan-based adsorbents. It is also recommended to study the scalability and economics of the adsorption process of heavy metals on the surface of chitosan or its derivatives [24].

2.4.3 Organic Pollutant Removal

Similar to heavy metals, organic pollutants are toxic for both humans and the aquatic system, which requires the attention of several researchers who suggest many technological approaches and material types [2, 137–146]. Chitosan could also be used to capture pollutants such as polycyclic aromatic hydrocarbons, pesticides, organic dyes, and endocrine disruptors, thanks to its adsorption properties and high specific surface area. However, its solubility in acidic media can be a drawback, but this can be overcome by cross-linking techniques. At low pH, chitosan undergoes protonation of the amine functions (to ammonium), which become positively charged. Chitosan then becomes a water-soluble cationic polyelectrolyte, and this protonation explains its swelling, dissolution, and reduced crystallinity. Generally, chitosan uses electrostatic interactions, hydrogen bonding, and π–π interactions for its adsorption of organic dyes. A positive charge usually develops in the protonated amine functional group of chitosan dissolved in an aqueous solution and draws negatively charged organic dyes toward its sites for appropriate adsorption to take place on its surface. The hydroxyl groups of chitosan (—OH) may form hydrogen bonds with functional groups of the organic dyes, for example, the azo group (—N=N—) and carbonyl groups (—C=O—), increasing the interaction between chitosan and dye molecules. Some organic dyes have conjugated π-bonds, which can interact by π–π interactions with aromatic groups that may be present in the chitosan structure, hence enhancing adsorption. Adsorption of organic dyes on chitosan may be represented by the following Eqs. (2.2–2.4):

$$\text{Chitosan} - \text{NH}_2^+ + \text{Dye}^- \rightarrow \text{Chitosan} - \text{Dye} + \text{H}^+ \tag{2.2}$$

$$\text{Chitosan} - \text{OH} + \text{Dye} - \text{C} = \text{O} \rightarrow \text{Chitosan} - \text{O} - \text{Dye} + \text{H}_2\text{O} \tag{2.3}$$

$$\text{Chitosan} - \pi + \text{Dye} - \pi \rightarrow \text{Chitosan} - \text{Dye} + \pi - \pi \text{ interactions} \tag{2.4}$$

The pH of the solution will be able to affect the surface charge of chitosan and organic dyes, hence affecting the electrostatic interaction between them. A higher concentration of organic dyes will saturate the active sites of chitosan, thereby reducing its adsorption efficiency. Smaller organic dyes with a flatter structure tend to be better adsorbed by chitosan due to their better accessibility to the active sites. Table 2.3 shows the adsorption capacities of different chitosan-based materials for specific organic dyes. Indeed, chitosan/*Pinus* spp. activated carbon plywood panel has an adsorption capacity of 30 mg g^{-1} for neutral red [147]. In addition, chitosan/cotton straw biochar has an adsorption capacity of 125.01 mg g^{-1} for reactive red 21 [148]. Peanut shell chitosan/activated

Table 2.3 Adsorption capacities of dyes on chitosan-based materials surfaces.

Dye	Chitosan-based materials	Adsorption capacity (mg g^{-1})	Reference
Neutral red	Chitosan/*Pinus* spp. plywood panel-activated carbon	30	[147]
Reactive red 21	Chitosan/cotton straw biochar	125.01	[148]
Indigo carmine	Chitosan/peanut shell activated carbon	208.33	[149]
Direct blue 78	Chitosan/date palm seeds activated carbon	33.5	[150]
Acid red 1	Chitosan-reinforced graphene oxide-hydroxyapatite	41.32	[151]
Congo red	Chitosan-reinforced graphene oxide-hydroxyapatite	43.06	[151]
Reactive red 2	Chitosan-reinforced graphene oxide-hydroxyapatite	40.03	[151]
Methylene blue	Chitosan/*Pinus* spp. plywood panel-activated carbon	84.89	[152]
Malachite green	Chitosan sulfur/tapioca peel biochar	53.35	[153]
Rhodamine B	Chitosan sulfur/tapioca peel biochar	40.86	[153]

carbon has an adsorption capacity of 208.33 mg g^{-1} for indigo carmine [149]. Chitosan/date palm seeds have an adsorption capacity of 33.5 mg g^{-1} for direct blue 78 [150]. Finally, reinforced graphene oxide-hydroxyapatite chitosan has an adsorption capacity of 41.32 mg g^{-1} for acid red 1, 43.06 mg g^{-1} for Congo red, and 40.03 mg g^{-1} for reactive red 2 [151]. Chitosan/*Pinus* spp. activated carbon plywood has an adsorption capacity of 84.89 mg g^{-1} for methylene blue [152]. Finally, chitosan sulfur/biocarbon from tapioca bark has an adsorption capacity of 53.35 mg g^{-1} for malachite green, and chitosan sulfide/biocarbon from tapioca bark has an adsorption capacity of 40.86 mg g^{-1} for rhodamine B [153].

Pharmaceutical by-products are chemical compounds present in pharmaceutical effluents and can pose a risk to the environment if not disposed of properly. Chitosan also shows important adsorption capacity for such compounds, based on interactions similar to those observed for organic dyes [115]. As for organic dyes, chitosan's positively charged sites can interact with negatively charged pharmaceutical by-products, promoting their adsorption. The hydrogen bonding between chitosan functional groups and the functional groups of pharmaceutical by-products also can take part in adsorption process [115]. Some pharmaceutical by-products have hydrophobic moieties, which might interact with chitosan alkyl groups via hydrophobic interactions, promoting adsorption. The chemical equations that describe the adsorption of pharmaceutical by-products on chitosan are similar to those for organic dyes and include hydrogen bonding, Eq. (2.5), electrostatic interactions, Eq. (2.6), and hydrophobic interactions, Eq. (2.7) [115].

$$\text{Chitosan} - \text{NH}_3^+ + \text{By-product}^- \rightarrow \text{Chitosan} - \text{By-product} + \text{H}^+ \tag{2.5}$$

$$\begin{aligned}&\text{Chitosan} - \text{OH} + \text{By-product} - \text{C}\\&\quad = \text{O} \rightarrow \text{Chitosan} - \text{O} - \text{By-product} + \text{H}_2\text{O}\end{aligned} \tag{2.6}$$

$$\begin{aligned}&\text{Chitosan} - \text{Alkyl} + \text{By-product} - \text{Alkyl}\\&\quad \rightarrow \text{Chitosan} - \text{By-product} + \text{hydrophobic interactions}\end{aligned} \tag{2.7}$$

Parameters influencing the adsorption efficiency of pharmaceutical by-products by chitosan are similar to those of organic dyes and include the pH, concentration, and molecular structure of the compounds. Organic pesticides are chemical compounds widely used in agriculture to protect crops from pests and diseases. However, if not disposed of properly, they contaminate soil and groundwater. Organic pesticides can also be adsorbed by chitosan, though the adsorption mechanisms may vary depending on the pesticide's specific chemical structure [154]. Positively charged organic pesticides may interact with the negatively charged sites of chitosan through electrostatic interactions, thus enhancing their adsorption. Hydrogen bonds between the functional groups of chitosan and the functional groups of organic pesticides can also contribute to adsorption.

Certain hydrophobic parts of organic pesticides can interact with chitosan alkyl groups via hydrophobic interactions, thereby promoting adsorption. Chemical equations describing the adsorption of organic pesticides on chitosan may include hydrogen bonding (Eq. 2.8), electrostatic interactions (Eq. 2.9), and hydrophobic interactions (Eq. 2.10), similar to those of pharmaceutical by-products [154].

$$\text{Chitosan} - \text{NH}_3^+ + \text{Pesticide}^- \rightarrow \text{Chitosan} - \text{By} - \text{Pesticide} + \text{H}^+ \tag{2.8}$$

$$\text{Chitosan} - \text{OH} + \text{Pesticide} - \text{C} = \text{O} \rightarrow \text{Chitosan} - \text{O} - \text{Pesticide} + \text{H}_2\text{O} \tag{2.9}$$

$$\begin{aligned}&\text{Chitosan} - \text{Alkyl} + \text{Pesticide} - \text{Alkyl}\\ &\quad\rightarrow \text{Chitosan} - \text{Pesticide} + \text{hydrophobic interactions}\end{aligned} \tag{2.10}$$

In photocatalysis, incident light energy higher than that corresponding to the band gap of a semiconductor initiates transitions of photogenerated electrons from the valence to the conduction band. A few electrons and holes recombine into the semiconductor with resultant heat and/or light. Other electrons and holes move to the surface of the semiconductor under the effect of an electric field. There, they react with oxidizable and reducible compounds present, forming highly oxidizing hydroxyl and dioxygen anion-free radicals that mineralize the organic contaminants into carbon dioxide and water. Functional groups in chitosan can contribute to the photocatalytic reaction. These groups, when combined into the composite photocatalyst, enhance light absorption and thus allow the energy conversion of light into excited electrons [155]. Moreover, the transfer of electrons to the catalyst surface can be supported by these groups, which increases their mobility. Chitosan functional groups also modulate the band structure of the composite by controlling the energy level, resulting in lower emission intensity and a narrower band gap [156]. In addition to influencing nanoparticle size and shape, chitosan's hydroxyl group enhances the material's stability and recyclability. Finally, the synergy between photocatalytic degradation and adsorption enhances the effectiveness of chitosan-based composite photocatalysts in water pollution control [157].

2.4.4 Prospects and Challenges

The use of biopolymer chitosan in organic pollutant and heavy metal adsorption has promising prospects in environmental remediation. Present research aims at the improvement of adsorption methodologies by chemically modifying chitosan to increase its capacity and selectivity. This biopolymer can be applied to different environments, from drinking water to industrial wastewater and contaminated soils, where it can be combined with other technologies, such as filtration and bioremediation, for better performance. Although it is of ecological origin, derived from crustacean chitin, its production cost is still a big bottleneck for large-scale application. The adsorption selectivity, regeneration, and reuse of chitosan also remain a challenge, and effective methods must be developed to maintain its performance. The complete environmental impact of this should be evaluated, covering production, use, and disposal. While chitosan can reduce clean-up costs, investment in production infrastructure and technology is needed for industrial use. Precautions should be done to ensure that it will not affect the environment and human health significantly by reducing exposure and handling wastes appropriately. Though there is a drawback in using chitosan for wastewater treatment, continued exploration offers opportunities to develop its efficiency and sustainability.

2.5 Conclusion

The chapter summarizes key findings and results concerning synthesis and characterizations of chitosan biopolymer material, followed by the summary of experimental methodology that was used; significant parameters tested have also been mentioned. Effectiveness regarding the subject of wastewater treatment applications was also presented for the chitosan biopolymer material in this chapter. It demonstrates the potential of chitosan in the adsorption of various types of pollutants, heavy metals, and organic compounds from wastewater for improving water quality. The chapter also presents an overview regarding the use of chitosan biopolymer material for the treatment of wastewater due to its biocompatibility, biodegradability, and reasonable cost. However, full recognition is given to the limitations and challenges in the use of chitosan-based adsorbents concerning limited adsorption capacity and regeneration issues, among others. Finally, it outlines the future research directions or areas for improvement that could be made in the field regarding chitosan-based wastewater treatments. This may include the investigation of new synthesis methods, optimization of the process conditions, and performance in research of chitosan composite or chemically modified chitosan for improved adsorption efficiency.

References

1 Wear, S.L., Acuña, V., McDonald, R., and Font, C. (2021). Sewage pollution, declining ecosystem health, and cross-sector collaboration. *Biological Conservation* 255: 109010.

2 El Messaoudi, N., Ciğeroğlu, Z., Şenol, Z.M. et al. (2024). Green synthesis of nanoparticles for remediation organic pollutants in wastewater by adsorption. *Advances in Chemical Pollution, Environmental Management and Protection* 10: 305–345.

3 Devlin, M. and Brodie, J. (2023). Nutrients and eutrophication. 75–100.

4 Akinnawo, S.O. (2023). Eutrophication: causes, consequences, physical, chemical and biological techniques for mitigation strategies. *Environmental Challenges* 12: 100733.

5 Amorim, C.A., Moura, A., and do N. (2021). Ecological impacts of freshwater algal blooms on water quality, plankton biodiversity, structure, and ecosystem functioning. *Science of the Total Environment* 758: 143605.

6 Islam, R., Kumar, S., Karmoker, J. et al. (2018). Bioaccumulation and adverse effects of persistent organic pollutants (POPs) on ecosystems and human exposure: a review study on Bangladesh perspectives. *Environmental Technology and Innovation* 12: 115–131.

7 Baudrot, V., Fritsch, C., Perasso, A. et al. (2018). Effects of contaminants and trophic cascade regulation on food chain stability: application to cadmium soil pollution on small mammals – Raptor systems. *Ecological Modelling* 382: 33–42.

8 Talbot, C.J., Bennett, E.M., Cassell, K. et al. (2018). The impact of flooding on aquatic ecosystem services. *Biogeochemistry* 141 (3): 439–461.

9 El Messaoudi, N., Ciğeroğlu, Z., Şenol, Z.M. et al. (2023). A comparative review of the adsorption and photocatalytic degradation of tetracycline in aquatic environment by g-C3N4-based materials. *Journal of Water Process Engineering* 55: 104150.

10 Ferreira, V., Bini, L.M., González Sagrario, M.D. et al. (2023). Aquatic ecosystem services: an overview of the special issue. *Hydrobiologia* 850 (12–13): 2473–2483.

11 Miller, J.D. and Hutchins, M. (2017). The impacts of urbanisation and climate change on urban flooding and urban water quality: a review of the evidence concerning the United Kingdom. *Journal of Hydrology: Regional Studies* 12: 345–362.

12 Bolan, S., Padhye, L.P., Jasemizad, T. et al. (2024). Impacts of climate change on the fate of contaminants through extreme weather events. *Science of the Total Environment* 909: 168388.

13 Kurniawan, S.B., Pambudi, D.S.A., Ahmad, M.M. et al. (2022). Ecological impacts of ballast water loading and discharge: insight into the toxicity and accumulation of disinfection by-products. *Heliyon* 8 (3): e09107.

14 El Mouden, A., El Messaoudi, N., El Guerraf, A. et al. (2023). Removal of cadmium and lead ions from aqueous solutions by novel dolomite-quartz@Fe3O4 nanocomposite fabricated as nanoadsorbent. *Environmental Research* 225: 115606.

15 Mayfield, A.E., Seybold, S.J., Haag, W.R. et al. (2021). Impacts of invasive species in terrestrial and aquatic systems in the United States. In: *Invasive Species in Forests and Rangelands of the United States: A Comprehensive Science Synthesis for the U.S. Forest Sector*, 5–39.

16 Obaideen, K., Shehata, N., Sayed, E.T. et al. (2022). The role of wastewater treatment in achieving sustainable development goals (SDGs) and sustainability guideline. *Energy Nexus* 7: 100112.
17 El Messaoudi, N., Ciğeroğlu, Z., Şenol, Z.M. et al. (2024). Green synthesis of CuFe2O4 nanoparticles from bioresource extracts and their applications in different areas: a review. *Biomass Conversion and Biorefinery* 2024: 1–22.
18 Silva, J.A. (2023). Wastewater treatment and reuse for sustainable water resources management: a systematic literature review. *Sustainability* 15 (14): 10940.
19 Englert, D., Zubrod, J.P., Schulz, R., and Bundschuh, M. (2013). Effects of municipal wastewater on aquatic ecosystem structure and function in the receiving stream. *Science of the Total Environment* 454–455: 401–410.
20 Garcia, X. and Pargament, D. (2015). Reusing wastewater to cope with water scarcity: economic, social and environmental considerations for decision-making. *Resources, Conservation and Recycling* 101: 154–166.
21 Tortajada, C. (2020). Contributions of recycled wastewater to clean water and sanitation sustainable development goals. *npj Clean Water* 3 (1).
22 Zubair, M., Arshad, M., and Ullah, A. (2020). Chitosan-based materials for water and wastewater treatment. In: *Handbook of Chitin and Chitosan*, vol. 3, 773–809.
23 Şenol, Z.M., Elma, E., El Messaoudi, N., and Mehmeti, V. (2023). Performance of cross-linked chitosan-zeolite composite adsorbent for removal of Pb2+ ions from aqueous solutions: experimental and Monte Carlo simulations studies. *Journal of Molecular Liquids* 391: 123310.
24 Omer, A.M., Dey, R., Eltaweil, A.S. et al. (2022). Insights into recent advances of chitosan-based adsorbents for sustainable removal of heavy metals and anions. *Arabian Journal of Chemistry* 15 (2): 103543.
25 Zang, X., Jiang, R., Zhu, H.Y. et al. (2024). A review on the progress of magnetic chitosan-based materials in water purification and solid-phase extraction of contaminants. *Separation and Purification Technology* 330: 125521.
26 Bhatt, P., Joshi, S., Urper Bayram, G.M. et al. (2023). Developments and application of chitosan-based adsorbents for wastewater treatments. *Environmental Research* 226: 115530.
27 Egorov, A.R., Kirichuk, A.A., Rubanik, V.V. et al. (2023). Chitosan and its derivatives: preparation and antibacterial properties. *Materials* 16 (18): 6076.
28 Morin-Crini, N., Lichtfouse, E., Torri, G., and Crini, G. (2019). Fundamentals and applications of Chitosan BT. In: *Sustainable Agriculture Reviews 35: Chitin and Chitosan: History, Fundamentals and Innovations* (ed. G. Crini and E. Lichtfouse), 49–123. Cham: Springer International Publishing.
29 Crini, G. (2019). Historical review on chitin and chitosan biopolymers. *Environmental Chemistry Letters* 17 (4): 1623–1643.

30 Peniston, Q.P. and Johnson, E.L. (1970). Method for treating an aqueous medium with chitosan and derivatives of chitin to remove an impurity.

31 Gooday, G.W. (1990). The ecology of chitin degradation. In: *Advances in Microbial Ecology*, 387–430. Springer.

32 Mohan, K., Ganesan, A.R., Muralisankar, T. et al. (2020). Recent insights into the extraction, characterization, and bioactivities of chitin and chitosan from insects. *Trends in Food Science and Technology* 105: 17–42.

33 Huq, T., Khan, A., Brown, D. et al. (2022). Sources, production and commercial applications of fungal chitosan: a review. *Journal of Bioresources and Bioproducts* 7 (2): 85–98.

34 Karthick Rajan, D., Mohan, K., Rajarajeswaran, J. et al. (2024). β-Chitin and chitosan from waste shells of edible mollusks as a functional ingredient. *Food Frontiers* 5 (1): 46–72.

35 Czechowska-Biskup, R., Jarosińska, D., Rokita, B. et al. (2012). Determination of degree of deacetylation of chitosan-comparision of methods. *Progress on Chemistry and Application of Chitin and its Derivatives* 17: 5–20.

36 Kozma, M., Acharya, B., and Bissessur, R. (2022). Chitin, chitosan, and nanochitin: extraction, synthesis, and applications. *Polymers (Basel)* 14 (19): 3989.

37 Li, X. and Xia, W. (2011). Effects of concentration, degree of deacetylation and molecular weight on emulsifying properties of chitosan. *International Journal of Biological Macromolecules* 48 (5): 768–772.

38 Dimzon, I.K.D. and Knepper, T.P. (2015). Degree of deacetylation of chitosan by infrared spectroscopy and partial least squares. *International Journal of Biological Macromolecules* 72: 939–945.

39 Hirai, A., Odani, H., and Nakajima, A. (1991). Determination of degree of deacetylation of chitosan by 1H NMR spectroscopy. *Polymer Bulletin* 26 (1): 87–94.

40 Jiang, X., Chen, L., and Zhong, W. (2003). A new linear potentiometric titration method for the determination of deacetylation degree of chitosan. *Carbohydrate Polymers* 54 (4): 457–463.

41 Dos Santos, Z.M., Caroni, A., Pereira, M.R. et al. (2009). Determination of deacetylation degree of chitosan: a comparison between conductometric titration and CHN elemental analysis. *Carbohydrate Research* 344 (18): 2591–2595.

42 Rinaudo, M., Pavlov, G., and Desbrières, J. (1999). Influence of acetic acid concentration on the solubilization of chitosan. *Polymer (Guildf)* 40 (25): 7029–7032.

43 Qin, C., Li, H., Xiao, Q. et al. (2006). Water-solubility of chitosan and its antimicrobial activity. *Carbohydrate Polymers* 63 (3): 367–374.

44 Qiao, C., Ma, X., Wang, X., and Liu, L. (2021). Structure and properties of chitosan films: effect of the type of solvent acid. *Lwt* 135: 109984.

45 Minh, N.C., Van Hoa, N., and Trung, T.S. (2020). Preparation, properties, and application of low-molecular-weight chitosan. In: *Handbook of Chitin and Chitosan*, 453–471. Elsevier.

46 Wang, W., Xue, C., and Mao, X. (2020). Chitosan: structural modification, biological activity and application. *International Journal of Biological Macromolecules* 164: 4532–4546.

47 de Farias, B.S., Grundmann, D.D.R., Rizzi, F.Z. et al. (2019). Production of low molecular weight chitosan by acid and oxidative pathways: effect on physicochemical properties. *Food Research International* 123: 88–94.

48 Budiati, T., Yudiastuti, S.O.N., Wardhani, N.K., and Wahyono, A. (2024). *Chitosan Extraction Methods of Black Soldier Fly Waste.* UNISMA Press.

49 Pakizeh, M., Moradi, A., and Ghassemi, T. (2021). Chemical extraction and modification of chitin and chitosan from shrimp shells. *European Polymer Journal* 159: 110709.

50 Abdou, E.S., Nagy, K.S.A., and Elsabee, M.Z. (2008). Extraction and characterization of chitin and chitosan from local sources. *Bioresource Technology* 99 (5): 1359–1367.

51 Wani, A.K., Akhtar, N., Mir, T. et al. (2023). Eco-friendly and safe alternatives for the valorization of shrimp farming waste. *Environmental Science and Pollution Research* 31 (1): 1–30.

52 Al Sagheer, F.A., Al-Sughayer, M.A., Muslim, S., and Elsabee, M.Z. (2009). Extraction and characterization of chitin and chitosan from marine sources in Arabian Gulf. *Carbohydrate Polymers* 77 (2): 410–419.

53 Islam, N., Hoque, M., and Taharat, S.F. (2022). Recent advances in extraction of chitin and chitosan. *World Journal of Microbiology and Biotechnology* 39 (1): 28.

54 Benhabiles, M.S., Drouiche, N., Lounici, H. et al. (2013). Effect of shrimp chitosan coatings as affected by chitosan extraction processes on postharvest quality of strawberry. *Journal of Food Measurement and Characterization* 7 (4): 215–221.

55 Hussain, M.R., Iman, M., and Maji, T.K. (2013). Determination of degree of deacetylation of chitosan and their effect on the release behavior of essential oil from chitosan and chitosan-gelatin complex microcapsules. *International Journal of Advanced Engineering Applications* 6 (4): 4–12.

56 Liu, Y., Xing, R., Yang, H. et al. (2020). Chitin extraction from shrimp (Litopenaeus vannamei) shells by successive two-step fermentation with Lactobacillus rhamnoides and Bacillus amyloliquefaciens. *International Journal of Biological Macromolecules* 148: 424–433.

57 Arbia, W., Arbia, L., Adour, L., and Amrane, A. (2013). Chitin extraction from crustacean shells using biological methods – a review. *Food Technology and Biotechnology* 51 (1): 12–25.

58 Tan, J.S., Abbasiliasi, S., Lee, C.K., and Phapugrangkul, P. (2020). Chitin extraction from shrimp wastes by single step fermentation with Lactobacillus acidophilus FTDC3871 using response surface methodology. *Journal of Food Processing & Preservation* 44 (11): e14895.

59 Pacheco, N., Garnica-Gonzalez, M., Gimeno, M. et al. (2011). Structural characterization of chitin and chitosan obtained by biological and chemical methods. *Biomacromolecules* 12 (9): 3285–3290.

60 Lagat, M.K., Were, S., Ndwigah, F. et al. (2021). Antimicrobial activity of chemically and biologically treated chitosan prepared from black soldier fly (Hermetia illucens) Pupal Shell Waste. *Microorganisms* 9 (12): 2417.

61 Jabeur, F., Mechri, S., Mensi, F. et al. (2022). Extraction and characterization of chitin, chitosan, and protein hydrolysate from the invasive pacific blue crab, portunus segnis (Forskål, 1775) having potential biological activities. *Environmental Science and Pollution Research* 29: 36023.

62 Manni, L., Ghorbel-Bellaaj, O., Jellouli, K. et al. (2010). Extraction and characterization of chitin, chitosan, and protein hydrolysates prepared from shrimp waste by treatment with crude protease from Bacillus cereus SV1. *Applied Biochemistry and Biotechnology* 162: 345–357.

63 Mohan, K., Ganesan, A.R., Ezhilarasi, P.N. et al. (2022). Green and eco-friendly approaches for the extraction of chitin and chitosan: a review. *Carbohydrate Polymers* 287: 119349.

64 Kou, S., Peters, L.M., and Mucalo, M.R. (2021). Chitosan: a review of sources and preparation methods. *International Journal of Biological Macromolecules* 169: 85–94.

65 Wang, J. and Zhuang, S. (2022). Chitosan-based materials: preparation, modification and application. *Journal of Cleaner Production* 355: 131825.

66 Gibson, H.W. and Bailey, F.C. (1980). Chemical modification of polymers. 13. Sulfonation of polystyrene surfaces. *Macromolecules* 13 (1): 34–41.

67 Li, B., Elango, J., and Wu, W. (2020). Recent advancement of molecular structure and biomaterial function of chitosan from marine organisms for pharmaceutical and nutraceutical application. *Applied Sciences* 10 (14): 4719.

68 Lamarque, G., Lucas, J.-M., Viton, C., and Domard, A. (2005). Physicochemical behavior of homogeneous series of acetylated chitosans in aqueous solution: role of various structural parameters. *Biomacromolecules* 6 (1): 131–142.

69 Shen, J., Jin, B., Qi, Y. et al. (2017). Carboxylated chitosan/silver-hydroxyapatite hybrid microspheres with improved antibacterial activity and cytocompatibility. *Materials Science and Engineering: C* 78: 589–597.

70 Chen, A.-H., Liu, S.-C., Chen, C.-Y., and Chen, C.-Y. (2008). Comparative adsorption of Cu (II), Zn (II), and Pb (II) ions in aqueous solution on the crosslinked chitosan with epichlorohydrin. *Journal of Hazardous Materials* 154 (1–3): 184–191.

71 Wang, W., Meng, Q., Li, Q. et al. (2020). Chitosan derivatives and their application in biomedicine. *International Journal of Molecular Sciences* 21 (2): 487.

72 Suchyta, D.J., Soto, R.J., and Schoenfisch, M.H. (2017). Selective monophosphorylation of chitosan via phosphorus oxychloride. *Polymer Chemistry* 8 (16): 2552–2558.

73 Hong, F., Qiu, P., Wang, Y. et al. (2024). Chitosan-based hydrogels: from preparation to applications, a review. *Food Chemistry: X* 21: 101095.

74 Takeshita, S., Zhao, S., Malfait, W.J., and Koebel, M.M. (2021). Chemistry of chitosan aerogels: three-dimensional pore control for tailored applications. *Angewandte Chemie, International Edition* 60 (18): 9828–9851.

75 Shi, W., Ching, Y.C., and Chuah, C.H. (2021). Preparation of aerogel beads and microspheres based on chitosan and cellulose for drug delivery: a review. *International Journal of Biological Macromolecules* 170: 751–767.

76 Takeshita, S., Zhao, S., Malfait, W.J., and Koebel, M.M. (2021). Chemie der Chitosan-Aerogele: Lenkung der dreidimensionalen Poren für maßgeschneiderte Anwendungen. *Angewandte Chemie* 133 (18): 9913–9938.

77 Mujtaba, M., Morsi, R.E., Kerch, G. et al. (2019). Current advancements in chitosan-based film production for food technology; A review. *International Journal of Biological Macromolecules* 121: 889–904.

78 Diblan, S., Gökkaya Erdem, B., and Kaya, S. (2020). Sorption, diffusivity, permeability and mechanical properties of chitosan, potassium sorbate, or nisin incorporated active polymer films. *Journal of Food Science and Technology* 57 (10): 3708–3719.

79 Arslan, D.Ş., Ertap, H., Şenol, Z.M. et al. (2024). Preparation of polyacrylamide titanium dioxide hybrid nanocomposite by direct polymerization and its applicability in removing crystal violet from aqueous solution. *Journal of Polymers and the Environment* 32: 573–587.

80 Amaral, I.F., Granja, P.L., and Barbosa, M.A. (2005). Chemical modification of chitosan by phosphorylation: an XPS, FT-IR and SEM study. *Journal of Biomaterials Science. Polymer Edition* 16 (12): 1575.

81 Kumari, S. and Singh, R.P. (2012). Glycolic acid-g-chitosan-gold nanoflower nanocomposite scaffolds for drug delivery and tissue engineering. *International Journal of Biological Macromolecules* 50 (3): 878.

82 Lee, J.W., Kim, S.Y., Kim, S.S. et al. (1999). Synthesis and characteristics of interpenetrating polymer network hydrogel composed of chitosan and poly(acrylic acid). *Journal of Applied Polymer Science* 73 (1): 113.

83 Şen, N.E. and Şenol, Z.M. (2023). Effective removal of Allura red food dye from water using cross-linked chitosan-diatomite composite beads. *International Journal of Biological Macromolecules* 253: 126632.

84 Eddya, M., Tbib, B., and EL-Hami, K. (2020). A comparison of chitosan properties after extraction from shrimp shells by diluted and concentrated acids. *Heliyon* 6 (2): e03486.

85 Corazzari, I., Nisticò, R., Turci, F. et al. (2015). Advanced physico-chemical characterization of chitosan by means of TGA coupled on-line with FTIR and GCMS: thermal degradation and water adsorption capacity. *Polymer Degradation and Stability* 112: 1.

86 Čerović, L.S., Milonjić, S.K., Todorović, M.B. et al. (2007). Point of zero charge of different carbides. *Colloids and Surfaces A: Physicochemical and Engineering Aspects* 297 (1–3): 1.

87 Şenol, Z.M., Şimşek, S., Özer, A., and Şenol Arslan, D. (2021). Synthesis and characterization of chitosan–vermiculite composite beads for removal of uranyl ions: isotherm, kinetics and thermodynamics studies. *Journal of Radioanalytical and Nuclear Chemistry* 327 (1): 159.

88 Bhatt, A.S., Krishna Bhat, D., and Santosh, M.S. (2010). Electrical and magnetic properties of chitosan-magnetite nanocomposites. *Physica B: Condensed Matter* 405 (8): 2078.

89 Sen, T., Ozcelik, B., Qiao, G.G., and Ozmen, M.M. (2017). Hierarchical porous hybrid chitosan scaffolds with tailorable mechanical properties. *Materials Letters* 209: 528–531.

90 Tasselli, F., Mirmohseni, A., Seyed Dorraji, M.S., and Figoli, A. (2013). Mechanical, swelling and adsorptive properties of dry-wet spun chitosan hollow fibers crosslinked with glutaraldehyde. *Reactive and Functional Polymers* 73 (1): 218–223.

91 Garnica-Palafox, I.M. and Sánchez-Arévalo, F.M. (2016). Influence of natural and synthetic crosslinking reagents on the structural and mechanical properties of chitosan-based hybrid hydrogels. *Carbohydrate Polymers* 151: 1073–1081.

92 Zhang, Z.H., Han, Z., Zeng, X.A. et al. (2015). Enhancing mechanical properties of chitosan films via modification with vanillin. *International Journal of Biological Macromolecules* 81: 638–643.

93 Prateepchanachai, S., Thakhiew, W., Devahastin, S., and Soponronnarit, S. (2017). Mechanical properties improvement of chitosan films via the use of plasticizer, charge modifying agent and film solution homogenization. *Carbohydrate Polymers* 174: 253–261.

94 Ma, Y., Xin, L., Tan, H. et al. (2017). Chitosan membrane dressings toughened by glycerol to load antibacterial drugs for wound healing. *Materials Science and Engineering: C* 81 (May): 522–531.

95 Sun, X., Wang, Z., Kadouh, H., and Zhou, K. (2014). The antimicrobial, mechanical, physical and structural properties of chitosan-gallic acid films. *Lwt* 57 (1): 83–89.

96 Haghighi, H., De Leo, R., Bedin, E. et al. (2018). (2019) Comparative analysis of blend and bilayer films based on chitosan and gelatin enriched with LAE (lauroyl arginate ethyl) with antimicrobial activity for food packaging applications. *Food Packaging and Shelf Life* 19: 31–39.

97 Kaya, M., Akyuz, L., Sargin, I. et al. (2017). Incorporation of sporopollenin enhances acid–base durability, hydrophobicity, and mechanical, antifungal and antioxidant properties of chitosan films. *Journal of Industrial and Engineering Chemistry* 47: 236.

98 Wang, X., Yong, H., Gao, L. et al. (2019). Preparation and characterization of antioxidant and pH-sensitive films based on chitosan and black soybean seed coat extract. *Food Hydrocolloids* 89: 56.

99 Kaim Billah, E.L. and R, Zaghloul A, Ahsaine HA, BaQais A, Khadoudi I, El Messaoudi N, Agunaou M, Soufiane A, Jugade R. (2022). Methyl orange adsorption studies on glutaraldehyde cross-linking chitosan/fluorapatite-based natural phosphate composite. *International Journal of Environmental Analytical Chemistry* 104 (17): 5840–5856.

100 El Messaoudi, N., El Khomri, M., Dbik, A. et al. (2016). Biosorption of Congo red in a fixed-bed column from aqueous solution using jujube shell: experimental and mathematical modeling. *Journal of Environmental Chemical Engineering* 4 (4): 3848–3855.

101 El Messaoudi, N., El Khomri, M., Bentahar, S. et al. (2016). Evaluation of performance of chemically treated date stones: application for the removal of cationic dyes from aqueous solutions. *Journal of the Taiwan Institute of Chemical Engineers* 67: 244–253.

102 El Khomri, M., El Messaoudi, N., Dbik, A. et al. (2020). Efficient adsorbent derived from Argania Spinosa for the adsorption of cationic dye: kinetics, mechanism, isotherm and thermodynamic study. *Surfaces and Interfaces* 20: 100601.

103 El Messaoudi, N., El Khomri, M., Fernine, Y. et al. (2022). Hydrothermally engineered Eriobotrya japonica leaves/MgO nanocomposites with potential applications in wastewater treatment. *Groundwater for Sustainable Development* 16: 100728.

104 Gil, A., Simon, S., Rodríguez-Santiago, L. et al. (2007). Influence of the side chain in the structure and fragmentation of amino acids radical cations. *Journal of Chemical Theory and Computation* 3 (6): 2210–2220.

105 He, X., Li, K., Xing, R. et al. (2016). The production of fully deacetylated chitosan by compression method. *Egyptian Journal of Aquatic Research* 42 (1): 75–81.

106 De Oliveira Pedro, R., Schmitt, C.C., and Neumann, M.G. (2016). Syntheses and characterization of amphiphilic quaternary ammonium chitosan derivatives. *Carbohydrate Polymers* 147: 97–103.

107 Feng, X., Long, R., Wang, L. et al. (2022). A review on heavy metal ions adsorption from water by layered double hydroxide and its composites. *Separation and Purification Technology* 284: 120099.

108 Mende, M., Schwarz, D., Steinbach, C. et al. (2016). Simultaneous adsorption of heavy metal ions and anions from aqueous solutions on chitosan—investigated by spectrophotometry and SEM-EDX analysis. *Colloids and Surfaces A: Physicochemical and Engineering Aspects* 510: 275–282.

109 Yuan, L., Yan, M., Huang, Z. et al. (2019). Influences of pH and metal ions on the interactions of oxytetracycline onto nano-hydroxyapatite and their co-adsorption behavior in aqueous solution. *Journal of Colloid and Interface Science* 541: 101–113.

110 Gomes, J.R.B., Jorge, M., and Gomes, P. (2014). Interaction of chitosan and chitin with Ni, Cu and Zn ions: a computational study. *The Journal of Chemical Thermodynamics* 73: 121–129.

111 Şenol, Z.M., El Messaoudi, N., Ciğeroglu, Z. et al. (2024). Removal of food dyes using biological materials via adsorption: a review. *Food Chemistry* 450: 139398.

112 El Mouden, A., El Guerraf, A., El Messaoudi, N. et al. (2022). Date stone functionalized with 3-aminopropyltriethoxysilane as a potential biosorbent for heavy metal ions removal from aqueous solution. *Chemistry Africa* 5 (3): 745–759.

113 Abdouss, M., Sakhaee, K., Renani, S.B., and Zahedi, P. (2024). Chitosan-based nanomaterials in decontamination of heavy metals. In: *Chitosan-Based Hybrid Nanomaterials*, 181–210.

114 Ahmed, M.A. and Mohamed, A.A. (2023). The use of chitosan-based composites for environmental remediation: a review. *International Journal of Biological Macromolecules* 242: 124787.

115 Dago-Serry, Y., Maroulas, K.N., Tolkou, A.K. et al. (2024). How the chitosan structure can affect the adsorption of pharmaceuticals from wastewaters: an overview. *Carbohydrate Polymer Technologies and Applications* 7: 100466.

116 Hsu, C.Y., Ajaj, Y., Mahmoud, Z.H. et al. (2024). Adsorption of heavy metal ions use chitosan/graphene nanocomposites: a review study. *Results in Chemistry* 7: 101332.

117 Ashraf, A., Dutta, J., Farooq, A. et al. (2024). Chitosan-based materials for heavy metal adsorption: recent advancements, challenges and limitations. *Journal of Molecular Structure* 1309: 138225.

118 Hamza, M.F., Hamad, N.A., Hamad, D.M. et al. (2021). Synthesis of eco-friendly biopolymer, alginate-chitosan composite to adsorb the heavy metals, Cd(II) and Pb(II) from contaminated effluents. *Materials (Basel)* 14 (9): 2189.

119 Croitoru, A.M., Ficai, A., Ficai, D. et al. (2020). Chitosan/graphene oxide nanocomposite membranes as adsorbents with applications in water purification. *Materials* 13 (7): 1687.

120 Anush, S.M., Chandan, H.R., Gayathri, B.H. et al. (2020). Graphene oxide functionalized chitosan-magnetite nanocomposite for removal of Cu(II) and Cr(VI) from waste water. *International Journal of Biological Macromolecules* 164: 4391–4402.

121 Zhang, D., Wang, L., Zeng, H. et al. (2020). Novel polyethyleneimine functionalized chitosan–lignin composite sponge with nanowall-network structures for fast and efficient removal of Hg(II) ions from aqueous solution. *Environmental Science. Nano* 7 (3): 793–802.

122 Jafarnejad, M., Asli, M.D., Taromi, F.A., and Manoochehri, M. (2020). Synthesis of multi-functionalized Fe_3O_4-NH_2-SH nanofiber based on chitosan for single and simultaneous adsorption of Pb(II) and Ni(II) from aqueous system. *International Journal of Biological Macromolecules* 148: 201–217.

123 Huang, Y., Zheng, H., Hu, X. et al. (2022). Enhanced selective adsorption of lead(II) from complex wastewater by DTPA functionalized chitosan-coated magnetic silica nanoparticles based on anion-synergism. *Journal of Hazardous Materials* 422: 126856.

124 Zhang, Y., Bian, T., Jiang, R. et al. (2021). Bionic chitosan-carbon imprinted aerogel for high selective recovery of Gd(III) from end-of-life rare earth productions. *Journal of Hazardous Materials* 407: 124347.

125 Bakar, S.A., Mohamed, R.M.S.R., Al-Gheethi, A. et al. (2023). Kinetics and isotherms of heavy metals removal from laundry greywater by chitosan ceramic beads. *Environmental Advances* 13: 100391.

126 Rosli, N., Yahya, W.Z.N., and Wirzal, M.D.H. (2022). Crosslinked chitosan/poly(vinyl alcohol) nanofibers functionalized by ionic liquid for heavy metal ions removal. *International Journal of Biological Macromolecules* 195: 132–141.

127 Patel, P.K., Pandey, L.M., and Uppaluri, R.V.S. (2024). Highly effective removal of multi-heavy metals from simulated industrial effluent through an adsorption process employing carboxymethyl-chitosan composites. *Environmental Research* 240: 117502.

128 Sharifi, M.J., Nouralishahi, A., and Hallajisani, A. (2023). Fe_3O_4-chitosan nanocomposite as a magnetic biosorbent for removal of nickel and cobalt heavy metals from polluted water. *International Journal of Biological Macromolecules* 248: 125984.

129 Zhuang, S., Zhu, K., and Wang, J. (2021). Fibrous chitosan/cellulose composite as an efficient adsorbent for Co(II) removal. *Journal of Cleaner Production* 285: 124911.

130 Rahaman, M.H., Islam, M.R., Islam, R. et al. (2024). Preparation, characterization, and adsorption kinetics of graphene oxide/chitosan/carboxymethyl cellulose composites for the removal of environmentally relevant toxic metals. *International Journal of Biological Macromolecules* 257: 128357.

131 Peralta, M.E., Nisticò, R., Franzoso, F. et al. (2019). Highly efficient removal of heavy metals from waters by magnetic chitosan-based composite. *Adsorption* 25 (7): 1337–1347.

132 Tang, S., Yang, J., Lin, L. et al. (2020). Construction of physically crosslinked chitosan/sodium alginate/calcium ion double-network hydrogel and its application to heavy metal ions removal. *Chemical Engineering Journal* 393: 124728.

133 Upadhyay, U., Sreedhar, I., Singh, S.A. et al. (2021). Recent advances in heavy metal removal by chitosan based adsorbents. *Carbohydrate Polymers* 251: 117000.

134 Al-Hazmi, G.A.A.M., Alayyafi, A.A.A., El-Desouky, M.G., and El-Bindary, A.A. (2024). Chitosan-nano CuO composite for removal of mercury (II): box-Behnken design optimization and adsorption mechanism. *International Journal of Biological Macromolecules* 261: 129769.

135 Al-Ahmed, Z.A., Alhasani, M., Aljohani, M.M. et al. (2024). Facile synthesis of new metal-organic framework/chitosan composite sponge for Hg(II) removal: characterization, adsorption efficiency, and optimization using Box-Behnken design. *International Journal of Biological Macromolecules* 259: 129282.

136 Khajavian, M., Kaviani, S., Piyanzina, I. et al. (2024). Chitosan-based adsorptive membrane modified by carboxymethyl cellulose for heavy metal ion adsorption: experimental and density functional theory investigations. *International Journal of Biological Macromolecules* 257: 128706.

137 Benjelloun, M., Miyah, Y., Evrendilek, G.A. et al. (2023). Synergistic effect of coupling ozonation/adsorption system for toxic dye efficient removal: chemometric optimization by Box–Behnken response surface methodology. *Desalination and Water Treatment* 306: 220–235.

138 Ciğeroğlu, Z., El Messaoudi, N., Şenol, Z.M. et al. (2024). Clay-based nanomaterials and their adsorptive removal efficiency for dyes and antibiotics: a review. *Materials Today Sustainability* 26: 100735.

139 Lagdali, S., Miyah, Y., El-Habacha, M. et al. (2023). Performance assessment of a phengite clay-based flat membrane for microfiltration of real-wastewater from clothes washing: characterization, cost estimation, and regeneration. *Case Studies in Chemical and Environmental Engineering* 8: 100388.

140 Mejbar, F., Miyah, Y., Benjelloun, M. et al. (2024). High-performance of Cu@eggshells for toxic dyes catalytic wet peroxide oxidation: kinetics, design of experiments, regeneration, and cost analysis. *Case Studies in Chemical and Environmental Engineering* 9: 100572.

141 Miyah, Y., Ssouni, S., Benjelloun, M. et al. (2023). DFT theoretical analysis and experimental approach combination to understand the toxic dye's adsorption mechanism on the corncob-activated carbon surface. *Journal of Molecular Structure* 1288: 135742.

142 Miyah, Y., Benjelloun, M., Salim, R. et al. (2022). Experimental and DFT theoretical study for understanding the adsorption mechanism of toxic dye onto innovative material Fb-HAp based on fishbone powder. *Journal of Molecular Liquids* 362: 119739.

143 Miyah, Y., Benjelloun, M., Lahrichi, A. et al. (2021). Highly-efficient treated oil shale ash adsorbent for toxic dyes removal: kinetics, isotherms, regeneration, cost analysis and optimization by experimental design. *Journal of Environmental Chemical Engineering* 9 (6): 106694.

144 Ssouni, S., Miyah, Y., Benjelloun, M. et al. (2023). High-performance of muscovite clay for toxic dyes' removal: adsorption mechanism, response surface approach, regeneration, and phytotoxicity assessment. *Case Studies in Chemical and Environmental Engineering* 8: 100456.

145 El Messaoudi, N., Franco, D.S.P., Gubernat, S. et al. (2024). Advances and future perspectives of water defluoridation by adsorption technology: a review. *Environmental Research* 252: 118857.

146 Georgin, J., Franco, D.S.P., Manzar, M.S. et al. (2024). A critical and comprehensive review of the current status of 17β-estradiol hormone remediation through adsorption technology. *Environmental Science and Pollution Research* 2024: 1–34.

147 de Freitas, F.P., Carvalho, A.M., Carneiro, A.D. et al. (2021). Adsorption of neutral red dye by chitosan and activated carbon composite films. *Heliyon* 7 (7): e07629.

148 Doondani, P., Gomase, V., Saravanan, D., and Jugade, R.M. (2022). Chitosan coated cotton-straw-biochar as an admirable adsorbent for reactive red dye. *Results in Engineering* 15: 100515.

149 Fatombi, J.K., Idohou, E.A., Osseni, S.A. et al. (2019). Adsorption of indigo carmine from aqueous solution by chitosan and chitosan/activated carbon composite: kinetics, isotherms and thermodynamics studies. *Fibers and Polymers* 20 (9): 1820–1832.

150 Sait, H.H., Hussain, A., Bassyouni, M. et al. (2022). Anionic dye removal using a date palm seed-derived activated carbon/chitosan polymer microbead biocomposite. *Polymers* 14 (12): 2503.

151 Sirajudheen, P., Karthikeyan, P., Ramkumar, K., and Meenakshi, S. (2020). Effective removal of organic pollutants by adsorption onto chitosan supported graphene oxide-hydroxyapatite composite: a novel reusable adsorbent. *Journal of Molecular Liquids* 318: 114200.

152 Trung, T.S., Minh, N.C., Cuong, H.N. et al. (2022). Valorization of fish and shrimp wastes to nano-hydroxyapatite/chitosan biocomposite for wastewater treatment. *Journal of Science: Advanced Materials and Devices* 7 (4): 100485.

153 Vigneshwaran, S., Sirajudheen, P., Nikitha, M. et al. (2021). Facile synthesis of sulfur-doped chitosan/biochar derived from tapioca peel for the removal of organic dyes: isotherm, kinetics and mechanisms. *Journal of Molecular Liquids* 326: 115303.

154 Gkika, D.A., Mitropoulos, A.C., Kokkinos, P. et al. (2023). Modified chitosan adsorbents in pharmaceutical simulated wastewaters: a review of the last updates. *Carbohydrate Polymer Technologies and Applications* 5: 100313.

155 Mohd Adnan, M.A., Bao, L.P., and Muhd Julkapli, N. (2020). Mitigation of pollutants by chitosan/metallic oxide photocatalyst: a review. *Journal of Cleaner Production* 261: 121190.

156 Cavallaro, G., Micciulla, S., Chiappisi, L., and Lazzara, G. (2021). Chitosan-based smart hybrid materials: a physico-chemical perspective. *Journal of Materials Chemistry B* 9 (3): 594–611.

157 Desai, N., Rana, D., Salave, S. et al. (2023). Chitosan: a potential biopolymer in drug delivery and biomedical applications. *Pharmaceutics* 15 (4): 1313.

3

Cellulose Nanomaterials: From Extraction to Wastewater Remediation

Nusrat Zahan Ramesha[1], Abdullah Al Jobair[2], and Md Enamul Hoque[3]

[1]*Department of Mechanical and Aerospace Engineering, Utah State University, Utah, United States*
[2]*Department of Civil and Environmental Engineering, Islamic University of Technology, Gazipur, Bangladesh*
[3]*Department of Mechanical Engineering, Faculty of Engineering, University of Tabuk, Tabuk, Saudi Arabia*

3.1 Introduction

Water contamination is a substantial worldwide issue threatening human well-being and natural environment. Wastewater originating from many sources, such as industrial and household sources, harbors a multitude of pollutants, including organic substances, pathogens, heavy metals, and suspended particles. These pollutants have the potential to inflict damage upon aquatic ecosystems and pose risks to human well-being. Water pollution leads to the dissemination of illnesses, such as diarrhea, dysentery, and cholera, because of the pollution of water sources with human waste, chemicals, and sewage. In addition, industrial effluent may contain poisonous, reactive, carcinogenic, or ignitable substances, which can endanger public health and the ecosystem [1]. Conventional methods for treating wastewater involve employing chemical, physical, and biological methods to eliminate contaminants/pollutants before releasing the water into the environment. These treatments may include processes such as solid–liquid separation, oxidation, reduction, precipitation reactions, and the use of microbes to break down and eliminate organic chemicals [2]. While these treatment approaches are generally efficient in removing most pollutants, their high energy consumption and expense and the creation of sludge and occasionally harmful by-products render them less desirable. Water pollution will continue to be a significant concern since over 5.5 billion individuals globally may be exposed to contaminated water by 2050 [3].

As water pollution is a burgeoning problem, the scientific community worldwide has shown significant interest in advancing affordable and renewable materials to promote sustainability. The prospective use of cellulose for wastewater remediation has been thoroughly examined in this context. Cellulose demonstrates improved benefits when its chains are grouped, leading to sufficiently structured regions that can be extracted as nanoscale particles referred to as cellulose nanomaterials (CNs). These materials are regarded as valuable futuristic substances due to their unique physicochemical properties [4]. They have attracted growing attention due to their desirable intrinsic characteristics, including many hydroxyl groups that facilitate functionalization, biodegradability, large surface area, and minimal toxicity. The ultimate nanocomposite characteristics can be readily customized through additional modifications of the nanocomposite, resulting in the creation of novel materials tailored for particular uses [5]. This chapter will focus on cellulose nanomaterials and their classification, as well as the potential application of cellulose nanomaterials in wastewater remediation.

3.2 Cellulose Nanomaterials

Anselme Payen discovered cellulose in 1838. He extracted it from plants and found its chemical formula to be $C_6H_{10}O_5$ [6]. Cellulose is widely regarded as the most plentiful renewable polymer on the planet. It is a fibrous structure naturally occurring in algae, microorganisms, plants (including specific bacteria), and other sources from nature. Indeed, its yearly output amounts to around 1.5×10^{12} tons [7]. This homopolysaccharide has a linear structure and consists of D-glucopyranose (glucose) units. These units are linked by β-(1,4) bonds. It also has a high degree of polymerization (DP). When they are connected together by β-(1,4)-glucosidic linkages, they form anhydroglucose units. The fundamental unit of the cellulose polymer, anhydrocellobiose, comprises two anhydroglucose units. However, the DP of cellulose is often expressed by the number of anhydroglucose units. The cellulose molecule exhibits significant functionality because each anhydroglucose unit comprises six carbon atoms, with three hydroxyl groups situated at the C2, C3, and C6 positions [8]. Cellulose fibers consist of a combination of arranged amorphous chains and densely compact highly crystalline regions. The arranged crystalline domains exhibit greater resistance to mechanical, enzymatic treatments, and chemicals resulting in higher degradation resistance than the amorphous domains. The crystalline areas are interconnected by hydrogen bonds, which confer stability to cellulose but reduce its solubility in water and other solvents [9]. Cellulose nanomaterials are characterized as those that have at least one dimension ranging from 1 to 100 nm in the nanoscale. During the past decade, multiple terminologies have been employed to classify different types of nanomaterials derived from cellulose, resulting in significant confusion and ambiguity. Standardizing word use according to

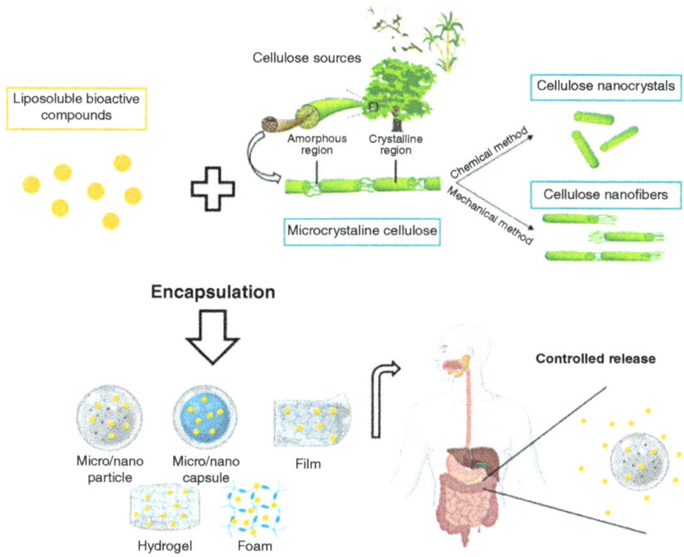

Figure 3.1 The layered arrangement of cellulose fibers and the molecular composition of cellulose. *Source:* Casanova et al. [11].

physical structure, size, and synthesis needs concerted work. Several years ago, a nanotechnology division was established by the Technical Association of the Pulp and Paper Industry (TAPPI) to standardize the terminology used for cellulose nanoparticles. An initial form of standard, TAPPI WI 3021: Standard Terms and Their Definitions for Cellulose Nanomaterials, has been created [10]. Thus, different forms of nanoscale cellulose can be produced through the breakdown of cellulose fibers; specifically, based on their origins, levels of crystallinity, and methods of extraction and production (i.e., top-down or bottom-up), three types of nanocellulose can be distinguished: (i) cellulose nanocrystals (CNCs), (ii) cellulose nanofibrils (CNFs), and (iii) bacterial nanocellulose (BNC). Figure 3.1 summarizes the layered arrangement of cellulose fibers and the molecular composition of cellulose.

3.2.1 Cellulose Nanocrystals (CNCs)

Cellulose nanocrystals, also referred to as crystalline nanocellulose (CNC), are elongated nanosized particles derived from diverse origins, such as wood pulp, cotton, and agricultural residues. This represents the predominant form of nanocellulose [12]. The width and length of these nanocrystals are influenced by factors such as time, temperature, and the purity of the cellulose source. Plant-derived cellulose nanocrystals have a diameter of 3–5 nm and a length of

100–300 nm. Conversely, the CNCs derived from tunicates exhibited a diameter of 10–20 nm and a length ranging from 500 to 2000 nm [13, 14]. Due to its high crystallinity, it exhibits less flexibility [15]. Cellulose nanocrystals possess numerous unique characteristics, including excellent stability, a broad surface area, a tensile strength of 10 GPa, an elastic modulus of 140–150 GPa, and various optical properties. The qualities of CNCs rely on the origin of cellulose, the extraction methodology, and its degree of crystallinity employed [11].

3.2.2 Cellulose Nanofibrils (CNFs)

Cellulose nanofibrils are known as nanofibrillated cellulose (CNF). These are typically derived from wood through a physical process involving wood pulp. However, these can additionally be recovered from other sources such as potato tubers, sugar beets, flax, and hemp [16]. CNFs have a width that falls within the 1–100 nm range, but their length can extend to several micrometers [17]. The properties of CNFs closely resemble those of cellulose nanocrystals, with the only difference being a reduced degree of crystallinity caused by the existence of an amorphous component [18]. CNFs possess notable characteristics, including their robustness and inflexibility, which render them a viable substitute for synthetic additives that enhance the durability of synthetic plastic composites. Furthermore, the minimal thermal expansion and restricted oxygen transfer rate allow for the use of these substances in food packaging, printing, and electronic devices [19]. The synthesis of nanocellulose (CNC and CNF) from cellulose microfibrils is illustrated in Figure 3.2.

3.2.3 Bacterial Nanocellulose (BNC)

There is one more type of nanocellulose generated by bacteria, known as bacterial nanocellulose (BNC) or microbial cellulose, which exhibits a higher level of

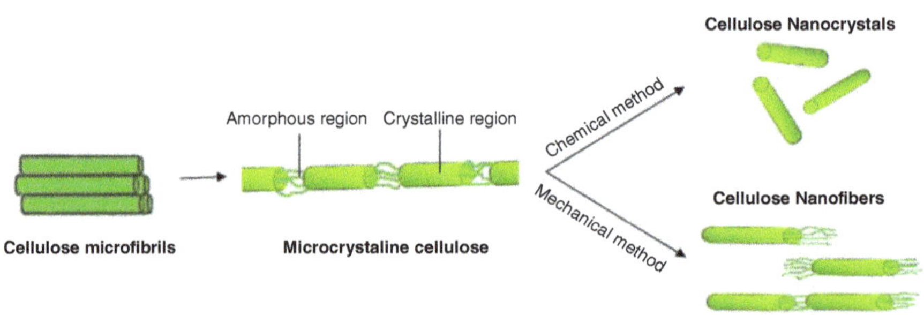

Figure 3.2 Nanocellulose (CNC and CNF) synthesis from cellulose microfibrils. *Source:* Casanova et al. [11].

Table 3.1 Comparison of the different types of nanocellulose [26].

	Cellulose nanocrystals (CNCs)	Cellulose nanofibrils (CNFs)	Bacterial nanocellulose (BNC)
Pictures	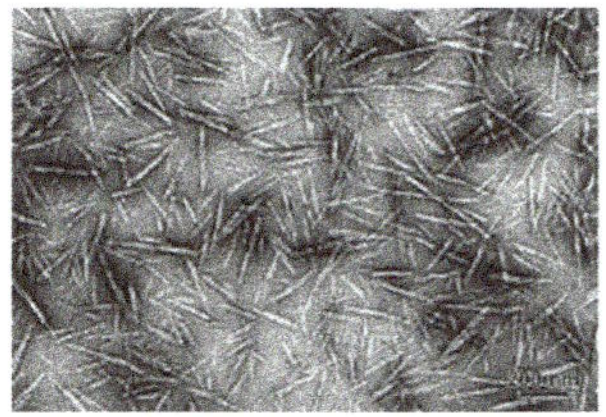 Transmission electron microscope (TEM) with a resolution of 100 nm.	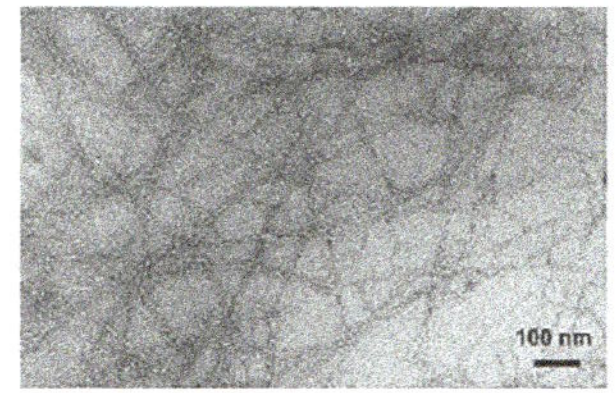 TEM with a resolution of 100 nm.	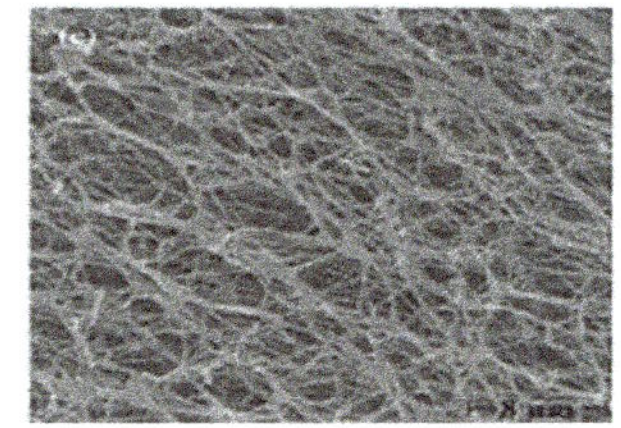 Scanning microscope with a resolution of 8 μm.
Equivalent terms	Rod-like cellulose microcrystals, whiskers, crystallites, nanocrystals, nanowhiskers.	Cellulose nanofibrils, nanofiber cellulose, microfibrillated cellulose, microfibrils.	Bacterial cellulose, microbial cellulose, biocellulose.
Typical origin	Swede root, kenaf fiber, black spruce and eucalyptus, sabdariffa fibers, palm oil, corncob, pea hull fiber, bacterial cellulose, wheat straw, avicel, softwood wood flour, wood fibers, ramie, pineapple leaf fiber, tunicate, rice husk, grass, tunicin, cotton, coconut fiber, rice straw, flax, microcrystalline cellulose (MCC), hemp, kraft wood, branch-bark of mulberry, chardonnay grape skins, spruce, pine, eucalyptus, and wood.	Mulberry, kenaf, recycled pulp, wheat straw, tunicate cellulose, cassava bagasse, banana, flax, sugar beet pulp, acacia pulp, bacterial cellulose, cotton, coconut husk fiber, hardwood and softwood pulp, pea hull, hemp, capimdourado, *luffacylindrica*, cotton linter, algae (valonia), date palm tree (rachis/leaflets), mengkuang leaves, sugarcane bagasse, curaua, and tunicate cellulose.	Pseudomonas, alcohols, Agrobacterium, Rhizobium, Sarcin, low molecular weight sugar, Gluconacetobacter, and various bacterial species.

(Continued)

Table 3.1 (Continued)

	Cellulose nanocrystals (CNCs)	Cellulose nanofibrils (CNFs)	Bacterial nanocellulose (BNC)
Production process	Enzymatic or chemical treatment applied after the delamination of wood pulp using mechanical pressure and acid hydrolysis.	Acid hydrolysis, the combination of mechanical shearing and high-pressure homogenization with enzymatic hydrolysis, and oxidation facilitated by TEMPO.	In aqueous culture media containing glucose, oxygen, and phosphate, bacterial synthesis and cultivation occur.
Typical dimensions	The diameter ranges from 5 to 80 nm, and the length ranges from 100 to 250 nm.	The diameter ranges from 5 to 70 nm, while the length varies from 100 to 250 nm for plant-based sources, and from 100 nm to several micrometers for cellulose derived from tunicates, algae, and bacteria.	The diameter ranges from 20 to 100 nm, and the length spans from 1.0 to 5.0 μm, forming various types of nanofiber networks.
Tensile strength	2–6 GPa	2–4 GPa	200–300 MPa
Young's modulus	50–143 GPa	15–150 GPa	15–35 Pa
Crystallinity index	54–88%	<50%	>88%

cleanliness compared to other varieties of nanocellulose [20, 21]. It is created by various bacterial species and microorganisms present in areas where carbohydrate fermentations occur [22–24]. The BNCs formed are usually quite crystalline, with ribbon-shaped fibrils 20–100 nm wide and many micrometers long. It is also possible to modify the size of BNCs by adjusting the experimental parameters [18]. BNC has many important characteristics, including a significant water content (about 99% in the hydrogel structures within the nanocellulose system), remarkable purity, impressive mechanical and thermal endurance, exceptional flexibility, and a high level of crystallinity (80–90%). Furthermore, BNC exhibits nontoxicity, biocompatibility, and bio-functionality, making it advantageous for medical implants, biotechnology, tissue reconstruction, and regenerative medicine [12]. Because of its production process, BNCs lack substances such as pectin, lignin, hemicellulose, and other substances commonly present in lignocellulosic biomass [25]. Nevertheless, the primary obstacle that needs to be addressed is the expansion of industrial production of BNC. This is a challenging task because of the limited productivity, costly culture media, and high fermentation expenses [12]. Table 3.1 presents a comparative analysis of several nanocellulose types.

3.3 Extraction of Cellulose Nanomaterials

3.3.1 Cellulose Nanocrystal Extraction

Cellulose nanocrystals are highly crystalline, which are less than half a micron [27]. Cellulose nanocrystals can be extracted by chemical, mechanical, and microbiological operations of cellulose microfibrils [28]. The nanocrystals are obtained from CNFs through the hydrolysis of amorphous domains using strong acid. The process of extracting cellulose nanocrystals involves three primary phases. The initial stage involves purification, which can be achieved through the use of potassium or sodium hydroxide, succeeded by bleaching with sodium chlorite. The amorphous areas are to be broken apart by a chemical process in the following step. Typically, acid hydrolysis is employed to carry out this particular procedure. The third phase entails the application of mechanical therapy [27]. The mechanical fibrillation process is hindered by two primary issues: fibril aggregation and fiber delamination, both of which contribute to increased energy consumption. Implementing an appropriate pretreatment may substantially reduce energy use. An optimal pretreatment enhances the internal surface area, disrupts hydrogen bonding, and augments the reactivity of cellulose. The energy required falls accordingly. Common pretreatments/common preliminary treatments include bleaching, pulping procedures, enzymatic, and alkaline-acid pretreatment [27]. The extraction of CNC involves specific processes to isolate these nanomaterials effectively.

3.3.1.1 Acid Hydrolysis

CNC preparation often involves the use of acid hydrolysis. This procedure entails the utilization of diverse high-concentration acids (6–8 M) under regulated temperature, in addition to other factors such as duration, agitation, and the ratio of acid to cellulosic material. These factors also have significant impacts on the result of the process [5]. Multiple researchers have documented the effect of process factors on the physical, thermal, chemical, and mechanical qualities. The particle size, mechanical, crystallinity, and thermal characteristics of the nanocrystals are significantly affected by the process operation duration, temperature, and acid content [29, 30]. The hydroxyl unit of the nanocrystallites reacts with sulfuric acid. This reaction results in the formation of a sulfonic group that carries a negative charge. By employing this method, it is possible to attain shorter chain segments without compromising the fundamental framework associated with the construction. The negatively charged nanocrystal surface's sulfate groups provide a strong repulsive force that successfully prevents agglomeration [31]. To neutralize the sulfate group and improve thermal stability, sodium hydroxide might be used [32]. More surface charge and a more constrained size are made possible by higher temperature, longer reaction times, and higher acid concentrations. However, thermal stability, crystallinity, and the manufacturing percentage of the produced CNCs are reduced [33]. The acid hydrolysis method has several drawbacks, including high water consumption, instrument damage, and increased waste creation. Increased interaction duration between cellulose and acids can lead to the degradation of crystallinity [34]. Cellulose nanocrystals have been processed from kinetics of high acid hydrolysis of a washed kraft pulp [35]. Cellulose nanocrystals were produced from cotton using ultrasound-assisted acid hydrolysis [36].

3.3.1.2 Hydrolysis with Solid Acid

There is another proposed method for extraction of cellulose nanocrystals. A study has documented a hydrolysis method that employs an acidic cation exchange resin as a solid catalyst. This method also incorporates high-power disintegration. It produces cellulose particles with an estimated yield of 50%. Furthermore, the solid acid can be restored to its original state by a posttreatment process. It has been reported that this technique yields cellulose nanocrystals at 50% with a crystallinity of 84% [37]. Liu et al. have synthesized CNC by hydrolyzing bleached pulp with solid phosphotungstic acid. Evidence was presented to demonstrate that the heat resistance of this cellulose nanocrystal is superior to that of the sulfuric acid hydrolyzed equivalent. By employing diethyl ether, this acid can be retrieved [38]. Hydrolysis using a solid acid offers significant benefits including convenient retrieval of the catalyst, little equipment corrosion, and a comparatively secure working environment. Additionally, the recovered acid can be reused multiple times for further cellulose hydrolysis without diminishing its effectiveness or reducing the final product yield [39]. Phosphotungstic acid is used in the formation of cellulose nanocrystals from cotton cellulose [40]. Ma et al. used synergistic effects of ionic liquid and reusable solid acid to remove cellulose nanocrystals [41].

3.3.1.3 Hydrolysis with Gaseous Acid

During this procedure, cellulose that contains moisture is brought into contact with a gaseous acid. The cellulose's moisture has a reaction with acid. Consequently, hydrolysis takes place at the noncrystalline regions and points of contact between fibrils [27]. CNCs are obtained through the process of mechanically grinding and/or subjecting hydrolyzed cellulose to ultrasonic treatment. Multiple forms of volatile acids can be employed in this process, including nitric acid, trifluoroacetic acid, and hydrochloric acid. This process is time-intensive and environmentally detrimental [39]. Through hydrolysis of metal salt catalysts, ionic liquids, and enzymes, cellulose nanocrystals can be synthesized [42]. There is no need for a lot of water, it is easy to recycle the acid, and the dialysis step can be avoided.

3.3.1.4 Hydrolysis with Metal Salt Catalyst

Metal salts are commonly used as catalysts due to their ability to enhance the hydrolysis effectiveness of cellulose. Researchers have shown that metal inorganic salts in the monovalent (NaCl, KCl), divalent ($CaCl_2$, $FeCl_2$, $FeSO_4$), and trivalent ($FeCl_3$, $Fe_2(SO_4)_3$, $Al(NO_3)_3$) categories can enhance the efficiency of cellulose hydrolysis [43, 44] and promote the synthesis of micro/nanocrystalline cellulose [45, 46]. The valence state of the metal ion plays a significant role in determining the effectiveness of hydrolysis in this process. During the interaction between metal ions and water molecules through polarization, an acidic solution (H^+) is generated [47]. The effectiveness of a transition metal ion in breaking down substances is largely influenced by its valence state [48].

3.3.2 Cellulose Nanofibril Extraction

CNFs are composed of elongated clusters (aggregates) of basic nanofibrils that are formed through the arrangement of amorphous and crystalline areas in an alternating pattern [39]. Comprising both primary and secondary layers, the cell wall of cellulose fiber includes the latter, which has three distinct components: S1, S2, and S3 [49]. The main layer consists of a disorganized network of microfibrils. The S1 layer has two lamellae arranged in a crossing ordering, close to the main layer. Within the S1 layer, microfibrils have a general alignment that is approximately parallel to one another. The S2 layer consists of almost parallel microfibrils that are organized in a helical pattern. The microfibrils have a spiral arrangement inside the S3 region. This coating is the innermost layer that covers the lumen. The microfibrils in these strata have distinct overall orientations [50]. CNF production comprises several methods. Different methods yield different types of CNFs. CNF can be obtained by strong mechanical decomposition. The energy required to overcome interfibrillar hydrogen bonding should be sufficient to delaminate nanofibrils rather than sever them. For maximum benefit from delaminated fibrils, reverse coalescence should be avoided [51]. This is because CNF is often employed in an aqueous solution to loosen interfibrillar hydrogen bonds. Because of its superior water retention capability, cellulose is dispersed at low concentrations, resulting in very viscous suspensions that are difficult to manage. Through the intense mechanical breakdown of cellulose fibers, CNFs or their aggregates can be generated.

3.3.2.1 High-Pressure Homogenization

The predominant technique for extracting CNF is high-pressure homogenization. This method is extensively employed for both industry and laboratory purposes. The process of homogenization led to a decrease in the size of output particles and a growth in particle quantity [52]. Many studies have been done to minimize the cellulose size by homogenization. The homogenization of cellulose has been incorporated in several ways, such as sugar beet [53] and wood pulp [54]. A technique for isolating nanocellulose from cotton cellulose utilizing high-pressure homogenization and an ionic liquid solution has been documented [55]. The solution is forced through a tiny aperture with a mechanical force at a high-pressure range of 50–2000 MPa [39]. Upon traversing a narrow aperture, the solution's velocity increased. Therefore, the dynamic pressure rises while the static pressure decreases. As a result, bubbles are created. Upon exiting the homogenization gap, these bubbles undergo collapse as a result of normal pressure, resulting in the creation of shock waves [39]. The solute has been broken down into smaller fragments because of the increased energy density in the cavitation area [56]. The particle size lowering method is a result of turbulent flow, interparticle collision, and shear stress [27]. The fabrication of nanofibrils by high-pressure homogenization is shown in Figure 3.3.

The size of cellulose microfibrils can be adjusted by altering the method parameters, the cycle count, and applied force [57]. Increasing the surface area of the output particle increases its physical stability. The homogenization gap width depends on the solution's force and viscosity. In a study, a high-pressure homogenizer is employed to fabricate CNFs from sugarcane bagasse, with a pressure of 15,000 psi and conducting 30 passes. The high-pressure homogenization was conducted by the use of xylanase pretreatment and a bleaching process [58]. Additionally, this method is used to generate cellulose nanocrystals from rice and oat husks [59]. Many mechanical pretreatments are possible preceding

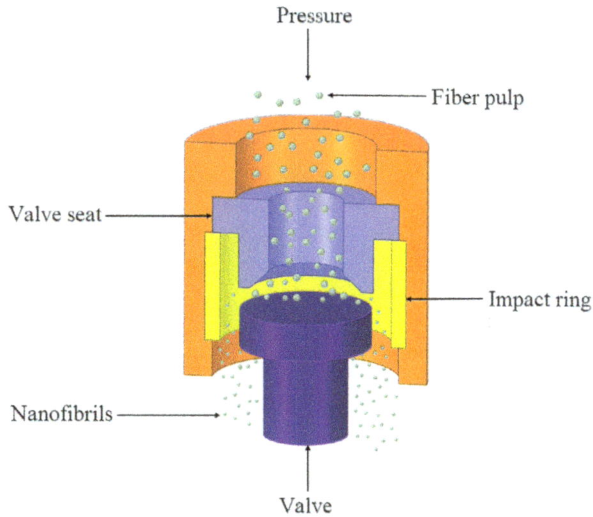

Figure 3.3 Fabrication of nanofibrils by high-pressure homogenizer. *Source:* Kalia et al. [57].

homogenization. Before homogenization, various mechanical pretreatments are employed. These are grinding, refining, and cryocrushing. Research has indicated that pretreatment methods like oxidation, enzymatic hydrolysis, acid hydrolysis, and others have been employed to address the issue of excessive energy usage.

3.3.2.2 Microfluidization

The microfluidizer is an important apparatus for generating cellulose microfibrils. The microfluidizer applies intense shear stress to break down cellulose particles in the suspension [27]. Since this uses a constant shear rate, this process minimizes the probability of clog formation. The suspension was transferred using a booster pump. As a result of this pump, the output pressure experiences a significant boost. The suspension is pumped into a contact chamber. The suspensions that undergo high shear stress can have a Y-shaped or Z-shaped configuration in this chamber. The interaction chamber contains distinct microchannels. The interaction between cellulose fiber and the microchannel wall generates shear force. A schematic of the microfluidizer used for microfluidization is provided in Figure 3.4.

The size of cellulose reduces until it reaches a specific number of cycles. It is typically important to increase the passing cycle to increase fibrillation. However, Lee et al. made an intriguing discovery by raising the quantity of consecutive phases. It was discovered that, after 10–15 cycles, the aspect ratio of the fiber's length to diameter expands, and the larger fibers divide into smaller strands. After 20 cycles of passing, there is a production of cellulose aggregates as a result of a rise in surface

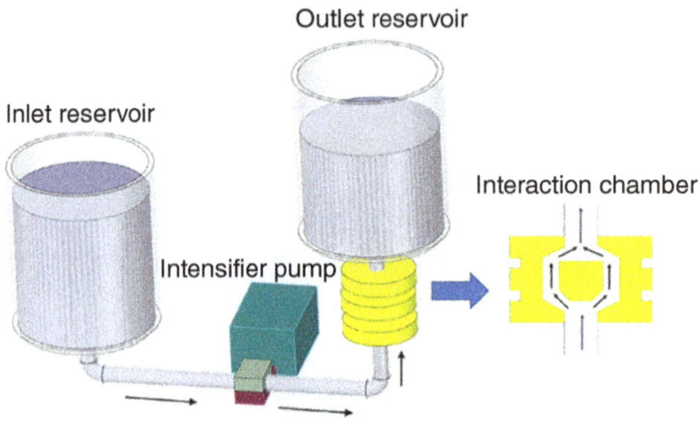

Figure 3.4 Microfluidizer. *Source:* Kalia et al. [57]; Khan et al. [60].

area and a larger concentration of hydroxyl groups [27]. The microfluidization procedure [61] can be used to achieve uniform particle size reduction in microfibrils. CNFs were extracted from agave bagasse using a microfluidizer [62]. The microfluidizer has also been utilized for the formation of CNFs from softwood kraft and paper mill sludge (PMS) [63, 64].

3.3.2.3 Grinding

Grinding is a mechanical method that has been employed to extract cellulose fibers from nanosized fibrils [5]. During the grinding process, the fiber fibrillation process employs a rotational speed of 1500 rpm to exert a shearing force on the fibers. The structural integrity of the cell wall is impaired by the shearing stress. The nanofibers in the cell wall are customized in a layered configuration derived from the pulp with the application of shear stresses. Furthermore, it should be emphasized that increasing the number of passes did not result in any substantial alterations in the fiber shape. The completion of pulp fiber fibrillation was inferred after only five runs through the grinder [5]. This treatment does not need any pretreatment. CNFs can be produced from wheat straw using the process of stone disk grinding [65]. The grinder system used for fibrillation is shown in Figure 3.5.

3.3.2.4 Cryocrushing

Cryocrushing is a mechanical fibrillation technique specifically employed for celluloses that have been frozen. Cellulose is pulverized using a mortar and pestle with liquid nitrogen combined with water-saturated cellulose [67]. The cell walls would rupture in the event of a high force being applied. These allow for the consistent production of nanofibers in water using a disintegrator for subsequent processing [68]. Even though this process can produce cellulose at the nanoscale, cryocrushing is expensive and has low productivity because it uses a

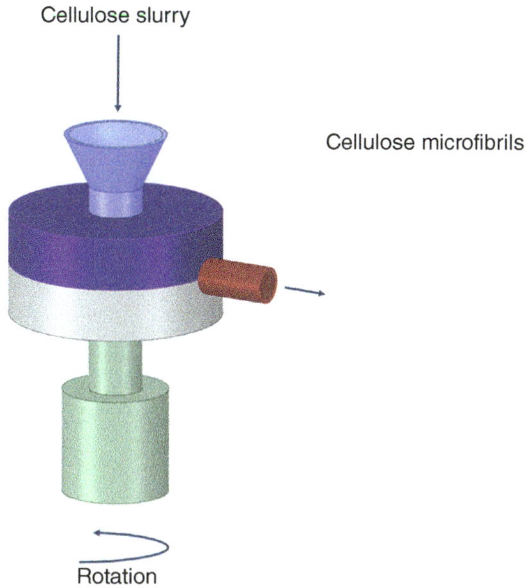

Figure 3.5 Grinder system. *Source:* Kargarzadeh et al. [39]; Missoum et al. [66].

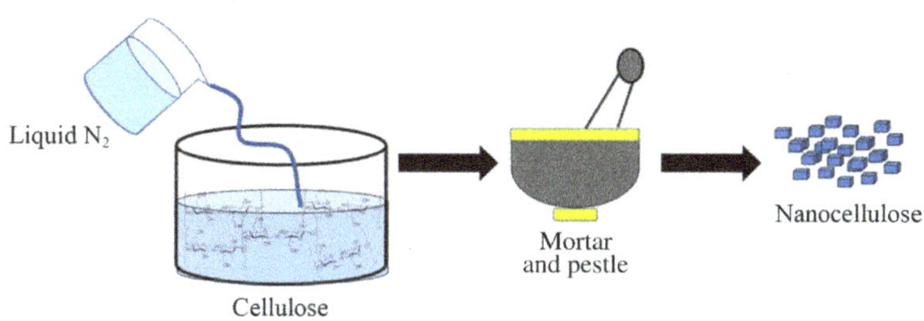

Figure 3.6 Illustration showing how cellulose is cyclocrushed to create nanocellulose. *Source:* Zinge and Kandasubramanian [15].

lot of energy [27]. Cryocrushing was used by Kaur et al. to extract the nanocrystals from rice husk [69]. The cryocrushing mechanism for producing nanocellulose is illustrated in Figure 3.6.

3.3.2.5 High-Intensity Ultrasonication

The high-intensity ultrasonication process can be utilized to extract CNFs. Ultrasonic frequencies (>20 kHz) are utilized in this method. A transducer in a high-intensity ultrasonicator converts mechanical or electrical energy into high-frequency acoustical energy, generating an ultrasound wave that produces ultrasonic vibrations. The shifting motion produces regular cycles of contracting and expanding in the slurry [70]. The occurrence of the cavitation process leads to the creation, expansion, and rapid implosion of small bubbles in the liquid. The collapse of the microbubbles leads to extreme circumstances, with a pressure exceeding 500 bar and a local temperature of around 5000 K [71]. The energy produced by this cavitation process is around 10–100 kJ mol^{-1}, falling within the range of hydrogen bond energy [72]. This is the cause of the fragmentation of cellulose fibers into smaller dimensions. The majority of cellulose fibrillation occurs near the needle tip of the sonicator. The concentration of the slurry affects how long fibrillation lasts [73]. The entire procedure is conducted within a container filled with water or ice in order to decrease the temperature. Wang et al. obtained fibrils from various cellulose sources. Their mixture consisted of fibrils at both the nanoscale and microscale levels. In the high-intensity ultrasonic approach, the temperature rose more rapidly, resulting in a greater amount of power. The fibrillation also demonstrated greater performance in terms of high power [74].

3.3.2.6 Refining

Refining is a mechanical technique that can cause permanent deformation in cellulose fibers and enhance their bonding energies. Hence, one might anticipate a variation in morphologies due to the disparity in applied forces and bonding energies. This technique can be considered a first step for homogenization [75]. A spring-loaded valve assembly is subjected to high pressure as diluted suspensions of processed cellulose fibers are forced through high pressure during the homogenization process. The quick actuation of this valve results in a considerable decrease in pressure as well as shearing and impact loads on the fibers. The interaction of these forces enhances the production of many fibrils in the cellulose fibers, leading to the continuous generation of CNFs [76].

3.3.3 Bacterial Nanocellulose Extraction

BNC is a biopolymer that has been studied a lot and has been known for more than 100 years. There are worries about its economic viability, especially when it comes to production, even though its structure and biology are very interesting. A variety of bacteria, including *Salmonella*, *Sarcina*, *Achromobacter*, *Pseudomonas*, *Acanthamoeba*, *Agrobacterium*, *Alcaligenes*, *Azotobacter*, *Rhizobium*, *Aerobacter*, *Zooglea*, and *Escherichia*, are used for the production of BNCs. Nevertheless, *Acetobacter xylinum*, a Gram-negative bacterium, stands out as the most effective species for BNC production [77]. The chemically manufactured medium can be used to make bacterial cellulose, but natural media can also be used [78]. The multi-stage production of bacterial cellulose is accurate [79]. Acetobacter is a top rod-shaped Gram-negative acetic bacterium. The current study suggests focusing on bacterial strains from this species as a prime candidate for biochemical, genetic, and molecular studies due to their fascinating behavior and high cellulose yield [80]. Biosynthesis represents a highly intricate and regulated technique. It requires precise coordination of protein and enzyme complexes with regulatory and catalytic functions [81]. The nanocellulose produced by various bacteria under varied culturing circumstances has a wide range of structural traits. The process determines their distinct physical and chemical properties and structures, rendering them convenient for various biological and manufacturing applications. Currently, BNC production can be achieved using three methods: agitated, static, and bioreactor-based bacterial culture [79].

3.3.3.1 Static Culture Method

This represents the most widely employed technique for producing BNC. Various containers of diverse shapes and sizes are loaded with a culture medium having a pH range of 4.5–6.5. These containers are then incubated for a period of 2–20 days at a temperature ranging from 25 °C to 30 °C. BNC generated using the static culture technique consists of a hydrogel film that develops at the interface between the air and the culture media [82, 83]. Cellulose ribbons that are self-assembling, overlapping, and interwoven in parallel-disorganized planes continuously crystallize into microfibrils and are extruded from linear pores at the bacteria's wall-membrane. Gelatinous cellulose membrane forms on the surface of the culture medium. Culture time determines BNC thickness. Washing the membrane in sodium hydroxide, neutralizing it with acetic acid, and rinsing it with deionized water until the BNC turns white purifies it. Following purification, the membranes can be dried or stored in a moist state. The pellicle possesses a 3D network structure characterized by strong stretch resistance, elasticity, and porosity. This static culture technique is simple and widely utilized for producing BNCs. Because of the high expense of synthetic medium and low BNC yield rates, this method is not ideal for large-scale industrial production. Insufficient oxygenation of static cultures of solely aerobic bacteria is a major issue. Researchers have explored the use of fruit juices, along with agricultural and industrial waste, as sources of carbon and nutrients [54, 84]. BNC shape, microstructure, and inherent properties are similar to conventional media as a control. Waste items can be used to make eco-friendly, low-cost BNCs, according to these studies.

3.3.3.2 Agitated Culture Method

This approach increases medium culture dissolved oxygen and production rates. In some cases, BNC production remains the same or decreases even after enhancing dissolved oxygen levels in the media for the same duration as static culture. BNC takes on morphologies such as spherical, cocoon-shaped, pellet-formed, or irregular cluster masses, differing from those in static culture methods [83]. The reduced polymerization and assembly of BNC into nanofibers lead to these morphological alterations, lower crystallinity, and diminished inherent mechanical properties. Additionally, the genetic instability of certain bacteria under agitation may impair BNC output, and this approach does not work for all bacterial strains.

3.3.3.3 Bioreactor Method

Bioreactor cultures that employ continuous cultivation techniques, such as rotating disc or airlift systems with a consistent oxygen transfer rate, have demonstrated elevated levels of BNC production. BNC crystallinity, elasticity, and polymerization degree were lower than agitated or static cultured bacterial cellulose [83]. The Wu and Li modified airlift bioreactor is one of several variants [85]. These BNC membranes hold more water than static-grown BNC. Changing the number of net plates changes the elastic modulus. Considering the pros and cons of the three strategies above, more research is needed to improve industrial production for increased productivity and lower costs.

3.4 Properties of Cellulose Nanomaterials

3.4.1 Physical Properties

Nanocellulose has a wide variety of structures and properties. The nanostructures, which are smaller than 100 nm, consist of nanofibrils, nanofibers, nanowhiskers, nanocrystals, nanorods, and nanoballs. These nanostructures can be transformed into 2D and 3D structures such as nanoplatelets, membranes, films, microparticles, and porous macroscopic matrices [86]. Cellulose nanomaterials have greater surface area, providing more binding sites for pollutant adsorption. Composite materials with CNs have demonstrated enhanced adsorption capability to remove a diverse range of contaminants effectively. CNs have a high aspect ratio which promotes the development of percolated networks in the case of CNCs and entangled networks in the case of CNFs at low concentrations, hence enhancing their efficacy as flocculating agents. Nanocomposites demonstrate favorable mechanical characteristics, with a tensile modulus of around 145–165 GPa and a tensile strength of 10,000 MPa [87]. BNC is highly stable and may be treated with hot acid and alkaline compounds that can boil for up to 120 minutes, depending on cellulose thickness. Nanocellulose fibers have a high water-holding capacity (>99 wt.%), forming a hydrogel, and a high DP with a molecular weight of approximately 10,000 repeating monomer cellulose units [86]. This enhancement allows for the use of these materials in various operating circumstances. The use of CNs in adsorbents and membranes benefits from their strong surface tension and stability in water, which improve wettability and reduce biofouling [88].

3.4.2 Biological Properties

Nanocellulose is biodegradable and biocompatible. Biocompatibility is the property of not causing harmful effects on biological systems and also requiring an acceptable reaction from the host [87]. CNs increase polymer membrane hydrophilicity and reduce surface protein adsorption, according to some research. CNF-doped cellulose acetate membranes had a 30% higher flux recovery ratio upon hydraulic washing than undoped membranes [89]. CNF materials have the potential to be employed in solvent separation processes, wherein the permeability of the CNF mats increases as the hydrophobicity of the solvent increases [90].

3.4.3 Thermal Properties

Cellulose nanoparticles possess remarkable thermal characteristics, rendering them highly attractive for a wide range of applications. Cellulose has a lower thermal stability compared to inorganic materials, which limits its application and the manufacturing conditions of its composites at high temperatures [91]. The issue of thermal stability is particularly crucial for sulfuric acid-hydrolyzed CNC. The determining element in this regard is the surface charge density, which leads to de-esterification and acid-catalyzed breakdown [92]. Cellulose exhibits a remarkably low thermal expansion coefficient (TEC) when it is in the form of a film at the nanoscale. The high crystallinity and strength of the cellulose nanomaterial network connections are both responsible for this feature [39].

3.4.4 Rheological Properties

CNs have a great ability of surface functionality in the presence of hydroxyl groups. These groups facilitate the straightforward modification of cellulose nanomaterials with a diverse range of functional groups, including amine, carboxyl, sulfate ester, and aldehyde groups. Additionally, CNs can be easily functionalized with small molecules, nanoparticles, or by polymer grafting [93]. Due to their great mechanical strength and crystallinity, CNs improve adsorbent and membrane stiffness and chemical resistance [89]. Fibrils create a cohesive and robust network that successfully opposes the movement of substances. The inclusion of cellulose nanomaterials at a low concentration results in the formation of a gel-like structure, for a 2% weight concentration of CNFs suspension. The force necessary to disrupt this network and commence fluid motion, known as the yield stress, can be substantial. Cellulose nanomaterial suspensions are effective stabilizers for emulsions or suspensions because they can entrap particles or droplets inside the network, preventing them from sinking or rising [91].

3.4.5 Optical Properties

Cellulose nanomaterials have some optical properties. Optical transparency can be achieved in pure CNF/CNC films by densely packing cellulose nanofibers and ensuring that the spaces between the fibers are sufficiently tiny to prevent light scattering. Transparent coatings or films can be produced from cellulose nanomaterial. The cellulose nanopapers, which are optically transparent, have great potential as flexible device substrates due to their lightweight nature, foldability, smooth surface, and high thermal dimensional stability. The dewatering technique is not suitable for producing films using CNC and CNF due to the particles' small size. Nanopapers are strong and clear, so they can be used to protect and restore cultural heritage. They can be used to replace lost paper on cultural items and keep the paper stable. CNC can create chiral nematic films that exhibit iridescence and color [91]. Mesoporous materials having a well-defined and organized chiral nematic structure, achieved by the use of CNC, can be manufactured by evaporation-induced self-assembly [39]. The cellulose component is subsequently removed through calcination [94].

3.5 Cellulose Nanomaterials for Wastewater Remediation

Cellulose nanomaterials are referred to as green materials. Nanocelluloses have been identified as a viable substitute for conventional materials used in wastewater treatment. Water contamination is a significant global issue that affects individuals worldwide. It is well-recognized that access to clean water is vital for the survival and well-being of the human population. Water bodies are getting polluted in many ways. As the world is developing gradually, the industries are getting bigger. Discharging industrial wastewater directly into water bodies leads to the contamination and pollution of natural water sources. The existence of many contaminants, including heavy metals, medicinal chemicals, dyes, oil and organic solvents, microorganisms, and other substances, can have detrimental impacts on the health of both animals and humans [95]. It has become a necessity to remove pollutants from wastewater. Though various techniques are already being used for wastewater remediation, the high energy consumption and expense, as well as the creation of sludge and sometimes harmful by-products, make it unfavorable [96, 97]. It is thus necessary to address water pollution problems and complex conventional treatment procedures by using nonselective sustainable materials and a straightforward pollutant removal approach. Cellulose nanomaterials have received significant attention in this regard.

3.5.1 Cellulose Nanomaterials as Adsorbents

Adsorption is a commonly used method for eliminating impurities from water. Adsorption is a surface process in which pollutants, known as adsorbates, accumulate on the solid surface, known as the adsorbent, from the surrounding medium. The imbalanced forces of attraction cause the preferential buildup of adsorbates on the surface of adsorbents. The process involves several sorts of interactions, and the specific bonding relies on the charge properties or surface functionalization of the adsorbate and adsorbent. When adsorption occurs from weak forces of contact, such van der Waals forces, it is referred to as physisorption; when a chemical bond forms between the molecules of the adsorbent and the adsorbate, it is referred to as chemisorption [98–100]. The process of adsorption is a cost-effective and efficient method for the removal of contaminants, including both inorganic and organic substances, which are found in wastewater. This is widely employed by materials based on nanocellulose for the purpose of wastewater cleanup, making it the predominant technology in this field. Key physicochemical features of nanocellulose and their effects on adsorption performance are summarized in Table 3.2.

This subsection will concentrate on cellulose nanomaterials as adsorbents for a range of pollutants, including heavy metals, oil, dyes, solvents, and other organic chemicals.

3.5.1.1 Heavy Metals

One of the biggest issues nowadays is wastewater containing heavy metals, which are harmful and nonbiodegradable pollutants. Manufacturing processes generate heavy metal ions and compounds, possibly contaminating water sources with naturally occurring harmful contaminants. It can be defined as any metal or metalloid with a density exceeding 5 $g\,cm^{-3}$ that exhibits deleterious effects [102]. A significant build-up of heavy metals can cause the development of neurological disorders, cancer, organ damage, and, in severe instances, death. Hence, the elimination of these heavy metals from aqueous solutions holds significant importance [103].

An increasing emphasis on sustainable materials has contributed to the growing popularity of bioadsorbents, such as nanocellulose, for their capability to adsorb heavy metals from wastewater. Heavy metals such as Cu^{2+}, Zn^{2+}, Hg^{+}, Cd^{2+}, Cr^{3+}, and Pb^{2+} are known as toxic. The nanocellulose showed a significant capacity for adsorbing Cr^{6+} and Pb^{2+} because of its characteristics of fibrous structure, high porosity, and significant surface area. The adsorption capacities reached up to 100 $mg\,g^{-1}$ for Cr^{6+} and 260 $mg\,g^{-1}$ for Pb^{2+} [17]. The CNFs-NAMH composite exhibits exceptional sensitivity to Co^{2+} ions in aqueous solutions, with a remarkable limit of detection (LOD) of 0.87 ppb. This high sensitivity allows naked-eye optical identification [104]. Carboxymethylcellulose nanofibers derived from

Table 3.2 The correlation between the physicochemical characteristics of nanocelluloses and their ability to adsorb harmful contaminants in water [101].

Nanocellulose features	Effect on adsorption
High surface area	Enhances the surface area from the microscale to the nanoscale, hence improving the adsorption capability of nanocellulose. The combined aerogel (1:3 ratio of CNC to CNF) provides a larger specific surface area compared to individual CNF or CNC.
High aspect ratio	The aspect ratio of CNC (10–80) typically is less compared to CNF (ranging up to 80–500), based on the origin of nanocellulose and the preparation procedure. Promotes the formation of interconnected CNC structures and interconnected CNF networks reinforced by stronger hydrogen bonding, hence improving the structural durability of the adsorbent.
High mechanical stiffness	The high mechanical stiffness (modulus) of CNC and CNF, approximately 130 and 70 GPa, respectively, improves the stiffness and cohesiveness of the adsorbent material. High crystalline forms (CNCs and CNFs) exhibit transparency, gas impermeability, and possess a tensile strength up to eight times that of steel.
High crystalline degree	The high crystallinity degree of nanocellulose (60–80%) improves the chemical resistance of the adsorbent and decreases cellulose solubility, even in highly polar solutions.
Susceptible to surface functionalization	Functionalization of hydroxyl groups (silylation, oxidation, etherification, radical grafting, and esterification) enhances the adsorption capacity of nanocellulose.
Stability in water	Decreases biofouling of nanocellulose-based adsorbents. The surface of cellulose-based water treatment materials possess a negative charge because of the higher abundance of carboxylate and hydroxyl groups, leading to stronger electrostatic repulsion between the surface layer and the majority of model foulants.
High surface tension	The higher surface tension of CN (about 60 mJ m^{-2}) enhances the wetting properties of adsorbents derived from nanocellulose in aqueous systems and decreases biofouling.

Australian spinifex grass were used for extracting cadmium from water. The findings of the study demonstrated that the generated CNFs exhibited effective cadmium removal capabilities in water. The interaction between nanocellulose and cadmium occurred through chelation and crystallization mechanisms at low and high concentrations, respectively [103]. The synthesis of nanocellulose particles is presented with a spherical shape, specifically designed for adsorbing mercury from aqueous solutions. In addition, the researchers introduced thiol groups to functionalize the spherical nanocellulose and subsequently conducted a comparative analysis of the adsorption activity exhibited by the two types of nanocellulose. The study revealed that the use of thiol-functionalized nanocellulose resulted in a remarkable adsorption efficiency of up to 97% for mercury within a time frame of 20 minutes, using a solution concentration of 100 ppm [105]. Nanocellulose-based materials are employed as radioactive adsorbents, including thorium and uranium. When graphene oxide-functionalized polymer grafted nanocellulose was utilized as an adsorbent for thorium, the outcome showed that up to 98% of the thorium was adsorbed [106]. A simple bonding and freeze-drying method was used to make CNFs aerogel. The ability of the CNFs aerogel to bind uranium showed that it can hold up to 195 $mg\,g^{-1}$ of uranium at a starting concentration level of 10 ppm [107].

Functionalized BNC was also tested for wastewater treatment. Untreated BNC lacks surface charges, hence it has low specific adsorption [108, 109]. Increased targeted adsorption and absorption of toxic heavy metals and various contaminants requires BNC fiber surface functionalization. In reaction with epichlorohydrin and diethylenetriamine, BNC becomes amino-functionalized. This BNC adsorbs 63 $mg\,g^{-1}$ Cu^{2+} and 87 $mg\,g^{-1}$ Pb^{2+} at pH 4.5. Adsorption decreases considerably when the amine group is protonated at lower pH. After filtering, aminated BNC may be rinsed in ethylenediaminetetraacetic acid to remove adsorbed metals [110]. BNC chemically altered with ammonium sulfamate ($NH_4OSO_2NH_2$) adsorbs 22.7 $mg\,g^{-1}$ Pb^{2+}. This number is five times unmodified BNC's adsorption capacity [111]. Similar to unmodified BNC, carboxymethylated BNC has higher adsorption capabilities for Cu^{2+} and Pb^{2+} (12.6 $mg\,g^{-1}$ for Cu^{2+} and 60.4 $mg\,g^{-1}$ for Pb^{2+} at pH = 4.5). The studies also shown that pH, contact length, and metal ion concentration effect Cu^{2+} and Pb^{2+} absorption [108].

The production of magnetically responsive nanocellulose adsorbents for efficient magnetic separation and reusability is a feasible method. This strategy enables the treatment of significant amounts of wastewater and the regeneration of the adsorbent material [112]. Hybrid Fe_3O_4/BNC nanocomposites have recently been utilized to specifically remove several poisonous metal ions from complicated wastewater mixes via magnetic separation. These nanocomposites have a significant adsorption capacity for Pb^{2+} (65 $mg\,g^{-1}$), Mn^{2+} (33 $mg\,g^{-1}$), and Cr^{3+} (25 $mg\,g^{-1}$) [105]. Similarly, BNC/Fe_3O_4 NPs that have been treated with amines show a strong ability to adsorb As^{5+} ions, with a rate of 90 mg g^{-1}. This is because of their strong attraction to magnetic Fe_3O_4 NPs and amines [113]. A research was conducted on BNC composites, where magnetite Fe_3O_4 nanoparticles were uniformly distributed in the BNC matrix, and it was discovered that the elimination of Cr^{6+} ions is significantly impacted by the pH of the medium. The investigation revealed that the best effectiveness of removal (5.13 mg g^{-1}) was attained at a pH of 4 [114]. The BNC/Fe_3O_4 particles, which are spherical in shape, were created by enclosing magnetite Fe_3O_4 nanoparticles with a diameter of 15 nm within BNC particles. These particles exhibited significant adsorption capabilities of 65, 33, and 25 mg g^{-1} for Pb^{2+}, Cr^{3+}, and Mn^{2+} ions, respectively [105]. Collectively, these investigations suggest that magnetic cellulose nanocomposites have superior adsorption efficacy in comparison to individual nanocelluloses.

3.5.1.2 Dyes

Dyes are pigmented compounds that undergo a chemical bonding process with the substrate onto which they are being put. The printing, rubber, paper, textile, leather, and cosmetic industries all utilize dyes to color their products. It is used to make products bright and lasting color. The rapid increase of the textile industry uses about 7×10^5 tons of dyes annually, making them a significant threat [115]. Dyes are often nonbiodegradable, posing a significant issue for plants and food sources. Direct dyes are water-soluble anionic dyes with a strong affinity for nanocelluloses. Acid dyes, also water-soluble anionic dyes, exhibit a very low binding affinity for nanocelluloses. The chromophores of acid and direct dyes are attracted to nanocelluloses by nonspecific hydrophobic contacts, dipolar forces, and hydrogen bonding. Generally, the process of modifying nanocelluloses with cationic substances enhances the absorption of direct and acid dyes. Vat dyes are not soluble in water and predominantly show a great tendency to be adsorbed onto nanocelluloses because of their high affinity, contrasting to direct and acid dyes, sulfur dyes (—S—, —S—S—, —Sn—), and azoic dyes (—N=N—) [101].

CNCs synthesized using sulfuric acid hydrolysis have the ability to bind cationic dyes, such as methylene blue, selectively. The studies showed that these pure CNCs naturally had a strong ability to adsorb substances. Additionally, by further modifying them with TEMPO-oxidation, their adsorption capability was increased by as much as 6.5 times. The surface of CNCs formed from the hydrolysis of wood pulp using sulfuric acid contains negative sulfate ester groups. Additionally, the main hydroxyl groups are further oxidized using TEMPO reagents to add extra negative charges. These carboxylated CNCs produced via the TEMPO-oxidation technique have a high concentration of carboxylic acid. The carboxylated CNCs were better at absorbing methylene blue (769 mg g^{-1} at pH 9) than the sulfated CNCs (118 mg g^{-1} at pH 9) [88].

Recently, research has been focused on making nanocellulose aerogels as an adsorbent because they have qualities like a large surface area. Researchers successfully fabricated aerogels with exceptional characteristics, including an extremely low density ranging from 1.7 to 8.1 mg cm^{-3} and an exceptionally high porosity ranging from 99.5% to 99.9%. These aerogels were generated by defibrillating CNFs derived from rice straws and subjecting them to a freeze-drying technique. CNF aerogels showed superior adsorption ability, capable of absorbing 210 times water and 375 times chloroform relative to their dry weight, respectively. Additionally, it has the potential to adsorb nonpolar hydrocarbons, polar aprotic solvents, and various oils at a rate that exceeds all earlier documented polymeric, carbonaceous, and cellulosic aerogels by a factor ranging from 2 to almost 20 times [116]. Reportedly, CNF aerogel was made from sago pith by three steps: dewaxing and delignification, ultra-sonication and homogenization, and freeze drying. The sample successfully removed up to 99.99% of methylene blue dye [117]. CNCs with high affinity for both anionic and cationic dye types can be used to create cross-linked microgels with polyvinyl amine. Its maximal adsorption uptake was 1469, 1250, and 896 mg g^{-1} for Congo red 4BS, reactive bright yellow K-4G, and acid red GR, respectively [118]. A composite made of nanocellulose and chitosan effectively removed Victoria Blue R dye. Adsorption followed Langmuir isotherm and pseudo-second-order kinetics [119].

3.5.1.3 Oil and Solvent

Oil-based chemicals resulting from discharges and oil spills are a significant contributor to environmental pollution. The presence of oil on the surface of water has a detrimental impact on the aquatic ecology and poses a contamination risk to drinking water sources for human consumption [103]. Typically, the development of effective water and oil separation involves the separation of water and oil, either through subsequent or simultaneous oil adsorption. Presently, the predominant techniques for separating oil and water, including centrifugation, hydrocyclone, and sedimentation, rely on disparities in gravity, or density [120, 121].

Owing to their highly porous structure, sponge-like nanocellulose gels and materials (CNCs, CNFs, and BNC) are widely employed for oil and water separation, as they can absorb significant amounts of oil [122]. One possible modification of CNFs involves the addition of silanes, which makes them hydrophobic, allowing them to absorb oil pollutants from the water surface [123]. Similarly, the use of sol-gel chemistry to cover CNF aerogels with a thin coating of TiO_2 NPs proves to be very effective in adsorbing oil [124]. In addition, silylated nanocellulose sponges provide a pliable, water-repellent, and very lightweight framework for extracting oil from water mixes [121]. The sponges made from oat straw cellulose pulp powder were very porous, with a porosity of up to 99%. These sponges were obtained by reacting the cellulose with methyltrimethoxysilane and utilizing a freeze-drying process. These sponges have more flexibility compared to typical inorganic porous materials, as shown by their ability to restore 96% of their original shape after being compressed. Their affinity for oil and resistance to water enable the effective and specific adsorption-based cleanup of dodecane spills from surface water. Similarly, hydrophobized cellulose microfibers may be used to produce lightweight carbon aerogels that have significant potential for selectively absorbing oil [125]. These sponges have a porosity of 99%, a density of 0.01 $g\,cm^{-3}$, excellent reusability, hydrophobicity, and an outstanding ability to adsorb paraffin oil (>86 $g\,g^{-1}$). The BNC membrane surface has been rendered water-repellent through treatment with trimethylchlorosilane. This treatment allows for the effective removal of plant oil from water. The process involves immersing BNC aerogels in liquid trimethylchlorosilane and then freeze-drying them. The surface-modified BNCs achieved a very high hydrophobicity, with a water/air contact angle reaching as high as 146.5°. Additionally, these BNCs exhibited a high level of porosity, with around 99.6% porosity. As a result, they demonstrated excellent efficiency in separating oil from water, with adsorption capacities reaching up to 185 $g\,g^{-1}$ [126].

The magnetic nanoparticles are combined with nanocelluloses through mixing or on-site hydrolysis of metallic precursors. These metal precursors are often used for the purpose of removing oil from wastewater. The magnetic nanocellulose aerobeads, produced via freeze-drying of iron oxide (Fe_3O_4)-incorporated spherical CNFs derived from recycled cardboard boxes, exhibit excellent oil removal selectivity. The aerobeads, with a density of 0.005 $g\,cm^{-3}$ and 99% porosity, exhibit exceptional absorption capabilities for a wide range of oils and organic solvents, particularly castor oil (279 $g\,g^{-1}$). Furthermore, these display outstanding reusability and can be utilized at least 10 times, showcasing an impressive ability to absorb diesel oil (101 $g\,g^{-1}$) [127]. Nanocellulose aerogels containing Fe_3O_4, produced using the freeze-drying technique, have the potential to absorb oil spills from water effectively. The density of these aerogels is 9.2 $mg\,cm^{-3}$ and possess an adsorption capacity of 68 $g\,g^{-1}$ toward cyclohexane. Additionally, they demonstrate effective absorption of ethyl acetate and vacuum pump oil [128].

3.5.1.4 Other Organic Pollutants

Humans may experience a number of health issues from organic pollutants in contaminated water, such as skin rashes, headaches, diarrhea, soreness in the muscles and joints, breathing difficulties, and irregular heartbeats. Emerging contaminants, including hormones, pesticides, pharmaceutically active substances, and personal care items, have raised significant concerns. As the global population grows, the demand for and use of these items rises, leading to higher levels of their presence in water. Inadequate disposal methods and the release of waste from both municipal and industrial sources are the main causes of this [129].

Pharmaceutical waste has the greatest detection rate in water among all other emerging contaminants. The usage of antibiotics is a significant problem in human as well as veterinary medicine. Their existence in the environment might potentially result in the emergence of antibiotic-resistant microorganisms [130, 131]. Nanocelluloses are used for the adsorption and removal of toxic pharmaceuticals, such as ciprofloxacin and diclofenac, from industrial effluent. The rate of adsorption for ciprofloxacin is 227 $mg\,g^{-1}$, and for diclofenac, it is 192 $mg\,g^{-1}$. In a separate study, nanocellulose was first modified before being used to adsorb vancomycin, ciprofloxacin, amoxicillin, and chlorhexidine. The adsorption process was observed to follow the second-order kinetic model, as shown by the adsorption data [132].

Other organic substances, such as herbicides, pesticides, and fertilizers, were examined as potential contaminants of wastewater. Several studies have shown that nanocellulose has the capability to function as an effective adsorbent for eliminating these. In recent times, there has been increasing attention in exploring the application of nanocelluloses as a possible adsorbent for the removal of pesticides, fertilizers, and pharmaceuticals as pollutants. CNFs aerogel was used to adsorb fertilizer by-products, including nitrate, nitrite, and phosphate. The achieved elimination percentages for nitrite, phosphate, and nitrate and were 73%, 98%, and 79%, respectively [133]. In a different study, the researchers used composites of polyvinyl alcohol (PVA) and nanocellulose to investigate the adsorption of fluoride. The findings showed that the adsorbent has an adsorption capacity of 11.36 $mg\,g^{-1}$ [134]. A study was carried out to analyze the process of adsorbing the pesticide chlorpyrifos utilizing nanocellulose that has been treated with polyvinylamine. The achieved elimination percentage was 93% [135].

3.5.2 Cellulose Nanomaterial as Flocculants

Flocculation is a widely applied technique for eliminating organic materials, metals, colors, and fine suspended solids from wastewater from both industrial and municipal sources. Flocculants are used in solutions to help colloidal particles collide and generate flocs—larger, unstable particles that ultimately precipitate out of the solution. Frequently, such flocculants stimulate the creation of flocs through electrostatic contact with particles of colloidal nature, resulting in the formation of bigger flocs [136, 137]. At present, water treatment methods rely on synthetic polymers like polyacrylamides and polyacrylic acids, which are originated from nonrenewable petroleum sources, for use as flocculants. Biological processes cannot easily break these substances down and often include unreacted monomers and compounds that have the potential to cause cancer and damage the nervous system [138, 139]. In this regard, an increasing amount of research is dedicated to creating biodegradable, biocompatible, and sustainable flocculants based on CNs. Nanocelluloses (CNFs and CNCs) possess distinct characteristics that distinguish them as excellent candidates for water treatment flocculants, in contrast with native cellulose: (i) The rod-like structure of CNFs and CNCs, together with their tiny size and large surface area, allows for percolation to occur at minimal concentrations. (ii) CNFs and CNCs have the ability to enhance the production of flocs in comparison to natural fibers. When comparing CNFs to CNCs, the greater electrostatic repulsion and stiffness of CNCs compared to CNFs will restrict physical and chemical bindings, reducing the likelihood of gelation [101].

Currently, there is a limited amount of research that discusses the uses of cellulose nanomaterials as flocculants in the treatment of wastewater. Suopajarvi et al. [139] produced carboxylated cellulose nanofibers (CNFs) with the purpose of using them as negatively charged flocculants in the treatment of municipal wastewater. Anionic CNFs exhibited remarkable durability and stability in water-based suspensions, which led to exceptional performance, achieving a turbidity reduction of 40–80%, and COD removal of 40–60%. These results were achieved under the optimal pH range of 6–8 [139]. In research carried out by Korhonen and Laine [140], the authors investigated the use of CNF/polyelectrolyte with varying charge density to determine its effectiveness in retaining and flocculating kaolin and calcium carbonate as fillers as additives in paper production. They demonstrated that when CNF/polyelectrolyte is used, the efficiency of flocculation increases from 80% to 95%. Polyelectrolytes facilitate the creation of CNF/polyelectrolyte connections between calcium carbonate and kaolin particles, resulting in effective flocculation [140]. A recent research evaluated the ability of hyperbranched cellulose modified with polyethyleneimine $(C_2H_5N)n$ to cause particles in kaolin-contaminated wastewater to clump together, facilitating their removal. The cellulose-based flocculant lessened the residual turbidity of the kaolin solution from an initial measurement of 490 NTU to 4 NTU upon the addition of 2.4 mg L^{-1} of the flocculant at pH 7.0 over a period of 30 minutes [141].

In their study, Yu et al. [142] showed that the CNCs obtained through a single-step hydrolysis process using citric acid and HCl on microcrystalline cellulose (MCC) exhibited superior ability to cause suspended kaolin clay particles to come together relative to two other kinds of CNCs produced through hydrolysis with sulfuric acid and formic acid/HCl. These CNCs can serve as a coagulant to efficiently remove positively charged dyes and kaolin particles from solutions, achieving a turbidity reduction capability of 99.5%. The superior performance of carboxylated CNCs relies on the existence of surface functional groups, which provide these CNCs with a higher number of anionic charges in comparison to other forms of CNCs [142]. Kemppainen et al. [143] developed dicarboxylic acicellulose (anionic) and sulfonic acid CNFs for quartz and hematite flocculation in contaminated water. At pH 8–9, 200–500 ppm of carboxylic modified CNFs flocculated hematite best. Sulfonated modified CNFs are as effective as carboxylic modified CNFs at 500 ppm as hematite flocculants with extended conditioning time and gentler stirring [143].

3.5.3 Cellulose Nanomaterials as Membranes

Membrane technology has become a popular way for processing and separating water from salty and polluted streams, allowing it to be reused by people. This technology is often used for filtering, extraction, and distillation purposes. Because of their great selectivity and low energy consumption, membranes based on nanocellulose are a very effective and ecologically friendly technique of purifying water. Two factors primarily dictate the functioning properties of membranes based on nanocellulose. They are the particular pollutant removal capacity and the water permeation flux value. The size of the pore and surface characteristics of nanocellulose play a crucial role in regulating the efficacy of the filter. Throughout time, several methods have been documented in the literature for the creation of composite membranes using nanocellulose and their subsequent usage in water treatment. The commonly used techniques are vacuum filtration, in situ polymerization, electrospinning, freeze-drying, and phase inversion. Elaborate analyses of each approach are provided in the following sections. In comparison to other approaches, the phases of phase inversion and in situ polymerization require additional fabrication steps. Nevertheless, they are significantly popular due to their straightforwardness and cost-efficiency. Frequently used methods for nanocellulose-based composite membrane production are summarized in Figure 3.7.

3.5.3.1 Phase Inversion

A simple process known as phase inversion (PI) uses a controlled mechanism to change a homogeneous fluid into a solid state [145]. The conversion may be initiated by processes such as evaporation, hydration, or thermal means [146, 147]. The wet PI procedure is the most commonly employed technique for producing membranes based on nanocellulose. During the process of synthesis, nanocellulose is typically introduced into a compatible solvent and evenly distributed by mechanical stirring or ultrasonication. Subsequently, a specific polymer is likewise dissolved in the same solvent. To get a uniform casting solution, the resultant mixture is agitated at an optimal temperature for a certain duration [148, 149]. In the last phase, membranes of a particular thickness are cast onto a petri dish and then submerged in a coagulation solution. The use of liquid with varying boiling levels alongside a substrate with small pores facilitates the regulation of the membrane's pore dimensions. Additionally, this approach enables regulation of the surface morphology. The substantial research on this preparation approach

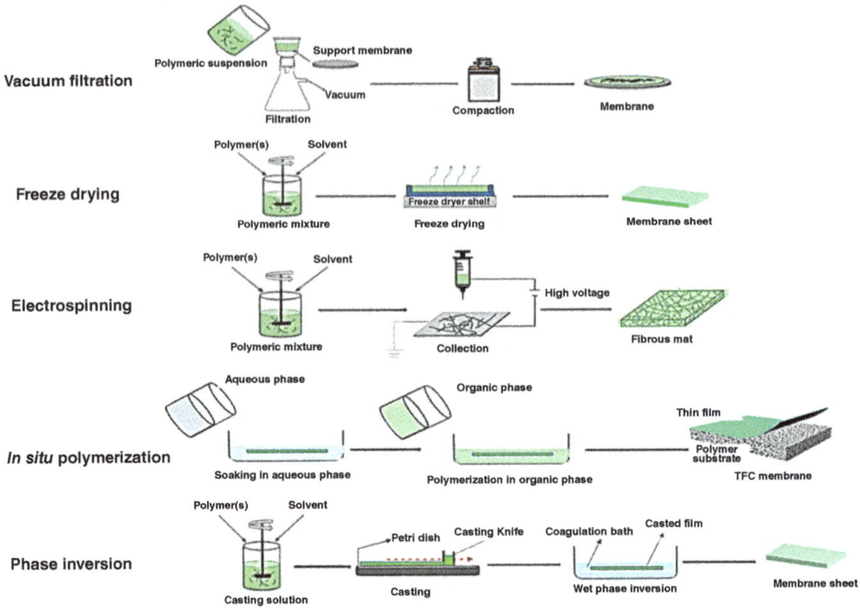

Figure 3.7 Graphical representation of frequently used methods for nanocellulose-based composite membrane production. *Source:* Mbakop et al. [144].

has focused on developing advanced composite membranes for low-pressure filtering due to its user-friendly nature [150, 151]. For example, CNCs was blended by using PI to make polyvinylidene difluoride (PVDF) composite ultrafiltration (UF) membranes [152]. The testing findings showed a significant improvement in permeate water flux (PWF; up to 47.5%) with the incorporation of CNCs. Furthermore, the study demonstrated improved mechanical characteristics and 92.5% bovine serum albumin (BSA) rejection. In their study, the intrinsic mechanical and hydrophilicity characteristics of polyethersulphone (PES) composite membranes that were mixed with carbon nanotubes (CNTs) and CNCs [153]. The hydrophilicity of the membranes based on CNCs was greater in comparison to the membranes based on CNTs. The membranes with CNCs demonstrated significantly more elasticity than the original PES membrane, even though both membranes had higher tensile strength and Young's modulus. Several supplementary studies illustrate the significant potential of the PI approach in creating composite membranes based on nanocellulose that possess exceptional qualities. Nevertheless, it is crucial to demonstrate a definitive link between the ratio of nanocellulose to polymer and the intrinsic features of the resultant composite membranes. This will provide enhanced regulation of the operation of the manufactured materials and lay the groundwork for industrial manufacturing.

3.5.3.2 Vacuum Filtration

The process of vacuum filtering, with the optional step of hot pressing, is often referred to as the "nanopaper approach." This method is used in both industrial and laboratory settings to create layered networks of membranes and nanopapers utilizing nanocellulose. After the procedure, nanocellulose is compactly arranged into the membrane and a film that has an array of pores with diameters ranging from 3 to 5 nm, similar to the diameter of the nanocellulose. Varied concentrations of nanocellulose solutions are often employed to customize permeability and the structure of the produced films. The pore size distribution of the membrane is significantly influenced by the length and diameter of nanocellulose utilized in the fabrication method [148–151]. Karim et al. [154] documented the method of vacuum filtering of solutions containing CNFs. After filtration, CNCs with sulfate or carboxyl functional groups were applied. The method produced multilayered membranes made of nanocellulose. The nanocellulose membranes demonstrated improved PWF and achieved a high removal efficiency (~100%) for Cu^{2+}, Fe^{2+}/Fe^{3+}, and Ag^{+} ions in mirror manufacturing discharge, regardless of the surface functioning. One typical problem with vacuum filtration technology is the likelihood of draining during filtering or backwashing operations, since the approach demands the mechanical incorporation of nanomaterials into the framework of polymeric material.

3.5.3.3 Electrospinning

This manufacturing method often generates nanofibrous membranes with many layers, which have exceptional active surface areas and interconnected porosity [155–157]. The membranes are created through integrating nanocellulose into the electrospun membrane substrate. This might lead to a rise in the membrane's effective surface area and surface chemistry for certain uses such as adsorptive filtration [148–151]. In another work conducted, PVDF electrospun membranes based on CNFs were created by Gopakumar [158]. The composite membranes demonstrated a high level of filtering efficiency against Fe_2O_3 nanoparticles. Notwithstanding the advancements made throughout time, the constrained manufacturing capability and the inadequate replicability are impeding the widespread use of this technology.

3.5.3.4 In Situ Polymerization

The in situ polymerization (IP) technique typically entails a polycondensation reaction between two monomers that are dispersed in solvents that do not mix. This reaction results in the formation of an extremely thin layer at the interface of a polymer substrate [159, 160]. First, polymeric support (namely, PES) is immersed in water-based solution for a short period. The organic phase is applied to the films to commence the IP technique. This is appropriate for the manufacturing of reverse osmosis (RO) and nanofiltration (NF) membranes. Factors including solvent types, monomer levels, and reaction times are often adjusted to regulate the morphological characteristics of the membranes that are generated. To enhance the qualities of the membranes that are produced, several studies that have shown the effective use of this technology over the years have also concentrated on enhancing compatibility between the organic and aqueous phases. In addition, the progress of contemporary research in the direction of more sustainable practices has resulted in the progressive use of ecologically compatible organic solvents for the purpose of large-scale manufacturing [148–151].

3.5.3.5 Freeze-Drying

The freeze-drying method is a simple procedure. First, the polymeric solution is poured onto a level surface. This is done at a temperature lower than room temperature to form a uniformly thick layer. A portion of the organic fluid rapidly evaporates. Afterward, the membrane is submerged in freezing water to fully form the membrane [148–151]. Karim et al. described using this process to create composite membranes made entirely from bio-based materials [161]. The membranes include CNCs as functional components embedded in a chitosan matrix. The observed reduced flux (64 $L\,m^{-2}\,h^{-1}$) was ascribed to the nanoporous nature of the membranes made of nanocomposite. Successful removal of dye molecules, including Victoria blue 2B (98%), methyl violet 2B (84%), and rhodamine 6G (70%), was obtained. Similarly, a freeze-extraction technique was used to fabricate ultrathin nanoporous membranes [162]. The membranes exhibited a clean water flow of $1.14 \times 104\,L\,m^{-2}\,h^{-1}\,bar^{-1}$. The manufactured membranes were deemed suitable for filtration tasks that include the separation of substrates and particles with sizes exceeding 10 nm from water-based solutions. Due to the early stage of development, advancements in technology are now confined to laboratory discoveries. Further research on the use of sustainable resources and optimization of processes could help overcome the limitations that prevent the widespread adoption of this technology.

3.5.4 Cellulose Nanomaterials for Photocatalytic Degradation

Photocatalysis has developed as an eco-friendly, cost-effective, and convenient process for the degradation of colors and organic contaminants in wastewater. Though nanocelluloses demonstrate limited visible light-UV photocatalytic activity, TiO_2 [55] and ZnO [121] have been included to augment this activity. Photocatalytic materials including individual particles, membranes, and thin films have been synthesized utilizing cellulose-based metal oxide nanostructures under UV and visible light irradiation. Photocatalysis utilizes photon energy to generate active radicals that facilitate the degradation of organic contaminants [163]. When photon energy exceeds the semiconductor energy gap, semiconductor materials can produce active hydroxyl radicals, resulting in the excitation of electrons from the valence band to the conduction band [164]. This forms a cavity that can produce hydroxyl radicals in an alkaline environment [165]. These radicals exhibit significant reactivity and can facilitate the oxidation of resistant compounds in wastewater, resulting in the complete breakdown of organic contaminants into nontoxic byproducts (i.e., CO_2 and H_2O) [166]. Active hydroxyl-free radicals interact with organic materials by four established mechanisms: radical addition, hydrogen abstraction, electron transfer, and radical combination [167]. The primary problem related to the application of semiconductor nanoparticles is their extraction from the solvent postreaction. Notably, nanocomposite containing nanocelluloses may be effortlessly extracted from wastewater for reuse and subsequent utilization, attributable to its distinctive magnetic characteristics. According to a study, magnetic Fe_3O_4/CNF composite aerogels catalyze Fenton-like Rhodamine B degradation with 100% efficiency [168]. The catalyst retained 97% removal effectiveness after six cycles [169]. New cellulose-based photocatalysts have an acceptable band gap (2.4 eV) and strong UV-Vis photocatalytic activity and stability [170]. Through hydrothermal treatment and thermal pyrolysis, photocatalyst-NPs cellulose aerogels and sponges may transform into photocatalytic carbonaceous materials [171]. Photocatalytic carbonaceous materials have increased photocatalytic characteristics because of their porosity, high surface area, and capacity to absorb organic pollutants [172].

3.5.5 Cellulose Nanomaterials as Absorbents

Absorption is a process employed in water treatment in which the pollutant disperses into the majority of the absorbent and is retained inside its porous matrix. Cellulose-based absorbents have garnered heightened interest owing to their distinctive features and sustainable nature. Previous studies have found that CNFs were used to make most CN-based absorbents, with few investigations utilizing CNCs as matrix materials. In a study, a porous cellulose scaffold that selectively absorbed organic solvents and oil has been created [173]. Freeze-drying and ice-templating acetate esterified CNCs were used to make the scaffold. Hydrophobic/oleophilic properties of this scaffold arise from the elevated concentration of acetate ester groups on the customized CNCs utilized in its preparation. The scaffolds absorbed 100–200 $g\,g^{-1}$ of organic solvents (toluene, chloroform, 1,2-dichloroethane, acetone, dichloromethane) and oils (silicone, paraffin). Chemically connected CNC aerogels suitable for superabsorbent applications in oil/water separations have been developed [174]. Aerogels were manufactured through the freeze-drying of hydrogels created by cross-linking hydrazide-modified CNCs and aldehyde-modified CNCs using hydrazone cross-linking chemistry. A hydrazone-carboxyl ligand-linked, CNC-based aerogel was synthesized that exhibits significant absorption capacity for water and other polar liquids [41]. The absorption capacities of these aerogels for water, toluene, dodecane, and ethanol were determined to be 133 ± 6, 34 ± 4, 54 ± 6, and $99 \pm 8\,g\,g^{-1}$, respectively. Aerogels are ideal for oil and water separation because of their extraordinary water and polar liquid absorption.

A simple and successful silylation method was created for hydrophobic CNF aerogels in water [116]. The silylation process involves freeze-drying the CNF water suspension in methyltrimethoxysilane solution in one step. These silylated CNF aerogels selectively and recycleably remove dodecane spills from water surfaces. Aerogels absorbed organic solvents (acetone, ethanol, toluene, chloroform, dodecane, and dichloromethane) and oils (silicone oil, motor oil, and mineral oil) with absorption capacities ranging from 49 to $102\,g\,g^{-1}$.

Vapor deposition with triethoxy(octyl)silane produced hydrophobic/oleophilic CNF aerogels by [116]. These aerogels demonstrated excellent absorption selectivity for nonpolar liquids and showed commendable recyclability. They exhibited remarkable absorption capabilities ranging between 139 and $356\,g\,g^{-1}$ for several nonpolar liquids, including aliphatic (octane, hexadecane, decane, hexan), cyclic (cyclohexane), and aromatic (toluene) hydrocarbons, as well as oils (soybean, pump).

3.5.6 Cellulose Nanomaterials as Disinfectants

Disinfection processes represent the final and crucial phase in treating wastewater, involving the elimination, deactivation, or destruction of pathogenic bacteria. To reduce harmful disinfection byproducts, numerous nanomaterials, including oxides of metal and polymer nanoparticles, have served as disinfectants in recent decades [34]. Nanocelluloses exhibit minimal impact on inhibition of microbes. This can offer a carbon supply for the growth of algae, fungi, and bacteria. Consequently, antibacterial and antimicrobial properties should be included into nanocellulose-based materials to suppress microbial development in contaminated water. Antibacterial properties can be imparted to nanocelluloses through surface modification using metal nanoparticles (e.g., Au, Ag), metal oxide nanoparticles (e.g., CuO, TiO_2, MgO, ZnO), and antimicrobial polymers [175]. Nanocelluloses can be surface modified by oxidation with oxidized groups (TEMPO oxidation), which results in their functionalized carboxyl groups that have a high affinity for *E. coli* and a rejection rate of approximately 96–99%. Both Gram-positive and Gram-negative bacteria are resistant when negative surface charges (carboxyl groups) are added. This also makes it easier to attach firmly to the negatively charged areas near the bacterial membrane. Although the surface carboxyl groups of nanocelluloses are not inherently antibacterial, their properties can be enhanced by the right antibacterial substances. An efficient method for introducing antibacterial activity is cationic surface functionalization with quaternary ammonium salts, including ammonium chlorides [176] and ammonium bromides [177].

Using mussel-inspired catechol chemistry, composites of carboxylated CNF with embedded silver nanoparticles and dopamine-linked (CCNF-DA) were developed. This has enhanced antibacterial activity, electrical conductivity, and mechanical properties. The composite exhibits anisotropic alignment due to stronger contact between catechol moieties and silver nanoparticles (Ag NPs), providing biomaterials that are capable of conducting electricity, are flexible, and have antimicrobial properties. These substances could be used as wastewater treatment disinfectants [178].

3.6 Challenges and Opportunities Associated with CN

One of the primary obstacles encountered when dealing with biomaterials, such as CNs, is their inherent vulnerability to water. One additional obstacle encountered when dealing with polymer-CN composites, similar to CNT-polymer composites, is the attainment of consistent and even dispersion across the polymer matrices [89]. The phenomenon of homo-aggregation induces regional nonuniformity within amorphous polymers, leading to disruptions in polymer alignment and thus affecting the crystalline structure of semi-crystalline polymers. Membranes made of homogeneously dispersed CNCs in polyethylene oxide (PEO) have a higher strain-at-failure compared to membranes containing cellulose nanofibers (CNF) doped with PEO [179]. The utilization of alternative functionalization strategies, distinct from the conventional two-phase system including nanocellulose and a uniform matrix, necessitates the development of a multiphase model. However, the building of such a model is challenging due to the incomplete understanding of the underlying reinforcing mechanisms. Therefore, further investigation is required to gain a comprehensive understanding of these mechanisms [110]. Another significant challenge that arises in the field is metrology.

The advancement of CN-metrology poses a significant obstacle that, once surmounted, will facilitate the rapid integration of CNs into the commercial sphere. One ultimate objective that remains is the identification of a viable approach to expedite the process of commercialization. The various components involved in the production of nanocellulose, including raw materials, procedures, and process parameters, can have varying degrees of influence on the resulting nanocellulose products. Consequently, these factors can also affect the qualities of composite products. It is imperative to establish research programs that are strategic, holistic, interdisciplinary, and adequately funded to facilitate the process of commercialization. These programs should be centered around specific applications and involve substantial collaboration with the product producer [180]. CNs can enhance membrane hydrophilicity for water diffusion but may compromise their mechanical characteristics. Using hydrophilic polymers avoids CN's mechanical qualities from decreasing, but it may reduce membrane water diffusion. A trade-off exists between improved hydrophilicity and mechanical strength from CNs in polymer membranes [181]. Although cellulose-based composite materials have promise for treating water by eliminating heavy metal ions and organic dyes, efficiently controlling the critical characteristics of porosity, permeability, and selective adsorption presents a considerable challenge. The issue of nanoscale cellulose agglomeration while the drying of films can be effectively addressed by applying a thin functional layer of nanocellulose over designated filters or membrane structures [101]. The implementation of research and development efforts focused on novel batch homogenization production techniques, as well as the establishment of comprehensive measurement standards, will significantly contribute to the advancement of precise measurement systems and standards for the mechanical characteristics of individual nanocellulose particles.

CN-based products are produced from sustainable materials, utilizing processes with a reduced carbon footprint, including sulfuric acid hydrolysis and mechanical disintegration. Klemm et al. found that producing CNFs from sulfite pulp required 27,000 $kWh\,t^{-1}$ of energy [182]. However, pretreatment with carboxymethylation improved processability, reducing this number to 500 $kWh\,t^{-1}$. In comparison, Roberts et al. found that biochar from late Stover, early Stover, and switchgrass from fossil fuels requires 208.1, 279.6, and 247.8 $kWh\,t^{-1}$, close to half the energy required for the production of CNFs following pulp carboxymethylation [183]. The implementation of continuous unit operations holds significant importance in the context of water treatment processes conducted on a large scale. However, the predominant focus of existing studies has been on batch procedures. Therefore, comprehensive investigations are required to gain a deeper understanding of the behavior and efficacy of carbon nanotube-based systems in diverse water treatment applications.

3.7 Environmental Impacts

Nanocellulose is widely regarded as a sustainable nanomaterial owing to its abundant availability, capacity for biodegradation, and compatibility with biological systems. Cellulose nanomaterials have the potential to decrease the environmental impact of various industries because of their biodegradability, low toxicity, and renewability [184]. Nonetheless, if not appropriately controlled, the manufacturing and disposal of cellulose nanoparticles may result in adverse environmental consequences. The technical difficulties associated with CNC pertain to the absence of developing effective and accurate pretreatment methods, extraction procedures, and surface functionalization techniques. The utilization of various hazardous chemicals during these processing stages can disturb the equilibrium of the ecosystem as a result of their adverse environmental effects and lack of sustainability [185]. It is imperative to take into account the life cycle assessment of cellulose nanomaterial manufacturing and disposal in order to optimize the process and mitigate the environmental consequences [186]. Like all other materials, cellulose nanomaterials have drawbacks. Though cellulose nanomaterials obtained from various biodegradable sources can be used in different applications from wastewater remediation to healthcare materials, they can bring toxicity to the environment.

3.8 Conclusions and Future Prospects

Cellulose nanomaterials are a potential material for wastewater remediation because of their abundant availability, biocompatibility, mechanical strength, and surface functionality. The adoption efficiency and functionality of any revolutionary technology are typically contingent upon two key factors: minimizing costs and considering potential toxicity. While nanoadsorbents have demonstrated high efficacy in certain water treatment scenarios, it is important to note that their production and deployment might be financially burdensome. The utilization of renewable and sustainable sources in the production of CNs expands the scope for prospective applications. However, it is necessary to address the issue of their pricing to enhance their competitiveness in the market. More research has been dedicated to making CNs economically sustainable and environmentally benign. The examination of several characteristics of CNs, including their shape, DP, chemical composition, rheological properties, and crystallinity, plays a crucial role in enhancing our comprehension of process optimization and facilitating their application in material creation. Though adsorbents based on CNs are widely employed, membrane fouling needs potential study for wastewater remediation. Cellulose nanomaterials are not yet widely used on a large scale. Research should be focused on new characterization methods for simultaneously using and producing CNs on a large scale. While several firms have claimed to successfully collect and reuse all substances utilized in the extraction process, further investigation is being conducted to enhance the extraction procedure to minimize labor intensity and minimize the use of harsh chemicals. Companies should emphasize the standard extraction method of CNs. Despite the fact that functionalized nanocellulose frequently exhibits adsorption capabilities that are comparable to, or even surpass, those of conventional adsorbents in the context of heavy metal and organic pollutant removal, it is imperative to investigate and address the issue of selectivity in complex water systems and large-scale experimental setups. This analysis may provide further elucidation on the efficacy and benefits of nanocellulose-based membranes in industrial contexts, particularly in comparison to alternative methods of water filtration. Future research should prioritize the development of sustainable and environmentally conscious approaches for the manufacturing and disposal of cellulose nanoparticles. Additionally, efforts should be made to establish standardized methodologies for characterizing cellulose nanomaterials. Furthermore, it is essential to explore novel applications for cellulose nanomaterials.

References

1 Lin, L., Yang, H., and Xu, X. (2022). Effects of water pollution on human health and disease heterogeneity: a review. *Frontiers in Environmental Science* 10: 880246. https://doi.org/10.3389/fenvs.2022.880246.

2 Kesari, K.K., Soni, R., Jamal, Q.M.S. et al. (2021). Wastewater treatment and reuse: a review of its applications and health implications. *Water, Air, & Soil Pollution* 232 (5): 208. https://doi.org/10.1007/s11270-021-05154-8.

3 Boretti, A. and Rosa, L. (2019). Reassessing the projections of the world water development report. *npj Clean Water* 2 (1): 15. https://doi.org/10.1038/s41545-019-0039-9.

4 Foster, E.J., Moon, R.J., Agarwal, U.P. et al. (2018). Current characterization methods for cellulose nanomaterials. *Chemical Society Reviews* 47 (8): 2609–2679. https://doi.org/10.1039/C6CS00895J.

5 Tusnim, J., Hoque, M.E., Hossain, S.A. et al. (2020). Nanocellulose and nanohydrogels for the development of cleaner energy and future sustainable materials. In: *Sustainable Nanocellulose and Nanohydrogels from Natural Sources* (ed. F. Mohammad, H.A. Al-Lohedan, and M. Jawaid), 81–113. Elsevier. https://doi.org/10.1016/B978-0-12-816789-2.00004-3.

6 Payen, A. (1838). *Mémoire sur la composition du tissu propre des plantes et du ligneux. (Memoir on the composition of the tissue of plants and of woody [material]). Comptes Rendus Hebdomadaires des Séances de l'Académie des Sci-ences* 7: 7. *References—Scientific Research Publishing*. (n.d.). Retrieved January 11, 2024, from https://www.scirp.org/reference/ReferencesPapers?ReferenceID=1640572&fbclid=IwAR20MZQ9E1KhMGTzu7UGSEIxSzYou0iSmEUhWwTB5u1MfwRoXOTC2KdZeQw.

7 Khalid, M.Y., Al Rashid, A., Arif, Z.U. et al. (2021). Recent advances in nanocellulose-based different biomaterials: types, properties, and emerging applications. *Journal of Materials Research and Technology* 14: 2601–2623. https://doi.org/10.1016/j.jmrt.2021.07.128.

8 Ching, Y.C., Ershad Ali, M., Abdullah, L.C. et al. (2016). Rheological properties of cellulose nanocrystal-embedded polymer composites: a review. *Cellulose* 23 (2): 1011–1030. https://doi.org/10.1007/s10570-016-0868-3.

9 Rajinipriya, M., Nagalakshmaiah, M., Robert, M., and Elkoun, S. (2018). Importance of agricultural and industrial waste in the field of nanocellulose and recent industrial developments of wood based nanocellulose: a review. *ACS Sustainable Chemistry & Engineering* 6 (3): 2807–2828. https://doi.org/10.1021/acssuschemeng.7b03437.

10 Dufresne, A. (2017). Cellulose nanomaterial reinforced polymer nanocomposites. *Current Opinion in Colloid & Interface Science* 29: 1–8. https://doi.org/10.1016/j.cocis.2017.01.004.

11 Casanova, F., Pereira, C.F., Ribeiro, A.B. et al. (2021). Novel micro- and nanocellulose-based delivery systems for liposoluble compounds. *Nanomaterials* 11 (10): 2593. https://doi.org/10.3390/nano11102593.

12 Mateo, S., Peinado, S., Morillas-Gutiérrez, F. et al. (2021). Nanocellulose from agricultural wastes: products and applications—a review. *Processes* 9 (9): 1594. https://doi.org/10.3390/pr9091594.

13 Mali, P. and Sherje, A.P. (2022). Cellulose nanocrystals: fundamentals and biomedical applications. *Carbohydrate Polymers* 275: 118668. https://doi.org/10.1016/j.carbpol.2021.118668.

14 Mishra, R.K., Sabu, A., and Tiwari, S.K. (2018). Materials chemistry and the futurist eco-friendly applications of nanocellulose: status and prospect. *Journal of Saudi Chemical Society* 22 (8): 949–978. https://doi.org/10.1016/j.jscs.2018.02.005.

15 Zinge, C. and Kandasubramanian, B. (2020). Nanocellulose based biodegradable polymers. *European Polymer Journal* 133: 109758. https://doi.org/10.1016/j.eurpolymj.2020.109758.

16 Omran, A.A.B., Mohammed, A.A.B.A., Sapuan, S.M. et al. (2021). Micro- and nanocellulose in polymer composite materials: a review. *Polymers* 13 (2): 231. https://doi.org/10.3390/polym13020231.

17 Trache, D., Tarchoun, A.F., Derradji, M. et al. (2020). Nanocellulose: from fundamentals to advanced applications. *Frontiers in Chemistry* 8: 392. https://doi.org/10.3389/fchem.2020.00392.

18 Ghasemlou, M., Daver, F., Ivanova, E.P. et al. (2021). Surface modifications of nanocellulose: from synthesis to high-performance nanocomposites. *Progress in Polymer Science* 119: 101418. https://doi.org/10.1016/j.progpolymsci.2021.101418.

19 Kumar, R., Rai, B., Gahlyan, S., and Kumar, G. (2021). A comprehensive review on production, surface modification and characterization of nanocellulose derived from biomass and its commercial applications. *Express Polymer Letters* 15 (2): 104–120. https://doi.org/10.3144/expresspolymlett.2021.11.

20 Abdullah, N.A., Rani, M.S.A., Mohammad, M. et al. (2021). Nanocellulose from agricultural waste as an emerging material for nanotechnology applications—an overview. *Polimery* 66 (3): 157–168. https://doi.org/10.14314/polimery.2021.3.1.

21 Thomas, P., Duolikun, T., Rumjit, N.P. et al. (2020). Comprehensive review on nanocellulose: recent developments, challenges and future prospects. *Journal of the Mechanical Behavior of Biomedical Materials* 110: 103884. https://doi.org/10.1016/j.jmbbm.2020.103884.

22 Shi, Z., Zhang, Y., Phillips, G.O., and Yang, G. (2014). Utilization of bacterial cellulose in food. *Food Hydrocolloids* 35: 539–545. https://doi.org/10.1016/j.foodhyd.2013.07.012.

23 Wu, C.-N. and Cheng, K.-C. (2017). Strong, thermal-stable, flexible, and transparent films by self-assembled TEMPO-oxidized bacterial cellulose nanofibers. *Cellulose* 24 (1): 269–283. https://doi.org/10.1007/s10570-016-1114-8.

24 Yano, H., Sugiyama, J., Nakagaito, A.N. et al. (2005). Optically transparent composites reinforced with networks of bacterial nanofibers. *Advanced Materials* 17 (2): 153–155. https://doi.org/10.1002/adma.200400597.

25 Phanthong, P., Reubroycharoen, P., Hao, X. et al. (2018). Nanocellulose: extraction and application. *Carbon Resources Conversion* 1 (1): 32–43. https://doi.org/10.1016/j.crcon.2018.05.004.

26 Ilyas, R.A. and Sapuan, S.M. (2018). Nanocellulose reinforced starch polymer composites: a review of preparation, properties and application.

27 Ali, S.Z., Nahian, M.K., and Hoque, M.E. (2023). Extraction of cellulose from agro-industrial wastes. In: *Extraction of Natural Products from Agro-Industrial Wastes* (ed. S.A. Bhawani, A. Khan, and F.B. Ahmad), 319–348. Elsevier. https://doi.org/10.1016/B978-0-12-823349-8.00013-7.

28 Domingues, R.M.A., Gomes, M.E., and Reis, R.L. (2014). The potential of cellulose nanocrystals in tissue engineering strategies. *Biomacromolecules* 15 (7): 2327–2346. https://doi.org/10.1021/bm500524s.

29 Al-Dulaimi, A.A. and Wanrosli, W.D. (2017). Isolation and characterization of nanocrystalline cellulose from totally chlorine free oil palm empty fruit bunch pulp. *Journal of Polymers and the Environment* 25 (2): 192–202. https://doi.org/10.1007/s10924-016-0798-z.

30 Dufresne, A. (2012). *Nanocellulose: From Nature to High Performance Tailored Materials*. In *Nanocellulose*. De Gruyter. https://doi.org/10.1515/9783110254600.

31 Lu, P. and Hsieh, Y.-L. (2010). Preparation and properties of cellulose nanocrystals: rods, spheres, and network. *Carbohydrate Polymers* 82: 329–336. https://doi.org/10.1016/j.carbpol.2010.04.073.

32 Kargarzadeh, H., Ishak, B., Ahmad, I. et al. (2012). Effects of hydrolysis conditions on the morphology, crystallinity, and thermal stability of cellulose nanocrystals extracted from kenaf bast fibers. *Cellulose* 19: 855–866. https://doi.org/10.1007/s10570-012-9684-6.

33 Ioelovich, M. (2012). Optimal conditions for isolation of nanocrystalline cellulose particles. *Journal of Nanoscience and Nanotechnology* 2: 9–13. https://doi.org/10.5923/j.nn.20120202.03.

34 Hassan, M., Abouzeid, R., Fadel, S. et al. (2014). Cellulose nanocrystals and carboxymethyl cellulose from olive stones and their use to improve paper sheets properties. *International Journal Of Nanoparticles* 7: https://doi.org/10.1504/IJNP.2014.067613.

35 Wang, Q., Zhao, X., and Zhu, J.Y. (2014). Kinetics of strong acid hydrolysis of a bleached kraft pulp for producing cellulose nanocrystals (CNCs). *Industrial & Engineering Chemistry Research* 53 (27): 11007–11014. https://doi.org/10.1021/ie501672m.

36 Pandi, N., Sonawane, S.H., and Anand Kishore, K. (2021). Synthesis of cellulose nanocrystals (CNCs) from cotton using ultrasound-assisted acid hydrolysis. *Ultrasonics Sonochemistry* 70: 105353. https://doi.org/10.1016/j.ultsonch.2020.105353.

37 Tang, L., Huang, B., Ou, W. et al. (2011). Manufacture of cellulose nanocrystals by cation exchange resin-catalyzed hydrolysis of cellulose. *Bioresource Technology* 102: 10973–10977. https://doi.org/10.1016/j.biortech.2011.09.070.

38 Liu, Y., Wang, H., Yu, G. et al. (2014). A novel approach for the preparation of nanocrystalline cellulose by using phosphotungstic acid. *Carbohydrate Polymers* 110: 415–422. https://doi.org/10.1016/j.carbpol.2014.04.040.

39 Kargarzadeh, H., Ioelovich, M., Ahmad, I. et al. (2017). *Methods for Extraction of Nanocellulose from Various Sources*, 1–49. Wiley-VCH Verlag GmBH & Co. KGaA. https://doi.org/10.1002/9783527689972.ch1.

40 Torlopov, M., Udoratina, E., Martakov, I., and Sitnikov, P. (2017). Cellulose nanocrystals prepared in H3PW12O40-acetic acid system. *Cellulose* 24: https://doi.org/10.1007/s10570-017-1256-3.

41 Ma, L., Xu, Y., Chen, J. et al. (2023). Preparation of cellulose nanocrystals by synergistic action of ionic liquid and recyclable solid acid under mild conditions. *Molecules* 28 (7): 7. https://doi.org/10.3390/molecules28073070.

42 Lu, Q., Tang, L., Lin, F. et al. (2014). Preparation and characterization of cellulose nanocrystals via ultrasonication-assisted $FeCl_3$-catalyzed hydrolysis. *Cellulose* 21: 3497–3506. https://doi.org/10.1007/s10570-014-0376-2.

43 López-Linares, J.C., Romero, I., Moya, M. et al. (2013). Pretreatment of olive tree biomass with $FeCl_3$ prior enzymatic hydrolysis. *Bioresource Technology* 128: 180–187. https://doi.org/10.1016/j.biortech.2012.10.076.

44 Zhang, Y., Li, Q., Su, J. et al. (2015). A green and efficient technology for the degradation of cellulosic materials: structure changes and enhanced enzymatic hydrolysis of natural cellulose pretreated by synergistic interaction of mechanical activation and metal salt. *Bioresource Technology* 177: 176–181. https://doi.org/10.1016/j.biortech.2014.11.085.

45 Abd Hamid, S.B., Chowdhury, Z., and Karim, M. (2014). Catalytic extraction of microcrystalline cellulose (MCC) from elaeis guineensis using central composite design (CCD). *BioResources* 9: https://doi.org/10.15376/biores.9.4.7403-7426.

46 Li, J., Zhang, X., Zhang, M. et al. (2015b). Ultrasonic enhance acid hydrolysis selectivity of cellulose with HCl-$FeCl_3$ as catalyst. *Carbohydrate Polymers* 117: 917–922. https://doi.org/10.1016/j.carbpol.2014.10.028.

47 Kamireddy, S.R., Li, J., Tucker, M. et al. (2013). Effects and mechanism of metal chloride salts on pretreatment and enzymatic digestibility of corn stover. *Industrial & Engineering Chemistry Research* 52 (5): 1775–1782. https://doi.org/10.1021/ie3019609.

48 Li, B., Xu, W., Kronlund, D. et al. (2015). Cellulose nanocrystals prepared via formic acid hydrolysis followed by TEMPO-mediated oxidation. *Carbohydrate Polymers* 133: 605–612. https://doi.org/10.1016/j.carbpol.2015.07.033.

49 Ioelovich, M. (2013). Plant biomass as a renewable source of biofuels and biochemicals.

50 Meier, H. (1962). Chemical and morphological aspects of the fine structure of wood. *Pure and Applied Chemistry* 5 (1–2): 37–52. https://doi.org/10.1351/pac196205010037.

51 Nechyporchuk, O., Belgacem, M.N., and Bras, J. (2016). Production of cellulose nanofibrils: a review of recent advances. *Industrial Crops and Products* 93: 2–25. https://doi.org/10.1016/j.indcrop.2016.02.016.

52 Comuzzo, P. and Calligaris, S. (2019). Potential applications of high pressure homogenization in winemaking: a review. *Beverages* 5 (3): 3. https://doi.org/10.3390/beverages5030056.

53 Dufresne, A., Cavaill, J.-Y., and Vignon, M.R. (1997). Mechanical behavior of sheets prepared from sugar beet cellulose microfibrils. *Journal of Applied Polymer Science* 64 (6): 1185–1194. https://doi.org/10.1002/(SICI)1097-4628(19970509)64:6<1185::AID-APP19>3.0.CO;2-V.

54 Siddiqui, N., Mills, R., Gardner, D., and Bousfield, D. (2011). Production and characterization of cellulose nanofibers from wood pulp. *Journal of Adhesion Science and Technology* 25: 709–721. https://doi.org/10.1163/016942410X525975.

55 Wang, Y., Wei, X., Li, J. et al. (2015). Study on nanocellulose by high pressure homogenization in homogeneous isolation. *Fibers and Polymers* 16 (3): 572–578. https://doi.org/10.1007/s12221-015-0572-1.

56 Müller, R.H., Jacobs, C., and Kayser, O. (2001). Nanosuspensions as particulate drug formulations in therapy. Rationale for development and what we can expect for the future. *Advanced Drug Delivery Reviews* 47 (1): 3–19. https://doi.org/10.1016/s0169-409x(00)00118-6.

57 Kalia, S., Boufi, S., Celli, A., and Kango, S. (2014). Nanofibrillated cellulose: surface modification and potential applications. *Colloid and Polymer Science* 292 (1): 5–31. https://doi.org/10.1007/s00396-013-3112-9.

58 Saelee, K., Yingkamhaeng, N., Nimchua, T., and Sukyai, P. (2016). An environmentally friendly xylanase-assisted pretreatment for cellulose nanofibrils isolation from sugarcane bagasse by high-pressure homogenization. *Industrial Crops and Products* 82: 149–160. https://doi.org/10.1016/j.indcrop.2015.11.064.

59 de Oliveira, J.P., Bruni, G.P., El Halal, S.L.M. et al. (2019). Cellulose nanocrystals from rice and oat husks and their application in aerogels for food packaging. *International Journal of Biological Macromolecules* 124: 175–184. https://doi.org/10.1016/j.ijbiomac.2018.11.205.

60 Khan, A., Vu, K.D., Chauve, G. et al. (2014). Optimization of microfluidization for the homogeneous distribution of cellulose nanocrystals (CNCs) in biopolymeric matrix. *Cellulose* 21 (5): 3457–3468. https://doi.org/10.1007/s10570-014-0361-9.

61 Bai, L. and McClements, D.J. (2016). Development of microfluidization methods for efficient production of concentrated nanoemulsions: comparison of single- and dual-channel microfluidizers. *Journal of Colloid and Interface Science* 466: 206–212. https://doi.org/10.1016/j.jcis.2015.12.039.

62 Palacios Hinestroza, H., Hernández Diaz, J.A., Esquivel Alfaro, M. et al. (2019). Isolation and characterization of nanofibrillar cellulose from *Agave tequilana* weber bagasse. *Advances in Materials Science and Engineering* 2019: 1–7. https://doi.org/10.1155/2019/1342547.

63 Du, H., Parit, M., Wu, M. et al. (2020). Sustainable valorization of paper mill sludge into cellulose nanofibrils and cellulose nanopaper. *Journal of Hazardous Materials* 400: 123106. https://doi.org/10.1016/j.jhazmat.2020.123106.

64 Wang, Q., Wei, W., Chang, F. et al. (2016). Controlling the size and film strength of individualized cellulose nanofibrils prepared by combined enzymatic pretreatment and high pressure microfluidization. *BioResources* 11 (1): 2536–2547. https://doi.org/10.15376/biores.11.1.2536-2547.

65 Bian, H., Gao, Y., Yang, Y. et al. (2018). Improving cellulose nanofibrillation of waste wheat straw using the combined methods of prewashing, p-toluenesulfonic acid hydrolysis, disk grinding, and endoglucanase post-treatment. *Bioresource Technology* 256: 321–327. https://doi.org/10.1016/j.biortech.2018.02.038.

66 Missoum, K., Belgacem, M., and Bras, J. (2013). Nanofibrillated cellulose surface modification: a review. *Materials* 6 (5): 1745–1766. https://doi.org/10.3390/ma6051745.

67 Frone, A.N., Panaitescu, D.M., and Donescu, D. (n.d.). Some aspects concerning the isolation of cellulose micro- and nano- fibers.

68 Chirayil, C.J., Mathew, L., and Thomas, S. (n.d.). Review of recent research in nano cellulose preparation from different lignocellulosic fibers.

69 Kaur, M., Kumari, S., and Sharma, P. (2018). Chemically modified nanocellulose from rice husk: synthesis and characterisation. *Advances in Research* 1–11. https://doi.org/10.9734/AIR/2018/38934.

70 Zou, H., Zhao, N., Sun, S. et al. (2020). High-intensity ultrasonication treatment improved physicochemical and functional properties of mussel sarcoplasmic proteins and enhanced the stability of oil-in-water emulsion. *Colloids and Surfaces A: Physicochemical and Engineering Aspects* 589: 124463. https://doi.org/10.1016/j.colsurfa.2020.124463.

71 Guangyin, Z. and Youcai, Z. (2017). Harvest of bioenergy from sewage sludge by anaerobic digestion. *Pollution Control and Resource Recovery for Sewage Sludge* 181–273. https://doi.org/10.1016/B978-0-12-811639-5.00005-X.

72 Tischer, P.C.S.F., Sierakowski, M.R., Westfahl, H., and Tischer, C.A. (2010). Nanostructural reorganization of bacterial cellulose by ultrasonic treatment. *Biomacromolecules* 11 (5): 1217–1224. https://doi.org/10.1021/bm901383a.

73 Chen, P., Yu, H., Liu, Y. et al. (2012). Concentration effects on the isolation and dynamic rheological behavior of cellulose nanofibers via ultrasonic processing. *Cellulose* 20: https://doi.org/10.1007/s10570-012-9829-7.

74 Wang, S. and Cheng, Q. (2009). A novel process to isolate fibrils from cellulose fibers by high-intensity ultrasonication, part 1: process optimization. *Journal of Applied Polymer Science* 113 (2): 1270–1275. https://doi.org/10.1002/app.30072.

75 Jonoobi, M., Oladi, R., Davoudpour, Y. et al. (2015). Different preparation methods and properties of nanostructured cellulose from various natural resources and residues: a review. *Cellulose* 22 (2): 935–969. https://doi.org/10.1007/s10570-015-0551-0.

76 Siró, I. and Plackett, D. (2010). Microfibrillated cellulose and new nanocomposite materials: a review. *Cellulose* 17 (3): 459–494. https://doi.org/10.1007/s10570-010-9405-y.

77 Skočaj, M. (2019). Bacterial nanocellulose in papermaking. *Cellulose* 26 (11): 6477–6488. doi:https://doi.org/10.1007/s10570-019-02566-y.

78 Hoque, E., Hossain, S., Mim, I.Z., and Chowdhury, J.I. (n.d.). Electroconductivity of bacterial cellulose and bacterial cellulose-based nanocomposites.

79 François, B. (2021). Bacterial nanocellulose production and biomedical applications. *The Journal of Biomedical Research* 35 (4): 310–317. https://doi.org/10.7555/JBR.35.20210036.

80 Castro, C., Cordeiro, N., Faria, M. et al. (2015). In-situ glyoxalization during biosynthesis of bacterial cellulose. *Carbohydrate Polymers* 126: 32–39. https://doi.org/10.1016/j.carbpol.2015.03.014.

81 Reshmy, R., Philip, E., Thomas, D. et al. (2021). Bacterial nanocellulose: engineering, production, and applications. *Bioengineered* 12 (2): 11463–11483. https://doi.org/10.1080/21655979.2021.2009753.

82 Czaja, W., Romanovicz, D., Brown, R., and malcolm. (2004). Structural investigations of microbial cellulose produced in stationary and agitated culture. *Cellulose* 11 (3): 403–411. https://doi.org/10.1023/B:CELL.0000046412.11983.61.

83 Wang, J., Tavakoli, J., and Tang, Y. (2019). Bacterial cellulose production, properties and applications with different culture methods—a review. *Carbohydrate Polymers* 219: 63–76. https://doi.org/10.1016/j.carbpol.2019.05.008.

84 Dubey, S., Singh, J., and Singh, R.P. (2018). Biotransformation of sweet lime pulp waste into high-quality nanocellulose with an excellent productivity using Komagataeibacter europaeus SGP37 under static intermittent fed-batch cultivation. *Bioresource Technology* 247: 73–80. https://doi.org/10.1016/j.biortech.2017.09.089.

85 Wu, S.-C. and Li, M.-H. (2015). Production of bacterial cellulose membranes in a modified airlift bioreactor by Gluconacetobacter xylinus. *Journal of Bioscience and Bioengineering* 120 (4): 444–449. https://doi.org/10.1016/j.jbiosc.2015.02.018.

86 Khatun, M.M., Hoque, M.E., Wadud, S.I., and Shawon, Z.B.Z. (2020). Recent developments in nanocellulose and nanohydrogel matrices—towards stem cell research and development. In: *Sustainable Nanocellulose and Nanohydrogels from Natural Sources* (ed. F. Mohammad, H.A. Al-Lohedan, and M. Jawaid), 315–328. Elsevier. https://doi.org/10.1016/B978-0-12-816789-2.00015-8.

87 Sadaf Afrin, Z.K. (2015). Nanocellulose as novel supportive functional material for growth and development of cells. *Cell & Developmental Biology* 4 (2): https://doi.org/10.4172/2168-9296.1000154.

88 Batmaz, R., Mohammed, N., Zaman, M. et al. (2014). Cellulose nanocrystals as promising adsorbents for the removal of cationic dyes. *Cellulose* 21 (3): 1655–1665. https://doi.org/10.1007/s10570-014-0168-8.

89 Carpenter, A.W., De Lannoy, C.-F., and Wiesner, M.R. (2015). Cellulose nanomaterials in water treatment technologies. *Environmental Science & Technology* 49 (9): 5277–5287. https://doi.org/10.1021/es506351r.

90 Mautner, A., Lee, K.-Y., Lahtinen, P. et al. (2014). Nanopapers for organic solvent nanofiltration. *Chemical Communications (Cambridge, England)* 50 (43): 5778–5781. https://doi.org/10.1039/c4cc00467a.

91 Dufresne, A. (2015). Cellulose nanomaterials. In: *Dekker Encyclopedia of Nanoscience and Nanotechnology*, 3e (ed. J.A. Schwarz, S.E. Lyshevski, and C.I. Contescu), 1–22. Taylor & Francis. https://doi.org/10.1081/E-ENN3-120053539.

92 Lin, N. and Dufresne, A. (2014). Surface chemistry, morphological analysis and properties of cellulose nanocrystals with gradiented sulfation degrees. *Nanoscale* 6 (10): 5384–5393. https://doi.org/10.1039/C3NR06761K.

93 Grishkewich, N., Mohammed, N., Tang, J., and Tam, K.C. (2017). Recent advances in the application of cellulose nanocrystals. *Current Opinion in Colloid & Interface Science* 29: 32–45. https://doi.org/10.1016/j.cocis.2017.01.005.

94 Wang, P.-X. and MacLachlan, M.J. (2018). Liquid crystalline tactoids: ordered structure, defective coalescence and evolution in confined geometries. *Philosophical Transactions of the Royal Society A: Mathematical, Physical and Engineering Sciences* 376 (2112): 20170042. https://doi.org/10.1098/rsta.2017.0042.

95 Singh, S., Kumar, V., Datta, S. et al. (2020). Current advancement and future prospect of biosorbents for bioremediation. *Science of the Total Environment* 709: 135895. https://doi.org/10.1016/j.scitotenv.2019.135895.

96 Saharan, V.K., Pinjari, D.V., Gogate, P.R., and Pandit, A.B. (2014). Advanced oxidation technologies for wastewater treatment. In: *Industrial Wastewater Treatment, Recycling and Reuse* (ed. V.V. Ranade and V.M. Bhandari), 141–191. Elsevier. https://doi.org/10.1016/B978-0-08-099968-5.00003-9.

97 Sahu, O. and Singh, N. (2019). Significance of bioadsorption process on textile industry wastewater. In: *The Impact and Prospects of Green Chemistry for Textile Technology* (ed. S.U. Islam and B.S. Butola), 367–416. Elsevier. https://doi.org/10.1016/B978-0-08-102491-1.00013-7.

98 Bhatnagar, A. and Sillanpää, M. (2010). Utilization of agro-industrial and municipal waste materials as potential adsorbents for water treatment—a review. *Chemical Engineering Journal* 157 (2–3): 277–296. https://doi.org/10.1016/j.cej.2010.01.007.

99 Dąbrowski, A. (2001). Adsorption—from theory to practice. *Advances in Colloid and Interface Science* 93 (1): 135–224. https://doi.org/10.1016/S0001-8686(00)00082-8.

100 Khajeh, M., Laurent, S., and Dastafkan, K. (2013). Nanoadsorbents: classification, preparation, and applications (with emphasis on aqueous media). *Chemical Reviews* 113 (10): 7728–7768. https://doi.org/10.1021/cr400086v.

101 Salama, A., Abouzeid, R., Leong, W.S. et al. (2021). Nanocellulose-based materials for water treatment: adsorption, photocatalytic degradation, disinfection, antifouling, and nanofiltration. *Nanomaterials* 11 (11): 3008. https://doi.org/10.3390/nano11113008.

102 Isa, E.D.M., Shameli, K., Ali, R.R., and Balasundram, V. (2022). Advances in nanocellulose based materials as adsorbent for wastewater treatment. *Journal of Research in Nanoscience and Nanotechnology* 5 (1): 43–64. https://doi.org/10.37934/jrnn.5.1.4364.

103 Gopakumar, D., Manna, S., Pasquini, D. et al. (2018). Nanocellulose: extraction and application as a sustainable material for wastewater purification. In: *New Polymer Nanocomposites for Environmental Remediation* (ed. C.M. Hussain and A.K. Mishra). https://doi.org/10.1016/B978-0-12-811033-1.00019-6.

104 Abualnaja, M.M., Almotairy, A.R.Z., Alorabi, A.Q. et al. (2024). Optimizing cellulose nanofiber composite for Co2+ ions recovery from lithium-ion battery wastes and removal from wastewater: a green environmental solution. *Journal of Water Process Engineering* 57: 104621. https://doi.org/10.1016/j.jwpe.2023.104621.

105 Zhu, H., Jia, S., Wan, T. et al. (2011). Biosynthesis of spherical Fe_3O_4/bacterial cellulose nanocomposites as adsorbents for heavy metal ions. *Carbohydrate Polymers* 86 (4): 1558–1564. https://doi.org/10.1016/j.carbpol.2011.06.061.

106 Sreenivasan, R., Suma Mahesh, S., and Sumi, V.S. (2019). Synthesis and application of polymer-grafted nanocellulose/graphene oxide nano composite for the selective recovery of radionuclides from aqueous media. *Separation Science and Technology* 54 (9): 1453–1468. https://doi.org/10.1080/01496395.2018.1534868.

107 Wang, Y., Li, Y., Zhang, Y. et al. (2021). Nanocellulose aerogel for highly efficient adsorption of uranium (VI) from aqueous solution. *Carbohydrate Polymers* 267: 118233. https://doi.org/10.1016/j.carbpol.2021.118233.

108 Chen, S., Zou, Y., Yan, Z. et al. (2009). Carboxymethylated-bacterial cellulose for copper and lead ion removal. *Journal of Hazardous Materials* 161 (2–3): 1355–1359. https://doi.org/10.1016/j.jhazmat.2008.04.098.

109 Oshima, T., Kondo, K., Ohto, K. et al. (2008). Preparation of phosphorylated bacterial cellulose as an adsorbent for metal ions. *Reactive and Functional Polymers* 68 (1): 376–383. https://doi.org/10.1016/j.reactfunctpolym.2007.07.046.

110 Shen, W., Chen, S., Shi, S. et al. (2009). Adsorption of Cu(II) and Pb(II) onto diethylenetriamine-bacterial cellulose. *Carbohydrate Polymers* 75 (1): 110–114. https://doi.org/10.1016/j.carbpol.2008.07.006.

111 Jin, X., Xiang, Z., Liu, Q. et al. (2017). Polyethyleneimine-bacterial cellulose bioadsorbent for effective removal of copper and lead ions from aqueous solution. *Bioresource Technology* 244: 844–849. https://doi.org/10.1016/j.biortech.2017.08.072.

112 Ambashta, R.D. and Sillanpää, M. (2010). Water purification using magnetic assistance: a review. *Journal of Hazardous Materials* 180 (1–3): 38–49. https://doi.org/10.1016/j.jhazmat.2010.04.105.

113 Nata, I.F., Sureshkumar, M., and Lee, C.-K. (2011). One-pot preparation of amine-rich magnetite/bacterial cellulose nanocomposite and its application for arsenate removal. *RSC Advances* 1 (4): 625. https://doi.org/10.1039/c1ra00153a.

114 Stoica-Guzun, A., Stroescu, M., Jinga, S.I. et al. (2016). Box-Behnken experimental design for chromium(VI) ions removal by bacterial cellulose-magnetite composites. *International Journal of Biological Macromolecules* 91: 1062–1072. https://doi.org/10.1016/j.ijbiomac.2016.06.070.

115 He, X., Male, K.B., Nesterenko, P.N. et al. (2013). Adsorption and desorption of methylene blue on porous carbon monoliths and nanocrystalline cellulose. *ACS Applied Materials & Interfaces* 5 (17): 8796–8804. https://doi.org/10.1021/am403222u.

116 Jiang, F. and Hsieh, Y.-L. (2014). Amphiphilic superabsorbent cellulose nanofibril aerogels. *Journal of Materials Chemistry A* 2 (18): 6337–6342. https://doi.org/10.1039/C4TA00743C.

117 Beh, J.H., Lim, T.H., Lew, J.H., and Lai, J.C. (2020). Cellulose nanofibril-based aerogel derived from sago pith waste and its application on methylene blue removal. *International Journal of Biological Macromolecules* 160: 836–845. https://doi.org/10.1016/j.ijbiomac.2020.05.227.

118 Jin, L., Li, W., Xu, Q., and Sun, Q. (2015). Amino-functionalized nanocrystalline cellulose as an adsorbent for anionic dyes. *Cellulose* 22 (4): 2443–2456. https://doi.org/10.1007/s10570-015-0649-4.

119 Frye, M., Vasisth, S.E., Atassi, A. et al. (2021). Effect of degree of sulfonation in nanocellulose/chitosan composite on adsorption of cationic dye as opioid simulant. *Cellulose Chemistry and Technology* 55 (1–2): 87–99. https://doi.org/10.35812/CelluloseChemTechnol.2021.55.09.

120 Wang, W., Lin, J., Cheng, J. et al. (2020). Dual super-amphiphilic modified cellulose acetate nanofiber membranes with highly efficient oil/water separation and excellent antifouling properties. *Journal of Hazardous Materials* 385: 121582. https://doi.org/10.1016/j.jhazmat.2019.121582.

121 Zhang, Z., Sèbe, G., Rentsch, D. et al. (2014). Ultralightweight and flexible silylated nanocellulose sponges for the selective removal of oil from water. *Chemistry of Materials* 26 (8): 2659–2668. https://doi.org/10.1021/cm5004164.

122 Sharma, P.R., Sharma, S.K., Lindström, T., and Hsiao, B.S. (2020). Nanocellulose-enabled membranes for water purification: perspectives. *Advanced Sustainable Systems* 4 (5): 1900114. https://doi.org/10.1002/adsu.201900114.

123 Zanini, M., Lavoratti, A., Lazzari, L.K. et al. (2017). Producing aerogels from silanized cellulose nanofiber suspension. *Cellulose* 24 (2): 769–779. https://doi.org/10.1007/s10570-016-1142-4.

124 Korhonen, J.T., Hiekkataipale, P., Malm, J. et al. (2011). Inorganic hollow nanotube aerogels by atomic layer deposition onto native nanocellulose templates. *ACS Nano* 5 (3): 1967–1974. https://doi.org/10.1021/nn200108s.

125 Meng, Y., Young, T.M., Liu, P. et al. (2015). Ultralight carbon aerogel from nanocellulose as a highly selective oil absorption material. *Cellulose* 22 (1): 435–447. https://doi.org/10.1007/s10570-014-0519-5.

126 Sai, H., Fu, R., Xing, L. et al. (2015). Surface modification of bacterial cellulose aerogels' web-like skeleton for oil/water separation. *ACS Applied Materials & Interfaces* 7 (13): 7373–7381. https://doi.org/10.1021/acsami.5b00846.

127 Karzar Jeddi, M., Laitinen, O., and Liimatainen, H. (2019). Magnetic superabsorbents based on nanocellulose aerobeads for selective removal of oils and organic solvents. *Materials & Design* 183: 108115. https://doi.org/10.1016/j.matdes.2019.108115.

128 Gu, H., Zhou, X., Lyu, S. et al. (2020). Magnetic nanocellulose-magnetite aerogel for easy oil adsorption. *Journal of Colloid and Interface Science* 560: 849–856. https://doi.org/10.1016/j.jcis.2019.10.084.

129 Sabouni, R. and Gomaa, H. (2019). Photocatalytic degradation of pharmaceutical micro-pollutants using ZnO. *Environmental Science and Pollution Research* 26 (6): 5372–5380. https://doi.org/10.1007/s11356-018-4051-2.

130 Elmolla, E.S. and Chaudhuri, M. (2010). Comparison of different advanced oxidation processes for treatment of antibiotic aqueous solution. *Desalination* 256 (1–3): 43–47. https://doi.org/10.1016/j.desal.2010.02.019.

131 Maniakova, G., Kowalska, K., Murgolo, S. et al. (2020). Comparison between heterogeneous and homogeneous solar driven advanced oxidation processes for urban wastewater treatment: pharmaceuticals removal and toxicity. *Separation and Purification Technology* 236: 116249. https://doi.org/10.1016/j.seppur.2019.116249.

132 Vismara, E., Bertolini, G., Bongio, C. et al. (2021). Nanocellulose from cotton waste and its glycidyl methacrylate grafting and allylation: synthesis, characterization and adsorption properties. *Nanomaterials* 11 (2): 476. https://doi.org/10.3390/nano11020476.

133 Darabitabar, F., Yavari, V., Hedayati, A. et al. (2020). Novel cellulose nanofiber aerogel for aquaculture wastewater treatment. *Environmental Technology & Innovation* 18: 100786. https://doi.org/10.1016/j.eti.2020.100786.

134 Das, L., Das, P., Bhowal, A., and Bhattacharjee, C. (2021). Enhanced biosorption of fluoride by extracted nanocellulose/polyvinyl alcohol composite in batch and fixed-bed system: ANN analysis and numerical modeling. *Environmental Science and Pollution Research* 28 (34): 47107–47125. https://doi.org/10.1007/s11356-021-14026-x.

135 Yang, J., Ma, C., Tao, J. et al. (2020). Optimization of polyvinylamine-modified nanocellulose for chlorpyrifos adsorption by central composite design. *Carbohydrate Polymers* 245: 116542. https://doi.org/10.1016/j.carbpol.2020.116542.

136 Lee, C.S., Robinson, J., and Chong, M.F. (2014). A review on application of flocculants in wastewater treatment. *Process Safety and Environmental Protection* 92 (6): 489–508. https://doi.org/10.1016/j.psep.2014.04.010.

137 Quinlan, P.J., Tanvir, A., and Tam, K.C. (2015). Application of the central composite design to study the flocculation of an anionic azo dye using quaternized cellulose nanofibrils. *Carbohydrate Polymers* 133: 80–89. https://doi.org/10.1016/j.carbpol.2015.06.095.

138 Rudén, C. (2004). Acrylamide and cancer risk—expert risk assessments and the public debate. *Food and Chemical Toxicology* 42 (3): 335–349. https://doi.org/10.1016/j.fct.2003.10.017.

139 Suopajärvi, T., Liimatainen, H., Hormi, O., and Niinimäki, J. (2013). Coagulation–flocculation treatment of municipal wastewater based on anionized nanocelluloses. *Chemical Engineering Journal* 231: 59–67. https://doi.org/10.1016/j.cej.2013.07.010.

140 Korhonen, M. and Laine, J. (2014). Flocculation and retention of fillers with nanocelluloses. *Nordic Pulp & Paper Research Journal* 29 (1): 119–128. https://doi.org/10.3183/npprj-2014-29-01-p119-128.

141 Zhang, L.-W., Hua, J.-R., Zhu, W.-J. et al. (n.d.). Flocculation performance of hyperbranched polyethyleneimine-grafted-cellulose in wastewater treatment.

142 Yu, H.-Y., Zhang, D.-Z., Lu, F.-F., and Yao, J. (2016). New approach for single-step extraction of carboxylated cellulose nanocrystals for their use as adsorbents and flocculants. *ACS Sustainable Chemistry & Engineering* 4 (5): 2632–2643. https://doi.org/10.1021/acssuschemeng.6b00126.

143 Kemppainen, K., Suopajärvi, T., Laitinen, O. et al. (2016). Flocculation of fine hematite and quartz suspensions with anionic cellulose nanofibers. *Chemical Engineering Science* 148: 256–266. https://doi.org/10.1016/j.ces.2016.04.014.

144 Mbakop, S., Nthunya, L.N., and Onyango, M.S. (2021). Recent advances in the synthesis of nanocellulose functionalized–hybrid membranes and application in water quality improvement. *Processes* 9 (4): 611. https://doi.org/10.3390/pr9040611.

145 Nthunya, L., Gutierrez, L., Nxumalo, E., and Mhlanga, S. (2019a). Water as the pore former in the synthesis of hydrophobic PVDF flat sheet membranes for use in membrane distillation*Hydro Science & Marine Engineering*, 1. 10.30564/hsme.v1i2.1346

146 Khumalo, N.P., Nthunya, L.N., De Canck, E. et al. (2019). Congo red dye removal by direct membrane distillation using PVDF/PTFE membrane. *Separation and Purification Technology* 211: 578–586. https://doi.org/10.1016/j.seppur.2018.10.039.

147 Masheane, M.L., Nthunya, L.N., Malinga, S.P. et al. (2017). Synthesis of Fe-Ag/f-MWCNT/PES nanostructured-hybrid membranes for removal of Cr(VI) from water. *Separation and Purification Technology* 184: 79–87. https://doi.org/10.1016/j.seppur.2017.04.018.

148 Gebreslase, G.A. (2018). Review on membranes for the filtration of aqueous based solution: oil in water emulsion. *Journal of Membrane Science & Technology* 8 (2): https://doi.org/10.4172/2155-9589.1000188.

149 Voisin, H., Bergström, L., Liu, P., and Mathew, A. (2017). Nanocellulose-based materials for water purification. *Nanomaterials* 7 (3): 57. https://doi.org/10.3390/nano7030057.

150 Dai, Z., Ottesen, V., Deng, J. et al. (2019). A brief review of nanocellulose based hybrid membranes for CO_2 separation. *Fibers* 7 (5): 5. https://doi.org/10.3390/fib7050040.

151 Khulbe, K.C. and Matsuura, T. (2018). Removal of heavy metals and pollutants by membrane adsorption techniques. *Applied Water Science* 8 (1): 19. https://doi.org/10.1007/s13201-018-0661-6.

152 Bai, H., Wang, X., Zhou, Y., and Zhang, L. (2012). Preparation and characterization of poly(vinylidene fluoride) composite membranes blended with nano-crystalline cellulose. *Progress in Natural Science: Materials International* 22 (3): 250–257. https://doi.org/10.1016/j.pnsc.2012.04.011.

153 Bai, L., Bossa, N., Qu, F. et al. (2017). Comparison of hydrophilicity and mechanical properties of nanocomposite membranes with cellulose nanocrystals and carbon nanotubes. *Environmental Science & Technology* 51 (1): 253–262. https://doi.org/10.1021/acs.est.6b04280.

154 Karim, Z., Claudpierre, S., Grahn, M. et al. (2016). Nanocellulose based functional membranes for water cleaning: tailoring of mechanical properties, porosity and metal ion capture. *Journal of Membrane Science* 514: 418–428. https://doi.org/10.1016/j.memsci.2016.05.018.

155 Nthunya, L.N., Derese, S., Gutierrez, L. et al. (2019b). Green synthesis of silver nanoparticles using one-pot and microwave-assisted methods and their subsequent embedment on PVDF nanofibre membranes for growth inhibition of mesophilic and thermophilic bacteria. *New Journal of Chemistry* 43 (10): 4168–4180. https://doi.org/10.1039/C8NJ06160B.

156 Nthunya, L.N., Masheane, M.L., Malinga, S.P. et al. (2017). Greener approach to prepare electrospun antibacterial β-cyclodextrin/cellulose acetate nanofibers for removal of bacteria from water. *ACS Sustainable Chemistry & Engineering* 5 (1): 153–160. https://doi.org/10.1021/acssuschemeng.6b01089.

157 Nthunya, L.N., Masheane, M.L., Malinga, S.P. et al. (2017). Thermally and mechanically stable β-cyclodextrin/cellulose acetate nanofibers synthesized using an environmentally benign procedure. *International Journal of Smart and Nano Materials* 8 (1): 1–19. https://doi.org/10.1080/19475411.2017.1289276.

158 Gopakumar, D.A., Pasquini, D., Henrique, M.A. et al. (2017). Meldrum's acid modified cellulose nanofiber-based polyvinylidene fluoride microfiltration membrane for dye water treatment and nanoparticle removal. *ACS Sustainable Chemistry & Engineering* 5 (2): 2026–2033. https://doi.org/10.1021/acssuschemeng.6b02952.

159 Nthunya, L.N., Gutierrez, L., Derese, S. et al. (2019). A review of nanoparticle-enhanced membrane distillation membranes: membrane synthesis and applications in water treatment. *Journal of Chemical Technology & Biotechnology* 94 (9): 2757–2771. https://doi.org/10.1002/jctb.5977.

160 Nthunya, L.N., Gutierrez, L., Nxumalo, E.N. et al. (2020). F-MWCNTs/AgNPs-coated superhydrophobic PVDF nanofibre membrane for organic, colloidal, and biofouling mitigation in direct contact membrane distillation. *Journal of Environmental Chemical Engineering* 8 (2): 103654. https://doi.org/10.1016/j.jece.2020.103654.

161 Karim, Z., Mathew, A.P., Grahn, M. et al. (2014). Nanoporous membranes with cellulose nanocrystals as functional entity in chitosan: removal of dyes from water. *Carbohydrate Polymers* 112: 668–676. https://doi.org/10.1016/j.carbpol.2014.06.048.

162 Zhang, Q.G., Deng, C., Soyekwo, F. et al. (2016). Sub-10 nm wide cellulose nanofibers for ultrathin nanoporous membranes with high organic permeation. *Advanced Functional Materials* 26 (5): 792–800. https://doi.org/10.1002/adfm.201503858.

163 Rehan, M., Barhoum, A., Khattab, T.A. et al. (2019). Colored, photocatalytic, antimicrobial and UV- decorated with Ag/Ag_2CO_3 and Ag/Ag_3PO_4 nanoparticles. *Cellulose* 26 (9): 5437–5453. https://doi.org/10.1007/s10570-019-02497-8.

164 Lcong, W.S., Luo, X., Li, Y. et al. (2015). Low resistance metal contacts to MoS_2 devices electrodes. *ACS Nano* 9 (1): 869–877. https://doi.org/10.1021/nn506567r.

165 Rehan, M., Barhoum, A., Van Assche, G. et al. (2017). Towards multifunctional cellulosic and in-situ synthesis of silver nanoparticles into cellulose fabrics. *International Journal of Biological Macromolecules* 98: 877–886. https://doi.org/10.1016/j.ijbiomac.2017.02.058.

166 Barhoum, A., Melcher, J., Van Assche, G. et al. (2017). Synthesis, growth photocatalytic activity of zinc oxide nanostructures: porous microparticles versus nonporous nanoparticles. *Journal of Materials Science* 52 (5): 2746–2762. https://doi.org/10.1007/s10853-016-0567-3.

167 Ren, C., Yang, B., Wu, M. et al. (2010). Synthesis of Ag/ZnO nanorods array performance. *Journal of Hazardous Materials* 182 (1):123–129. https://doi.org/10.1016/j.jhazmat.2010.05.141.

168 Jiao, Y., Wan, C., Bao, W. et al. (2018). Facile hydrothermal synthesis of Fe_3O_4@cellulose aerogel application in Fenton-like degradation of Rhodamine B. *Carbohydrate Polymers* 189: 371–378. https://doi.org/10.1016/j.carbpol.2018.02.028.

169 Yanfen, F., Yingping, H., Jing, Y. et al. (2011). Unique ability of BiOBr to decarboxylate d-Glu and d-degradation of microcystin-LR in water. *Environmental Science and Technology* 45 (4): 1593–1600. https://doi.org/10.1021/es103422j.

170 Li, Q., Zhou, D., Zhang, P. et al. (2016). The BiOBr/regenerated cellulose composite film as a degradation of phenol. *Colloids and Surfaces A: Physicochemical and Engineering Aspects* 501: 132–137. https://doi.org/10.1016/j.colsurfa.2016.04.054.

171 Gaur, M., Misra, C., Yadav, A.B. et al. (2021). Biomedical applications nanomaterials: fullerenes, quantum dots, nanotubes, nanofibers, and graphene. *Materials (Basel)* 14 (20). https://doi.org/10.3390/ma14205978.

172 Wang, Z., Ma, H., Chu, B., and Hsiao, B.S. (2017). Fabrication of cellulose nanofiber-based ultrafiltration membranes by spray. *Polymer Science* 134 (11). https://doi.org/10.1002/app.44583.

173 Anirudhan, T.S. and Deepa, J.R. (2017). Nano-zinc oxide incorporated graphene oxide/nanocellulose composite for the adsorption degradation of ciprofloxacin hydrochloride from aqueous solutions. *Journal of Colloid Interface Science* 490: 343–356. https://doi.org/10.1016/j.jcis.2016.11.042.

174 Yang, X. and Cranston, E.D. (2014). Chemically cross-linked cellulose nanocrystal aerogels with shape recovery and superabsorbent. *Chemistry of Materials* 26 (20): 6016–6025. https://doi.org/10.1021/cm502873c.

175 Li, J., Cha, R., Mou, K. et al. (2018). Nanocellulose-based antibacterial materials. *Advanced Healthcare Materials* 7 (20):1800334. https://doi.org/10.1002/adhm.201800334.

176 Fei, P., Liao, L., Meng, J. et al. (2018). Non-leaching antibacterial cellulose triacetate reverse osmosis immobilization of quaternary ammonium cations. *Carbohydrate Polymers* 181: 1102–1111. https://doi.org/10.1016/j.carbpol.2017.11.036.

177 Abitbol, T., Marway, H., and Cranston, E.D. (2014). Surface modification of cellulose nanocrystals with cetyltrimethylammonium bromide. *Nordic Pulp and Paper Research Journal* 29 (1): 46–57. https://doi.org/10.3183/npprj-2014-29-01-p046-057.

178 Fu, F., Gu, J., Cao, J. et al. (2018). Reduction of silver ions using an alkaline cellulose dope: straightforward access to Ag/ZnO decorated cellulose nanocomposite film with enhanced antibacterial activities. *ACS Sustainable Chemistry and Engineering* 6 (1):738–748. https://doi.org/10.1021/acssuschemeng.7b03059.

179 Xu, X., Liu, F., Jiang, L. et al. (2013). Cellulose nanocrystals vs. cellulose nanofibrils: a comparative study on their microstructures and effects as polymer reinforcing agents. *ACS Applied Materials and Interfaces* 5 (8): 2999–3009. https://doi.org/10.1021/am302624t.

180 Moon, R.J., Schueneman, G.T., and Simonsen, J. (2016). Overview of cellulose nanomaterials, their capabilities and applications. *JOM* 68 (9): 2383–2394. https://doi.org/10.1007/s11837-016-2018-7.

181 Moon, R.J., Martini, A., Nairn, J. et al. (2011). Cellulose nanomaterials review: structure, properties and nanocomposites. *Chemical Society Reviews* 40 (7): 3941–3994. https://doi.org/10.1039/C0CS00108B.

182 Klemm, D., Kramer, F., Moritz, S. et al. (2011). Nanocelluloses: a new family of nature-based materials. *Angewandte Chemie International Edition* 50 (24): 5438–5466. https://doi.org/10.1002/anie.201001273.

183 Roberts, K.G., Gloy, B.A., Joseph, S. et al. (2010). Life cycle assessment of biochar systems: estimating the energetic, economic, and climate change potential. *Environmental Science and Technology* 44 (2): 827–833. https://doi.org/10.1021/es902266r.

184 Shak, K.P.Y., Pang, Y.L., and Mah, S.K. (2018). Nanocellulose: recent advances and its prospects in environmental remediation. *Beilstein Journal of Nanotechnology* 9 (1): 2479–2498. https://doi.org/10.3762/bjnano.9.232.

185 Zuliat, F.A., Suardi, M., and Djamaan, A. (n.d.). *CNC (Cellulose Nanocrystals) Isolation from Various Agriculture and Industrial Waste Using Acid Hydrolysis Methods. Cellulose Nanocrystals.*

186 Krexner, T., Bauer, A., Zollitsch, W. et al. (2022). Environmental life cycle assessment of nano-cellulose and biogas production from manure. *Journal of Environmental Management* 314: 115093. https://doi.org/10.1016/j.jenvman.2022.115093.

4

Biochar-Based Biocomposites for Wastewater Treatment

Aziz Ihammi[1], Nadia Zari[1], and Rachid Bouhfid[2]

[1] *Moroccan Foundation for Advanced Science, Innovation and Research (MAScIR), Mohammed VI Polytechnic University, Lot 660-Hay Moulay Rachid, 43150 Ben Guerir, Morocco*
[2] *International Water Research Institute (IWRI), Mohammed VI Polytechnic University, Lot 660 Hay Moulay Rachid, 43150 Ben Guerir, Morocco*

4.1 Introduction

Water is vital for life and is crucial for the sustenance of communities, businesses, and societies. Water is essential for life, and access to potable water is equivalent to health. The quality and availability of water have become significant concerns for sustainability [1].

The increasing need for clean water, exacerbated by swift industrialization and urbanization, has rendered wastewater treatment a crucial global concern [2]. Conventional treatment systems, while somewhat effective, exhibit significant limitations, including elevated operational costs, the production of secondary pollutants, and a restricted ability to eliminate new contaminants. Considering these issues, there is a growing interest in sustainable and eco-friendly materials as substitutes for conventional treatment approaches [3, 4].

Biochar, a carbon-dense substance generated from the pyrolysis of biomass, has surfaced as a viable adsorbent owing to its elevated surface area, porosity, and functional groups adept for sequestering diverse pollutants [5, 6]. To improve its adsorption capability and mechanical durability, biochar is frequently integrated with other materials to create biocomposites [7, 8].

Biochar-based biocomposites alongside polymers, metal oxides, and biogenic elements provide significant advantages for environmental remediation [9]. By leveraging the properties of these materials, we may create multifunctional systems that efficiently tackle various contaminants.

This chapter seeks to constructively examine the promising potential of biochar-based biocomposites in wastewater treatment [10]. We will explore their unique characteristics, novel synthesis methods, and various applications. Furthermore, we will investigate the adsorption mechanisms and the interactions of these biocomposites with various contaminants, including heavy metals, organic pollutants, and new contaminants such as pharmaceuticals and endocrine disruptors. We will emphasize the environmental and economic benefits of utilizing these biocomposites, illustrating their role in sustainable water treatment solutions.

Biocomposites using biochar utilize the beneficial qualities of both biochar and numerous complementing components, including polymers [9], metal oxides [10–12], and biologically generated compounds [13]. This combination produces multifunctional systems capable of effectively tackling a wide range of contaminants [8, 14]. This chapter will critically analyze the potential of biochar-based biocomposites for wastewater treatment. We shall delineate their unique characteristics, novel synthesis methods, and multifaceted uses. Additionally, we will examine the adsorption mechanisms and interactions of these biocomposites with various contaminants, including heavy metals, organic pollutants, and emerging contaminants such as pharmaceuticals and endocrine disruptors. Ultimately, we will emphasize the environmental and economic benefits of biocomposites, illustrating their significance in sustainable wastewater management.

4.2 Wastewater and Wastewater Treatment

The swift expansion of contemporary society, propelled by industrialization and urbanization, has emerged as a principal factor in water resource contamination. The deterioration of water quality arises from several origins, which can be categorized into two primary types: direct, or point sources, and indirect, or diffuse sources. Direct sources encompass effluent discharges from industrial operations, refineries, wastewater treatment plants, and other human infrastructures that emit pollutants directly into rivers, lakes, or seas. Conversely, indirect sources consist of pollutants that permeate groundwater systems, soils, or are conveyed into streams via atmospheric precipitation. These contaminants frequently originate from agricultural practices, mining operations, or air pollutants that ultimately deposit onto terrestrial and aquatic environments (Figure 4.1) [11].

Industrial effluents encompass a diverse array of deleterious compounds, including dyes, heavy metals, surfactants, and byproducts from mining operations and wastewater. Agricultural chemicals, including pesticides and fertilizers, by-products from energy extraction, radioactive waste, and the effects of expanding urban development all contribute to pollution [12]. Contaminants

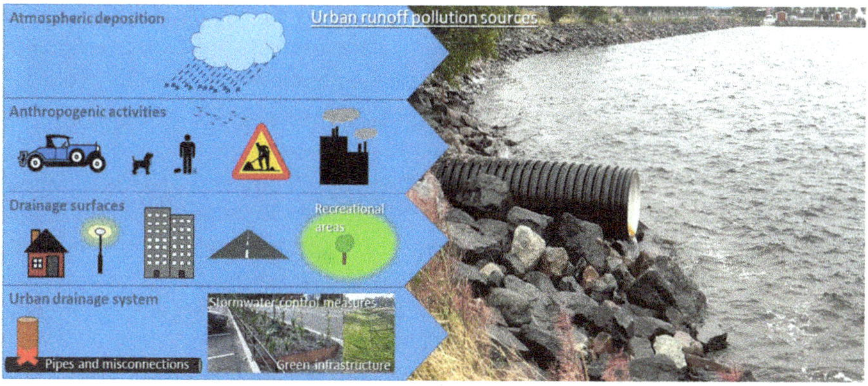

Figure 4.1 Major sources of water pollution [11] / Elsevier/ CC BY-NC-ND 4.0.

are typically categorized into three primary types based on their composition: organic compounds, encompassing hydrocarbons and industrial solvents; inorganic compounds, including heavy metals like lead and mercury; and microorganisms, which contribute to the biological contamination of water resources [13]. The diversity and complexity of these pollutants underscore the need to implement more effective and sustainable treatment solutions to safeguard water quality.

The increasing scarcity of freshwater resources, intensified by climate change, fast population increase, and water-related environmental hazards, renders wastewater management a global need. In response to these issues, numerous treatments have been devised to diminish wastewater discharges and alleviate the adverse environmental effects of contaminants. These interventions include both conventional therapy methods and more novel contemporary technologies [14]. Nonetheless, wastewater treatment facilities are intricate systems that must continually operate at peak efficiency, regardless of the fluctuations in wastewater flow that occur hourly, daily, monthly, and seasonally. Life cycle assessment (LCA) has been extensively utilized to evaluate the comprehensive environmental impact of wastewater treatment and to encourage more sustainable management practices. This tool evaluates the environmental consequences of each treatment stage, providing a more integrated approach to the sustainability of wastewater treatment systems [15].

4.3 Biochar: Definition and Properties

Biochar (BC) is a carbon-dense organic substance generated from the pyrolysis of biomass by-products in a high-temperature, low-oxygen environment. The pyrolysis process stabilizes organic matter, converting biological waste into a solid, porous substance with qualities appropriate for many applications. Biochar is distinguished by the plentiful availability of its antecedents, typically sourced from agricultural or forestry byproducts, with the ease and low expense of its manufacturing. Its elevated surface area and porosity, along with the simplicity of chemical or physical alteration, render it a promising material for various applications. It is extensively utilized in areas such as soil enhancement, where it enhances water retention and fertility; carbon sequestration, which contributes to the mitigation of greenhouse gas emissions; and, most significantly, in pollution management for the elimination of organic and inorganic pollutants from wastewater [16].

However, raw biochar typically exhibits constrained adsorption capacities, hence limiting its efficacy in addressing complex contaminants. To address this constraint, further adjustments are frequently necessary to improve their physicochemical qualities. This includes increasing surface area, improving pore structure, and enriching surface functional groups, which can interact more effectively with contaminants. In this context, activation processes, whether physical (such as thermal activation) or chemical (through impregnation with reactive agents), are often explored to improve biochar's affinity for a wide range of pollutants, from heavy metals to complex organic compounds [17].

Thus, modified biochar emerges as a competitive alternative to other carbon-based materials such as activated carbon, carbon nanotubes, or graphene, which are often more expensive to produce. Activated biochar can provide comparable or even superior pollutant removal performance in certain environmental applications, while offering both economic and ecological benefits. Moreover, its production serves as a viable solution for managing biomass waste, which poses an increasing challenge in the context of natural resource management. By transforming these residues into a valuable product, biochar helps alleviate environmental pressures while opening new perspectives for sustainable pollutant treatment [18].

4.4 Methods of Biochar Production

Biochar production takes place under anaerobic conditions, meaning with a limited oxygen supply. Although numerous production methods are described in the literature, no standardized classification exists to date. This section aims to highlight the various approaches to biochar production. These methods can be primarily divided into two categories: traditional and modern approaches, based on their level of technological advancement and modernization (Table 4.1).

Table 4.1 Production of biochar using various methodologies and its application.

Production techniques	Production conditions	Applications	References
Slow pyrolysis	T: 300–700 °C; heating rate: 5 to 7 °C min^{-1}	Soil enhancement, bio-oil, and syngas used as energy sources	[19]
Fast pyrolysis	$T > 500$ °C; heating rate: 300 °C min^{-1}	Adsorbent, soil enhancement, bio-oil, and syngas	[20, 21]
Gasification	$T > 700$ °C with gasifying agents	Dyestuff removal, adsorption, carbon sequestration, soil improvement, and syngas production	[22]
Torrefaction	T: 230–300 °C	Regarded as a pretreatment step to improve the physical, chemical, and biochemical characteristics of raw biomass	[23]
Flash pyrolysis	T: 900–1200 °C; heating rate: 800–1000 °C s^{-1}	Soil enhancement, bio-oil, and syngas	[24]
Hydrothermal carbonization	P: 2–10 MPa and temperature: 220–240 °C	Retention of nutrients high calorific value, better grindability, and improved hydrophobicity	[25, 26]
Microwave pyrolysis	T: 450–800 °C; microwave power: 400–600 W	Soil amendment, bio-oil, and syngas	[27, 28]

4.4.1 Traditional Methods

The production and use of biochar date back over 2500 years, notably with the discovery of Terra Preta in the Amazon basin [29]. At that time, methods involved slowly burning wood in pits with limited oxygen supply or burying partially carbonized biomass underground [30]. In addition to improving soil quality, biochar was used for other purposes, such as preservation and construction [31]. With technological advancements, traditional methods were gradually replaced by modern techniques, such as the use of steel kilns, which optimize production and the recovery of volatile compounds [31].

4.4.1.1 Pyrolysis

Pyrolysis is a thermal decomposition process of biomass under anaerobic conditions or with limited oxygen, producing gases, liquids, and biochar. Agricultural biomass, composed of lignin, cellulose, hemicelluloses, and silica, decomposes at temperatures ranging from 300 to 700 °C [32]. Various pyrolysis techniques exist depending on process parameters, such as slow and fast pyrolysis (traditional), and flash, vacuum, and microwave pyrolysis (modern), which are more recent and technologically advanced.

Slow Pyrolysis Slow pyrolysis, also known as conventional pyrolysis, is a process that takes several hours and occurs at temperatures ranging from 300 to 700 °C, with a moderate heating rate of 5–7 °C min^{-1}. This process primarily produces biochar (35–45%), as well as bio-oil and syngas. The biomass utilized may comprise diverse components including wood, deinking sludge, nut shells, and maize. Studies demonstrate that both temperature and biomass type influence the chemical composition and yield of biochar, with yields generally diminishing at elevated temperatures. A gradual heating rate is typically more efficacious for enhancing biochar synthesis [33, 34].

Fast Pyrolysis Fast pyrolysis is an effective thermochemical method for producing biofuels from biomass, primarily yielding bio-oil and syngas, along with 20% biochar [21, 35]. It transpires at temperatures exceeding 500 °C with a rapid heating rate of 15 °C min^{-1}, facilitating biochar synthesis within seconds. Research indicates that this biochar is especially proficient at adsorbing pollutants like salicylic acid and ibuprofen, facilitating preesterification in biodiesel synthesis, and enhancing soil quality [20].

4.4.2 Modern Methods

4.4.2.1 Gasification

Gasification is a widely utilized technology for manufacturing syngas from solid fuels, yielding greater syngas output with less emissions in comparison to methods such as pyrolysis or combustion. Although hydrogen is the principal product, biochar is concurrently produced, frequently seen as waste; however, it possesses significant value for applications including dye removal, chemical adsorption, carbon sequestration, and soil enhancement. The process functions at elevated temperatures (exceeding 700 °C), employing steam and oxygen to transform materials into hydrogen, carbon monoxide, and carbon dioxide. Biochar is produced during the breakdown of biomass, often at temperatures ranging from 400 to 500 °C and can subsequently be transformed into syngas at higher temperatures [36].

4.4.2.2 Flash Pyrolysis

Flash pyrolysis is an enhanced form of fast pyrolysis, in which biomass breaks down at very high temperatures (900–1200 °C) in a matter of seconds, typically under a minute. This approach achieves heating rates above 1000°C s^{-1}, resulting in a substantial yield of bio-oil while markedly reducing biochar production [37].

4.4.2.3 Hydrothermal Carbonization

Hydrothermal carbonization (HTC) is a technique that converts moisture-laden raw materials, such as sewage sludge and animal manure, into biochar without requiring drying. The method involves heating wet biomass to 220–240 °C under high pressure (2–10 MPa) for several hours. HTC takes place in rotating drums, ovens or stoves, producing carbon-rich materials with a range of properties. The biochar generated by HTC preserves essential elements including nitrogen and phosphorus, enhancing soil fertility. In comparison to torrefaction and pyrolysis, the HTC process presents advantages including a reduced oxygen/carbon ratio, enhanced energy content, improved grindability, and superior water resistance [38].

4.4.2.4 Microwave Pyrolysis

Microwave heating is a sophisticated method of biochar production that offers several advantages over conventional techniques. It uses dielectric heating to generate thermal energy, which is introduced into the reactor without direct contact. This technique enables rapid, selective and uniform heating, reducing the temperature required to produce biochar by 200 °C while achieving comparable results. Optimum biochar production is achieved at 450 °C with a microwave power of 400 W and a residence time of 4–6 minutes, while the highest hydrogen yield is obtained at 700 °C [28].

4.4.3 Biochar Activation

The properties of biochar obtained from thermochemical conversion of biomass can be optimized using physical, chemical, and impregnation techniques prior to its use in specific environments [39]. Producing biochar from biomass under controlled conditions results in an environmentally friendly, carbon-rich material with a polycyclic aromatic structure [40]. Its high porosity, large specific surface area and functional groups make it an attractive option for many applications. However, modified or activated biochar demonstrates a higher adsorption capacity to organic and inorganic contaminants compared to unmodified biochar (Figure 4.2) [42].

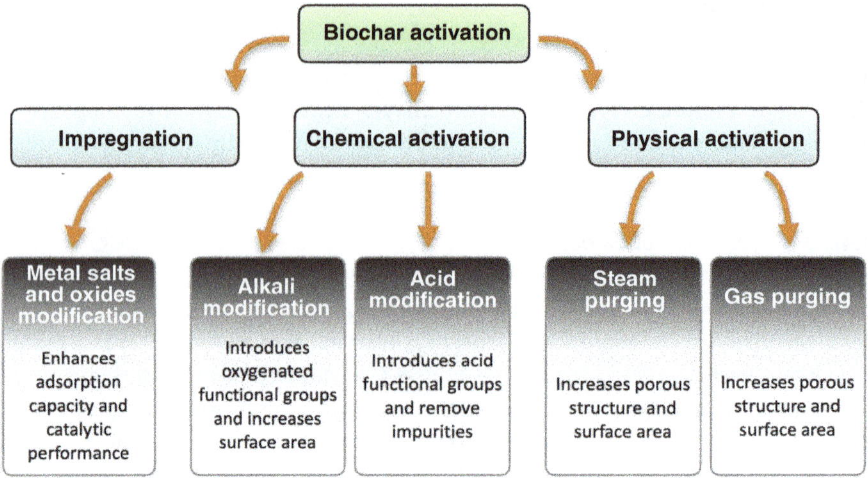

Figure 4.2 Biochar activation methods [42] / Springer Nature.

4.4.3.1 Physical Activation

Biochar is activated at higher temperatures, reaching up to 900 °C for 60 minutes, under partial pressures of up to 53 kPa [43]. This process involves a reaction with steam, air, ozone or carbon dioxide, creating pores in the structure by converting fixed carbon and volatiles into CO and CO_2. The main objective of this physical activation is to increase pore volume, specific surface area and modify morphology while reducing polarity and aromaticity [41].

4.4.3.2 Chemical Activation

This method is employed to produce chemically modified biochar and can be executed through either a one-step or two-step process. In the one-step procedure, carbonization and activation occur concurrently with the assistance of a modifying agent. The two-step procedure entails initially carbonizing the biomass, succeeded by chemical activation, which can be executed either by applying chemical agents postcarbonization or by incorporating the agents with the biomass prior to pretreatment and carbonization [41].

Acid Activation Acid modification enhances the physicochemical properties of biochar by adding acidic functional groups on its surface for high-value applications. This procedure mostly eliminates metal-based contaminants and modifies the structural and functional properties of the biochar. Common acidifying agents comprise hydrochloric, sulfuric, oxalic, nitric, citric, and phosphoric acids [44]. The surface area and pore volume of acid-modified biochar are affected by the acid type, impregnation ratio, and activation temperature, enhancing its surface area and mesoporous volume. Phosphoric acid notably enhances biochar's thermal resilience and introduces oxygen-phosphorus groups, hence augmenting surface acidity [45].

Alkali Activation Alkaline activation enhances the surface and functional characteristics of biochar, especially its oxygenation. The predominant alkaline agents are sodium hydroxide and potassium hydroxide. The biochar is initially immersed and agitated in various agents at different concentrations, at ambient temperature (20–22 °C), for a duration of 6–24 hours, contingent upon the chemical agent employed [41]. The material is thereafter cleaned, dried, and subjected to pyrolysis in a nitrogen environment at temperatures ranging from 300 to 700 °C for a duration of 1–2 hours, facilitating functional change. The primary objective of alkaline activation is to generate a positive charge on the biochar's surface, hence enhancing the removal of anionic pollutants from contaminated water. Biochar activated with KOH is regarded as a cost-effective and eco-friendly adsorbent, particularly proficient in the removal of heavy metals from wastewater treatment facilities [46]. The pore volume and surface area of activated biochar are influenced by the type of raw material, the alkaline agent employed, the preparation method, and the treatment conditions.

Impregnation Activation The impregnation step in biochar processing improves its physical and chemical properties by forming composites, which increases the adsorption capacity and mass yield of activated biochar [41]. Impregnation with metal salts or oxides optimizes the adsorption, catalysis, and magnetic properties of biochar. In recent years, there has been growing interest in plant and modified biochar's impregnated with metal salts such as $MgCl_2$, $AlCl_3$, $FeCl_3$, $ZnCl_2$, $CaCl_2$, etc., and metal oxides such as CaO, MgO, MnO, FeO, ZnO, and TiO_2, to improve anion adsorption efficiency [47].

4.5 Biochar Biocomposites Application in Wastewater Treatment

4.5.1 Biochar-Polymer Biocomposites

Biocomposites made of biochar and polymers represent a promising advancement in the creation of sustainable materials for wastewater treatment. Biochar, derived from the pyrolysis of plant biomass, is a porous, carbon-rich material with a high specific surface area, a variety of functional groups (—OH, —COOH, etc.), and excellent adsorption capacity. However, when used alone, biochar presents certain constraints, notably insufficient mechanical stability and material loss during filtering processes. To address these drawbacks, it is increasingly incorporated into natural or synthetic polymer matrices to create biochar-polymer biocomposites with optimized performance [48, 49].

These biocomposites combine the adsorption capacity of biochar with the mechanical properties and flexibility of polymers. For example, the integration of biochar into polymers such as polyacrylamide (PAM), polylactic acid (PLA), or polyvinyl acetate (PVA) improves mechanical strength, thermal stability, and the adsorption efficiency of pollutants such as heavy metals (Pb^{2+}, Cd^{2+}), organic dyes (methylene blue, Congo red), or pharmaceuticals. A recent study by Bartoli et al. (2022) revealed that incorporating 20% biochar into a polyamide matrix increases tensile strength by 31% and impact resistance by over 200%, while maintaining satisfactory adsorption capacity [50, 51].

The performance of these biocomposites is significantly influenced by several factors, including the pyrolysis temperature of the biochar (ideally between 400 and 600 °C to preserve the functional groups), the particle size (reduced by grinding to <300 μm to improve dispersion), and the chemical compatibility between the biochar and the polymer. Preliminary treatments, such as the chemical activation of biochar or the addition of functional groups, also optimize the interfacial interaction and distribution within the polymer matrix [52].

These materials can be used in the form of films, spheres, or foams in wastewater treatment systems, offering notable advantages in terms of reusability, cost reduction, and organic waste valorization. Biochar-polymer biocomposites fit perfectly into circular economy strategies and sustainable water management.

4.5.2 Metal Oxide–Biochar Hybrids

Biocomposites, which combine metal oxides with biochar, represent a category of multifunctional materials essential for wastewater treatment. By combining the excellent adsorption capacity of biochar with the catalytic, magnetic, and photocatalytic properties of metal oxides such as (Fe_3O_4, ZnO, TiO_2), these hybrids enhance the efficiency of purification processes while facilitating the recovery of the material (Figure 4.3) [54–56].

For example, the hybrid material Fe_3O_4@biochar, derived from biomass such as *Ascophyllum nodosum* algae, demonstrated a 20% increase in its Cu^{2+} removal efficiency, from 36.99 to 45.13 $mg\,g^{-1}$. This improvement is mainly due to a chemisorption adsorption process, as explained by Langmuir models ($Q_{max} \approx 53.19\,mg\,g^{-1}$), as well as easy posttreatment magnetic recovery [9, 47].

In addition, a Fe_3O_4@biochar composite made from activated wood waste was employed as a catalyst for persulfate, enabling efficient degradation of Azurant Orange 7 at room temperature. The study revealed a joint mechanism of initial adsorption and persulfate-induced radical oxidation, supported by electron

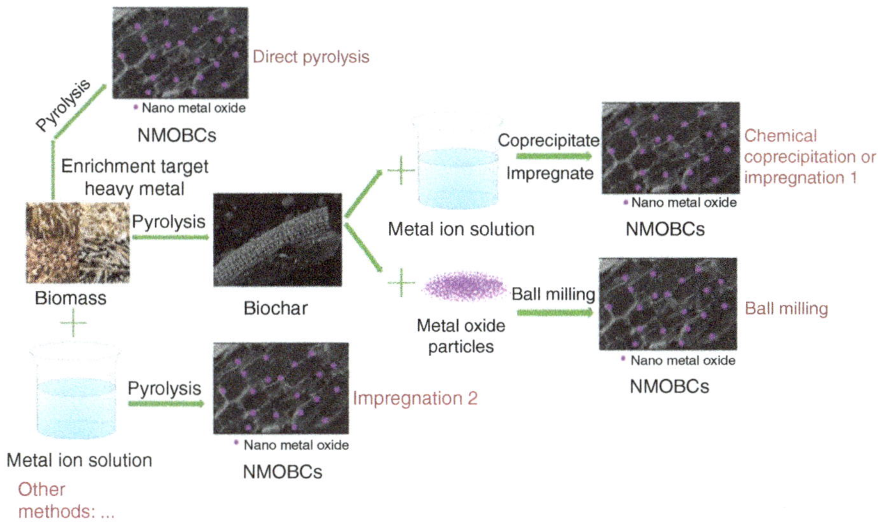

Figure 4.3 Diverse methodologies in the manufacture of nanoscale metal oxide-impregnated biochar's [61] / with permission of Elsevier.

paramagnetic resonance (EPR) spectroscopic analyses and density functional theory (DFT) calculations [57].

Research by Weidner and al (2022) highlights the use of systems containing titanium dioxide, zinc oxide, iron oxide, and other oxides. They point out that the uniform distribution of metal nanoparticles in the biochar matrix leads to improved catalytic activities, greater ease of recovery by sedimentation or magnetization, and treatment efficiencies ranging from 80% to 100% depending on the pollutants targeted [47].

Recent research by Dhila and Bhame (2025), published in the journal *Desalination and Water Treatment*, has demonstrated removal rates of up to 99% for textile dyes such as methylene blue and reactive orange. This removal was achieved using biochar nanocomposites functionalized with metal oxides such as ZnO, which integrate adsorption and photodegradation mechanisms when exposed to UV or visible light. These compounds are characterized by their strength, their ability to be reused cyclically, and their manufacture from abundant, recyclable materials [58].

As an example, a ternary ZnO-Sm_2O_3-MgO/biochar hybrid material, synthesized in a single step, exhibited an exceptional adsorption capacity of 979 mg g^{-1} for the red dye Rhodamine B, with a removal efficiency reaching almost 99.5% at an initial concentration of 410 mg g^{-1}. The process involved a synergy between adsorption, photonic capture, and catalysis by active sites located at the interface of oxides and biochar [59].

4.6 Removal of Contaminants from Wastewater by Biochar-Based Biocomposites

4.6.1 Adsorption of Pharmaceutical Contaminants

4.6.1.1 Preparation of Biochar-Based Biocomposites for Elimination of Pharmaceuticals

In the past decade, extensive research has been conducted on the elimination of pharmaceuticals from water and wastewater. Traditional treatment approaches, including flocculation, filtering, phytoremediation, activated sludge, and chlorination, have demonstrated ineffectiveness in eliminating these chemicals [60]. Despite the potential of improved oxidation methods, they are constrained by high energy requirements and limited mineralization efficiency. Biological techniques, while employed, are constrained by the protracted degradation of pharmaceuticals, the requirement for extensive surface areas, and the inadequately biodegradable characteristics of these pollutants [61].

In light of these challenges, adsorption techniques have surfaced as a viable alternative owing to their adaptability, user-friendliness, low initial cost, and efficacy. Among the utilized adsorbents, including activated carbons, clay minerals, and polymeric materials, biochar and its modified composites emerge as compelling alternatives for the removal of pharmaceutical pollutants (Figure 4.4). Biochar is esteemed for its affordability and efficacy in eliminating contaminants, especially through its interactions with hydrophobic medicines [63].

To enhance the effectiveness of adsorption processes, it is essential to have a deeper understanding of the fundamental mechanics. The mechanisms encompass π–π interactions between the electrons of the pharmaceuticals' aromatic rings and the adsorbent surface, the formation of donor-acceptor complexes involving functional groups like carbonyls, hydroxyls, or amines, along with

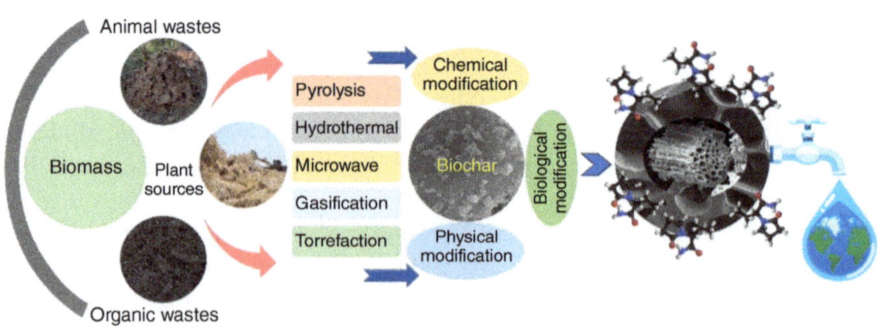

Figure 4.4 Biochar production and modification from various biomasses for removal of pharmaceutical pollutants in wastewater [64] / with permission of Elsevier.

Table 4.2 Biochar-based biocomposites for the removal of pharmaceuticals from wastewater.

Feedstock	Modified biochar	Pharmaceuticals	Capacity ($mg\,g^{-1}$)	Reference
Corn stover	$CoFe_2O_4$/biochar	Ciprofloxacin (5 $mg\,L^{-1}$)	11.38	[67]
Coffee grounds	$ZnCl_2$-biochar	Amoxicillin (200 $mg\,L^{-1}$)	178.57	[68]
Date palm	Fe_3O_4/chitosan/biochar	Ibuprofen (20 $mg\,L^{-1}$)	72.20	[69]
Walnut shell	Fe/N-biochar	Tetracycline (200 $mg\,L^{-1}$)	238.94	[70]
Rice husk	Fe_3O_4/ZnO-biochar	Diclofenac (20 $mg\,L^{-1}$)	24.33	[71]
Corncob	Fe-loaded/biochar	Tetracycline (50 $mg\,L^{-1}$)	606.52	[72]
Banana pseudo stem	$CoFe_2O_4$/modified biochar	Amoxicillin (50 $mg\,L^{-1}$)	99.99	[56]
Date palm	MIL–53(Fe)/biochar	• Ofloxacin (75 $mg\,L^{-1}$) • Ciprofloxacin (75 $mg\,L^{-1}$)	218.29 223.89	[66]

electrostatic interactions and hydrogen bonds. Hydrophobic medications exhibit a greater affinity for surface adsorption, whereas hydrophilic drugs are more challenging to eliminate [64, 65].

Despite numerous studies on the subject, research still needs to delve deeper into adsorption mechanisms and the maximum adsorption capacity of adsorbents to optimize processes [56, 66]. Table 4.2 shows the main biochar-based adsorbent composites used for various pharmaceutical pollutants.

4.6.1.2 Adsorption Mechanism of Pharmaceutical Compounds

Pharmaceutical contaminants in water can engage with adsorbent surfaces via both physical and chemical interactions. Physical adsorption, which generally entails nonpolar medicines adhering to nonpolar surfaces such as carbonaceous materials or minerals, is less common; yet its efficacy is predominantly determined by the adsorbent's surface area. Pharmaceutical compounds frequently participate in interaction sorption processes influenced by their functional groups that contain heteroatoms like sulfur, nitrogen, and oxygen. These functional groups facilitate several chemical interactions, such as van der Waals forces, electrostatic attractions, and hydrogen bonding, hence augmenting the binding affinity between the contaminant and the adsorbent material [73].

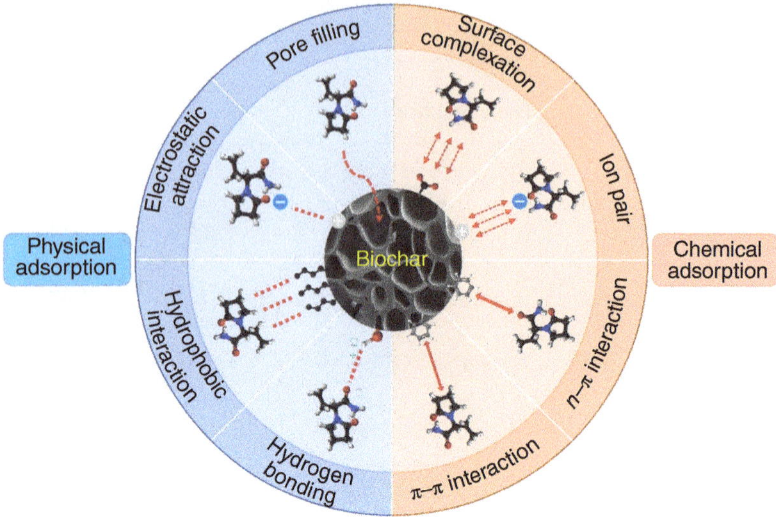

Figure 4.5 Adsorption mechanisms of pharmaceutical from aqueous solutions utilizing biochar and modified biochar [64] / with permission of Elsevier.

Biochar exhibits significant potential for the adsorption of pharmaceutical contaminants via multiple processes, including hydrophobic interactions, pore-filling, electrostatic attraction, π–π and n–π interactions, ion-pair formation, and hydrogen bonding (Figure 4.5). The adsorption processes are inherently connected to the unique surface properties of biochar, influenced by its carbonized and noncarbonized components. Electrostatic interactions and π–π stacking are predominantly acknowledged as the principal mechanisms for pharmaceutical removal, although other interactions also play a significant role, contingent upon the molecular structure of the pollutants and the physicochemical characteristics of the biochar [74].

The adsorption of pharmaceuticals onto charcoal entails various critical interactions, especially for compounds with aromatic rings. One primary mechanism is π–π dispersion interactions, which transpire between the aromatic rings of medicinal compounds and the aromatic framework of carbon-based biochar. These interactions are essential in the adsorption process, as they form robust connections between the two [73].

Another crucial mechanism involves the n–π interaction, in which oxygen ions present on the biochar surface transfer electrons to the electron-deficient aromatic rings of pharmaceutical compounds, leading to the formation of a donor-acceptor complex. Thus, oxygen-containing functions on the biochar surface, such as —COOH and —OH groups, act as electron donors, while the aromatic rings present in the pharmaceuticals act as electron acceptors. This interaction is of vital importance in enhancing the adsorption phenomenon [75].

Hydrogen bonding plays an important role in the adsorption process of pharmaceutical compounds. The oxygen and nitrogen atoms present in pharmaceutical molecules can act as H-receptors, while the —OH groups located on the biochar surface act as H-donors. This dipole–dipole interaction has the capacity to generate significant hydrogen bonds, particularly with pharmaceutical compounds such as paracetamol. Two categories of hydrogen bonds have been identified: the first is formed between the hydrogen atoms of the —OH groups present on the biochar and the oxygen/nitrogen atoms of the pharmaceutical compounds, while the second, known as the Yoshida hydrogen bond, is created between the aromatic ring of the pharmaceutical compounds and the —OH groups located on the biochar surface [76].

Finally, hydrophobic interactions are particularly important for nonpolar pharmaceutical molecules. In aqueous environments, nonpolar groups tend to cluster together to reduce their exposure to water, making them more likely to bind to biochar adsorbents. This clustering process represents a further essential mechanism for removing pharmaceutical pollutants from water [77].

4.6.2 Adsorption of Heavy Metals

4.6.2.1 Biochar-Based Composites for Adsorption of Heavy Metals

Anthropogenic activities in the agricultural and industrial sectors have led to a significant increase in waste containing heavy metals being discharged into soil and water, causing severe environmental pollution. Removal of these heavy metals is essential to ensure a safe and healthy environment. Adsorption stands out from other pollutant treatment techniques due to its simplicity, low energy consumption, and ability to adapt to various types of pollutants [78].

Various adsorbent categories are listed, including biosorbents, magnetic adsorbents, and industrial adsorbents. Biosorbents, such as biochar, are recognized for their renewable nature; however, they face challenges such as soil degradation and the complexity of post-use recovery. Magnetic adsorbents, recognized for their ability to be recycled, their moderate cost, and their compatibility with living organisms, offer interesting prospects despite the constraints associated with their manufacturing process. Industrial adsorbents, derived from industrial waste, are widely available but require modification to optimize their adsorption capacity, resulting in higher costs [79].

Biochar-based adsorbents, derived from renewable raw materials such as agricultural residues, can be modified to enhance their adsorption performance. These materials are mainly made up of cellulose, hemicellulose, and lignin, which undergo thermochemical transformation (pyrolysis) to generate biochar.

Biochar characteristics, such as specific surface area, functional groups, pore distribution, and mineral composition, are influenced by pyrolysis parameters. Higher pyrolysis temperatures increase the carbon content of the biochar, which improves its ability to adsorb contaminants by altering its chemical structure, reducing its hydrophilic character, and increasing its degree of aromaticity [80].

Thus, the effectiveness of using biochar to remove contaminants is closely linked to the temperature conditions during its production, with higher temperatures resulting in biochar with more favorable adsorption properties [81].

The incorporation of organic compounds into biochar enhances its effectiveness in adsorbing heavy metals (HMs) from water, by exploiting a variety of mechanisms. Organic compounds increase the surface area of the composite material and incorporate functionalities such as —COOH, —OH, and —NH_2, thus promoting the creation of complexes with metal ions and improving the efficiency of the adsorption process. These functional groups encourage the formation of reactive sites on the biochar surface, promoting chemical interactions such as redox reactions and surface complexation, leading to improved heavy metal removal [82].

Chitosan, a natural polymer from crustacean shells, is particularly effective due to its —NH_2, C=O, and —OH groups, which bind metal ions. For example, chitosan-modified magnetic loofah biochar exhibited significantly higher adsorption capacities for Cr(VI) and Cu(II) compared to unmodified biochar (Figure 4.6), due to the addition of —COOH groups [83]. This shows that modifying biochar with chitosan or other organic materials can greatly improve its effectiveness in removing heavy metals from contaminated water (Table 4.3).

4.6.2.2 Adsorption Mechanism of Biochar for Heavy Metals

To optimize biochar's use in water treatment, understanding its heavy metal (HM) adsorption mechanisms is crucial. Biochar interacts with HM ions through various processes, including electrostatic attraction, ion exchange, surface complexation, chemical precipitation, and hydrogen bonding. These mechanisms depend on factors such as the specific pollutant, biochar's pore structure, and chemical bonding between the sorbate (HM ions) and the sorbent (biochar) [92].

Biochar's porous structure, with both macro- and micro-pores, offers a large surface area for adsorption. HM ions penetrate biochar's pores and bind to functional groups, such as —COOH and —OH, on the biochar surface. These functional groups may change after adsorption as they form complexes with HM ions. Additionally, biochar's oxygen-containing functional groups can undergo ion exchange, where metal ions replace hydrogen ions (Figure 4.7) [94].

The pH of the water significantly affects adsorption due to biochar's surface charge. At a pH above the point of zero charge (pHpzc), biochar becomes negatively charged, enhancing the electrostatic attraction between positively charged metal ions (e.g., Zn^{2+}, Pb^{2+}) and the biochar surface. Conversely, at a pH below

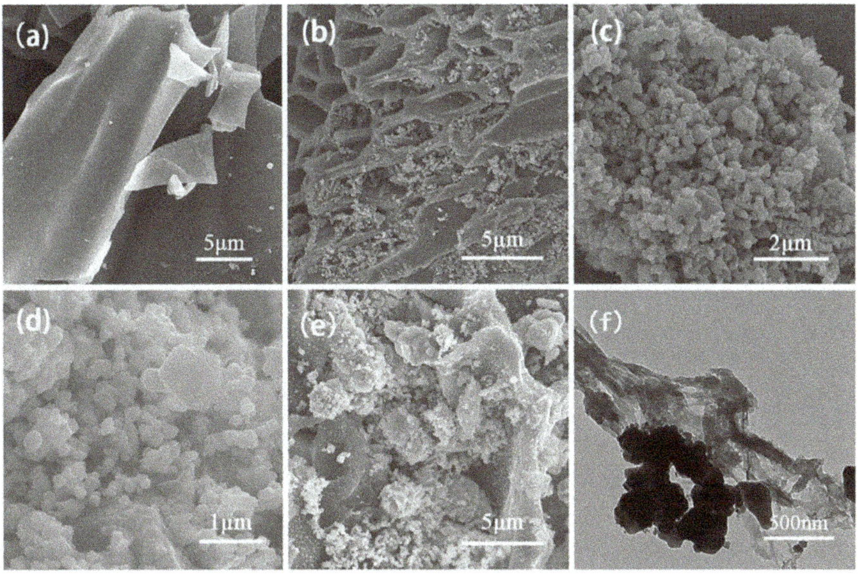

Figure 4.6 Scanning electron microscopy (SEM) images: (a) show SEM of loofah biochar, magnetic loofah biochar at different magnification levels (b) to (d), and (e) the 40 wt.% chitosan-modified magnetic biochar composite. In addition, figure (f) shows a transmission electron microscopy (TEM) image of the same 40 wt.% chitosan-modified magnetic biochar composite [86] / Elsevier / CC BY-NC 4.0.

the pHpzc, the biochar surface becomes positively charged, promoting attraction between negatively charged ions ($HCrO_4^-$, $Cr_2O_7^{2-}$) and the biochar.

Specific studies support these mechanisms. For example, As(V) adsorption on peanut shell biochar's involves surface complexation and electrostatic interactions with functional groups like C=O and —OH. Similarly, at low pH, Cr(VI) is adsorbed via electrostatic interactions with protonated functional groups on the biochar surface ($—NH_3^+$, $—OH_2^+$, $—COOH_2^+$) [95].

Modifying biochar with materials like ZnO and MnO_2 can further enhance HM adsorption. ZnO increases biochar's surface charge and promotes photocatalytic reduction of Cr(VI) to Cr(III), while MnO_2 increases the composite's surface area and pore volume, improving its ability to form complexes with HM ions [88].

4.7 Factors Affecting the Adsorption Performance of Biochar Biocomposites

The effective use of biochar as an adsorbent for pollutants, particularly in the treatment of organic and inorganic contaminants present in wastewater, relies on a

Table 4.3 Modifications of biochar on adsorption of heavy metal from contaminated water.

Biomass	Composite	Surface AREA ($m^2 g^{-1}$)	Heavy metals adsorbed	Adsorption capacity ($mg g^{-1}$)	Reference
Rice husk	Polyethyleneimine/ modified biochar	—	Cr (VI) ($100 mg L^{-1}$)	435.7	[84]
Wood	Fe_3O_4/ biochar	320	As (III) ($10 mg L^{-1}$)	5.49	[85]
Peanut shell	FeO/ biochar	—	Cd(II) ($50 mg L^{-1}$)	29.9	[86]
Swine manure	MnO_2/ biochar	70.9	Pb (II) ($210 mg L^{-1}$)	268	[87]
Water hyacinth	MnO_2/ biochar	120.2	Cd(II) ($100 mg L^{-1}$)	232.5	[88]
Corn cobs	TiO_2/ biochar	450.4	Cd(II) ($300 mg L^{-1}$)	72.5	[89]
Corn stover	ZnO/ZnS/ biochar	397.4	Cr(VI) ($400 mg L^{-1}$)	24.5	[90]
Banana pseudostem	Modified biochar	25.7	Cd(II) ($30.2 mg L^{-1}$)	12.5	[91]

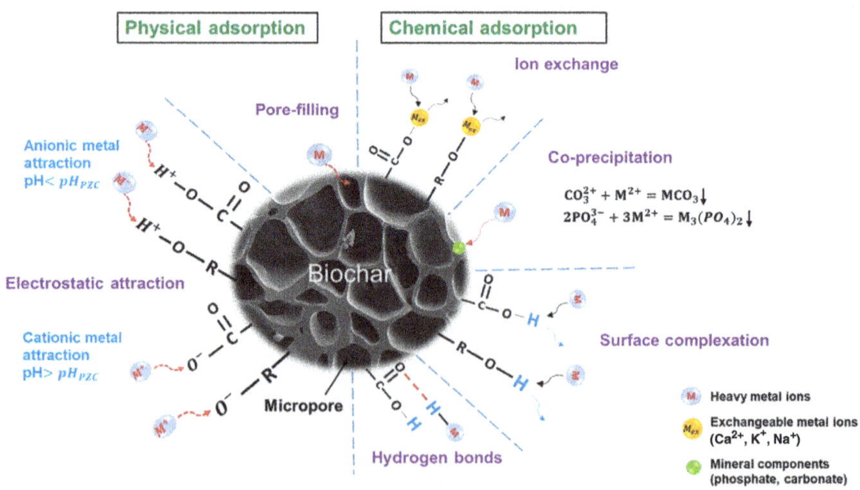

Figure 4.7 Mechanisms of adsorption of heavy metals ions by biochar [97] / Elsevier / CC BY-NC 4.0.

number of essential characteristics such as specific surface area, functional groups, porosity, carbon and ash content, and cation exchange capacity. These properties are influenced by various parameters such as pyrolysis temperature, the nature of the raw material and the pyrolysis process. Recent advances have increased the efficiency of biochar as a pollutant adsorbent [96].

The rise in temperature during the pyrolysis process results in an increase in the carbon content and ash levels of the biochar. This enhances the surface and microporosity of the material, which is beneficial for the adsorption of heavy metals and organic pollutants. On the other hand, biochar manufactured at lower temperatures retains the oxygen-containing functional groups required for sorption of polar inorganic and organic pollutants by electrostatic attraction [97].

The adsorption of pharmaceuticals by biochar is influenced by various factors, including pH, temperature, contact time, initial pollutant concentration, and the chemical structure of the pharmaceutical. In real water systems, ions such as Na^+, K^+, carbonate, sulfate, and phosphate can compete with pharmaceutical pollutants for adsorption sites. Studies have shown that the presence of carbonate and bicarbonate reduces the effectiveness of biochar by creating alkaline conditions that lead to electrostatic repulsion.

Adsorption of pharmaceutical compounds by biochar is subject to the influence of several parameters, such as pH, temperature, contact time, initial contaminant concentration, as well as the chemical composition of the pharmaceutical compound. In real aquatic systems, ions such as Na^+, K^+, carbonate, sulfate, and phosphate can compete with pharmaceutical pollutants for adsorption sites. Research has shown that the presence of carbonate and bicarbonate decreases biochar efficiency by inducing alkaline conditions favoring electrostatic repulsion [98].

To optimize pharmaceutical adsorption, the biochar-to-water ratio is critical, as an initial increase in adsorbent enhances adsorption until equilibrium is attained. The duration of contact influences efficiency, characterized by fast initial adsorption succeeded by a decelerated rate when the active sites reach saturation [99].

Biochar can be generated from diverse feedstocks, including agricultural residues, algal biomass, and activated sludge, with the selection of feedstock significantly influencing adsorption capacity. Numerous studies have concentrated on individual pharmaceutical contaminants in synthetic water, underscoring the necessity for more study to assess biochar efficacy in actual wastewater systems and large-scale industrial applications [100].

4.8 Challenges and Limitations

The biochar biocomposites present a promising range of uses; however, it is crucial to carefully evaluate their potential environmental concerns. The integrity of their carbon-based structure is fundamental to safe utilization within our ecosystems. Biochar's durability is seen in its elevated aromatic content in dissolved organic matter, influenced by aromaticity and the extent of aromatic condensation, which might differ based on the source material. The introduction of biochar can increase carbon levels in aquatic systems, but it is crucial to monitor its impact on microorganisms. Studies indicate that modest doses generally exhibit negligible toxicity, however elevated amounts may raise cytotoxic and genotoxic issues [90].

The future direction involves a more thorough investigation of biochar's stability and environmental consequences, especially as we assess the implications of various pyrolysis settings. By conducting rigorous toxicity assessments on different biochar varieties and a range of creatures, such as fish, plants, and algae, we can facilitate the safe and effective utilization of biochar, thereby fostering a more sustainable future [91].

4.9 Conclusion

Biochar-based biocomposites have developed new materials for wastewater treatment, providing a sustainable and effective answer to the growing difficulties of environmental pollution. This chapter highlights the optimization of the structural and chemical properties of biochar, obtained from diverse biomass sources, by modifications utilizing both organic and inorganic elements. These biocomposites demonstrate an extraordinary ability to adsorb various contaminants, including heavy metals and pharmaceutical residues. The efficacy of biochar biocomposites is determined by several critical criteria, including pyrolysis parameters, raw material selection, and the incorporation of functional additives. High-temperature pyrolysis markedly increases the specific surface area, microporosity, and carbon concentration of biochar, hence augmenting its effectiveness in adsorbing nonpolar organic contaminants and heavy metals. In contrast, low-temperature pyrolysis preserves oxygenated functional groups that facilitate the elimination of inorganic pollutants through mechanisms including electrostatic attraction, ion exchange, and surface complexation. The integration of organic components such as chitosan with inorganic substances like metal oxides (ZnO, MnO_2) has shown particularly successful in enhancing the adsorption capacity and stability of biochar in actual wastewater treatment applications. These alterations improve surface reactivity by generating new active sites and optimizing interactions with contaminants via mechanisms such as ion exchange, electrostatic interactions, and redox reactions. Moreover, biochar-based magnetic biocomposites enable enhanced recovery and regeneration, rendering them particularly appropriate for large-scale use.

References

1 Tyagi, S., Sharma, B., Singh, P., and Dobhal, R. (2020). Water quality assessment in terms of water quality index. *American Journal of Water Resources* 1 (3): 34–38. https://doi.org/10.12691/ajwr-1-3-3.

2 Meena, M., Sonigra, P., Yadav, G., and Barupal, T. (2021). Wastewater treatment techniques: an introduction. In: *Removal of Emerging Contaminants Through Microbial Processes* (ed. M.P. Shah), 161–182. Springer, Singapore https://doi.org/10.1007/978-981-15-5901-3_8.

3 Kathi, S. and El Din Mahmoud, A. (2024). Trends in effective removal of emerging contaminants from wastewater: a comprehensive review. *Desalination and Water Treatment* 317: https://doi.org/10.1016/j.dwt.2024.100258.

4 Jafarinejad, S. and Jafarinejad, S.A. (2017). Comprehensive study on the application of reverse osmosis (RO) technology for the petroleum industry wastewater treatment. *Journal Of Water And Environmental Nanotechnology* 2 (4): 243–264. https://doi.org/10.22090/jwent.2017.04.003.

5 Samuel Olugbenga, O., Goodness Adeleye, P., Blessing Oladipupo, S. et al. (2024). Biomass-derived biochar in wastewater treatment- a circular economy approach. *Waste Management Bulletin* 1 (4): 1–14. https://doi.org/10.1016/j.wmb.2023.07.007.

6 Amalina F, Abd Razak AS, Krishnan S, Zularisam AW, Nasrullah M. A comprehensive assessment of the method for producing biochar, its characterization, stability, and potential applications in regenerative economic sustainability—a review. *Cleaner Materials* 2022 3:100045. https://doi.org/10.1016/j.clema.2022.100045.

7 Kurniawan TA, Ali S, Mohyuddin A, Haider A, Riaz M, Khan S, Othman MH, Goh HH, Anouzla A, Aziz F, Ali I. Cultivating sustainability: harnessing biochar-derived composites for carbon-neutral wastewater treatment. *Process Safety and Environment Protection* 2024 187:665-97. https://doi.org/10.1016/j.psep.2024.04.040.

8 Shukla, P., Giri, B.S., Mishra, R.K. et al. (2021). Lignocellulosic biomass-based engineered biochar composites: a facile strategy for abatement of emerging pollutants and utilization in industrial applications. *Renewable and Sustainable Energy Reviews* 152: https://doi.org/10.1016/j.rser.2021.111643.

9 Dzoujo HT, Shikuku VO, Tome S, Simo AC, Ng'eno EC, Getenga ZM, Etoh MA, Dina DD. Recent advances in metal oxide-biochar composites for water and soil remediation: a review. *Hybrid Advances* 2024, 7:100292. https://doi.org/10.1016/j.hybadv.2024.100292.

10 Gao N, Du W, Zhang M, Ling G, Zhang P. Chitosan-modified biochar: preparation, modifications, mechanisms and applications. *International Journal of*

Biological Macromolecules 2022;209:31-49. https://doi.org/10.1016/j.ijbiomac.2022.04.006.

11 Müller A, Österlund H, Marsalek J, Viklander M. The pollution conveyed by urban runoff: a review of sources. *Science of the Total Environment* 2020;709:136125. https://doi.org/10.1016/j.scitotenv.2019.136125.

12 Crini G, Lichtfouse E. Advantages and disadvantages of techniques used for wastewater treatment. *Environmental Chemistry Letters* 2019;17(1):145-55. https://doi.org/10.1007/s10311-018-0785-9.

13 Singh, P. Singh S, Sinha R.et al. (2021). Pollutants and water management: resources, strategies and scarcity, first edition. Edited microbial degradation of organic contaminants in water bodies technological advancements.

14 Xu, Z., Wei, C., Jin, J. et al. (2018). Development of a novel mixed titanium, silver oxide polyacrylonitrile nanofiber as a superior adsorbent and its application for MB removal in wastewater treatment. *Journal of the Brazilian Chemical Society* 29 (3): 560–570. https://doi.org/10.21577/0103-5053.20170168.

15 Corominas, L., Garrido-Baserba, M., Villez, K. et al. (2018). Transforming data into knowledge for improved wastewater treatment operation: a critical review of techniques. *Environmental Modelling and Software* 106: 89–103. https://doi.org/10.1016/j.envsoft.2017.11.023.

16 Maniraj, J., Ramesh, M., Kumar, S.G., and Sahayaraj, A.F. (2023). Introduction of biochar: sources, composition, and recent updates. *Materials Horizons: From Nature to Nanomaterials* Part F1442: 1–17. https://doi.org/10.1007/978-981-99-5239-7_1/FIGURES/5.

17 Ambaye TG, Vaccari M, van Hullebusch ED, Amrane A, Rtimi SJ. Mechanisms and adsorption capacities of biochar for the removal of organic and inorganic pollutants from industrial wastewater. *International Journal of Environmental Science and Technology* 2021 18(10):3273-94. https://doi.org/10.1007/s13762-020-03060-w.

18 Ravindiran G, Rajamanickam S, Janardhan G, Hayder G, Alagumalai A, Mahian O, Lam SS, Sonne C. Production and modifications of biochar to engineered materials and its application for environmental sustainability: a review. *Biochar* 2024;6(1):62. https://doi.org/10.1007/s42773-024-00350-1.

19 Lai, W.Y., Lai, C.M., Ke, G.R. et al. (2013). The effects of woodchip biochar application on crop yield, carbon sequestration and greenhouse gas emissions from soils planted with rice or leaf beet. *Journal of the Taiwan Institute of Chemical Engineers* 44 (6): 1039–1044. https://doi.org/10.1016/j.jtice.2013.06.028.

20 Essandoh, M., Kunwar, B., Pittman, C.U. et al. (2015). Sorptive removal of salicylic acid and ibuprofen from aqueous solutions using pine wood fast pyrolysis biochar. *Chemical Engineering Journal* 265: 219–227. https://doi.org/10.1016/j.cej.2014.12.006.

21 Dai, L., Fan, L., Liu, Y. et al. (2017). Production of bio-oil and biochar from soapstock via microwave-assisted co-catalytic fast pyrolysis. *Bioresource Technology* 225: 1–8. https://doi.org/10.1016/j.biortech.2016.11.017.

22 Al-Rahbi, A.S. and Williams, P.T. (2017). Hydrogen-rich syngas production and tar removal from biomass gasification using sacrificial tyre pyrolysis char. *Applied Energy* 190: 501–509. https://doi.org/10.1016/j.apenergy.2016.12.099.

23 Sarker TR, Nanda S, Dalai AK, Meda V. A review of torrefaction technology for upgrading lignocellulosic biomass to solid biofuels. *BioEnergy Research* 2021;14(2):645-69. https://doi.org/10.1007/s12155-020-10236-2/Published.

24 Li, L., Rowbotham, J.S., Christopher Greenwell, H., and Dyer, P.W. (2013). An introduction to pyrolysis and catalytic pyrolysis: versatile techniques for biomass conversion. In: *New and Future Developments in Catalysis: Catalytic Biomass Conversion* (ed. L.S. Steven), 173–208. Elsevier B.V https://doi.org/10.1016/B978-0-444-53878-9.00009-6.

25 Hu B, Wang K, Wu L, Yu SH, Antonietti M, Titirici MM. Engineering carbon materials from the hydrothermal carbonization process of biomass. *Advanced Materials* 2010;22(7):813-28. https://doi.org/10.1002/adma.200902812.

26 Lynam, J.G., Coronella, C.J., Yan, W. et al. (2011). Acetic acid and lithium chloride effects on hydrothermal carbonization of lignocellulosic biomass. *Bioresource Technology* 102 (10): 6192–6199. https://doi.org/10.1016/j.biortech.2011.02.035.

27 Arafat Hossain, M., Ganesan, P., Jewaratnam, J., and Chinna, K. (2017). Optimization of process parameters for microwave pyrolysis of oil palm fiber (OPF) for hydrogen and biochar production. *Energy Conversion and Management* 133: 349–362. https://doi.org/10.1016/j.enconman.2016.10.046.

28 Huang, Y.F., Te Chiueh, P., and Lo, S.L. (2019). CO_2 adsorption on biochar from co-torrefaction of sewage sludge and leucaena wood using microwave heating. *Energy Procedia* 4435–4440. https://doi.org/10.1016/j.egypro.2019.01.772.

29 Glaser B. Prehistorically modified soils of central Amazonia: a model for sustainable agriculture in the twenty-first century. *Philosophical Transactions of the Royal Society, B: Biological Sciences* 2007;362(1478):187-96. https://doi.org/10.1098/rstb.2006.1978.

30 Barrow, C.J. (2012). Biochar: potential for countering land degradation and for improving agriculture. *Applied Geography* 34: 21–28. https://doi.org/10.1016/j.apgeog.2011.09.008.

31 Reilly, J. (1925). The technology of wood distillation: with special reference to the methods of obtaining the intermediate and finished products from the primary distillate. *Nature* 116: 779–780. https://doi.org/10.1038/116779a0.

32 Suárez-Abelenda, M., Kaal, J., and McBeath, A.V. (2017). Translating analytical pyrolysis fingerprints to thermal stability indices (TSI) to improve biochar characterization by pyrolysis-GC-MS. *Biomass and Bioenergy* 98: 306–320. https://doi.org/10.1016/j.biombioe.2017.01.021.

33 Roberts, D.A., Cole, A.J., Paul, N.A., and de Nys, R. (2015). Algal biochar enhances the re-vegetation of stockpiled mine soils with native grass. *Journal of Environmental Management* 161: 173–180. https://doi.org/10.1016/j.jenvman.2015.07.002.

34 Morali, U. and Şensöz, S. (2015). Pyrolysis of hornbeam shell (Carpinus betulus L.) in a fixed bed reactor: characterization of bio-oil and bio-char. *Fuel* 150: 672–678. https://doi.org/10.1016/j.fuel.2015.02.095.

35 Tomczyk A, Sokołowska Z, Boguta P. Biochar physicochemical properties: pyrolysis temperature and feedstock kind effects. *Reviews in Environmental Science and Bio/Technology* 2020;19(1):191-215. https://doi.org/10.1007/s11157-020-09523-3.

36 Guan G, Kaewpanha M, Hao X, Abudula A. Catalytic steam reforming of biomass tar: prospects and challenges. *Renewable and Sustainable Energy Reviews* 2016;58:450-61. https://doi.org/10.1016/j.rser.2015.12.316.

37 Tripathi, M., Sahu, J.N., Ganesan, P. et al. (2015). Effect of microwave frequency on dielectric properties of oil palm shell (OPS) and OPS char synthesized by microwave pyrolysis of OPS. *Journal of Analytical and Applied Pyrolysis* 112: 306–312. https://doi.org/10.1016/j.jaap.2015.01.007.

38 Regmi, P., Garcia Moscoso, J.L., Kumar, S. et al. (2012). Removal of copper and cadmium from aqueous solution using switchgrass biochar produced via hydrothermal carbonization process. *Journal of Environmental Management* 109: 61–69. https://doi.org/10.1016/j.jenvman.2012.04.047.

39 Cha, J.S., Park, S.H., Jung, S.C. et al. (2016). Production and utilization of biochar: a review. *Korean Society of Industrial Engineering Chemistry* https://doi.org/10.1016/j.jiec.2016.06.002.

40 Collard FX, Blin J. A review on pyrolysis of biomass constituents: mechanisms and composition of the products obtained from the conversion of cellulose, hemicelluloses and lignin. *Renewable and Sustainable Energy Reviews* 2014;38:594-608. https://doi.org/10.1016/j.rser.2014.06.013.

41 Panwar, N.L. and Pawar, A. (2022). Influence of activation conditions on the physicochemical properties of activated biochar: a review. *Biomass Conversion and Biorefinery* 12 (3): 925–947. https://doi.org/10.1007/s13399-020-00870-3/Published.

42 Nair, V. and Vinu, R. (2016). Peroxide-assisted microwave activation of pyrolysis char for adsorption of dyes from wastewater. *Bioresource Technology* 216: 511–519. https://doi.org/10.1016/j.biortech.2016.05.070.

43 Bardestani, R. and Kaliaguine, S. (2018). Steam activation and mild air oxidation of vacuum pyrolysis biochar. *Biomass and Bioenergy* 108: 101–112. https://doi.org/10.1016/j.biombioe.2017.10.011.

44 Peng, P., Lang, Y.H., and Wang, X.M. (2016). Adsorption behavior and mechanism of pentachlorophenol on reed biochars: PH effect, pyrolysis temperature, hydrochloric acid treatment and isotherms. *Ecological Engineering* 90: 225–233. https://doi.org/10.1016/j.ecoleng.2016.01.039.

45 Cordero-Lanzac, T., Hita, I., Veloso, A. et al. (2017). Characterization and controlled combustion of carbonaceous deactivating species deposited on an activated carbon-based catalyst. *Chemical Engineering Journal* 327: 454–464. https://doi.org/10.1016/j.cej.2017.06.077.

46 Bashir, S., Zhu, J., Fu, Q., and Hu, H. (2018). Comparing the adsorption mechanism of Cd by rice straw pristine and KOH-modified biochar. *Environmental Science and Pollution Research* 25 (12): 11875–11883. https://doi.org/10.1007/s11356-018-1292-z.

47 Weidner E, Karbassiyazdi E, Altaee A, Jesionowski T, Ciesielczyk F. Hybrid metal oxide/biochar materials for wastewater treatment technology: a review. *ACS Omega* 2022 7(31):27062-78. https://doi.org/10.1021/acsomega.2c02909.

48 Bahsaine, K., Benzeid, H., Zari, N. et al. (2024). Biochar-alginate beads derived from argan nutshells for effective methylene blue removal: a sustainable approach to wastewater treatment. *International Journal of Biological Macromolecules* 282: https://doi.org/10.1016/j.ijbiomac.2024.136853.

49 Tan, X.F., Liu, Y.G., Gu, Y.L. et al. (2016). Biochar-based nano-composites for the decontamination of wastewater: a review. *Bioresource Technology* 1 (212): 318–333. https://doi.org/10.1016/j.biortech.2016.04.093.

50 Yasim-Anuar, T.A., Yee-Foong, L.N., Lawal, A.A. et al. (2022). Emerging application of biochar as a renewable and superior filler in polymer composites. *RSC Advances* 12 (22): 13938–13949. https://doi.org/10.1039/d2ra01897g.

51 Bartoli M, Arrigo R, Malucelli G, Tagliaferro A, Duraccio D. Recent advances in biochar polymer composites. *Polymers* 2022 14(12):2506. https://doi.org/10.3390/polym14122506.

52 Aryanfar, Y., Keçebaş, A., Blanco Fernandez, J. et al. (2024). Biochar polymer composites for wastewater treatment. *Reference Module in Materials Science and Materials Engineering* https://doi.org/10.1016/B978-0-323-95486-0.00034-X.

53 Zhao, C., Wang, B., Theng, B.K. et al. (2021). Formation and mechanisms of nano-metal oxide-biochar composites for pollutants removal: a review. *Science of the Total Environment* 767: 145305. https://doi.org/10.1016/J.SCITOTENV.2021.145305.

54 Chakhtouna H, Mekhzoum ME, Zari N, Benzeid H, Qaiss AE, Bouhfid R. "*Biochar-Supported Materials for Wastewater Treatment.*" In: Inamuddin, M.I. Ahamed, R. Boddula and T.A. Rangreez (eds) *Applied Water Science*. 2021, https://doi.org/10.1002/9781119725237.ch7.

55 Chakhtouna, H., Ouhssain, A., Kadmiri, I.M. et al. (2023). Photocatalytic and bactericidal behaviors of Ag/TiO_2 doped biochar through Ball–milling approach. *Journal of Photochemistry and Photobiology A: Chemistry* 444: https://doi.org/10.1016/j.jphotochem.2023.114971.

56 Chakhtouna, H., Benzeid, H., Zari, N. et al. (2021). Functional $CoFe_2O_4$-modified biochar derived from banana pseudostem as an efficient adsorbent for the removal of amoxicillin from water. *Separation and Purification Technology* 266: https://doi.org/10.1016/j.seppur.2021.118592.

57 Zhu, Q., Zhang, K., Xu, J. et al. (2023). Performance and mechanism of Fe3O4 loaded biochar activating persulfate to degrade acid orange 7. *Water (Switzerland)* 15 (10): https://doi.org/10.3390/w15101849.

58 Dhila, H., Bhapkar, A., and Bhame, S. (2025). Metal oxide/biochar hybrid nanocomposites for adsorption and photocatalytic degradation of textile dye effluents: a review. *Desalination and Water Treatment* https://doi.org/10.1016/j.dwt.2025.101004.

59 Zhou, Y., Wen, J., Dong, J. et al. (2024). Efficient adsorption-photocatalysis activity of ternary ZnO-Sm_2O_3-MgO co-modified biochar for dye removal. *Applied Organometallic Chemistry* 38 (3): e7349.

60 Calisto, V., Ferreira, C.I.A., Oliveira, J.A.B.P. et al. (2015). Adsorptive removal of pharmaceuticals from water by commercial and waste-based carbons. *Journal of Environmental Management* 152: 83–90. https://doi.org/10.1016/j.jenvman.2015.01.019.

61 Hama Aziz, K.H., Mahyar, A., Miessner, H. et al. (2018). Application of a planar falling film reactor for decomposition and mineralization of methylene blue in the aqueous media via ozonation, Fenton, photocatalysis and non-thermal plasma: a comparative study. *Process Safety and Environmental Protection* 113: 319–329. https://doi.org/10.1016/j.psep.2017.11.005.

62 Aziz, K.H., Mustafa, F.S., Hassan, M.A. et al. (2024). Biochar as green adsorbents for pharmaceutical pollution in aquatic environments: a review. *Desalination* 583: 117725. https://doi.org/10.1016/j.desal.2024.117725.

63 Anoob, F., Arachchi, S., Azamathulla, H.M. et al. (2024). Nanoadsorbents as an effective wastewater treatment candidate for pharmaceutical contaminants; towards sustainable policy development. *Case Studies in Chemical and Environmental Engineering* 9: https://doi.org/10.1016/j.cscee.2024.100639.

64 Lladó, J., López, F., Rossell, J.M. et al. (2024). Multivariate analysis of pharmaceutical pollutants adsorption in aqueous media with tailored waste-based carbonaceous adsorbent materials and commercial activated carbons. *Sustainable Chemistry and Pharmacy* 38: https://doi.org/10.1016/j.scp.2024.101453.

65 Sanga P, Al-mashriqi HS, Chen J, Qiu H. A mechanistic view of removing pharmaceutical-induced pollutants by MXene and MXene-functionalized composites via adsorption and advanced oxidation process. *Journal of Environmental Chemical Engineering* 2024 12(1):111685. https://doi.org/10.1016/j.jece.2023.111685.

66 Chakhtouna, H., Benzeid, H., Zari, N. et al. (2023). Microwave-assisted synthesis of MIL–53(Fe)/biochar composite from date palm for ciprofloxacin and ofloxacin antibiotics removal. *Separation and Purification Technology* 308: https://doi.org/10.1016/j.seppur.2022.122850.

67 Yang, J., Wang, L., Yang, J. et al. Magnetic biochar coupled with bismuth tungstate for multiple antibiotic removal from contaminated water: characteristics, performance, and competitive adsorption synergistic photocatalysis mechanism. *Journal of Environmental Chemical Engineering* 12 (1): https://doi.org/10.1016/j.jece.2023.111768.

68 Choi, S.W., Hong, J., Youn, S., and Kim, I. (2023). Removal of amoxicillin by coffee grounds biochar with different pretreatment methods. *Environmental Advances* 14: https://doi.org/10.1016/j.envadv.2023.100446.

69 Shaheen, J.F., Eniola, J.O., and Sizirici, B. (2024). Adsorption of ibuprofen from aqueous solution by modified date palm biochar: performance, optimization, and life cycle assessment. *Bioresource Technology Reports* 25: https://doi.org/10.1016/j.biteb.2023.101696.

70 Li, X. and Shi, J. (2022). Simultaneous adsorption of tetracycline, ammonium and phosphate from wastewater by iron and nitrogen modified biochar: kinetics, isotherm, thermodynamic and mechanism. *Chemosphere* 293: https://doi.org/10.1016/j.chemosphere.2022.133574.

71 Hinsene, H., Bhawawet, N., and Imyim, A. (2024). Rice husk biochar doped with deep eutectic solvent and Fe_3O_4/ZnO nanoparticles for heavy metal and diclofenac removal from water. *Separation and Purification Technology* 339: https://doi.org/10.1016/j.seppur.2024.126638.

72 Deng Y, Chen J, She A, Ni F, Chen W, Ao T, Zhang Y. A novel Fe-loaded porous hydrothermal biochar for removing tetracycline from wastewater: performance, mechanism, and fixed-bed column. *Journal of Environmental Chemical Engineering* 2024;12(2):112256. https://doi.org/10.1016/j.jece.2024.112256.

73 Pauletto, P.S., Lütke, S.F., Dotto, G.L., and Salau, N.P.G. (2021). Adsorption mechanisms of single and simultaneous removal of pharmaceutical compounds onto activated carbon: isotherm and thermodynamic modeling. *Journal of Molecular Liquids* 336: https://doi.org/10.1016/j.molliq.2021.116203.

74 Nasiri, A., Golestani, N., Rajabi, S., and Hashemi, M. (2024). Facile and green synthesis of recyclable, environmentally friendly, chemically stable, and cost-effective magnetic nanohybrid adsorbent for tetracycline adsorption. *Heliyon* 10 (2): https://doi.org/10.1016/j.heliyon.2024.e24179.

75 Oladoye PO, Kadhom M, Khan I, Aziz KH, Alli YA. Advancements in adsorption and photodegradation technologies for Rhodamine B dye wastewater treatment: fundamentals, applications, and future directions. *Green Chemical Engineering* 2024 5(4):440-60. https://doi.org/10.1016/j.gce.2023.12.004.

76 Nguyen, D.T., Tran, H.N., Juang, R.S. et al. (2020). Adsorption process and mechanism of acetaminophen onto commercial activated carbon. *Journal of Environmental Chemical Engineering* 8 (6): https://doi.org/10.1016/j.jece.2020.104408.

77 Salahshoori, I., Jorabchi, M.N., Ghasemi, S. et al. (2023). An in silico study of sustainable drug pollutants removal using carboxylic acid functionalized-MOF nanostructures (MIL-53 (Al)-(COOH)2): towards a greener future. *Desalination* 559: https://doi.org/10.1016/j.desal.2023.116654.

78 Abdel Maksoud MI, Fahim RA, Bedir AG, Osman AI, Abouelela MM, El-Sayyad GS, Elkodous MA, Mahmoud AS, Rabee MM, Al-Muhtaseb AA, Rooney DW. Engineered magnetic oxides nanoparticles as efficient sorbents for wastewater remediation: a review. *Environmental Chemistry Letters* 2022 20(1):519-62. https://doi.org/10.1007/s10311-021-01351-3.

79 Osman AI, El-Monaem EM, Elgarahy AM, Aniagor CO, Hosny M, Farghali M, Rashad E, Ejimofor MI, Lopez-Maldonado EA, Ihara I, Yap PS. Methods to prepare biosorbents and magnetic sorbents for water treatment: a review. *Environmental Chemistry Letters* 2023 21(4):2337-98. https://doi.org/10.1007/s10311-023-01603-4.

80 Tan, H., Lee, C.T., Ong, P.Y. et al. (2021). A review on the comparison between slow pyrolysis and fast pyrolysis on the quality of lignocellulosic and lignin-based biochar. *IOP Conference Series: Materials Science and Engineering* 1051 (1): 012075. https://doi.org/10.1088/1757-899x/1051/1/012075.

81 Sizmur T, Fresno T, Akgül G, Frost H, Moreno-Jiménez E. Biochar modification to enhance sorption of inorganics from water. *Bioresource Technology* 2017 246:34-47. https://doi.org/10.1016/j.biortech.2017.07.082.

82 Zhou YF, Haynes RJ. Sorption of heavy metals by inorganic and organic components of solid wastes: significance to use of wastes as low-cost adsorbents and immobilizing agents. *Critical Reviews in Environmental Science and Technology* 2010 40(11):909-77. https://doi.org/10.1080/10643380802586857.

83 Xiao, F., Cheng, J., Cao, W. et al. (2019). Removal of heavy metals from aqueous solution using chitosan-combined magnetic biochars. *Journal of Colloid and Interface Science* 540: 579–584. https://doi.org/10.1016/j.jcis.2019.01.068.

84 Ma, Y., Liu, W.J., Zhang, N. et al. (2014). Polyethylenimine modified biochar adsorbent for hexavalent chromium removal from the aqueous solution. *Bioresource Technology* 169: 403–408. https://doi.org/10.1016/j.biortech.2014.07.014.

85 Navarathna, C.M., Karunanayake, A.G., Gunatilake, S.R. et al. (2019). Removal of Arsenic(III) from water using magnetite precipitated onto Douglas fir biochar. *Journal of Environmental Management* 250: https://doi.org/10.1016/j.jenvman.2019.109429.

86 Li, Y., Gao, L., Lu, Z. et al. (2020). Enhanced removal of heavy metals from water by hydrous ferric oxide-modified biochar. *ACS Omega* 5 (44): 28702–28711. https://doi.org/10.1021/acsomega.0c03893.

87 Liang J, Li X, Yu Z, Zeng G, Luo Y, Jiang L, Yang Z, Qian Y, Wu H. Amorphous MnO_2 modified biochar derived from aerobically composted swine manure for adsorption of Pb (II) and Cd (II). *ACS Sustainable Chemistry & Engineering* 2017 5(6):5049-58. https://doi.org/10.1021/acssuschemeng.7b00434.

88 Zhang, H., Xu, F., Xue, J. et al. (2020). Enhanced removal of heavy metal ions from aqueous solution using manganese dioxide-loaded biochar: behavior and mechanism. *Scientific Reports* 10 (1): https://doi.org/10.1038/s41598-020-63000-z.

89 Luo, M., Lin, H., He, Y. et al. (2019). Efficient simultaneous removal of cadmium and arsenic in aqueous solution by titanium-modified ultrasonic biochar. *Bioresource Technology* 284: 333–339. https://doi.org/10.1016/j.biortech.2019.03.108.

90 Li, C., Zhang, L., Gao, Y., and Li, A. (2018). Facile synthesis of nano ZnO/ZnS modified biochar by directly pyrolyzing of zinc contaminated corn stover for Pb(II), Cu(II) and Cr(VI) removals. *Waste Management* 79: 625–637. https://doi.org/10.1016/j.wasman.2018.08.035.

91 Bahsaine, K., Chakhtouna, H., Mekhzoum, M.E. et al. (2024). Efficient cadmium removal from industrial phosphoric acid using banana pseudostem-derived biochar. *Biomass Conversion and Biorefinery* 14 (15): 17745–17759. https://doi.org/10.1007/s13399-023-04130-y.

92 Yin, G., Chen, X., Sarkar, B. et al. (2023). Co-adsorption mechanisms of Cd(II) and As(III) by an Fe-Mn binary oxide biochar in aqueous solution. *Chemical Engineering Journal* 466: https://doi.org/10.1016/j.cej.2023.143199.

93 Ullah MH, Rahman MJ. Adsorptive removal of toxic heavy metals from wastewater using water hyacinth and its biochar: a review. *Heliyon* 2024 10(17). https://doi.org/10.1016/j.heliyon.2024.e36869.

94 Wang, Y., Li, J., Xu, L. et al. (2024). EDTA functionalized Mg/Al hydroxides modified biochar for Pb(II) and Cd(II) removal: adsorption performance and mechanism. *Separation and Purification Technology* 335: https://doi.org/10.1016/j.seppur.2023.126199.

95 Sattar, M.S., Shakoor, M.B., Ali, S. et al. (2019). Comparative efficiency of peanut shell and peanut shell biochar for removal of arsenic from water. *Environmental Science and Pollution Research* 26 (18): 18624–18635. https://doi.org/10.1007/s11356-019-05185-z.

96 Wang, C., Ling, X., Sun, W. et al. (2024). Elaboration of cost-effective adsorbents derived from sewage sludge through hydrothermal carbonization: influence of process parameters and feedstock components on the adsorption performance. *Journal of Cleaner Production* 434: https://doi.org/10.1016/j.jclepro.2023.140358.

97 Sotoudehnia, F., Baba Rabiu, A., Alayat, A., and McDonald, A.G. (2020). Characterization of bio-oil and biochar from pyrolysis of waste corrugated cardboard. *Journal of Analytical and Applied Pyrolysis* 145: https://doi.org/10.1016/j.jaap.2019.104722.

98 Zeng ZW, Tan XF, Liu YG, Tian SR, Zeng GM, Jiang LH, Liu SB, Li J, Liu N, Yin ZH. Comprehensive adsorption studies of doxycycline and ciprofloxacin antibiotics by biochars prepared at different temperatures. *Frontiers in Chemistry* 2018 6:80. https://doi.org/10.3389/fchem.2018.00080.

99 Zhang, Z., Zhang, M., Zhao, X., and Cao, J. (2024). High-efficient removal and adsorption mechanism of organic dyes in wastewater by KOH-activated biochar from phenol-formaldehyde resin modified wood. *Separation and Purification Technology* 330: https://doi.org/10.1016/j.seppur.2023.125542.

100 Amalina F, Abd Razak AS, Krishnan S, Sulaiman H, Zularisam AW, Nasrullah M. Biochar production techniques utilizing biomass waste-derived materials and environmental applications—a review. *Journal of Hazardous Materials Advances* 2022 1;7:100134. https://doi.org/10.1016/j.hazadv.2022.100134.

5

Lignin Valorization as Biopolymer in Wastewater Treatment

Brener Felipe Melo Lima Gomes[1,2], *André Luís Rossoni Ferreira*[1,2], *and Leandro Vinícius Alves Gurgel*[1]

[1] *Physical Organic Chemistry Group, Department of Chemistry, Institute of Exact and Biological Sciences, Federal University of Ouro Preto, Ouro Preto, Brazil*
[2] *Graduate Program in Environmental Engineering, School of Mines, Federal University of Ouro Preto, Ouro Preto, Brazil*

5.1 Introduction

With the advances in technology, the composition of wastewaters from production processes has become increasingly complex, necessitating the use of treatment methods that are highly efficient for the removal of various pollutants, while at the same time being inexpensive and easy to operate. Adsorption is one of the commonest techniques employed in advanced treatments, since it enables the removal of target species present in the fluid phase, without generating byproducts that might be more toxic than the initial target species [1]. This treatment technology involves the separation of a contaminant (adsorbate) from the fluid phase onto a solid phase (adsorbent). For this, the adsorbent should have suitable physicochemical surface characteristics that ensure high affinity for the target pollutants.

Many highly efficient adsorbents have been studied by scientists worldwide, including materials based on graphene [2], chitosan [3], and zeolite [4], among others. However, although these materials have been shown to be effective for wastewater treatment on a laboratory scale, difficulties remain concerning technological scale-up for applications at pilot or industrial levels. The production of materials with exceptional adsorption capacities is associated with greater consumption of energy and/or reagents, resulting in higher production costs. This can bring both economic and environmental disadvantages, including higher carbon dioxide equivalent (CO_2-eq) emissions [5].

One strategy to overcome this problem is to use lignin as a precursor raw material for the manufacture of new biogenic polymers. Globally, lignin is the second most abundant organic material, after cellulose. Given its presence in plant tissues, lignin is frequently produced in high volumes as an unintended byproduct during industrial processes. The main contributor is the pulp and paper industry, responsible for the generation of around 70 million tons of lignin annually [6]. This volume is expected to increase to around 225 million tons annually by 2030, due to the contribution from second-generation lignocellulosic biorefineries, currently expanding due to the increased demand for biofuels and other bioproducts [7]. Although a fraction of the lignin generated in the paper and cellulose industry, in the form of black liquor, is used in energy cogeneration systems that also enable the recovery of chemical reagents for the process, 95% of the industrial lignin produced is still not fully utilized [6]. This industrial residue with varied chemical composition, but rich in lignin (also known as technical lignin), has desirable physicochemical characteristics for the synthesis of new adsorbent materials, including the presence of oxygenated functional groups and a high carbon content. These are attractive features for high-performance activated carbons used in technological and environmental applications [5].

Recent studies have described the use of lignin in the development of new biopolymers showing potential for use in environmental applications, especially wastewater treatment [8–10]. This chapter presents the main lignin-based biopolymers that have been developed and used as adsorbents in the treatment of water containing different organic and inorganic pollutants. The physical and chemical characteristics of these biopolymers are discussed, as well as their performance and the implications of their technology readiness levels (TRLs) concerning the possible transfer of the technologies to the market. The perspectives for future research are also considered.

5.2 Lignin Structure and Characteristics

Lignin is the second most abundant biopolymer on Earth, with an annual photosynthetic production of around 200×10^8 tons [11]. Lignocellulosic biomasses typically contain between 10% and 30% of lignin in their compositions [12], representing a potential renewable source of aromatic compounds. However, lignin is not found in its isolated form in nature, while its rigid structure and the presence of different chemical bonds, including the formation of carbohydrate-lignin complexes, have hindered its exploitation for practical applications [13]. Hence, research is needed to obtain a better understanding of the structure of lignin, to develop effective techniques to isolate it and reduce its molar mass, and to obtain high value-added products from this macromolecule and its basic units, in addition to identifying relevant applications for the resulting materials [14].

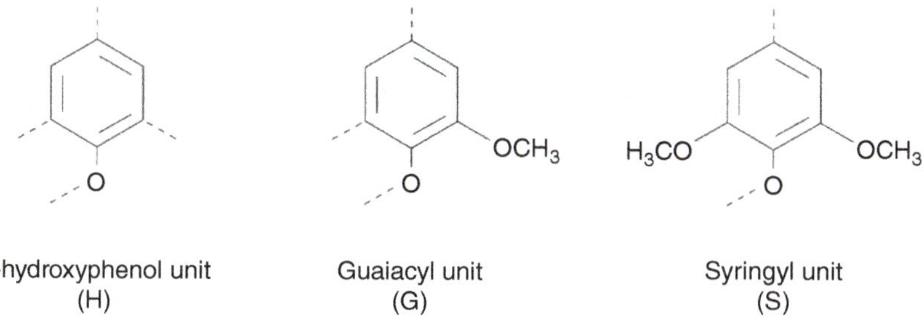

Figure 5.1 H, G, and S subunits of lignin.

Lignin is a three-dimensional macromolecule with high molar mass, whose basic structure is formed by *p*-coumaryl, coniferyl, and sinapyl monolignol alcohols, which originate *p*-hydroxyphenyl (H), guaiacyl (G), and syringyl (S) phenylpropanoid subunits (Figure 5.1) [15]. Functional groups linked to these subunits include phenolic and aliphatic hydroxyls, as well as methoxy and carbonyl groups (of ketones, aldehydes, and carboxylic acids) [11].

According to the proportions of G, S, and H subunits present, lignin may be classified into three types: softwood (G-type lignin), hardwood (GS-type lignin), or herbaceous (GSH-type lignin) [16]. Softwood lignin typically contains 0–5% H units and 95–100% G units, while hardwood lignin contains 0–10% H units, 25–50% G units, and 45–74% S units [17]. Herbaceous lignin typically presents 5–35% H units, 35–80% G units, and 20–55% S units [17]. Additionally, herbaceous lignin generally has a lower degree of "polymerization," while softwood and hardwood lignins have a higher degree of "polymerization" [18–20]. Among other factors, this is partly related to the quantity of free positions in the aromatic ring available for polymerization reactions, together with the ratio among the G, S, and H subunits. Lignins with a greater quantity of G subunits tend to present more crosslinks, since the G subunits can form crosslinking bonds with up to three other subunits, while the S subunits can only form two crosslinking bonds with other subunits [21]. The GSH type lignin (G: 35–80%; S: 20–55%; H: 5–35%) presents more free positions on the aromatic ring. On the other hand, although lignin types G (G: 95–100%) and GS (G: 25–50%; S: 45–74%) possess fewer free positions on the aromatic ring, they are more reactive and consequently present higher degrees of "polymerization," compared to GSH-type lignin. The different proportions among the lignin subunits mean that the lignins differ substantially in terms of reactivity, molar mass, and quantities of specific functional groups, consequently leading to different potential applications. For example, the greater quantity of free positions in the aromatic ring of GSH-type lignin, when accompanied by a smaller number of carbon–carbon bonds and a greater number of α- and β—O—4 bonds, can make it an attractive raw material for producing fine chemicals by catalytic fractionation of the lignin into low molar mass lignols, followed by chemical modification and/or isolation [5]. In contrast, the greater degree of "polymerization" of G and GS lignins, with a large quantity of the lignin units cross-linked by the original functional groups, accompanied by a greater number of carbon–carbon bonds, can make these types of lignin more suitable as raw materials for obtaining products such as activated carbons [17].

5.3 Lignin-Based Materials for Removal of Pollutants from Wastewater

In recent years, there have been many studies concerning the development of lignin-based materials for wastewater treatment, due to the low cost of lignin, its high availability, and the diverse functional groups present in its composition (which is a highly desirable characteristic for a purification process such as adsorption). However, crude lignin typically has a low specific surface area (0.5–21 $m^2\ g^{-1}$), which is normally insufficient for use in adsorption [22]. Therefore, it is necessary to develop new lignin-based materials that are highly porous and have high availability of functional groups on their surfaces, to obtain high-performance adsorbents for practical applications. In this way, it is possible to transform lignin, a renewable and widely available raw material, into higher value-added materials, promoting a circular economy in the sectors involved and potentially reducing CO_2-eq emissions. These new materials can be employed as adsorbents to improve wastewater purification, effectively removing many organic and inorganic pollutants. Some examples of lignin-based materials commonly used in wastewater treatment processes are discussed below.

5.3.1 Lignin-Based Hydrogels

Hydrogels are hydrophilic polymers formed by three-dimensional bonds (Figure 5.2) that may be chemical or physical, depending on the interactions in the material structure. For example, hydrogels with chemical bonds have a structure interconnected by covalent bonding, while the three-dimensional structure of hydrogels with physical bonds is maintained by hydrogen bonds, electrostatic interactions, and van der Waals forces [23]. Hydrogels can absorb water and liquid solvents in quantities greater than thousands of times their own weight, making them promising materials for application in adsorption processes. The use of lignin, a natural aromatic macromolecule, to produce hydrogels offers the advantages of biocompatibility and biodegradability [24], as well as the presence of diverse surface functional groups, which are highly relevant characteristics from both chemical and environmental perspectives, for applications in water treatment systems.

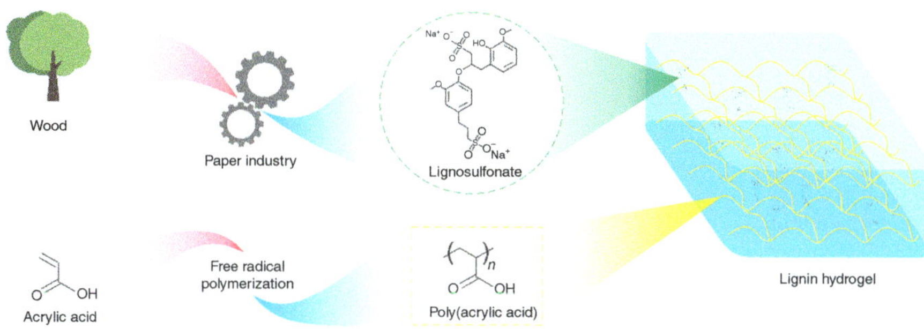

Figure 5.2 Illustration of a lignin-based hydrogel.

Kraft, lignosulfonate, and organosolv lignins are the types that have so far been most widely used to obtain hydrogels, due to the large volumes of these materials produced in pulping processes [24]. However, high volumes of isolated lignin can also be produced in the pretreatment processes used to disrupt lignocellulosic biomass and obtain the carbohydrate fraction, as recently implemented in second-generation biorefineries for manufacturing biofuels and bioproducts [5]. These lignins, with different degrees of purity and composition, can be used to produce hydrogels.

The commonest methods used to synthesize lignin-based hydrogels involve interlinking by copolymerization and graft copolymerization, although methods such as interpenetrating polymer network, atom transfer radical polymerization (ATRP), reversible addition-fragmentation chain transfer (RAFT), hydrothermal, and wet spinning techniques are also used. These methods result in hydrogels with different structural characteristics, as summarized in Table 5.1.

Lignin-based hydrogels have been investigated for the removal of different organic and inorganic species from wastewater, due to their surface characteristics including chemical stability, porous structure, high specific surface area, and the presence of oxygenated functional groups. Wang et al. [25] synthesized a hydrogel denoted CKLA (cationic kraft lignin acrylate) using cationic kraft lignin acrylate, acrylamide, and *N*,*N*′-methylenebisacrylamide. The presence of hydroxyl groups and 3.15% nitrogen in the composition of this hydrogel resulted in high adsorption capacity ($125.63\,mg\,g^{-1}$) for Congo red dye. Yu et al. [26] obtained a hydrogel by inserting acrylic acid (AA) into the lignosulfonate (LS) structure, in the presence of *N*,*N*′-methylenebisacrylamide, initiated by laccase/*tert*-butyl hydroperoxide. The hydrogel, denoted LS-g-AA (lignosulfonate-g-acrylic acid), presented a strong presence of surface hydroxyl and carboxylic acid groups and had a high adsorption capacity ($2013\,mg\,g^{-1}$) for methylene blue dye.

The production of composites based on lignin and other fine materials for the purpose of adsorption has also attracted attention. Li et al. [27] produced a

Table 5.1 Main preparation methods and characteristics of lignin-based hydrogels[a].

Synthesis method	Structural characteristics
Crosslinking polymerization	Highly porous structure and heterogeneous surface
Graft copolymerization	High absorption capacity and smooth surface
Semi-interpenetrating	High swelling rate
Atom transfer radical polymerization (ATRP)	Well-aligned interlinking
Reversible addition-fragmentation chain transfer (RAFT)	Well-aligned interlinking
Hydrothermal	Wide pores
Wet-spinning	Larger fiber diameters
Oxidation	Complex porous structure

a) Adapted from Liu et al. [6].

hydrogel by a hydrothermal process using LS impregnated with graphene oxide. The hydrogel, denoted LS-GH (lignosulfonate-modified graphene), contained sulfonic acid groups and presented a high adsorption capacity (1210 $mg\,g^{-1}$) for Pb(II). This material had a high specific surface area (459 $m^2\,g^{-1}$), a porous structure, and oxygenated functional groups of lignin and graphene oxide, which provided active sites for adsorption of Pb(II). Elsewhere, Liu et al. [28] used the intermatrix synthesis (IMS) method to obtain a composite hydrogel denoted nZVI@LH (hydrogel-supported nanoscale zero-valent iron composite), with incorporation of Fe(II) and subsequent reduction of Fe^{2+} to Fe^0 by the addition of $NaBH_4$. The nZVI@LH had S and Fe^0 contents of 1.05% and 7.82%, respectively, and showed a good adsorption capacity for Cr(VI) (370.37 mg Cr^{6+} g^{-1} Fe^0). The authors found that the efficiency of the Cr(VI) adsorption process was significantly influenced by pH and the amount of nano zero-valent iron (nZVI). The adsorption capacity of nZVI@LH for Cr(VI) was substantially higher (67.65%) under more acidic conditions, compared to more alkaline conditions (~18%), indicating that the adsorption of Cr(VI) was favored in an acidic medium. Considering the effect of the amount of nZVI incorporated into the composite, it was found that better adsorption capacity was obtained using $FeSO_4$ concentrations up to 0.1 mol L^{-1}, compared to nZVI@LH obtained with higher concentrations of $FeSO_4$. This was suggested to be due to the aggregation of nZVI particles present in larger amounts, which was reflected in a reduced adsorption capacity of the hydrogel.

Figure 5.3 Illustration of a lignin-based aerogel produced by freeze-drying of a hydrogel.

Lignin-based hydrogels typically have a complex porous structure with micro-, meso-, and macropores, where the three-dimensional biopolymer rich in oxygenated functional groups is dispersed throughout the gel, providing active sites for the adsorption of various organic and inorganic substances. The heterogeneous nature of this material commonly results in multilayer adsorption processes [26, 27, 29].

5.3.2 Lignin-Based Aerogels

Aerogels are three-dimensional materials obtained by the freeze-drying of hydrogels (Figure 5.3), usually presenting low density, high porosity, and high specific surface area [6, 30]. These characteristics, combined with the physicochemical properties of lignin, make aerogels highly suitable for use in wastewater treatment. For example, Elsayed et al. [31] obtained an aerogel by the chemical interlinking of amino-functionalized cellulose nanofibers (Amf-CNF) and LS. This material, denoted Amf-CNF/LS (amino-functionalized cellulose nanofibers and lignosulfonate), presented surface phenolic hydroxyl and sulfonic acid functional groups and was shown to be highly versatile, with high adsorption capacities for the dyes methylene blue (170.94 $mg\,g^{-1}$) and rhodamine B (147.28 $mg\,g^{-1}$), as well as for Cd(II) (129.87 $mg\,g^{-1}$), under optimal adsorption conditions.

In addition to the removal of dyes and toxic metals, lignin-based aerogels have been shown to be effective in treatments for other contaminants. Zou et al. [32] produced an aerogel by chemical interlinking of a gel composed of lignocellulose nanofibrils with epichlorohydrin and tetrabutyl titanate ($Ti(OC_4H_9)_4$). This material, denoted C-SLCNF-TNP (sulfated lignocellulose nanofibrils with TiO_2 nanoparticles), was used for the removal of tetracycline by simultaneous adsorption and photocatalysis processes. The C-SLCNF-TNP presented surface sulfonic acid groups and had a maximum adsorption capacity of 70 $mg\,g^{-1}$, providing approximately 90% removal of the antibiotic in 40 minutes. Wang et al. [29]

reported the high adsorption capacity of a magnetic composite aerogel, obtained using a single-step hydrothermal process applied to a mixture of graphene oxide, ethylenediamine, Fe_3O_4, and polyvinyl alcohol. This aerogel, denoted FLGA (magnetic lignin/graphene aerogel), presented excellent adsorption capacities for sunflower oil (140 $g\,g^{-1}$), benzene (65 $g\,g^{-1}$), *n*-hexane (70 $g\,g^{-1}$), and carbon tetrachloride (115 $g\,g^{-1}$), reaching 135.2 times the weight of the aerogel itself. Although still at the laboratory bench level, the studies described above demonstrate the potential of composite aerogels for use in recovering relevant chemical compounds of economic and environmental interest.

Several recent studies have focused on the production of porous aerogels that can provide high performance. For this, conventional aerogel production, involving the formation of composites by interlinking with the lignin structure, is followed by a carbonization step in the presence of activation reagents [33]. The use of this type of material in water treatment was reported by Lv et al. [34], who synthesized an activated carbon (NaLS-kACA) from a sodium lignin sulfonate and κ-carrageenan aerogel (NaLS-kA), which was used to remove methylene blue from water. In comparison to NaLS-kA, NaLS-kACA presented a better-developed porous structure and higher specific surface area (595 $m^2\,g^{-1}$), which was reflected in an excellent dye adsorption capacity of 421.94 $mg\,g^{-1}$. The adsorption isotherm for methylene blue on NaLS-kACA followed the Freundlich model, indicating the heterogeneity of the surface of the material. The adsorption process was suggested to be governed by π–π electron donor/acceptor interactions, electrostatic forces, and hydrogen bonding.

A trend observed in the literature has been toward the synthesis of lignin biopolymers interlinked with graphene oxide. For example, Yan et al. [3] proposed an aerogel obtained from the interlinking of graphene oxide, chitosan, and LS (GLCA), with a specific surface area of 75 $m^2\,g^{-1}$ and the presence of surface sulfonic acid and amine groups, resulting in a very high adsorption capacity for methylene blue (1023.9 $mg\,g^{-1}$). The adsorption of the dye on GLCA was strongly pH-dependent, reaching 99% removal efficiency at pH above 4, suggesting that the adsorption process was governed by hydrogen bonding and π–π type interactions. However, in this case, the Langmuir model provided the best fit to the adsorption isotherm, indicating that adsorption occurred homogeneously in the form of a monolayer, probably due to the low porosity of the GLCA structure. Wu et al. [33] proposed a composite aerogel composed of lignin alkali and graphene oxide (GO-AL), which showed an excellent maximum adsorption capacity for methylene blue of 1185.98 $mg\,g^{-1}$. High performance was maintained even after five adsorption/desorption cycles (1083.84 $mg\,g^{-1}$), with the loss of only 8.6% in adsorption capacity indicating the high reuse potential of GO-AL.

Despite the favorable structural properties of aerogels for adsorption processes, the hydrogel freeze-drying step required to produce the aerogel involves high energy consumption, while long times may be required for complete drying of the aerogel [6], which could be economically disadvantageous and hinder practical application on an industrial scale.

5.3.3 Lignin-Based Porous Carbon Materials

Activated carbons (ACs) are extensively used in water treatment systems, where the removal of pollutants is favored by their surface and textural properties including complex porous structure, high specific surface area, and presence of oxygenated functional groups. Since lignin has a wide variety of oxygenated functional groups, is abundant in nature, and has a high carbon content (>60%), it can be considered an ideal precursor raw material for producing activated carbons [5]. Various methodologies can be used to synthesize activated carbons, with pyrolysis being the commonest carbonization method. Other processes can be applied simultaneously with pyrolysis, in a subsequent step (such as physical or chemical activation) (Figure 5.4) or before pyrolysis (such as a hydrothermal process, among others), to obtain activated carbons with different textural properties [6]. The hydrothermal process results in the formation of a hydrochar with a greater quantity of functional groups and higher carbon content, but with low specific surface area. Therefore, a second activation step may be applied to obtain an activated carbon with improved textural properties [35].

Pyrolysis is usually the basic technique employed to produce AC, involving carbonization of the precursor material at high temperatures in an inert atmosphere. The activation can occur simultaneously during the initial carbonization, or be performed in a subsequent step. The purpose of activation, which may be physical or chemical, is to develop a structure that is porous and has an increased specific surface area [36]. For example, Miliotti et al. [37] produced a nonactivated biochar, as well as chemically activated carbons, from enzymatic hydrolysis

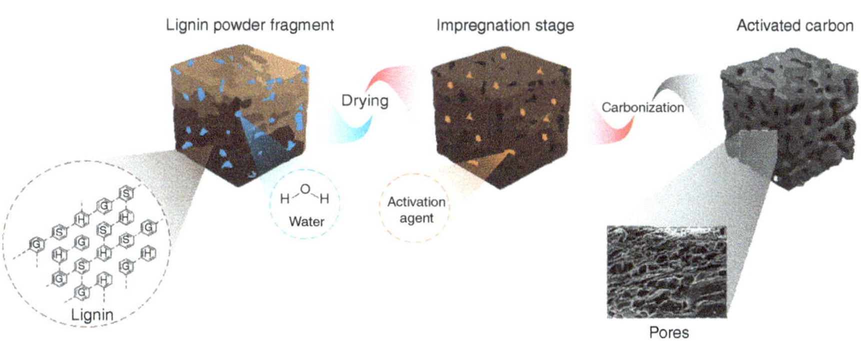

Figure 5.4 Illustration of the stages of the lignin carbonization process to obtain chemically activated carbons.

lignin (EHL), where the nonactivated material had much lower specific surface area (370 m^2 g^{-1}) and total pore volume (0.19 cm^3 g^{-1}), compared to the activated carbon (1484 m^2 g^{-1} and 1.39 cm^3 g^{-1}), demonstrating the essential role of the activation step in obtaining materials with superior textural characteristics. However, the authors did not evaluate the potentials of the developed activated carbons for the treatment of contaminated water.

Some of the lignin-based materials, including ACs, that have been reported in the literature for wastewater treatment are listed in Table 5.2. Examples of high-performing lignin-based ACs used for water treatment were described by Kamwilaisak et al. [41] and Li et al. [8]. Kamwilaisak et al. [41] produced an AC from hydrolyzed black liquor lignin derived from an acid hydrolysis process with H_2SO_4, for use in Cd(II) adsorption. The material (ACHL-66-1) presented high specific surface area (1646 m^2 g^{-1}) and total pore volume (0.8595 cm^3 g^{-1}), with maximum Cd(II) adsorption capacity of 9.98 mg g^{-1}, where the adsorption occurred uniformly over the adsorbent surface. Li et al. [8] synthesized ACs from corn stover lignin (CSLAC), under different temperature conditions, for the removal of phenol. It was found that increasing the carbonization temperature, up to 800 °C, led to increases of the specific surface area and pore volume. However, when the temperature was increased to 900 °C, the specific surface area decreased from 1679 to 695 m^2 g^{-1}, the pore volume decreased from 0.369 to 0.128 cm^3 g^{-1}, and the AC yield decreased from 26.8% to 3.3%. The CSLAC obtained at 800 °C (CSLAC-800) was used for the removal of phenol, showing a maximum adsorption capacity of 612.3 mg g^{-1} and attainment of equilibrium within 60 minutes.

Although ACs are usually obtained by carbonization processes, new materials can be produced by combining carbonization with other emerging techniques. Example of such hybrid methodologies are aerogel carbons and materials obtained using templates, which have been investigated in electrochemistry to produce supercapacitors [48–50]. However, initial studies have also indicated the potential of these materials for the removal of contaminants from water, as in the case of the NaLS-kA material developed by Lv et al. [34], which was used to remove methylene blue dye. Meng et al. [23] produced an advanced material denoted LCAGO (lignin-based carbon aerogel enhanced by graphene oxide), capable of separating oil from water, by pyrolysis of an aerogel interlinked with aminated lignin, graphene oxide, *N*,*N*′-methylenebisacrylamide, and acrylic acid. LCAGO presented adsorption capacities of 32 and 34 g g^{-1} for gasoline and chloroform, respectively. Xie et al. [47] produced an innovative porous carbon composed of lignin nanosheets, denoted LPCNS (lignin-based porous carbon nanosheets), using NaCl as a template and activation with KOH, which was used for the adsorption of tetracycline. LPCNS showed superior surface and textural properties, compared to a conventional lignin-based carbon activated

Table 5.2 Literature reports of adsorbents synthesized from lignin for wastewater treatment.

Material	Specific surface area ($m^2\ g^{-1}$)[a]	Purpose	Species of interest	Adsorption capacity ($mg\ g^{-1}$)	Isotherm model	Reference
			Hydrogel			
CKLA	—	Adsorption	Congo red	125.63	—	[25]
LS-g-AA	—	Adsorption	Methylene blue	2013	Freundlich	[26]
LS-GH	459	Adsorption	Pb(II)	1210	Langmuir	[27]
nZVI@LH	—	Adsorption	Cr(VI)	310.86	—	[28]
AN-L/M	—	Adsorption	Methylene blue	9646.92	Langmuir and Freundlich	[38]
SL-g-PAA	—	Adsorption	Co(II)	130	Langmuir	[39]
			Cu(II)	115	Langmuir	
			Ni(II)	140	Langmuir	
			Cd(II)	255	Langmuir	
			Pb(II)	340	Langmuir	
			Aerogel			
FLGA	65	Adsorption	Sunflower seed oil	140,000	—	[29]
			Benzene	65,000	—	
			n-Hexane	70,000	—	
			Carbon tetrachloride	115,000	—	
C-SLCNF-TNPs	95	Adsorption	Tetracycline	70	—	[32]
LN/MMT @cellulose	1	Adsorption	Ciprofloxacin	140.73	Langmuir	[40]

(Continued)

Table 5.2 (Continued)

Material	Specific surface area ($m^2\ g^{-1}$)[a]	Purpose	Species of interest	Adsorption capacity ($mg\ g^{-1}$)	Isotherm model	Reference
Amf-CNF/LS	12	Adsorption	Methylene blue	170.94	Langmuir	[31]
			Rhodamine B	147.28	Langmuir	
			Cd(II)	129.87	Langmuir	
GLCA	75	Adsorption	Methylene blue	1023.9	Langmuir	[3]
GO-AL	41	Adsorption	Methylene blue	1185.98	Langmuir	[33]
Porous carbon						
Alkali kraft AC	2031	Adsorption	Methylene blue	850.00	—	[35]
Organosolv AC	2538			870.00	—	
ACHL-66-1	1646	Adsorption	Cd(II)	9.98	Langmuir	[41]
CSLAC-800	1679	Adsorption	Phenol	612.3	Freundlich	[8]
NaLS-kA	595	Adsorption	Methylene blue	421.94	Freundlich	[34]
LES 1:2	886	Adsorption	Methyl orange	262	Langmuir	[42]
L-HPC	749	Adsorption	Cr(VI)	887.8	Freundlich	[43]
SLCs	296	Adsorption	Cd(II)	845.84	Freundlich	[44]
ILAC-N	1411	Adsorption	Tetracycline	1396	Redlich-Peterson	[45]
LCAGO	—	Adsorption	Petrol	32000	—	[23]
LCAGO	—		Chloroform	34000	—	
DKLAAC	1305	Adsorption	Phenol	223	Langmuir	[46]
			2-Nitrophenol	333	Langmuir	
			4-Nitrophenol	319	Langmuir	
LPCNS	3505	Adsorption	Tetracycline	1613	Langmuir	[47]
LCA	2554			1139	Langmuir	

a) The specific surface area values in the table are given as approximate whole numbers, without decimal points.

by KOH (LCA). LPCNS exhibited high specific surface area (3505 m^2 g^{-1}) and pore volume (2.0 cm^3 g^{-1}), resulting in efficient tetracycline removal (95.47%) and an adsorption capacity of 1613 mg g^{-1}. The specific surface area (2554 mg g^{-1}) and pore volume (2.02 cm^3 g^{-1}) of LCA could be considered satisfactory for an AC, but its adsorption capacity (1139 mg g^{-1}) and drug removal performance (75.92%) were lower than found for LPCNS, indicating the benefit of using the NaCl template for obtaining a high-performance advanced adsorbent for the removal of tetracycline.

A variety of different possible methodologies can be used to obtain adsorbent materials with high adsorption capacities for target pollutants. However, a drawback of including additional synthesis processes, such as those employing templates to obtain aerogels, is the increased consumption of energy and reagents. This can be limiting from both economic and technological maturity perspectives, considering the challenges of scaling up technologies for applications on commercial and industrial scales. For this reason, in addition to their attractive characteristics, ACs obtained by pyrolytic conversion are still the adsorbent materials of greatest commercial interest.

5.4 Technology Readiness Level and Techno-Economic Implications of Producing Activated Carbon from Lignin

As discussed in the previous sections, the physicochemical characteristics of lignin make this material suitable for obtaining ACs with excellent performance in the removal of organic and inorganic pollutants from water. This potential could be key to adding value to lignin by its use as a raw material to obtain biopolymers for water treatment, while also contributing to a circular economy. However, the economic success of lignin conversion depends on the technology used and the value of the final product, considering the yield of its conversion into AC, the waste generated during the production and use of lignin-based AC, and the associated energy consumption [5]. It is also necessary to consider environmental aspects of the developed technology, ensuring that both the process and the product have the lowest possible environmental impacts. For this purpose, it is recommended to perform techno-economic analysis (TEA) and life cycle assessment (LCA) of the processes and products.

For both economic and environmental reasons, new lignin-based materials with higher added value should ideally be obtained using mature technologies that already have a TRL at the commercial scale. To this end, the production of ACs from lignocellulosic biomasses is already well established using rotating tubular furnaces and fluidized beds in industrial plants [51]. According to Zhu et al. [52], the use of fluidized bed reactors to produce activated carbons from lignocellulosic

Table 5.3 Techno-economic and environmental aspects of some lignin-based materials[a].

Material	Emissions (kg CO_2 per kg of product)	Possible gross profit (US$ kg^{-1})
AC from lignin	12	0.38
Carbon fiber from lignin	20	9.76
PU foam from lignin	5	0.12
Lignin hydrogel	2.8	−0.72 to 0.14

a) Adapted from Gomes et al. [5].

wastes can reduce greenhouse gas (GHG) emissions by up to −1839 t CO_2-eq. In other work, Li et al. [53] performed LCA of activated carbon production from lignocellulosic materials, showing that among the raw materials evaluated, lignin had the lowest global warming impact (2.11×10^2 kg CO_2-eq). These studies suggested that the production of AC in a single step had a lower environmental impact and was more appropriate. However, the global warming impact values can vary depending on the scope of the analysis, the location, the type of waste, and the data available to perform the LCA.

Many technologies are emerging to add value to lignin, where the costs associated with the production of new products may be higher than those of conventional routes. In such cases, it may be necessary to consider the potential market size and volatility of the product price, prior to scale-up of these emerging technologies [7]. Table 5.3 lists some lignin-based materials, together with the corresponding CO_2 emissions and financial data. Evidently, materials such as carbon fibers (21.78 US$ kg^{-1}) or polyurethane (PU) foam (8.38 US$ kg^{-1}) have higher market values than AC (1.75–13.6 US$ kg^{-1}). However, considering GHG emissions, the production of AC (12 kg CO_2 kg^{-1}) has a much lower environmental impact, compared to carbon fibers (20–27 kg CO_2 kg^{-1}). Although hydrogel has the lowest environmental impact, it provides a low financial return (−0.72–0.14 US$ kg^{-1}) and has major technical and economic limitations for applications on an industrial scale. Despite presenting excellent performance at the laboratory scale, both hydrogels and aerogels have very low TRL, indicating the existence of technical and economic disadvantages of these materials for practical applications [54].

Since this is a relatively new and expanding area of knowledge, there is a lack of studies concerning LCA and TEA of advanced materials obtained from lignin. Furthermore, while many investigations have focused on the production of high-performance materials for electrochemical applications, the area of water treatment has been largely ignored. However, it is notable that many of the lignin-based materials reported for electrochemical applications in the literature have favorable textural and surface properties that could be exploited in applications for water treatment.

5.5 Research Opportunities and Future Perspectives

Bibliometric analysis is an important way of identifying hotspots and trends in research reported in the literature. Use of the search string "((lignin OR lignin-based OR lign*) AND (hydrogel OR aerogel OR *carbon) AND (adsor*) AND (water* OR wastewater OR *removal))" in the Web of Science database (Clarivate®, London, UK) resulted in the identification of 46 articles published between 1998 and 2024. Figure 5.5 shows a cloud of the keywords with the highest occurrence among these studies. High density around the terms "adsorption" and "activated carbon" indicates that activated carbon was the most widely studied lignin-based material. The temporal evolution of the keywords showed that until 2019, the main focus was on the chemical modification of ACs (clusters in blue), while currently there is a greater focus on the production of composite materials (clusters in yellow), such as hydrogels and nanocomposites, and their application for the removal of organic (methylene blue, rhodamine B, and phenol) and inorganic (Pb(II), Cr(VI), and cadmium) species from contaminated water.

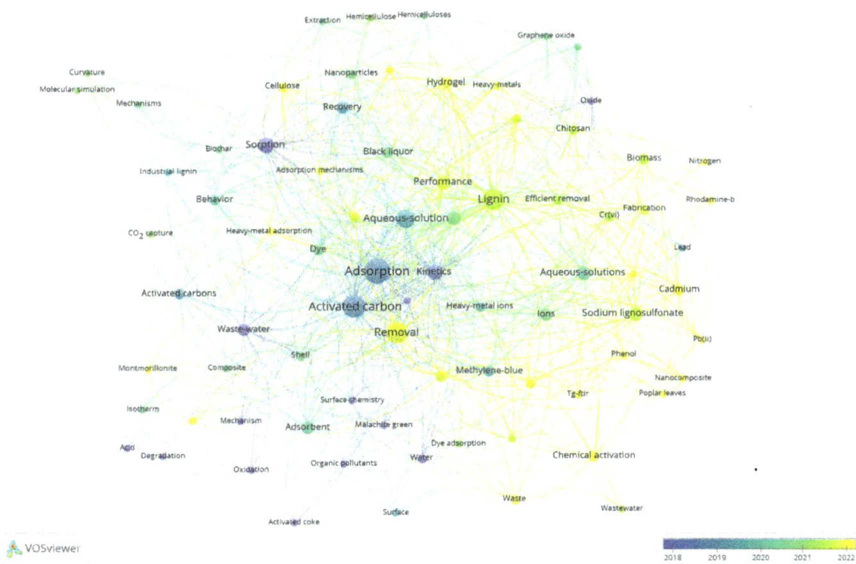

Figure 5.5 Keywords cloud from publications retrieved from the Web of Science database for the period 1998–2024.

Considering that the studies found in the literature are relatively recent, with most being in the initial stages, it is evident that there are excellent opportunities for further research, examples being: (i) use of materials, including composites, based on lignin and other compounds for the adsorption of various contaminants, especially emerging pollutants; (ii) application of the developed materials in continuous wastewater treatment systems; and (iii) scale-up of the technologies to pilot and industrial (commercial) scales.

It is also important that the production of lignin-based materials for application in wastewater treatment should consider the market dynamics of supply and demand, so that the technologies are viable, both technically and economically, and competitive with existing commercial products [55]. The production of advanced materials with high adsorption performance must be accompanied by application of TEA and LCA, to further ensure the economic and environmental viability of transferring the technology from the bench scale (TRL 1–5) to the market (TRL 8 and 9), avoiding the so-called "valley of death."

Several relevant opportunities for research work in this area are as follows:

1) Investigation of advanced lignin-based materials with outstanding textural and surface properties for applications in electrochemistry and adsorption.
2) Analysis of the potentials of new advanced lignin-based materials for the removal of other organic, inorganic, and emerging contaminants in single- and multi-component systems, using both batch and continuous modes, evaluating aspects including the versatility, selectivity, and reusability of the material.
3) Application of new lignin-based biopolymers in pilot-scale continuous treatment systems to validate LCA and TEA data.
4) Development of new environmentally and economically viable materials, by applying insights from analyses such as TEA and LCA.
5) Scale-up of production of these materials to prototype and commercial levels.

References

1 Baêta, B.E.L., Lima, D.R.S., Silva, S.Q., and Aquino, S.F. (2015). Evaluation of soluble microbial products and aromatic amines accumulation during a combined anaerobic/aerobic treatment of a model azo dye. *Chemical Engineering Journal* 259: 936–944. https://doi.org/10.1016/j.cej.2014.08.050.

2 Nascimento, B.F., Silva, L.F.O., Araujo, C.M.B. et al. (2022). Synthesis and application of ferromagnetic graphene oxide nanocomposite as an effective adsorbent for Clonazepam: batch experiments, modeling, regeneration, and phytotoxicity. *Journal of Environmental Chemical Engineering* 10 (5): 108331. https://doi.org/10.1016/j.jece.2022.108331.

3 Yan, M.F., Huang, W.X., and Li, Z.L. (2019). Chitosan cross-linked graphene oxide/lignosulfonate composite aerogel for enhanced adsorption of methylene blue in water. *International Journal of Biological Macromolecules* 136: 927–935. https://doi.org/10.1016/j.ijbiomac.2019.06.144.

4 Gawel, A., Sühnholz, S., Georgi, A. et al. (2023). Fe-zeolites for the adsorption and oxidative degradation of nitroaromatic compounds in water. *Journal of Hazardous Materials* 459 (11): 132125. https://doi.org/10.1016/j.jhazmat.2023.132125.

5 Gomes, B.F.M.L., Vaz Júnior, S., and Gurgel, L.V.A. (2024). Production of activated carbons from technical lignin as a promising pathway towards carbon emission neutrality for second-generation (2G) ethanol plants. *Journal of Cleaner Production* 450: 141648. https://doi.org/10.1016/j.jclepro.2024.141648.

6 Liu, Y.L., Jin, C., Yang, Z.Z. et al. (2021). Recent advances in lignin-based porous materials for pollutants removal from wastewater. *International Journal of Biological Macromolecules* 187: 880–891. https://doi.org/10.1016/j.ijbiomac.2021.07.152.

7 Bajwa, D.S., Pourhashem, G., Ullah, A.H., and Bajwa, S.G. (2019). A concise review of current lignin production, applications, products and their environmental impact. *Industrial Crops and Products* 139 (11): 111526. https://doi.org/10.1016/j.indcrop.2019.111526.

8 Li, M.P., Mu, J.C., Liu, Y.X. et al. (2023). Removal of phenol by lignin-based activated carbon as an efficient adsorbent for adsorption of phenolic wastewater. *Research on Chemical Intermediates* 49 (5): 2209–2232. https://doi.org/10.1007/s11164-023-04958-z.

9 Chen, W., Xie, H.J., Jiang, N. et al. (2023). Synthesis of magnetic sodium lignosulfonate hydrogel(Fe3O4@LS) and its adsorption behavior for Cd2+in wastewater. *International Journal of Biological Macromolecules* 245 (17): 125498. https://doi.org/10.1016/j.ijbiomac.2023.125498.

10 Gul, E., Alrawashdeh, K.A., Masek, O. et al. (2021). Production and use of biochar from lignin and lignin-rich residues (such as digestate and olive stones) for wastewater treatment. *Journal of Analytical and Applied Pyrolysis* 158 (17): 105263. https://doi.org/10.1016/j.jaap.2021.105263.

11 Poveda-Giraldo, J.A., Solarte-Toro, J.C., and Alzate, C.A.C. (2021). The potential use of lignin as a platform product in biorefineries: a review. *Renewable & Sustainable Energy Reviews* 138: 110688. https://doi.org/10.1016/j.rser.2020.110688.

12 Carrott, P.J. and Carrott, M.R. (2007). Lignin—from natural adsorbent to activated carbon: a review. *Bioresource Technology* 98 (12): 2301–2312. https://doi.org/10.1016/j.biortech.2006.08.008.

13 Sivagurunathan, P., Raj, T., Mohanta, C.S. et al. (2021). 2G waste lignin to fuel and high value-added chemicals: approaches, challenges and future outlook for

sustainable development. *Chemosphere* 268: 129326. https://doi.org/10.1016/j.chemosphere.2020.129326.

14 Lobato-Peralta, D.R., Duque-Brito, E., Villafan-Vidales, H.I. et al. (2021). A review on trends in lignin extraction and valorization of lignocellulosic biomass for energy applications. *Journal of Cleaner Production* 293: 126123. https://doi.org/10.1016/j.jclepro.2021.126123.

15 Sheng, Y.Q., Lam, S.S., Wu, Y.J. et al. (2021). Enzymatic conversion of pretreated lignocellulosic biomass: a review on influence of structural changes of lignin. *Bioresource Technology* 324: 124631. https://doi.org/10.1016/j.biortech.2020.124631.

16 Wang, T.P., Li, H., Yuan, J.M. et al. (2021). Structures and pyrolytic characteristics of organosolv lignins from typical softwood, hardwood and herbaceous biomass. *Industrial Crops and Products* 171: 113912. https://doi.org/10.1016/j.indcrop.2021.113912.

17 Patel, R., Dhar, P., Babaei-Ghazvini, A. et al. (2023). Transforming lignin into renewable fuels, chemicals, and materials: a review. *Bioresource Technology Reports* 22: 101463. https://doi.org/10.1016/j.biteb.2023.101463.

18 Arruda, M.D.M., Alves, S.D.L., da Cruz, I.J. et al. (2021). Characterization of a lignin from *Crataeva tapia* leaves and potential applications in medicinal and cosmetic formulations. *International Journal of Biological Macromolecules* 180: 286–298. https://doi.org/10.1016/j.ijbiomac.2021.03.077.

19 Mottiar, Y., Vanholme, R., Boerjan, W. et al. (2016). Designer lignins: harnessing the plasticity of lignification. *Current Opinion in Biotechnology* 37: 190–200. https://doi.org/10.1016/j.copbio.2015.10.009.

20 Qin, Z., Yang, Q.L., Cheng, X.C. et al. (2021). Structural features, chemical composition, antioxidant activities of organosolv lignins extracted from black and white sesame capsules and stalks. *Industrial Crops and Products* 169: 113677. https://doi.org/10.1016/j.indcrop.2021.113677.

21 Kačík, F., Ďurkovič, J., and Kačíková, D. (2014). Structural characterization of lignin by syringyl to guaiacyl ratio and molecular mass determination. In: *Lignin: Structural Analysis, Applications in Biomaterials and Ecological Significance* (ed. F. Lu), 67–89. *Hauppauge, NY*: *Nova Science Publishers.*

22 Liu, D., Gu, W.Y., Zhou, L. et al. (2023). From biochar to functions: lignin induced formation of Fe_3C in carbon/Fe composites for efficient adsorption of tetracycline from wastewater. *Separation and Purification Technology* 304: 10. https://doi.org/10.1016/j.seppur.2022.122217.

23 Meng, Y., Liu, T.L., Yu, S.S. et al. (2020). A lignin-based carbon aerogel enhanced by graphene oxide and application in oil/water separation. *Fuel* 278: 118376. https://doi.org/10.1016/j.fuel.2020.118376.

24 Meng, Y., Lu, J., Cheng, Y. et al. (2019). Lignin-based hydrogels: a review of preparation, properties, and application. *International Journal of Biological Macromolecules* 135: 1006–1019. https://doi.org/10.1016/j.ijbiomac.2019.05.198.

25 Wang, C., Feng, X.Z., Li, W.B. et al. (2023). Fabricating cationic lignin hydrogels for dye adsorption. *Journal of Renewable Materials* 11 (4): 1793–1805. https://doi.org/10.32604/jrm.2023.024521.

26 Yu, C.H., Wang, F., Zhang, C.H. et al. (2016). The synthesis and absorption dynamics of a lignin-based hydrogel for remediation of cationic dye-contaminated effluent. *Reactive & Functional Polymers* 106: 137–142. https://doi.org/10.1016/j.reactfunctpolym.2016.07.016.

27 Li, F.F., Wang, X.L., Yuan, T.Q., and Sun, R.C. (2016). A lignosulfonate-modified graphene hydrogel with ultrahigh adsorption capacity for Pb(II) removal. *Journal of Materials Chemistry A* 4 (30): 11888–11896. https://doi.org/10.1039/c6ta03779h.

28 Liu, X.Y., Zhang, S.Y., Zhang, X.Y. et al. (2022). A novel lignin hydrogel supported nZVI for efficient removal of Cr(VI). *Chemosphere* 301: 134781. https://doi.org/10.1016/j.chemosphere.2022.134781.

29 Wang, Q., Wang, S.L., Yao, K., and Li, H.H. (2022). Preparation of magnetic lignin/graphene aerogel and its adsorption properties to oil and organic solvents. *Industrial Crops and Products* 187: 115372. https://doi.org/10.1016/j.indcrop.2022.115372.

30 Joul, P., Ho, T.T., Kallavus, U. et al. (2022). Characterization of organosolv lignins and their application in the preparation of aerogels. *Materials* 15 (8): 2861. https://doi.org/10.3390/ma15082861.

31 Elsayed, I., Schueneman, G.T., El-Giar, E.M., and Hassan, E. (2023). Amino-functionalized cellulose nanofiber/lignosulfonate new aerogel adsorbent for the removal of dyes and heavy metals from wastewater. *Gels* 9 (2): 154. https://doi.org/10.3390/gels9020154.

32 Zou, X.Y., Yao, L.Y., Zhou, S.Q. et al. (2022). Sulfated lignocellulose nanofibril based composite aerogel towards adsorption-photocatalytic removal of tetracycline. *Carbohydrate Polymers* 296: 119970. https://doi.org/10.1016/j.carbpol.2022.119970.

33 Wu, Z.J., Huang, W.X., Shan, X.Y., and Li, Z.L. (2020). Preparation of a porous graphene oxide/alkali lignin aerogel composite and its adsorption properties for methylene blue. *International Journal of Biological Macromolecules* 143: 325–333. https://doi.org/10.1016/j.ijbiomac.2019.12.017.

34 Lv, D., Li, Y., and Wang, L.J. (2020). Carbon aerogels derived from sodium lignin sulfonate embedded in carrageenan skeleton for methylene-blue removal. *International Journal of Biological Macromolecules* 148: 979–987. https://doi.org/10.1016/j.ijbiomac.2020.01.136.

35 Correa, C.R., Stollovsky, M., Hehr, T. et al. (2017). Influence of the carbonization process on activated carbon properties from lignin and lignin-rich biomasses. *ACS Sustainable Chemistry & Engineering* 5 (9): 8222–8233. https://doi.org/10.1021/acssuschemeng.7b01895.

36 González-García, P. (2018). Activated carbon from lignocellulosics precursors: a review of the synthesis methods, characterization techniques and applications. *Renewable & Sustainable Energy Reviews* 82: 1393–1414. https://doi.org/10.1016/j.rser.2017.04.117.

37 Miliotti, E., Rosi, L., Bettucci, L. et al. (2020). Characterization of chemically and physically activated carbons from lignocellulosic ethanol lignin-rich stream via hydrothermal carbonization and slow pyrolysis pretreatment. *Energies* 13 (16): 4101. https://doi.org/10.3390/en13164101.

38 Wang, Y., Xiong, Y., Wang, J.Y., and Zhang, X.D. (2017). Ultrasonic-assisted fabrication of montmorillonite-lignin hybrid hydrogel: highly efficient swelling behaviors and super-sorbent for dye removal from wastewater. *Colloids and Surfaces a-Physicochemical and Engineering Aspects* 520: 903–913. https://doi.org/10.1016/j.colsurfa.2017.02.050.

39 Jiao, G.J., Ma, J.L., Li, Y.C. et al. (2021). Enhanced adsorption activity for phosphate removal by functional lignin-derived carbon-based adsorbent: optimization, performance and evaluation. *Science of the Total Environment* 761: 143217. https://doi.org/10.1016/j.scitotenv.2020.143217.

40 Li, N., Tao, K.K., Xia, W. et al. (2023). A novel cellulose/lignin/montmorillonite ternary hybrid aerogel for efficiently adsorptive removal of antibiotics from water. *Chemical Engineering Journal* 466: 143265. https://doi.org/10.1016/j.cej.2023.143265.

41 Kamwilaisak, K., Siripanee, S., Jutakridsada, P. et al. (2022). Porosity enhancement of activated carbon by hydrolyzed lignin from black liquor. *Clean Technologies and Environmental Policy* 24 (8): 2517–2530. https://doi.org/10.1007/s10098-022-02331-z.

42 Saini, K., Kumari, A., Sahoo, A. et al. (2023). Eggshell-lignin carbon composite for methyl orange removal from wastewater. *Industrial & Engineering Chemistry Research* 62 (38): 15499–15510. https://doi.org/10.1021/acs.iecr.3c02029.

43 Liang, H.X., Ding, W., Zhang, H.W. et al. (2022). A novel lignin-based hierarchical porous carbon for efficient and selective removal of Cr(VI) from wastewater. *International Journal of Biological Macromolecules* 204: 310–320. https://doi.org/10.1016/j.ijbiomac.2022.02.014.

44 Li, X., Shuai, K., Zhang, Y. et al. (2023). Removal of Cd2+ from wastewater to form a three-dimensional fiber network using Si-Mg doped industrial lignin-based carbon materials. *International Journal of Biological Macromolecules* 229: 62–69. https://doi.org/10.1016/j.ijbiomac.2022.12.274.

45 Zhou, H.J., Li, X.Z., Jin, H.T., and She, D. (2022). Mechanism of a double-channel nitrogen-doped lignin-based carbon on the highly selective removal of tetracycline from water. *Bioresource Technology* 346: 126652. https://doi.org/10.1016/j.biortech.2021.126652.

46 Chaudhary, M., Suhas, K., Chaudhary, S. et al. (2022). Studies on the removal of phenol and nitrophenols from water by activated carbon developed from demineralized kraft lignin. *Agronomy* 12 (10): 2564. https://doi.org/10.3390/agronomy12102564.

47 Xie, A.T., Dai, J.D., Chen, Y. et al. (2019). NaCl-template assisted preparation of porous carbon nanosheets started from lignin for efficient removal of tetracycline. *Advanced Powder Technology* 30 (1): 170–179. https://doi.org/10.1016/j.apt.2018.10.020.

48 Fu, F.B., Zhao, B.W., Yang, D.J. et al. (2021). Insights into gas-exfoliation and the in-situ template mechanism of zinc compound for lignin-derived supercapacitive porous carbon. *ACS Applied Energy Materials* 4 (12): 13617–13626. https://doi.org/10.1021/acsaem.1c02283.

49 Karaaslan, M.A., Lin, L.T., Ko, F., and Renneckar, S. (2022). Carbon aerogels from softwood kraft lignin for high performance supercapacitor electrodes. *Frontiers in Materials* 9: 894061. https://doi.org/10.3389/fmats.2022.894061.

50 Yang, B.S., Kang, K.Y., and Jeong, M.J. (2017). Preparation of lignin-based carbon aerogels as biomaterials for nano-supercapacitor. *Journal of the Korean Physical Society* 71 (8): 478–482. https://doi.org/10.3938/jkps.71.478.

51 Rashidi, N.A., Chai, Y.H., Ismail, I.S. et al. (2022). Biomass as activated carbon precursor and potential in supercapacitor applications. *Biomass Conversion and Biorefinery* 15: https://doi.org/10.1007/s13399-022-02351-1.

52 Zhu, X.F., Labianca, C., He, M.J. et al. (2022). Life-cycle assessment of pyrolysis processes for sustainable production of biochar from agro-residues. *Bioresource Technology* 360: 127601. https://doi.org/10.1016/j.biortech.2022.127601.

53 Li, C., Sun, K., Sun, Y.F. et al. (2024). Sustainable production of porous carbon from biomass: influence of pre-carbonization on pore evolution and environment impact. *Chemical Engineering Journal* 480: 148176. https://doi.org/10.1016/j.cej.2023.148176.

54 Tofanica, B.M., Belosinschi, D., and Volf, I. (2022). Gels, aerogels and hydrogels: a challenge for the cellulose-based product industries. *Gels* 8 (8): 497. https://doi.org/10.3390/gels8080497.

55 Rinaldi, R., Jastrzebski, R., Clough, M.T. et al. (2016). Paving the way for lignin valorisation: recent advances in bioengineering, biorefining and catalysis. *Angewandte Chemie, International Edition* 55 (29): 8164–8215. https://doi.org/10.1002/anie.201510351.

6

Alginate-TiO_2 Composite: Formulation and Optimization for Wastewater Treatment

Sadia Khalid[1,], Anca Mazare[2,*], and Amara Nasir[3,*]*

[1]*Nanoscience & Technology Department, National Centre for Physics, Quaid-i-Azam University, Islamabad, Pakistan*
[2]*Department of Materials Science WW4-LKO, Friedrich Alexander University of Erlangen Nürnberg, Erlangen, Germany*
[3]*Department of Physics, The University of Lahore, 53700, Pakistan*

6.1 Introduction

Water is essential to our daily lives and its quality must be preserved. Sources of contamination affect our water resources, with primary pollutants coming from industrial, domestic, and agricultural activities, as well as other environmental and global changes. Water pollutants can be categorized as inorganic, organic, and biological. Common inorganic pollutants include heavy metals, which are highly toxic and carcinogenic, as well as nitrates, sulfates, phosphates, fluorides, chlorides, and oxalates, all of which pose significant health hazards. Organic pollutants include a wide range of toxic substances, such as pesticides (including insecticides, herbicides, and fungicides), polynuclear aromatic hydrocarbons, phenols, polychlorinated biphenyls, halogenated aromatic hydrocarbons, formaldehyde, polybrominated biphenyls, biphenyls, detergents, oils, and greases. In addition, wastewater contains normal hydrocarbons, alcohols, aldehydes, ketones, proteins, lignin, and pharmaceuticals. Biological pollutants in wastewater include harmful microbes such as bacteria, fungi, algae, plankton, amoebas, viruses, and other worms that can cause various diseases. These pollutants are present in solvated, colloidal, or suspended forms [1]. In many areas of the world, water is contaminated to levels that are unfit for human consumption. As the world's population grows, the likelihood of significant

*All authors have made an equal contribution in this manuscript.

freshwater shortages and the need for wastewater treatment increases [2]. In an estimate, half of the world's population will be living in water-stressed areas by 2025 [3]. In light of these issues, various wastewater treatment methods have been used to mitigate contaminated water resources and ensure the availability of clean water. They can be broadly categorized into physical (e.g., adsorption, and membrane separation), chemical (e.g., chlorination, ion exchange, ozonation, coagulation, photo- and chemical oxidation), and biological processes (e.g., enzyme bioreactor, biodegradation, and membrane bioreactor). The classification of commonly used water treatment technologies are presented in Figure 6.1a [19].

The application of wastewater treatment technologies using natural materials is a viable way to reduce secondary harm to humans and the environment. In recent years, there has been an increasing focus on the application of biodegradable and environmentally benign materials to reduce toxicity and reduce treatment costs. Natural materials have attracted considerable interest due to their intrinsic advantages, such as widespread availability, cost-effectiveness, environmental degradability, high reactivity, and low sludge production. Among these, sodium alginate (SA), an abundant bioactive material in nature, has been extensively used in water treatment. Its remarkable water absorption and retention capabilities, reversible swelling properties, and effective pollutant adsorption performance make it particularly valuable for these applications [4].

Alginate-TiO_2 composites are advantageous for wastewater treatment due to their improved photocatalytic degradation of pollutants, increased stability, and ease of handling in continuous flow systems, making them an ideal material for large-scale wastewater treatment processes [5–8]. However, TiO_2 has several advantages such as high photocatalytic activity, low cost, and nontoxicity, which make it a popular material for various applications, including wastewater treatment. Two of the key processes that play a crucial role in advanced oxidation processes that target wastewater treatment using biopolymers with TiO_2 are adsorption and photocatalysis [5]. Adsorption to incorporate contaminants into the composite and effectiveness depends on the adsorbent (activated carbon, zeolites, bioadsorbents, etc.) characteristics, such as surface area, pore size distribution, and functionality with limitations such as periodic regeneration and potential saturation [6, 9]. Photocatalysis is based on light-activated catalysts that degrade organic pollutants, mainly by generating electron-hole pairs upon illumination, which participate in further redox reactions. The free hydroxyl radicals (OH·) formed during the reactions can then attack the organic contaminants (without regard to selectivity), resulting in lower molecular weight species (intermediary level) or in final products in the form of water and carbon dioxide (complete mineralization level).

Titanium dioxide (TiO_2) remains a benchmark material for photocatalysis due to its high photocatalytic activity, chemical stability, nontoxicity, and affordability. Titania can effectively degrade a wide range of organic pollutants, which makes it a popular choice for wastewater treatment. A schematic of the main steps in the mechanism of photocatalysis is outlined in Figure 6.1b [10, 11], when upon suitable illumination electrons (e^-) are transferred from the valence band (VB) to the conduction band (CB), resulting also in hole (h^+) formation, further participating in reduction and oxidation reactions. The produced hydroxyl radicals are effective in the mineralization of organic pollutants. Extensive research studies have evaluated the use of TiO_2 in complex systems (cocatalysts, heterojunctions, composites, etc.) for environmental remediation, including air purification, water disinfection or purification, and hazardous waste remediation [12–14]. However, the use of TiO_2 has some limitations. For instance, TiO_2 has a narrow absorption range for visible light (bandgap of 3.2 eV for anatase), leading to low efficiency under solar illumination [15–18]. The rapid recombination of photogenerated electron-hole pairs further reduces its photocatalytic efficiency [15–17].

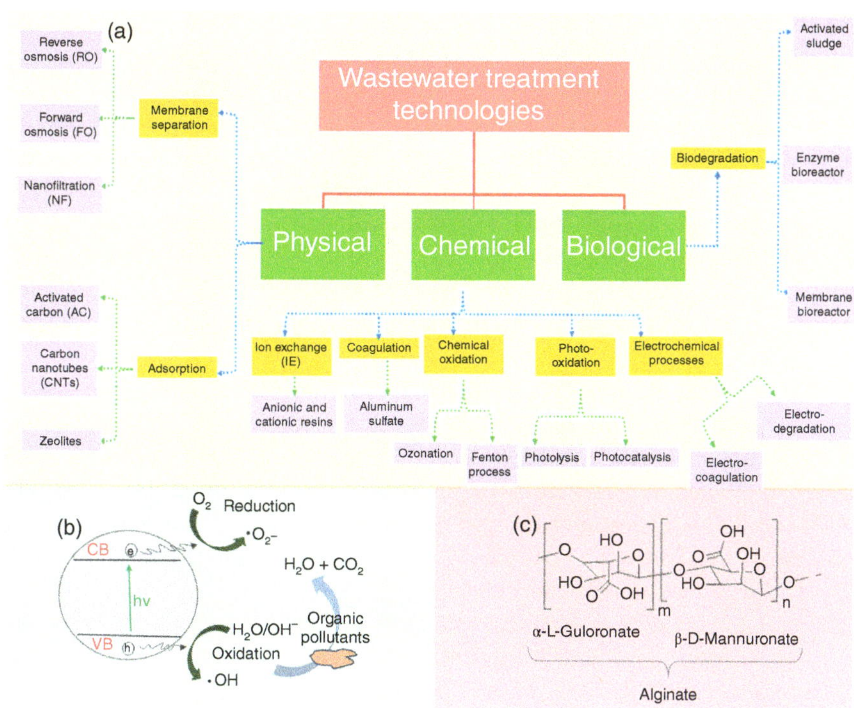

Figure 6.1 (a) Overview of water treatment technologies [19] / Intechopen Limited / CC BY-NC-ND 4.0, (b) schematic of photocatalytic degradation on TiO_2, and (c) chemical structure of alginate.

Strategies to overcome these limitations include doping with metals or nonmetals, and forming composites with other materials. Moreover, its use in suspensions requires posttreatment separation steps, which can complicate its application in large-scale processes.

One of the most widely used nanomaterial forms of TiO_2 is Degussa (Evonik) P25, a commercial form that is widely used as a standard in photocatalysis research. P25 is a mixed-phase TiO_2 consisting of approximately 80% anatase and 20% rutile, which enhances photocatalytic performance owing to the synergistic effect between the two phases. This standardization facilitates comparisons across studies, providing a consistent benchmark for evaluating new photocatalysts [5].

To overcome the TiO_2 disadvantages of nanoparticle-based photocatalysts in wastewater treatment, which have a high agglomeration tendency and low adsorption capacity, their use in composites with the addition of adsorbents brings into play the adsorption/photocatalysis synergistic effect. This means, for example, the use of TiO_2-polymer composite photocatalysts, such as polyvinyl alcohol, polystyrene, alginate, and so on [20–24]. In recent years, biopolymers have received considerable attention in scientific research due to their nontoxic properties, versatility, adaptability to different materials, flexibility, distinctive surface morphology, porosity, and exceptional mechanical strength [25]. This is also the case for alginate-TiO_2 composites, which provide a stable, recoverable photocatalyst that maintains high photocatalytic activity while simplifying the recovery process.

Alginate, a biocompatible and biodegradable polymer, is a linear block copolymer formed of α-L-guluronate and 1,4-linked β-D mannuronate (as shown in Figure 6.1c) [26]. Alginate can effectively immobilize TiO_2 nanoparticles, facilitating their recovery and reuse while preventing aggregation. This, coupled with its biocompatibility, nontoxicity, and cost-effectiveness, makes alginate an ideal material for use in alginate-based composites for various industrial and environmental applications. The general efficacy of alginate in adsorbing pollutants, such as dyes, heavy metals, and antibiotics, is well documented. This characteristic makes alginate an ideal material for use in alginate-based composites in various industrial and environmental applications, where incorporating materials such as carbonaceous substances and engineered nanoparticles such as TiO_2 can further enhance their efficiency.

In the following subsections, we further discuss alginate-based nanobiosorbents, as well as practical applications and case studies of various alginate-TiO_2 composites as a function of their design and formulation, emphasizing the current limitations and further optimization.

6.2 Alginate-Based Nano-Biosorbents (ANBs)

Alginate is a naturally derived polysaccharide primarily extracted from brown algae [27]. It has gained significant attention in the field of environmental remediation due to its unique physicochemical properties. For instance, the presence of carboxyl functional groups along the polymer backbone facilitates ionic cross-linking with divalent cations (e.g., Ca^{2+}) which leads to the formation of stable hydrogel networks. The chemical structure of alginate possesses functional groups that play pivotal roles in biosorption processes. The presence of carboxyl (—COOH) and hydroxyl (—OH) groups along the polymer backbone makes alginate highly effective in adsorbing various pollutants from aqueous environments. Carboxyl groups serve as primary sites for metal ion binding, facilitating chelation or complexation reactions with metal pollutants. They also contribute to ion exchange processes by forming electrostatic interactions with charged species, enhancing alginate's affinity toward contaminants. Hydroxyl groups, on the other hand, enhance alginate's hydrophilicity, promoting its dispersion and interaction with aqueous pollutants. Additionally, hydroxyl groups participate in hydrogen bonding interactions with water molecules and functional groups of pollutants, further enhancing their adsorption onto alginate surfaces. The arrangement and density of these functional groups along the alginate polymer chain significantly influence its biosorption capacity, selectivity, and overall effectiveness as a biosorbent. Understanding the role of these functional groups is paramount for designing and optimizing alginate-based biosorbents for various environmental remediation applications. Moreover, alginate exhibits biocompatibility, biodegradability, facile molding into diverse geometries, gelation capacity, and superior encapsulation properties, which feature its potential as an ideal biomaterial for the preparation of biocomposites [28]. These properties facilitate the development of versatile and sustainable biocomposite materials suitable for various environmental applications.

In recent years, the advent of nanotechnology has revolutionized the application potential of alginate, enhancing its effectiveness as a biosorbent at the nanoscale. Nanoadsorbents offer significant promise for wastewater treatment owing to their inherent characteristics, including their minute size, extensive surface area, pronounced porosity, catalytic activity, reactivity, adjustable molecular dimensions, hydrophobic nature, ease of separation, and regenerability. These advantageous attributes render nanoadsorbents highly suitable for use as pollutant removal. To harness multiple functionalities within a single platform, researchers are actively investigating the engineering of nanotechnology, aiming to fabricate binary, ternary, polymer-based, and diverse support-mediated nanocomposites for enhanced pollutant adsorption capabilities.

Alginate-based nano-biosorbents are synthesized through various techniques, including ionotropic gelation, emulsification, and electrospinning, which allow for precise control over particle size and surface area. These nano-biosorbents exhibit superior adsorption capacities due to their high surface-to-volume ratio and the enhanced availability of active binding sites. Alginate-based composites are prepared with various materials such as chitosan, cellulose nanofibers, casein, activated carbon, biochar, graphene oxide, carbon nanotube, magnetic materials, and carbon and non-carbon-based nanomaterials, as well as microorganisms in hydrogels/alginate granules. The ionic bond formation, emulsion formation, electrostatic complexation, and self-assembly due to hydrogen bonding result in the preparation of alginate-based composites [29]. Alginate composites are tailored into specific geometries to fulfill practical demands in water treatment, including spherical beads, millimeter-sized monoliths, and functional films, such as aerogels/foams, hydrogels, and thin films. The advancement of alginate composites presents a promising approach to mitigate dependence on nonrenewable resources, thereby curbing their ecological footprint [30].

6.2.1 Adsorption Mechanisms of ANBs for Pollutants

In the interaction between pollutants and alginate-based adsorbents, various mechanisms come into play, encompassing electrostatic attraction, surface complexation, physical adsorption, and ion exchange. While a dominant mechanism may emerge in certain scenarios, in many cases, adsorption is governed by a combination of these mechanisms. The specific mechanism's contribution to heavy metal adsorption varies significantly, influenced by factors such as metal speciation, ionic composition of the solution, and the chemical properties of alginate materials. Notably, ion exchange stands out as a major mechanism for heavy metal ion adsorption onto alginate-based adsorbents. This process typically involves the exchange of heavy metal ions (M^{2+}) with cross-linking cations (Ca^{2+}), leading to the formation of a stable egg-box structure on the surface of the adsorbent [31]. The adsorption mechanism for methylene blue dye involved electrostatic attraction between TiO_2 or COO^- groups and the cationic dye, as well as hydrogen bonding between hydroxyl groups and the imine groups of the dye [32, 33]. Figure 6.2 includes the possible adsorption mechanisms in alginate-based adsorbents [31].

6.2.2 Modification of ANBs

Traditional adsorbent materials applied in wastewater treatment demonstrate common problems such as limited adsorption capacity, poor selectivity toward pollutants, reusability, and stability. Numerous modification strategies have been

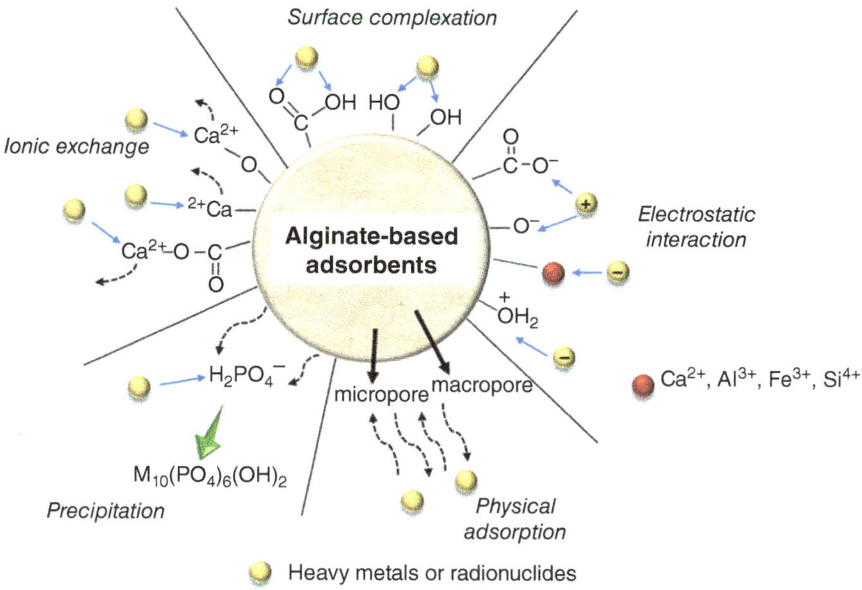

Figure 6.2 Schematic adsorption mechanisms of heavy metals and radionuclides onto alginate-based adsorbents [31] / with permission of Elsevier.

applied to enhance the adsorption properties of the alginates such as cross-linking, grafting, surface functionalization, and composite preparation. These modifications increase the porosity, specific surface area, ion exchange capacity, chemical and thermal stability, flexibility, mechanical strength and durability, structural tenability, wettability, and affinity toward pollutants. Incorporation or integration of suitable materials into alginate matrices allow easier and safer operation for practical applications. The approaches to modify alginates through composite formation include incorporation of the following nanomaterials into the alginate matrix [7, 34, 35]:

- Polymers: chitosan, polyvinyl alcohol (PVA), polyaniline (PANi)
- Clays: montmorillonite (Mt), bentonite (BT), kaolin (KN)
- Carbon materials: biochar (BC), activated carbon (AC), carbon nanotubes (CNTs), graphene oxide (GO)
- Layered double hydroxides (LDHs): Mg–Al LDH, Zn–Fe LDHs
- Metals: lanthanum (La), zirconium (Zr), iron (Fe)
- Metal oxides: magnesium oxide (MgO), zinc oxide (ZnO), cadmium oxide (CdO), titanium dioxide (TiO_2), manganese dioxide (MnO_2)
- Magnetic: spinel ferrites ($MgFe_2O_4$, Fe_3O_4)

Alginate-based composites can be fabricated through various approaches, e.g., gelation method, freeze-drying method, in situ growth method, dip-coating method, and carbonization. The structural properties, morphological features, surface functionalities, and sorption properties of the nanocomposites significantly depend on synthetic approaches [30].

6.3 Alginate-TiO_2 Composites (ATCs): Design and Application

Titania or TiO_2 is a wide bandgap semiconductor (3.2 eV – ultraviolet region for anatase) photocatalyst. It is the most promising and extensively used photocatalyst among all photocatalysts owing to its nontoxicity, photostability, chemical inertness, and better oxidation efficiency. It is widely employed to eliminate organic and inorganic pollutants. But there are some important limitation factors that restrict its practical applications:

1) Limited utilization of solar light for activation of photocatalyst
2) Low adsorption of organic pollutants
3) Limited photocatalyst recovery and reusability
4) Aggregation of TiO_2 nanoparticles
5) Difficulty in uniform distribution of catalyst

Doping and co-doping with metals and nonmetals, heterojunction construction, surface organic modifications, and immobilization of support structures are the key strategies to tackle above-mentioned issues. An effective strategy is immobilization of support structures such as polymers, clay, silica, and perlite for the TiO_2 photocatalyst. This approach improves the recovery, reusability, and degradation efficiency to a greater extent. Use of alginate as support aids some exciting features such as environmentally friendly nature, low cost, durability, enhanced adsorption, availability, and biodegradability [25].

6.3.1 Beads

Alginate TiO_2 beads are typically prepared by an ionotropic gelation technique. In this technique, TiO_2 nanoparticles are encapsulated with the alginate polymer, followed by cross-linking alginate with multivalent ions in the presence of TiO_2 particles. This method forms a stable gel network where TiO_2 is evenly distributed. Some studies have used gas-forming agents ($CaCO_3$, $NaHCO_3$) during the gelation process. The prepared solution is dropped into $CaCl_2$ solution. These beads are used for photocatalytic degradation, in addition to adsorption, to treat water and wastewater [36]. The beads are spherical, possess size in the range of micrometers, and have a porous, rough surface [36]. Typically, they are used for photocatalytic dye degradation in addition to adsorption. This spherical shape gives good contact between the light source, dyes, and beads [37].

Wan et al. [37] loaded commercial TiO_2 (P25) into the alginate salt network and cross-linked to form spherical beads. The cross-linking was performed using divalent ions. Bead formation occurs by a typical gelation process during which TiO_2 particles uniformly disperse and immobilize in the alginate. The content of TiO_2 in SA was 2 wt.%. Beads were 3.2–3.7 mm and were used for methylene orange degradation. Lam et al. [38] reported 90% removal of reactive black 5 in less than 200 minutes by photocatalysis using such beads. Alhattab et al. [39] developed alginate/bentonite-TiO_2 composite beads for water treatment by adsorption. The beads were used for amoxicillin and 4-chlorophenol removal and showed 87.56% efficiency for amoxicillin and 82.56% for 4-chlorophenol. Recyclability and desorption studies showed excellent reuse of composite beads, with 100% regeneration efficiency. Fadl et al. [40] formed TiO_2/alginate beads cross-linked with $CaCl_2$ and loaded with Ag^+ ions and reduced to Ag nanoparticles by irradiation. These beads showed 98% 2-nitrophenol reduction to 2-aminophenol, which was 76% for alginate/TiO_2 beads. Figure 6.3 shows alginate-TiO_2 beads prepared by ionotropic gelation which exhibit much higher adsorption of methylene blue compared to TiO_2 powder due to their porous structure, which allows dye molecules to diffuse into internal macropores. This shows a practical advantage of beads in handling and recovery of the catalyst in wastewater treatment. However, under

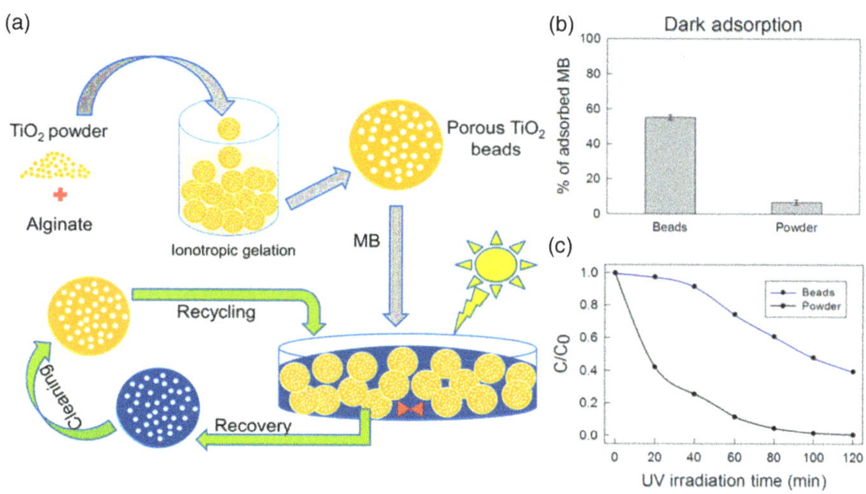

Figure 6.3 (a) General mechanism of ionotropic gelation and wastewater treatment by alginate-TiO_2 beads, and (b) adsorption, and (c) photoactivity of beads versus anatase powder [57] / MDPI / CC BY 4.0.

UV irradiation, the photocatalytic degradation rate is lower for the beads because part of the dye is trapped inside the pores and remains less exposed to light, while the alginate matrix partially reduces the accessible TiO_2 surface area [41].

To extend the light absorption range and photoactivity of TiO_2, a recent approach is modification of TiO_2 by rare earth elements. For example, Guetni et al. [42] doped 1% lanthanum in anatase nanoparticles and obtained 96.49% in only 105 minutes to degrade Orange G dye under 300–800 nm light source. Some other approaches, such as elemental doping or metal or metal oxide nanoparticles modifications, have also been performed to obtain enhanced photoactivity. Azeroul et al. [43] used doped TiO_2 nanowires in SA matrix to enhance surface area and, therefore, adsorption properties of the composite. Nanowires were doped with zirconium, and the beads were cross-linked by calcium chloride. These composites were used to remove safranin and methylene blue dyes and 29 and 26 $mg\,g^{-1}$ adsorption capacity, respectively. They were effective for both single or binary organic dyes. Idris et al. [44] immobilized C and N co-doped TiO_2 nanoparticles on alginate beads for diazinon degradation under solar light. This floating photocatalyst showed 80.61% in 8 hours under solar light and reusable up to five cycle.

Interestingly, Dalponte et al. [36] prepared a floating photocatalyst by modifying TiO_2 and alginate polymer for the first time. During ionic gelation, a gas-forming agent ($CaCO_3$, $NaHCO_3$) in an acidic medium (CH_3COOH) is used. The catalyst had enhanced light absorption and was easier to separate after treatment. The optimal formulation for the 100% floating of the beads after 24 hours was 2.3 wt.% $CaCO_3$, 6.4 wt.% CH_3COOH, and 1.1 wt.% TiO_2. The photocatalytic activity of the beads for tartrazine dye removal was up to 89% and was maintained even after seven cycles. This floating photocatalyst is easily recoverable and is an efficient approach for industrial wastewater treatment without the need for mechanical stirring.

6.3.2 Hydrogels

Hydrogels are three-dimensional polymeric matrices that possess multifunctional hydrophilic groups which facilitate the absorption and subsequent release of water or other fluids containing various substances, such as dyes, heavy metals, and radionuclides. This reversible process is governed by diffusion, polymer relaxation, and environmental stimuli, such as temperature, solvent composition, and pH variations. The adsorption and desorption mechanisms of hydrogels in wastewater treatment are complex, requiring careful consideration of synthesis parameters to optimize their performance. Factors such as the functional groups present in the polymers, chemical properties of the contaminants, and experimental conditions like temperature, pH, and salt concentration influence the adsorption processes [45]. Hydrogel-based composites are tailored to enhance their functional properties, such as maximum equilibrium swelling, high absorption rates, minimal residual content, porosity, uniform softness, high durability, and photostability. These composites are reusable and multifunctional [46]. The incorporation of TiO_2 nanoparticles into the hydrogels improves their water absorption ability [32]. An interesting aspect of hydrogels is that they can retain substantial amounts of water within their structures without dissolving in water [46].

Thakur et al. developed a titania-incorporated sodium alginate/acrylic acid hydrogel nanocomposite for the adsorption of methylene blue dye from aqueous solutions. The hydrogel nanocomposite was synthesized through free radical polymerization of acrylic acid onto sodium alginate, using potassium persulfate as an initiator and *N*,*N*′-methylenebisacrylamide as a cross-linker. The maximum adsorption capacity was 2257.36 $mg\,g^{-1}$ [47]. A sodium alginate/polyacrylamide/titanium dioxide double network composite hydrogel was synthesized using sunlight-induced polymerization and ionic cross-linking. Interestingly, TiO_2 served as an initiator and was uniformly immobilized on the hydrogel matrix to support photocatalysis. The hydrogel's dual network structure and hydrophilic functional groups improved its adsorption capacities and photocatalytic performance by preventing TiO_2 agglomeration. The hydrogel was effective as a reusable adsorbent and photocatalyst for dye treatment [48]. Nanocomposite hydrogels composed of barium ion cross-linked alginate and carboxymethyl cellulose encapsulating TiO_2 nanoparticles were synthesized by a dissipative convective method followed by freeze-drying. The composite hydrogels demonstrated excellent photocatalytic activity (91.5% in 240 minutes) toward Congo Red dye under direct solar irradiation and had high photostability as a photocatalyst [23].

6.3.3 Aerogels

Aerogels are ultralight, highly porous materials that consist of up to 99% air, giving them an extremely low density and a unique nanostructure. This structure results in exceptional properties, such as superior thermal insulation, transparency, and high surface area [49]. Polymer aerogels are typically produced through a sol-gel process followed by drying techniques such as supercritical drying, freeze-drying, or ambient pressure drying. These methods ensure that the gel's structure is preserved while the liquid is removed, maintaining the aerogel's porosity and low density. With the advent of nanotechnology, these aerogels facilitate embedding nanomaterials, to enhance performance and obtain multifunctional nanoarchitectures [50].

The preparation of aerogels with immobilized TiO_2 nanoparticles for wastewater remediation is another application of alginate and TiO_2. Utilizing homogeneous dispersion and ionic cross-linking techniques, Dai et al. [51] developed TiO_2-alginate aerogels for oil/water separation. A hybrid sodium alginate-TiO_2 sol was created using ultrasonication, followed by freeze-drying.

The resulting hybrid aerogel was then immersed in a $CaCl_2$ solution for cross-linking and subsequently freeze-dried again after washing. The aerogels had a 3D porous structure, rough surface and high water affinity. A reduced graphene oxide-TiO_2/sodium alginate aerogel was synthesized for the degradation of ibuprofen and sulfamethoxazole. The aerogel had three-dimensional interconnected macroporous structure, large surface area that facilitated TiO_2 nanoparticle immobilization, efficient charge separation, and enhanced adsorption properties. The degradation was over 99% for both contaminants within 45–90 minutes under UV-A irradiation. The optimal mass ratio of TiO_2 nanoparticles to GO in the aerogel was 2:1 in the presence of 1 wt.% alginate solution. It was easy to separate and recyclable which reduced the energy cost associated with sample recovery from effluents [52]. The mixed silver and nickel sulfide (Ni_3S_4/Ag_2S) synthesized over the TiO_2 surface is a potential photocatalyst for microbial and pharmaceutical drug decontamination from wastewater. The immobilization of this nanocomposite in alginate show higher adsorption and microbial decontamination properties, and its aerogel can be easily isolated from the wastewater after use. This $Ni_3S_4/Ag_2S/TiO_2$/calcium alginate aerogel was used to remove the oxytetracycline drug and promote microbial decontamination of wastewater from the beverage industry. The maximum adsorption was $29.57\,mg\,g^{-1}$ in 180 minutes. The aerogel had superior cytotoxicity within microbial cells and effectively destroyed nucleic acids under solar light illumination [53]. Figure 6.4 shows the schematics of aerogel synthesis and microbe degradation. Mixed phase TiO_2 is obtained by treating it with concentrated NaOH including rutile or anatase and sodium titanate. The Na^+ ions in sodium titanate were replaced by Ni^{2+} and Ag^+ ions, forming stable sulfides on the TiO_2 surface, and the final aerogel photo after freeze drying is also present in Figure 6.4. The microbial deactivation mechanism of the $Ni_3S_4/Ag_2S/TiO_2$/calcium alginate aerogel involves both physical and chemical interactions. In darkness, the aerogel traps microbes on its surface, where metallic moieties can disrupt cell respiration and cellular pathways, inhibiting biofilm formation and further microbial growth. Under solar light, the aerogel is more effective due to the generation of reactive oxygen species and active charged species like e^-, h^+, $^-O_2$, OH, and H_2O_2. The synergistic effect between all three constituents enhances the production of Reactive oxygen species (ROS), increasing cytotoxicity and destroying microbial nucleic acids [53].

6.3.4 Membranes

Free-standing Ca-alginate hydrogel filtration membranes can be synthesized starting from Na-alginate as the membrane substrate, $CaCl_2$ as the ionic cross-linker agent, and polyethylene glycol (PEG) as the pore-forming agent [54]; the resulting membranes showed excellent antifouling performance. The overall synthesis

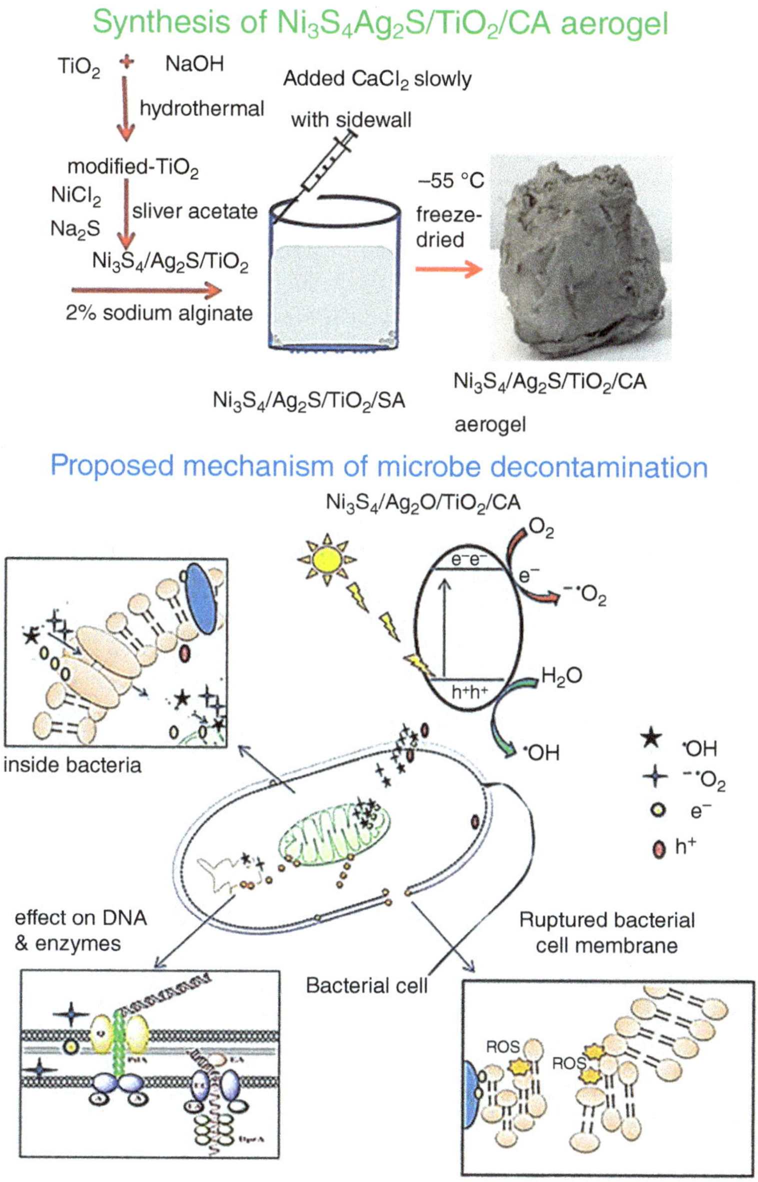

Figure 6.4 Schematic synthesis of $Ni_3S_4/Ag_2S/TiO_2$/calcium alginate aerogel and its presumed mechanism of photocatalytic microbe degradation [44] / with permission of Elsevier.

process is simple, inexpensive, and does not produce organic wastewater [54]. Various metal ions were initially evaluated as cross-linking agents (Ca^{2+}, Zn^{2+}, Mn^{2+}, Co^{2+}, Fe^{2+}, Al^{3+}), and Ca^{2+} cross-linking was found to be optimal in terms of separation performance [55]. Recent studies have also investigated mixed cross-linking metal ions (Al^{3+} and Ca^{2+}) [56].

To obtain alginate-TiO_2 composite membranes [57–59], an ionic cross-linking mechanism similar to that of alginate membranes can be used. This means that sodium alginate is used as the substrate material, TiO_2 as the modifier, $CaCl_2$ as the cross-linking agent, and a pore-forming agent, such as PEG, can be employed. The process itself is simple; briefly, this consists of dispersing TiO_2 (P25 or nanoparticles with different functionalities) and sodium alginate in water with a specific concentration of the modifier (i.e., TiO_2) to sodium alginate under stirring to ensure homogeneity [58], followed by casting/spreading on a glass plate. The obtained film was then immersed in a $CaCl_2$ aqueous solution for cross-linking. Elution in warm water must also be performed to remove pore-forming agents [59].

Wang et al. [58] designed uniformly dispersed TiO_2 nanoparticles (by using carboxylated TiO_2, TiO_2-COOH) in calcium alginate membranes, with the 20 wt.% TiO_2-alginate membrane having a tensile strength value of 2.44 MPa, threefold higher compared to the alginate membrane, and the highest rejection rations for the tested dyes (Brilliant blue G250–98.4%, Direct black 38–96.8%, and Congo red–95.9%) and lowest for the tested inorganic salts, and finally the 24-hour filtration test indicated a permeate flux of 14.1 $L\,m^{-2}\,h^{-1}$, thus confirming their applicability to dye/salt separation. Nevertheless, as no pore-forming agent was employed in the synthesis, the porosity was low and with a thicker membrane, thus resulting in a lower flux.

Interestingly, further investigations by Gao et al. [59] report a hydrophilic TiO_2-alginate filtration membrane with good antifouling and chlorine resistance (enabling disinfection with sodium hypochlorite), which targets the protein purification by alginate-based membranes (when coupled with polyhydroxybutyrate nanofibers, the membrane has high flux (136.7 $L\,m^{-2}$ h) and high bovine serum albumin (BSA) rejection (98.0%) at 0.1 MPa). The authors reported that with increasing TiO_2 content, the membrane flux gradually increased with a decreasing rejection rate (attributed to the increased looseness upon adding TiO_2, with a 4.5 MPa elastic modulus for the 50 wt.% TiO_2 membrane). While such a membrane has been applied in protein purification and separation (using BSA as a model), it can also be used in other media remediation applications owing to the increased control of the pore size and flux.

The preparation conditions, in terms of the cross-linking method and amount or cross-linking time, can also influence the adsorption performance of the prepared materials [26, 56]. The surface chemistry of the TiO_2 nanoparticles and the membrane surface significantly influence and control their interaction and arrangement in the final composite. For the surface chemistry of the TiO_2 nanoparticles, we have discussed in the previous chapter that it is a key factor controlling the dispersibility of the nanoparticles [66], which was also successfully incorporated into TiO_2-alginate membranes [58]. Functionalization or chemical modification by grafting with hydrophilic polymers (e.g., polyacrylamide-grafted sodium alginate copolymeric membranes [69]) can influence the physicochemical and transport properties of the resulting membrane. Mansourpanah et al. [70] showed that the -OH functionalization (polyethersulfone (PES)/polyimide (PI)) led to excellent adhesion of TiO_2 nanoparticles to the alginate surface, increased reversible deposition, and significantly decreased irreversible fouling in the case of membrane. For a more detailed discussion on alginate-based membrane modification approaches, please see the dedicated reviews in [7, 9, 67].

In addition to surface chemistry, an additional influencing factor that can affect the morphology of the resulting membrane is related to the synthesis composition; the amount of TiO_2 nanoparticles, the use of a pore-forming agent, is crucial for the morphology, and if used, its amount can dictate the pore size and resulting flux. As shown by Wang et al. [58], when carboxylated TiO_2 was used, the optimal TiO_2 loading amount was 20 wt.%. Nevertheless, when using a pore-forming agent such as PEG [59], it was shown that the molecular weight of PEG and the amount used significantly influenced the performance (flux and rejection of BSA) of the membrane in the filtration tests.

In most water purification studies, methylene blue (MB) is used as a model pollutant to evaluate its efficacy in dye degradation, irrespective of photocatalyst morphology. Considering that TiO_2 decorated with noble-metal cocatalysts is well-known for enhancing dye degradation, it stands to reason that the use of such noble-metal-decorated TiO_2 in the alginate membranes would be advantageous. Note that Au nanoparticle decoration on TiO_2 nanotubular structures was shown to efficiently photoreduce Hg(II) to Hg(I)/Hg(0) and accumulate the compounds under sunlight irradiation (this can be followed by regenerating the photocatalyst and recovering the Hg waste) [71]. Nevertheless, there is a limited amount of data on noble-metal-decorated TiO_2-alginate composites, for example, Au-TiO_2/alginate films for active food packaging (with degradable and antibacterial properties) [72]. This approach was also successfully reported for Ag-TiO_2 on polyvinylidene fluoride membranes, which degraded MB and ensured bacterial inactivation under visible light illumination [73].

6.3.5 Coatings

Thin films or coatings of TiO_2-alginate composites can be achieved by coating or dispersing various formulations (hydrogels, a cross-linked mixture of Na-alginate, and TiO_2 used in membrane manufacturing, etc.). A composite hydrogel coating consisting of TiO_2 deposited polyacrylamide-sodium alginate was obtained, in which the use of polydopamine/TiO_2 ensured adhesion to the substrates and a photodegradation ability reaching ~98% MB [60]. The resulting hydrogel coating was deposited on a copper mesh or polyvinylidene fluoride membrane and showed superhydrophilic properties and high oil/water separation performance in a long-term separation test.

For example, TiO_2-alginate films can be obtained in a similar manner as membranes by immobilizing TiO_2 nanoparticles on sodium alginate and spreading on a glass plate, controlling the thickness, and subsequently cross-linking with $CaCl_2$ [61] or without cross-linking and drying in air [62]. The former TiO_2-alginate thin films (0.1–0.4 mm) [61] showed ~82% degradation of MB and good reusability while maintaining the mechanical properties and MB degradation rates. Interestingly, the photocatalyst nanoparticles were recovered by dissolving alginate. Omar et al. [62] investigated the optical properties of various nanocomposite films, including TiO_2-alginate.

Reveendran and Ong [63] investigated the use of casted TiO_2-alginate thin films for Congo Red dye removal in an aqueous environment, and the optimal conditions for dye degradation were found to be a pH of 8 for 5 mg L^{-1} Congo Red dye concentration and a contact time of 5 hours. Thin films of TiO_2-alginate with polyacrylamide (0.3 mm thickness) were synthesized by a combination of UV irradiation and cross-linking, and showed good degradation of methyl orange [64].

Similarly, the surface chemistry and loading amount of coatings are expected to significantly influence their properties. For example, more complex films can be obtained via casting using more than two polymers and TiO_2. Some examples include (i) coatings with antifogging, self-healing, and antifouling properties based on polyvinyl alcohol, sodium alginate, glycerin, and TiO_2 [74] and (ii) novel flexible polymer nanocomposite films containing polyvinyl alcohol, Na-alginate, TiO_2, and graphene oxide, which have shown promise in solar cells and electrical energy storage applications [75].

6.4 Limitations and Stability Concerns of ATCs

A major challenge with TiO_2 in alginate-TiO_2 composites is its wide bandgap, which tends to hamper its practical application only in the UV region. Pristine TiO_2 possesses very low adsorption capacity due to low porosity and surface area [33]. Despite this fact, a common disadvantage of nanofiller-alginate composites is ensuring uniform dispersion of the nanofiller, in our case, TiO_2 nanoparticles. For alginate composites, this is related to the tendency of TiO_2 nanoparticles to agglomerate or aggregate, which is attributed to their large surface-area-to-particle-size ratio. The degradation of material properties is widely recognized as a result of the combination and incompatibility of inorganic particles with polymer substrates (e.g., polyvinylidene fluoride) [65]. Reports also show that carboxylation of TiO_2 nanoparticles reduced agglomeration phenomena, as the modification improved the dispersibility and interfacial adhesion of the nanoparticles [66]. Wang et al. [58] have recently shown that such carboxylated TiO_2 can also be used in the synthesis of TiO_2-alginate membranes, and that the carboxylated TiO_2 nanoparticles were uniformly dispersed on the membrane surface and enhanced its mechanical performance. Moreover, the selective removal and recovery of targeted pollutants are complicated in the presence of other substances in wastewater, leading to unwanted competition and interference.

One of the significant challenges in hydrogels is their recovery from solutions, particularly for micro- or nano-sized hydrogels. These stability concerns for hydrogel composites impact their performance and reusability [46]. This issue has also been observed for pristine sodium alginate membranes, especially when in contact with high water content solutions [67]. However, literature reports indicate that cross-linked alginate membranes show exceptional chemical stability toward organic solvents (dimethylformamide, dimethylsulfoxide, etc.) [67, 68] and good resistance to alkali or high salt content environments [56, 67]. Despite their remarkable properties, aerogels also face challenges of fragility and high production costs. The scanning electron microscopy (SEM) analysis of the dehydrated alginate beads shows enhancement in structural stability and presence of a crack-free surface after TiO_2 loading [37]. This is because TiO_2 is stable at elevated temperatures, and its incorporation into sodium alginate increases the decomposition temperature of alginate at 55.6% TiO_2 content in the composite [38].

Literature shows that although the potential of alginate-nanofiller composites in the fields of environmental cleanup, advancements in manufacturing techniques continue to address the present issues in stability, selectivity, homogeneous dispersion of filler, low production cost, and easy application and processing of composite for wastewater cleanup.

6.5 Conclusion and Future Perspectives

The future of wastewater treatment by sodium alginate-TiO_2 composites appears highly promising and offers innovative and effective solutions to address the water pollution challenges and water scarcity. Below are the key areas to shape the outlook of sodium alginate-TiO_2 composites for efficient wastewater treatment:

1) Continuous advent in nanotechnology, particularly in material synthesis and functionalization, is expected to lead to the development of more efficient and

versatile sodium alginate-TiO_2 composites, which can remove a broader range of contaminants. Various polysaccharides for wastewater treatment are the focus of researchers to discover new materials with novel properties, and this will expand the applications of sodium alginate-TiO_2 composites.

2) Sodium alginate-TiO_2 composites may become a priority in pharmaceutical and chemical industries to seek reliable and cost-effective solutions for sustainable wastewater treatment. This is due to the increased concerns over antibiotic waste and other harmful chemical discharges.
3) Sodium alginate-TiO_2 composites can enhance biological processes in wastewater treatment by removing heavy metals and pollutants that inhibit microbial activity, thereby improving overall treatment efficiency and reducing environmental impacts.
4) With the increase in use of sodium alginate-TiO_2 composites, a comprehensive environmental impact assessment will be essential to ensure the safe and sustainable utilization of these composites in wastewater treatment by understanding potential risks associated with their use and disposal.
5) In order to transit sodium alginate-TiO_2 composites from lab-scale to large-scale application, challenges in production, cost-effectiveness, and integration into existing infrastructure must be addressed.
6) The selectivity and stability of the sodium alginate-TiO_2 composites must be addressed to improve their efficiency in wastewater treatment.
7) Suitable modifications or metal-doping in TiO_2 can increase the photoactivity and thereby utilization of composites in natural sunlight more effectively.
8) A complete investigation and understanding of the mechanisms of biopolymer incorporation in TiO_2 will help to understand the impact on photodegradation and bandgap reduction.

The future of sodium-TiO_2 composites for water and wastewater treatment is promising and has the potential to develop the industry. The current research and their implementation could help mitigate water pollution, safeguard public health, and preserve ecosystems globally. However, management and budgeting considerations must be taken into account for successful application.

References

1 Gupta, V.K., Ali, I., Saleh, T.A. et al. (2012). Chemical treatment technologies for waste-water recycling—an overview. *RSC Advances* 2 (16): 6380–6388. https://doi.org/10.1039/C2RA20340E.

2 Gusain, R., Gupta, K., Joshi, P., and Khatri, O.P. (2019). Adsorptive removal and photocatalytic degradation of organic pollutants using metal oxides and

their composites: a comprehensive review. *Advances in Colloid and Interface Science* 272: 102009. https://doi.org/10.1016/J.CIS.2019.102009.

3 UNICEF (n.d.) Water scarcity|UNICEF. https://www.unicef.org/wash/water-scarcity (accessed 10 June 2024).

4 Zheng, D., Wang, K., and Bai, B. (2024). A critical review of sodium alginate-based composites in water treatment. *Carbohydrate Polymers* 331: 121850. https://doi.org/10.1016/J.CARBPOL.2024.121850.

5 Chkirida, S., Zari, N., Achour, R. et al. (2021). Highly synergic adsorption/photocatalytic efficiency of alginate/bentonite impregnated TiO_2 beads for wastewater treatment. *Journal of Photochemistry and Photobiology A: Chemistry* 412: 113215. https://doi.org/10.1016/j.jphotochem.2021.113215.

6 Shabbir, A., Akbar, M., Yaqoob, A., and Ahmad, A. (2023). Recent advances and development of blended sodium alginate wastewater management. In: *Sodium Alginate-Based Nanomaterials for Wastewater Treatment*, (ed. Ahmad A., Ahmad I., Kamal T., Asiri A.M., Tabassum S.), 315–330. Elsevier. https://doi.org/10.1016/B978-0-12-823551-5.00015-X.

7 Wang, B., Wan, Y., Zheng, Y. et al. (2019). Alginate-based composites for environmental applications: a critical review. *Critical Reviews in Environmental Science and Technology* 49 (4): 318–356. https://doi.org/10.1080/10643389.2018.1547621.

8 Su, C., Sun, X., Mu, Y. et al. (2021). Multilayer calcium alginate beads containing Diatom Biosilica and Bacillus subtilis as microecologics for sewage treatment. *Carbohydrate Polymers* 256: 117603. https://doi.org/10.1016/j.carbpol.2020.117603.

9 Radoor, S., Karayil, J., Jayakumar, A. et al. (2024). Recent advances in cellulose- and alginate-based hydrogels for water and wastewater treatment: a review. *Carbohydrate Polymers* 323: 121339. https://doi.org/10.1016/j.carbpol.2023.121339.

10 Schneider, J., Matsuoka, M., Takeuchi, M. et al. (2014). Understanding TiO_2 photocatalysis: mechanisms and materials. *Chemical Reviews* 114 (19): 9919–9986. https://doi.org/10.1021/CR5001892.

11 Hoffmann, M.R., Martin, S.T., Choi, W., and Bahnemann, D.W. (1995). Environmental applications of semiconductor photocatalysis. *Chemical Reviews* 95 (1): 69–96. https://doi.org/10.1021/cr00033a004.

12 Tsang, C.H.A., Li, K., Zeng, Y. et al. (2019). Titanium oxide based photocatalytic materials development and their role of in the air pollutants degradation: overview and forecast. *Environment International* 125: 200–228. https://doi.org/10.1016/j.envint.2019.01.015.

13 Purabgola, A., Mayilswamy, N., and Kandasubramanian, B. (2022). Graphene-based TiO_2 composites for photocatalysis & environmental remediation: synthesis and progress. *Environmental Science and Pollution Research* 29 (22): https://doi.org/10.1007/S11356-022-18983-9.

14 Linsebigler, A.L., Lu, G., and Yates, J.T. (1995). Photocatalysis on TiO_2 surfaces: principles, mechanisms, and selected results. *Chemical Reviews* 95 (3): 735–758. https://doi.org/10.1021/cr00035a013.

15 Fujishima, A. and Honda, K. (1972). TiO_2 photoelectrochemistry and photocatalysis. *Nature* 213 (1998): 8656.

16 Carp, O., Huisman, C.L., and Reller, A. (2004). Photoinduced reactivity of titanium dioxide. *Progress in Solid State Chemistry* 32 (1–2): 33–177. https://doi.org/10.1016/j.progsolidstchem.2004.08.001.

17 Kang, X., Liu, S., Dai, Z. et al. (2019). Titanium dioxide: from engineering to applications. *Catalysts* 9 (2): 191. https://doi.org/10.3390/catal9020191.

18 Etacheri, V., Di Valentin, C., Schneider, J. et al. (2015). Visible-light activation of TiO_2 photocatalysts: advances in theory and experiments. *Journal of Photochemistry and Photobiology C Photochemistry Reviews* 25: 1–29. https://doi.org/10.1016/J.JPHOTOCHEMREV.2015.08.003.

19 El-Gendy, A., El-Shafie, A.S., Issa, A. et al. (2020. https://doi.org/10.5772/INTECHOPEN.91750). Carbon-based materials (cbms) for determination and remediation of antimicrobials in different substrates: wastewater and infant foods as examples. In: *Carbon-Based Material for Environmental Protection and Remediation*, (ed. Bartoli M., Frediani M., Rosi L.), 1–20 IntechOpen.

20 Benhabiles, O., Mahmoudi, H., Lounici, H., and Goosen, M.F.A. (2016). Effectiveness of a photocatalytic organic membrane for solar degradation of methylene blue pollutant. *Desalination and Water Treatment* 57 (30): 14067–14076. https://doi.org/10.1080/19443994.2015.1061954.

21 Lei, P., Wang, F., Gao, X. et al. (2012). Immobilization of TiO_2 nanoparticles in polymeric substrates by chemical bonding for multi-cycle photodegradation of organic pollutants. *Journal of Hazardous Materials* 227–228: 185–194. https://doi.org/10.1016/j.jhazmat.2012.05.029.

22 Vavilapalli, D.S., Rosen, J., and Singh, S. (2023). Immobilization of a TiO_2–PEDOT:PSS hybrid heterojunction photocatalyst for degradation of organic effluents. *RSC Advances* 13 (5): 3095–3101. https://doi.org/10.1039/D2RA06729C.

23 Thomas, M., Naikoo, G.A., Sheikh, M.U.D. et al. (2016). Effective photocatalytic degradation of Congo red dye using alginate/carboxymethyl cellulose/TiO_2 nanocomposite hydrogel under direct sunlight irradiation. *Journal of Photochemistry and Photobiology A: Chemistry* 327: 33–43. https://doi.org/10.1016/j.jphotochem.2016.05.005.

24 Ren, M., Frimmel, F.H., and Abbt-Braun, G. (2015). Multi-cycle photocatalytic degradation of bezafibrate by a cast polyvinyl alcohol/titanium dioxide

(PVA/TiO_2) hybrid film. *Journal of Molecular Catalysis A: Chemical* 400: 42–48. https://doi.org/10.1016/j.molcata.2015.02.004.

25 Balakrishnan, A., Appunni, S., Chinthala, M., and Vo, D.-V.N. (2022). Biopolymer-supported TiO_2 as a sustainable photocatalyst for wastewater treatment: a review. *Environmental Chemistry Letters* 20 (5): 3071–3098. https://doi.org/10.1007/s10311-022-01443-8.

26 Xie, F., Gao, C., and Avérous, L. (2024). Alginate-based materials: enhancing properties through multiphase formulation design and processing innovation. *Materials Science & Engineering R: Reports* 159: 100799. https://doi.org/10.1016/j.mser.2024.100799.

27 Mostolizadeh, S. and Mostolizadeh, S. (2024, . https://doi.org/10.5772/INTECHOPEN.112666). Alginate, polymer purified from seaweed. In: *Alginate-Applications and Future Perspectives*, (ed. Severo I.A., Mariano A.B., Vargas J.V.C.), 1–16. New York, NY: IntechOpen.

28 Venkatesan, J., Bhatnagar, I., Manivasagan, P. et al. (2015). Alginate composites for bone tissue engineering: a review. *International Journal of Biological Macromolecules* 72: 269–281. https://doi.org/10.1016/J.IJBIOMAC.2014.07.008.

29 Du, K., Cárdenas-Arenas, A., Martínez, C.A. et al. (2017). Fabrication of transparent TiO_2 nanotube-based photoanodes for CdS/CdTe quantum co-sensitized solar cells. *Journal of Physics Conference Series* 786 (1): 012044. https://doi.org/10.1088/1742-6596/786/1/012044.

30 Wang, Y. and Lu, Y. (2023). Sodium alginate-based functional materials toward sustainable applications: water treatment and energy storage. *Industrial and Engineering Chemistry Research* 62 (29): 11279–11304. https://doi.org/10.1021/ACS.IECR.3C01082/ASSET/IMAGES/LARGE/IE3C01082_0015.JPEG.

31 Sutirman, Z.A., Sanagi, M.M., and Wan Aini, W.I. (2021). Alginate-based adsorbents for removal of metal ions and radionuclides from aqueous solutions: a review. *International Journal of Biological Macromolecules* 174: 216–228. https://doi.org/10.1016/J.IJBIOMAC.2021.01.150.

32 Pourjavadi, A., Doulabi, M., and Doroudian, M. (2014). Adsorption characteristics of malachite green dye onto novel kappa-carrageenan-g-polyacrylic acid/TiO_2-NH2 hydrogel nanocomposite. *Journal of the Iranian Chemical Society* 11 (4): 1057–1065. https://doi.org/10.1007/S13738-013-0374-6/METRICS.

33 Mishra, A., Mehta, A., and Basu, S. (2018). Clay supported TiO_2 nanoparticles for photocatalytic degradation of environmental pollutants: a review. *Journal of Environmental Chemical Engineering* 6 (5): 6088–6107. https://doi.org/10.1016/J.JECE.2018.09.029.

34 Eltaweil, A.S., Abd el-Monaem, E.M., Elshishini, H.M. et al. (2022). Recent developments in alginate-based adsorbents for removing phosphate ions from wastewater: a review. *RSC Advances* 12 (13): 8228–8248. https://doi.org/10.1039/D1RA09193J.

35 Qamar, S.A., Qamar, M., Basharat, A. et al. (2022). Alginate-based nano-adsorbent materials – bioinspired solution to mitigate hazardous environmental pollutants. *Chemosphere* 288: 132618. https://doi.org/10.1016/J.CHEMOSPHERE.2021.132618.

36 Dalponte, I., de Sousa, B.C., Mathias, A.L., and Jorge, R.M.M. (2019). Formulation and optimization of a novel TiO_2/calcium alginate floating photocatalyst. *International Journal of Biological Macromolecules* 137: 992–1001. https://doi.org/10.1016/J.IJBIOMAC.2019.07.020.

37 Wan, S., Zhao, W., Xiong, D. et al. (2022). Novel alginate immobilized TiO_2 reusable functional hydrogel beads with high photocatalytic removal of dye pollutions. *Journal of Polymer Engineering* 42 (10): 978–985. https://doi.org/10.1515/POLYENG-2022-0017/MACHINEREADABLECITATION/RIS.

38 Lam, W.H., Tee, L.H., and Ban, Z.H. (2024). A facile and rapid immobilization method of titanium dioxide-alginate composite for the photocatalytic removal of reactive black-5. *Bulletin of Chemical Reaction Engineering & Catalysis* 19 (2): 230–241. https://doi.org/10.9767/BCREC.20133.

39 Alhattab, Z.D., Aljeboree, A.M., Jawad, M.A. et al. (2023). Highly adsorption of alginate/bentonite impregnated TiO_2 beads for wastewater treatment: optimization, kinetics, and regeneration studies. *Caspian Journal of Environmental Sciences* 21 (3): 657–664. https://doi.org/10.22124/CJES.2023.6943.

40 El Fadl, F.I.A. and Taleb, M.F.A. (2023). Hybrid alginate/TiO_2/Ag bio-nanocomposite beads for catalytic hydrogenation of 2-nitrophenol. *Journal of Inorganic and Organometallic Polymers and Materials* 33 (7): 2142–2153. https://doi.org/10.1007/S10904-023-02651-7/METRICS.

41 Gjipalaj, J. and Alessandri, I. (2017). Easy recovery, mechanical stability, enhanced adsorption capacity and recyclability of alginate-based TiO_2 macrobead photocatalysts for water treatment. *Journal of Environmental Chemical Engineering* 5 (2): 1763–1770. https://doi.org/10.1016/J.JECE.2017.03.017.

42 Guetni, I., Belaiche, M., Ferdi, C.A. et al. (2024). Novel modified nanophotocatalysts of TiO_2 nanoparticles and TiO_2/Alginate beads with lanthanides [La, Sm, Y] to degrade the Azo dye Orange G under UV-VIS radiation. *Materials Science in Semiconductor Processing* 174: 108193. https://doi.org/10.1016/J.MSSP.2024.108193.

43 Azeroual, S., Khatib, K., Belfkira, A. et al. (2024). A novel approach for adsorption of organic dyes from aqueous solutions using a sodium alginate/titanium dioxide nanowire doped with zirconium cryogel beads. *Frontiers in Chemistry* 12: 1285230. https://doi.org/10.3389/FCHEM.2024.1285230/BIBTEX.

44 Mohamad Idris, N.H., Cheong, K.Y., Smith, S.M., and Lee, H.L. (2023). C,N-codoped TiO_2 nanoparticles immobilized on floating alginate beads for diazinon removal under solar light irradiation. *ACS Applied Nano Materials* https://doi.org/10.1021/ACSANM.3C03622/SUPPL_FILE/AN3C03622_SI_001.PDF.

45 Malik, R., Saxena, R., and Warkar, S.G. (2023). Organic hybrid hydrogels: a sustenance technique in waste-water treatment. *ChemistrySelect* 8 (5): e202203670. https://doi.org/10.1002/SLCT.202203670.

46 Thakur, S., Sharma, B., Verma, A. et al. (2018). Recent progress in sodium alginate based sustainable hydrogels for environmental applications. *Journal of Cleaner Production* 198: 143–159. https://doi.org/10.1016/J.JCLEPRO.2018.06.259.

47 Thakur, S., Pandey, S., and Arotiba, O.A. (2016). Development of a sodium alginate-based organic/inorganic superabsorbent composite hydrogel for adsorption of methylene blue. *Carbohydrate Polymers* 153: 34–46. https://doi.org/10.1016/J.CARBPOL.2016.06.104.

48 Wu, T., Sawut, A., Simayi, R. et al. (2023). Green synthesis and environmental applications of alginate/polyacrylamide/titanium dioxide composite hydrogel. *Journal of Applied Polymer Science* 140 (37): e54394. https://doi.org/10.1002/APP.54394.

49 Idumah, C.I., Ezika, A.C., and Okpechi, V.U. (2021). Emerging trends in polymer aerogel nanoarchitectures, surfaces, interfaces and applications. *Surfaces and Interfaces* 25: 101258. https://doi.org/10.1016/J.SURFIN.2021.101258.

50 Idumah, C.I. (2021). Novel trends in polymer aerogel nanocomposites. *Polymer-Plastics Technology and Materials* 60 (14): 1519–1531. https://doi.org/10.1080/25740881.2021.1912092.

51 Dai, J., Tian, Q., Sun, Q. et al. (2019). TiO_2-alginate composite aerogels as novel oil/water separation and wastewater remediation filters. *Composites. Part B, Engineering* 160: 480–487. https://doi.org/10.1016/J.COMPOSITESB.2018.12.097.

52 Nawaz, M., Khan, A.A., Hussain, A. et al. (2020). Reduced graphene oxide–TiO_2/sodium alginate 3-dimensional structure aerogel for enhanced photocatalytic degradation of ibuprofen and sulfamethoxazole. *Chemosphere* 261: 127702. https://doi.org/10.1016/J.CHEMOSPHERE.2020.127702.

53 Kumar, R., Oves, M., Ansari, M.O. et al. (2022). Biopolymeric Ni_3S_4/Ag_2S/TiO_2/calcium alginate aerogel for the decontamination of pharmaceutical drug and microbial pollutants from wastewater. *Nanomaterials* 12 (20): 3642. https://doi.org/10.3390/NANO12203642/S1.

54 Zhao, K., Zhang, X., Wei, J. et al. (2015). Calcium alginate hydrogel filtration membrane with excellent anti-fouling property and controlled separation performance. *Journal of Membrane Science* 492: 536–546. https://doi.org/10.1016/j.memsci.2015.05.075.

55 Huang, R.Y.M., Pal, R., and Moon, G.Y. (1999). Characteristics of sodium alginate membranes for the pervaporation dehydration of ethanol–water and isopropanol–water mixtures. *Journal of Membrane Science* 160 (1): 101–113. https://doi.org/10.1016/S0376-7388(99)00071-X.

56 Liu, F., Zhou, Z., Tu, Y. et al. (2022). Preparation of green sodium alginate adsorption membrane and its high adsorption performance for fluoroquinolones antibiotics. *Journal of Water Process Engineering* 49: 103124. https://doi.org/10.1016/j.jwpe.2022.103124.

57 Zhao, K., Feng, L., Lin, H. et al. (2014). Adsorption and photocatalytic degradation of methyl orange imprinted composite membranes using TiO_2/calcium alginate hydrogel as matrix. *Catalysis Today* 236: 127–134. https://doi.org/10.1016/j.cattod.2014.03.041.

58 Wang, X., Qin, W., Wang, L.X. et al. (2020). Desalination of dye utilizing carboxylated TiO_2/calcium alginate hydrogel nanofiltration membrane with high salt permeation. *Separation and Purification Technology* 253: 117475. https://doi.org/10.1016/j.seppur.2020.117475.

59 Gao, N., Xie, W., Xu, L. et al. (2023). Characterization of a chlorine resistant and hydrophilic TiO_2/calcium alginate hydrogel filtration membrane used for protein purification maintaining protein structure. *International Journal of Biological Macromolecules* 253: 126367. https://doi.org/10.1016/j.ijbiomac.2023.126367.

60 Liu, Y., Huang, J., and Guo, Z. (2024). TiO_2 deposited dual functional hydrogel coatings with superhydrophilic and photocatalytic properties for efficient oil/water separation and dye photodegradation. *Journal of Environmental Chemical Engineering* 12 (4): 113133. https://doi.org/10.1016/j.jece.2024.113133.

61 Zhao, K., Feng, L., Li, Z. et al. (2014). Preparation, characterization and photocatalytic degradation properties of a TiO_2/calcium alginate composite film and the recovery of TiO_2 nanoparticles. *RSC Advances* 4 (93): 51321–51329. https://doi.org/10.1039/C4RA08102A.

62 Omar, A. et al. (2022). Enhancing the optical properties of chitosan, carboxymethyl cellulose, sodium alginate modified with nano metal oxide and graphene oxide. *Optical and Quantum Electronics* 54 (12): 806. https://doi.org/10.1007/s11082-022-04107-7.

63 Reveendran, G.a./p. and Ong, S.-T. (2018). Application of experimental design for dyes removal in aqueous environment by using sodium alginate-TiO_2 thin film. *Chemical Data Collections* 15–16: 32–40. https://doi.org/10.1016/j.cdc.2018.03.002.

64 Wei, S., Zhang, X., Zhao, K. et al. (2016). Preparation, characterization, and photocatalytic degradation properties of polyacrylamide/calcium alginate/TiO_2 composite film. *Polymer Composites* 37 (4): 1292–1301. https://doi.org/10.1002/pc.23295.

65 Madaeni, S.S., Zinadini, S., and Vatanpour, V. (2011). A new approach to improve antifouling property of PVDF membrane using in situ polymerization of PAA functionalized TiO_2 nanoparticles. *Journal of Membrane Science* 380 (1–2): 155–162. https://doi.org/10.1016/j.memsci.2011.07.006.

66 Rajaeian, B., Heitz, A., Tade, M.O., and Liu, S. (2015). Improved separation and antifouling performance of PVA thin film nanocomposite membranes incorporated with carboxylated TiO_2 nanoparticles. *Journal of Membrane Science* 485: 48–59. https://doi.org/10.1016/j.memsci.2015.03.009.

67 Ehsan, M., Razzaq, H., Razzaque, S. et al. (2022). Recent advances in sodium alginate-based membranes for dehydration of aqueous ethanol through pervaporation. *Journal of Polymer Science* 60 (16): 2435–2453. https://doi.org/10.1002/pol.20220190.

68 Aburabie, J.H., Puspasari, T., and Peinemann, K.-V. (2020). Alginate-based membranes: paving the way for green organic solvent nanofiltration. *Journal of Membrane Science* 596: 117615. https://doi.org/10.1016/j.memsci.2019.117615.

69 Kurkuri, M.D., Kumbar, S.G., and Aminabhavi, T.M. (2002). Synthesis and characterization of polyacrylamide-grafted sodium alginate copolymeric membranes and their use in pervaporation separation of water and tetrahydrofuran mixtures. *Journal of Applied Polymer Science* 86 (2): 272–281. https://doi.org/10.1002/app.10948.

70 Mansourpanah, Y., Madaeni, S.S., Rahimpour, A. et al. (2009). Formation of appropriate sites on nanofiltration membrane surface for binding TiO_2 photocatalyst: performance, characterization and fouling-resistant capability. *Journal of Membrane Science* 330 (1–2): 297–306. https://doi.org/10.1016/j.memsci.2009.01.001.

71 Spanu, D., Bestetti, A., Hildebrand, H. et al. (2019). Photocatalytic reduction and scavenging of Hg(II) over templated-dewetted Au on TiO_2 nanotubes. *Photochemical & Photobiological Sciences* 18 (5): 1046–1055. https://doi.org/10.1039/c8pp00424b.

72 Tang, S., Wang, Z., Li, P. et al. (2018). Degradable and photocatalytic antibacterial Au-TiO_2/sodium alginate nanocomposite films for active food packaging. *Nanomaterials* 8 (11): 930. https://doi.org/10.3390/nano8110930.

73 Li, J.-H., Yan, B.-F., Shao, X.-S. et al. (2015). Influence of Ag/TiO_2 nanoparticle on the surface hydrophilicity and visible-light response activity of polyvinylidene fluoride membrane. *Applied Surface Science* 324: 82–89. https://doi.org/10.1016/j.apsusc.2014.10.080.

74 Xu, X., Zhu, T., Zheng, W. et al. (2023). A robust and transparent hydrogel coating for sustainable antifogging with excellent self-cleaning and self-healing ability. *Chemical Engineering Journal* 451: 137879. https://doi.org/10.1016/j.cej.2022.137879.

75 Alsalmah, H.A., Almehmadi, S.J., Farea, M.O. et al. (2023). Hybrid GO/TiO_2 nanoparticles reinforced NaAl/PVA blend: nanocomposites for high-performance energy storage devices. *Polymers for Advanced Technologies* 34 (9): 2831–2840. https://doi.org/10.1002/pat.6105.

7

Cellulose Nanocrystalline Biocomposites Applications in Wastewater Treatment

Sadia Khalid[1], Asiya Gul[1,2], and Tania Ali[1,2]

[1]*Nanosciences & Technology Department (NS&TD), National Centre for Physics (NCP), Quaid-i-Azam University, Islamabad, Pakistan*
[2]*Department of Chemistry, Hazara University, Mansehra, Pakistan*

7.1 Introduction

Our planet is endowed with vast resources that support life and foster growth. Among these, water stands out as a fundamental component, predating the emergence of life on Earth. As a crucial natural resource, water plays a pivotal role in shaping the trajectory of human society [1]. Water pollution from industrial waste is a significant environmental concern that has become alarming and has a significant impact on freshwater resources [2]. Access to safe drinking water is a paramount humanitarian objective, persisting as a pressing worldwide issue in modern times [3, 4]. According to Gupta et al. [5], 97.5% of the total water content on this planet is salt water, which is not useable without purification. Pure water is 2.5%, but almost 70% of it is in the form of polar ice. Less quantity of the total amount of fresh water is available to us, which is 1% or 0.007% of the total volume of water on the planet [6] (Figure 7.1).

Water quality is progressively deteriorating because of industrialization, population growth, global warming, and others [8]. The large amount of pollutants produced and discharged into water due to various industrial activities is a major problem for human and aquatic life [9]. A wide variety of pollutants can contaminate freshwater, including potentially hazardous elements [10], dyes [11], phenolic derivatives [12], agricultural chemicals [13], herbicides [14], medications [15], and cosmetics [16]. These contaminants are responsible for bioaccumulation, carcinogenesis, and mutagenesis in aquatic species and are harmful to aquatic plants, animals, and human life [17]. The management of water pollution is still a main task for scientists due to its increasing toxicity to humans and the ecosystem [18].

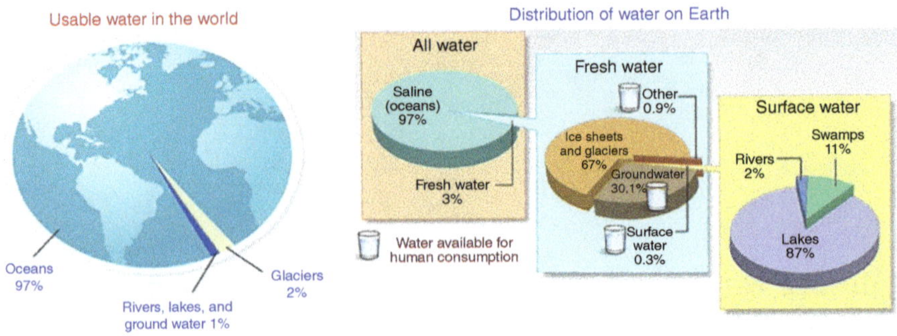

Figure 7.1 Water distribution in the world. *Source:* Reprinted with permission from [7]. Copyright 2018 ELSEVIER.

Therefore, the removal of water pollutants using environmentally friendly materials is necessary for sustainable development [19]. Cellulose is a semi-crystalline biopolymer. It is nonthermoplastic with excellent thermal stability in the range of 300–350 °C and decomposes without melting [20]. Nanocellulose is the best promising, maintainable, and eco-friendly biodegradable sorbent which can be used for environmental remediation after various surface modifications [21]. The extraction of nanocellulose from biomass waste allows its production on an industrial scale. This chapter comprehensively summarizes recent reports on wastewater treatment using nanocellulose-based nanocomposites. The core of this chapter is to highlight the importance of wastewater treatment by nanocellulose-based materials. Limitations, drawbacks, and remediation of nanocellulose-based composites are also comprehensively discussed.

Why Nanocellulose for Wastewater Treatment?

- Renewable, nontoxic, and abundant natural biopolymer
- Biocompatibility, adaptability, and low cytotoxicity
- Biodegradable and sustainable nanofiller
- High aspect ratio and high surface area
- Tunable surface chemistry
- Excellent mechanical properties and strong networking
- Promising antifouling properties
- Promising pollutant adsorption properties

7.2 Nanocellulose

Cellulose, a key constituent of lignocellulosic biomass, is the most plentiful natural polymer globally, with annual production exceeding 7.5 trillion tons [22].

Cellulose is an insoluble, tough, fibrous biomass that is essential for sustaining the erection and structure of plant cell walls. Since its extraction in 1938, its properties, structural characteristics, and biosynthesis have been extensively studied [22]. Numerous reviews have been published on this renewable material. Cellulose is characterized by its high molecular weight and uniform composition, consisting of a long chain of identical glucose molecules of repeating β-1,4 glucose units linked by glycosidic linkages, with each unit rotated 180° relative to its neighbor. The two ends of a cellulose chain are different; one reducing end has a hemiacetal unit and the other nonreducing end has a hydroxyl group. Typically, the degree of polymerization of cellulose chains contains about 20,000 glucose units, and the chains might be shorter in plant cell wall.

The cellulose chain comprises β-D-glucopyranose rings, each assuming a chair-like C1 conformation, where hydroxyl groups occupy equatorial sites and hydrogen atoms are positioned axially. Intramolecular hydrogen bonding between these hydroxyl groups and hydrogen atoms stabilizes this conformation [23]. Cellulose does not exist as a unimolecule, but as a fiber composed of multiple cellulose molecules. During biosynthesis, individual cellulose molecules twist together in an ordered fashion to form elementary fibrils, each consisting of about 36 cellulose molecules. These elementary fibrils aggregate into microfibrils, which then form cellulose fibers. The arrangement of fibrils within cellulose fibers varies. Cellulose microfibrils exhibit varying diameters (2–20 nm) depending on their source and biosynthetic conditions. Their structure is influenced by hydrogen bonding, van der Waals forces, and hydrophobic interactions, leading to the formation of microfibrils with distinct crystalline and amorphous regions [22].

7.2.1 Classification and Properties

Cellulose, an environmentally friendly biopolymer, provides the primary structural framework for plant and bacterial cell walls. Its molecular structure comprises anhydroglucose units interconnected via β-glucosidic bonds at carbon positions 1 and 4, with hydroxyl groups located at positions 2, 3, and 6 [24]. Nanocellulose is obtained from various raw materials by hydrolysis. The unique combination of properties, including biodegradability, biocompatibility, renewability, and nontoxicity, along with its tunable surface chemistry, high specific surface area, and excellent reactivity, makes it an exceptional filler material for developing sustainable biocomposites. It is used in industries such as paper [25], food packaging [26], medical [27], and hygiene products [28].

Nanocellulose exists in several forms (Figure 7.2) depending on its source: cellulose nanocrystals (CNCs), tunicate cellulose nanocrystals (TCNCs), cellulose nanofibers (CNFs), bacterial nanocellulose (BNC), and the newly identified hairy cellulose nanocrystalloids (HNC). CNFs are typically spaghetti-like structures

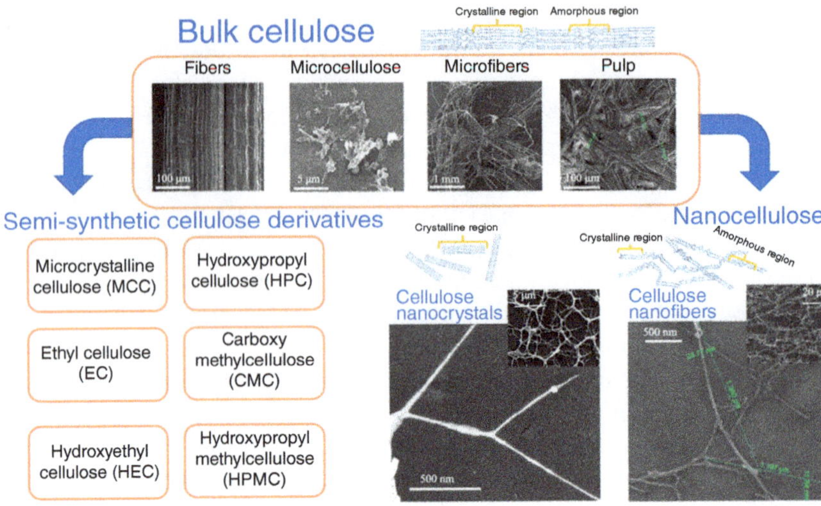

Figure 7.2 Cellulose derivatives: chemically modified celluloses (MCC, EC, HEC, HPC, CMC, and HPMC) and nanocelluloses (SEM images of diluted cellulose nanocrystals and nanofibers). *Source:* Reprinted with permission from [29]. Copyright 2021 American Chemical Society.

with widths of 3–50 nm and lengths of 0.5–5 μm. CNCs resemble rice-like needles, with lengths of 200–600 nm and widths of 3–50 nm. BNCs have a ribbon-like morphology, with widths of 20–100 nm and lengths of 1–5 μm. Various methods have been developed and optimized for the extraction of nanocellulose, although commercial production has not yet been achieved. Cellulose nanocrystals are extracted from different sources through acid hydrolysis using sulfuric acid (H_2SO_4), which leaves sulfate groups ($-SO4^{2-}$) on the CNC surface (30–90 mmol kg^{-1}) through esterification with hydroxyl groups, imparting negative charges to the CNC surface [30].

CNFs are derived from the cell walls of lignocellulosic biomass. In its natural form, cellulose exists as fibers composed of chains of cellulose molecules. These chains are tightly packed into protofibrils, because of strong intramolecular and intermolecular hydrogen bonding. The protofibrils further aggregate into microfibrils, which then form the larger cellulose fibers. These fibers display a dual-phase composition, featuring amorphous regions that confer flexibility and crystalline areas that enhance strength and stiffness [31].

Basically, cellulose is a semi-crystalline material with a tunable degree of crystallinity that varies with its origin, extraction method, extraction conditions (acid, time, etc.), and pretreatment process. Plant extraction results in a degree of crystallinity in the range of 40–60%, whereas cellulose extracted from bacteria and tunicin exhibits a degree of crystallinity in the range of 80–100%. For example, the acid hydrolysis process yields CNCs with higher crystallinity than CNFs, primarily due to the dissolution of amorphous cellulose components [32].

7.2.2 Preparation Methods

Nanocellulose extraction typically involves two main approaches: sulfuric acid hydrolysis and TEMPO-mediated oxidation. The acid hydrolysis method employs proton-mediated degradation to break down amorphous cellulose regions, yielding CNCs. Conversely, TEMPO-mediated oxidation selectively targets the primary hydroxyl groups at the C6 position, facilitating the disintegration of cellulose fibers into CNFs [33].

Ultrasonic cavitation is emerging as an effective and environmentally friendly technology for both the production and surface functionalization of nanocellulose. Generally, the extraction of nanocellulose from plant sources involves a multi-step top-down approach that includes the following methods:

- Physical: refining, mechanical grinding, ultrasonic grinding, and thermal treatment
- Chemical: acid hydrolysis, alkali treatment, and chemical modification
- Biological: enzymatic hydrolysis and the production of CNFs from bacteria
- Hybrid methods

Despite these methods, challenges such as low yield, by-product toxicity, high water usage, and significant energy consumption remain. A schematic representation of typical nanocellulose production methods is provided in Figure 7.3 [34].

7.2.3 Surface Modification

Bio-based sorbents are primarily constituted of three key biomass components: cellulose, hemicellulose, and lignin. Due to its hydrophilic nature, cellulose tends to have low buoyancy. Various methods are used to functionalize and modify the surface properties of nanocellulosic materials, including chemical, physical, and enzymatic functionalization (Figure 7.4) [35].

Hydrophilicity of cellulose is due to the presence of surface hydroxyl groups. The wettability of a solid substrate is usually examined by surface-free energy and surface roughness [36]. Surface of CNCs can be modified by noncovalent physical interactions or adsorption of molecules (e.g., surfactants—surface active agents) and macromolecules (poly-oxyethylene—PEO). Surfactants are typically amphiphilic in nature, consisting of hydrophobic (oil-loving tails) and hydrophilic (water-loving heads) groups. The head of the surfactant can attach to the surface of the CNCs, and the tail can protrude to provide a nonpolar surface that reduces

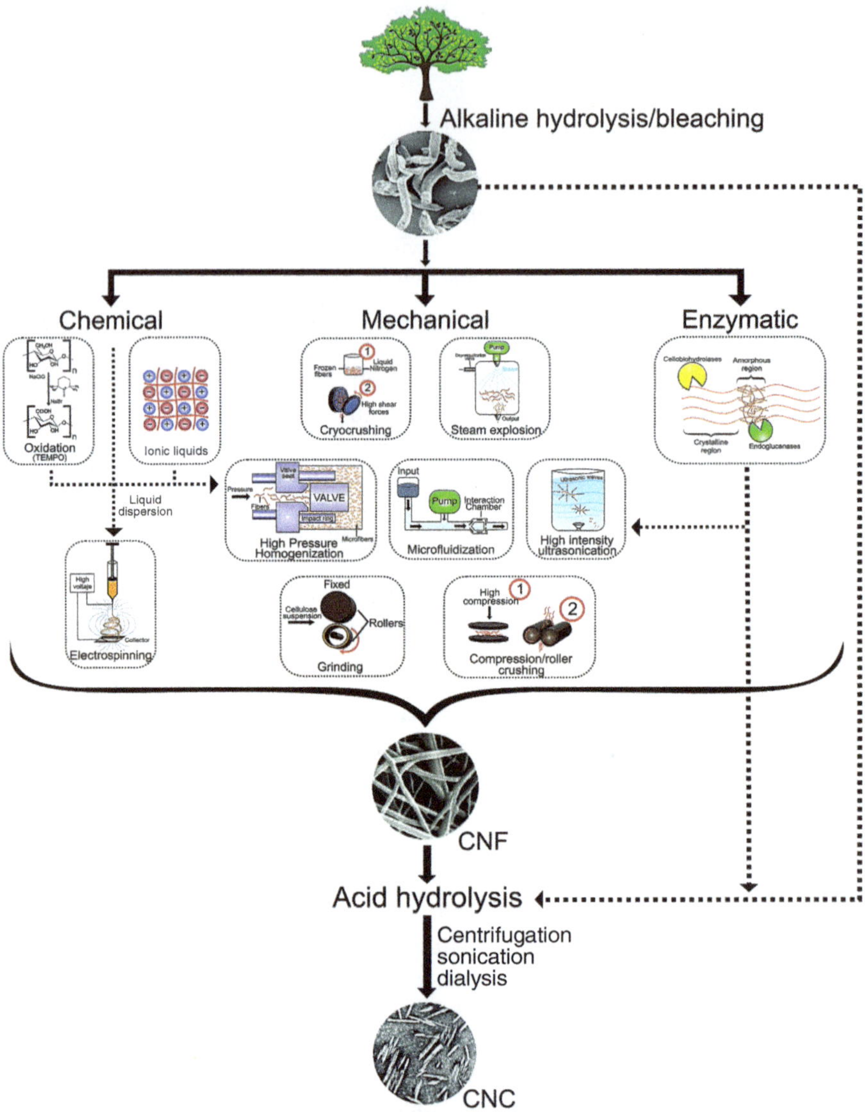

Figure 7.3 Conventional treatments to obtain nanocellulose. *Source:* Reprinted with permission from [34]. Copyright 2015 under Creative Commons Attribution 3.0 License.

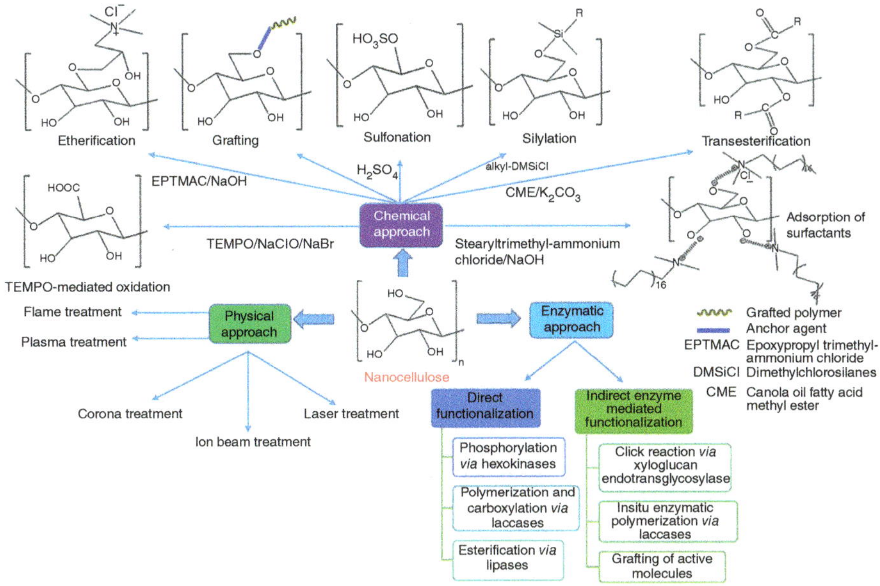

Figure 7.4 Schematic representation of the most commonly used surface modification routes of nanocellulose. *Source:* Reprinted with permission from [35]. Copyright 2021 under Creative Commons Attribution License (CC BY).

surface tension. Coating CNCs with PEO increases the thermal stability as well as the dispensability of CNCs [37].

Chemical modification of cellulose and hemicellulose involves replacing their functional groups with hydrophobic moieties through various catalytic and noncatalytic processes. As a result, the contact angle increases significantly, exceeding 150° (Figure 7.4) [35]. Notably, the adsorption-free energy reaches its maximum value when the surface is in a uniformly wet state, corresponding to a contact angle of 90°. Hydrophobization of nanocellulose surfaces is commonly achieved through esterification, a process that replaces hydroxyl groups with hydrophobic moieties, such as carbonyl groups (C=O), thereby enhancing hydrophobicity and oil sorption capacity. Alternatively, acetylation involves the reaction of acetyl groups (CO—CH_3) from acetic anhydride with hydroxyl groups (OH) on cellulose and hemicellulose, resulting in the formation of oleophilic acetate groups (O—CO—CH_3). Research has consistently shown that this method effectively increases the hydrophobicity of cellulosic materials. Silylation, a chemical modification technique, substitutes silyl groups (R3Si) onto the nanocellulose surface, resulting in enhanced hydrophobicity and improved mechanical properties. Furthermore, the TEMPO-mediated oxidation process facilitates the substitution of hydroxyl groups on CNCs with carboxyl groups (R—COOH).

Various chemical modification techniques have been employed to enhance the adsorption properties of cellulosic materials, including acid treatment, surfactant modification, polymer grafting, and the incorporation of metal oxide nanoparticles. Additionally, thermal treatments, such as hydrothermal carbonization, have been utilized to improve sorption capacity. This process involves subjecting the material to high temperatures and pressures in subcritical or supercritical water, resulting in reduced oxygen-containing functional groups and volatile components, increased carbon content, and enhanced hydrophobicity [24]. Moreover, magnetic nanoadsorbents have garnered significant interest due to their ease of recovery after use, which enhances operational efficiency and reduces operational costs during the separation process [38].

7.2.4 Adsorption and Adsorbents

Adsorption is a surface phenomenon where atoms, ions, or molecules from a liquid or gas phase bind to a solid surface, forming a thin layer of adhered species. The process of adsorption [39] was first discovered by C. W. Scheele in 1773 during his experiment in which gases were exposed to activated carbon. Further observations were made by Larvitz in 1792 and Khel in 1793 to study the decolorizing effects of activated carbon (charcoal) [40]. However, the concept of adsorption was initially introduced by Kayser in 1881 [5], who differentiated it from intermolecular penetration and emphasized that the core principle of adsorption lies in the surface accumulation of substances.

According to the *Handbook of Clay Science* [41], there are four different mechanisms for adsorption process, namely surface adsorption [42], partitioning [43], surface precipitation, and structural incorporation [44, 45]. During the surface adsorption process, contaminants accumulate on the surface or within the pores of the adsorbent material. This attachment can occur through physical adsorption, which is facilitated by relatively weak van der Waals forces [46] or chemisorption, as a result of chemical bond formation between pollutants and adsorbent surface [44]. The physically adsorbed pollutants can be easily removed, while the chemisorbed pollutants can be removed by specific treatments.

Many adsorbents have been identified for wastewater treatment. The ideal adsorbent should possess a unique set of characteristics, including high surface area, mechanical strength, porosity, fast adsorption rates, high capacity, and low cost [47, 48]. Furthermore, it should exhibit low water affinity, good recyclability, biodegradability, and resistance to physical and chemical stress. Adsorbents can be classified into five distinct categories: (i) natural materials in their raw form; (ii) natural materials modified for enhanced performance; (iii) manufactured materials, such as polymers and zeolites; (iv) waste-derived materials; and (v) biosorbents, including chitosan and microbial biomass [49]. Biopolymer-based materials and derivatives sourced from agricultural and industrial waste streams are attracting interest due to renewability, sustainability, and nontoxicity [24]. Examples of such materials include peat moss, kapok, sawdust, plant fibers, milkweed, and straw [50]. Nanocellulose, a bio-derived adsorbent, holds great promise for environmental remediation due to its ease of processing, customizable surface functionality, reactive properties, reversibility, and biodegradability [51].

7.2.5 Sorption Properties of Nanocellulose

Nanocellulose is favorable for wastewater treatment because of its exceptional physicochemical properties such as higher surface area, high degree of crystallinity, larger aspect ratio, outstanding mechanical stiffness and surface tension, tunable surface functionalization, and water stability. Nanocellulose sorbents can be presented in six categories [52]:

- Powders, beads, droplets
- Capsules
- Continuous fibers
- Films and membranes
- Hydrogels
- Aerogels and dried foams

The most common form of nanocellulose adsorbent is aerogel. They offer the advantage of easy recovery of the adsorbent. Aerogels constitute a novel class of nanoscale mesoporous materials, distinguished by their exceptionally high porosity (95–99%), low density (approximately 10 $mg\,mL^{-1}$), and expansive surface area (exceeding 100 $m^2\,g^{-1}$), surpassing the characteristics of xerogels and cryogels. The production of aerogels involves the replacement of the liquid solvent within the gel with air, without altering the network structure or gel volume, typically achieved through supercritical drying, although this method's scalability is limited despite its cost-effectiveness [53]. Adsorption processes used for contaminant removal are typically simple and highly efficient.

Recently, environmental and ecological concerns have prompted the development of innovative, environmentally friendly, and cost-effective sorbent materials. The integration of nanotechnology and biotechnology has led to the development of nanocellulose-based biosorbents, which have shown impressive sorption capacities [54]. Adsorption properties of CNCs, CNFs, and HNC-based sorbents are summarized in Figure 7.5. Key parameters that affect the efficiency of these biosorbents include surface functionality, chemical composition, synthesis procedure, and experimental conditions [55].

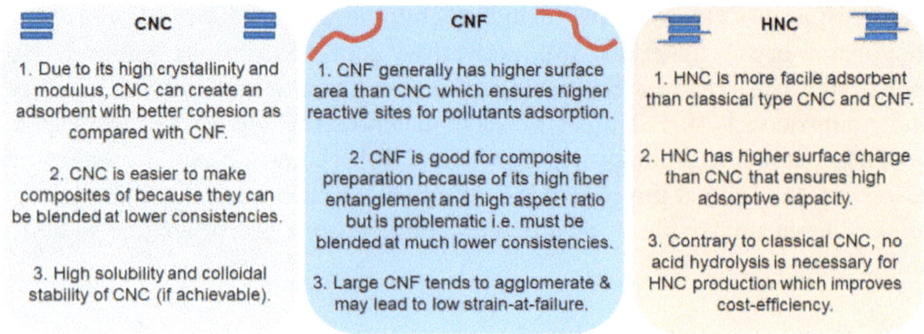

Figure 7.5 Adsorption characteristics of CNC, CNF, and HNC-based sorbents for water purification. *Source:* Reprinted with permission from [21]. Copyright 2022 American Chemical Society.

7.2.6 Biocomposites of Nanocellulose

One of the most notable features of nanocrystalline cellulose is their ability to act as nanofillers, reinforcing various synthetic and natural polymers. The wide range of applications for CNCs significantly improves the mechanical properties of materials such as adhesives, tires, gloves, and coatings. Here are some common applications of nanocrystalline-based biocomposites:

i) Nanocellulose-based biocomposites extensively play an important role in food packaging, agriculture, and hygiene products. Examples include CNC-β-chitosan (CS) [56], CNC-gelatin-corn starch (ST) [57], polylactide-co-polyglycolide (PLGA)-alginate (Alg)-CS-CNCs [58], CS/CNCs [59], CNCs-PVA (poly(vinyl alcohol)/carboxymethyl cellulose (CMC)) [60], CNCs-lignin [61], CNCs-natural rubber (NR) [62], soy protein/CNC/pine needle extract (PNE) [63], CNCs-acrylonitrile-butadiene-styrene [64], CMC/ST/polysaccharide-CNCs [65], gluten-cellulose nanofibrils (CNF)-CNCs [66], nitrile rubber (NBR)-CNCs [67], CNCs-oxidize [68], and carboxylated styrene butadiene rubber/CNCs [69].
ii) The CNCs-based bionanocomposites are significantly used in bioscience, biomedical, bioprinting, and biotechnology, e.g., CNCs-poly (vinyl alcohol) (PVA)/sodium alginate (Na-Alg) use in scaffold for tissue engineering [70], CNCs-soy protein-glycerol-ethylene glycol diglycidyl ether (EGDE) [71], Gelatin/glycerin-CNCs [65], CNCs-CMC [72], CS-CNCs [73], Alg/CNC 3D [74], CNC/whey protein/alginate [75], CNC/calcium alginate [76], protein/C [56] NC [77], and whey protein-CNC [78].
iii) CNCs-based composites are widely employed in energy and environmental applications, e.g., multi-walled carbon nanotubes (MWCNT)/CS/CNCs

[79], CS/polyethylene oxide (PEO)-CNCs [80], CNCs/CS [81], and PLA/polybutylene succinate (PBS)/CNC [82].

CNCs are highly regarded as premium nanofillers, capable of enhancing the mechanical strength and barrier properties of biopolymer-based materials. The combination of CNCs with chitosan (CS) has garnered significant attention, largely due to the inherent compatibility between these two polysaccharide-based components [6, 18–24]. Chitosan-cellulose nanocrystal (CS/CNC) bio-nanocomposites display a unique combination of beneficial properties, including biocompatibility, biodegradability, stability in biological environments, and bioactive functionality [25, 26]. The exceptional properties of CS/CNC nanocomposites render them versatile materials for various applications, spanning biomedical, packaging, agricultural, and environmental fields. The subsequent section delves into the application of nanocellulose-based composites in wastewater treatment.

7.3 Wastewater Treatment by Nanocellulose Biocomposites

Water pollutants can be divided into inorganic and organic contaminants. Industrial and agricultural pollutants include petrochemicals, herbicides, insecticides, detergents, chlorinated solvents, hydrocarbons, and cosmetics industry waste.

Emerging organic contaminants (EOCs) refer to a collection of synthetic organic compounds, including pharmaceuticals, personal care products, hormones, pesticides, flame retardants, surfactants, and plasticizers, which have raised concerns due to their potential environmental and health impacts [83].

These nanocellulose-based composites have shown great promise in removing a broad range of pollutants from water, with a particular affinity for capturing cationic species such as heavy metal ions. Moreover, with strategic modifications, such as the incorporation of ammonium groups, they can also efficiently remove anionic contaminants like nitrates and phosphates, as well as various organic pollutants, including dyes, oils, pesticides, and pharmaceutical residues. Hydroxyl groups on nanocellulose adsorbents prepared by acid hydrolysis appeal to dyes and cationic heavy metal ions [84]. Surface functionalization of nanocellulose increases the affinity toward charged moieties and hence improves the adsorption capacities. TEMPO-modified nanocellulose fibers also exhibit anionic carboxylic groups that attract cations. An alternative approach involves flocculation, which aggregates colloidal particles through electrostatic interactions, facilitating the removal of suspended contaminants from wastewater. Additionally, nanocellulose-based photocatalysts have been explored for catalytic degradation of organic pollutants and oil–water separation, offering a more environmentally friendly option compared to synthetic materials. Details of wastewater treatment applications based on nanocellulose have been described in subsections of this heading [85].

7.3.1 Oil Spills Cleanup

In recent years, the increasing extent of oil pollution because of expansion of offshore oil industries, oil transportation, and frequent oil spills has raised significant environmental concerns globally, particularly impacting freshwater resources. Groundwater contamination has led to severe health issues [86–88]. The devastating impact of oil spills on marine ecosystems is a growing concern, with an estimated 100 million gallons of crude oil leaking into seawater annually. Effective management of oil-contaminated water is crucial to mitigate its harmful effects on aquatic life, human health, and the environment [89].

Various methods are employed to clean up oil spills, broadly classified into physical, biological, and chemical approaches. Physical methods utilize adsorbents, booms, and skimmers. Biological methods involve bioremediation processes utilizing specific bacteria. Chemical methods employ chemical coagulation, electrocoagulation strategies, and dispersants with surfactants. Each method has its advantages and limitations, influencing its selection based on specific conditions and objectives. Some methods may become obsolete because of some aspects, e.g., cost effectiveness, inadequate oil sorption and retention, poor biodegradability, and difficulties in recovery [90].

Adsorption using adsorbent nanomaterials has garnered significant attention as a physical method for oil spill cleanup. Choosing the appropriate adsorbent is crucial for effective adsorption processes. Numerous studies have investigated various aspects, including adsorption capacity and mechanisms, for separating oil from rivers. It's been demonstrated that adsorbents based on nanomaterials possess substantial oil uptake capacity from water surfaces, thereby reducing treatment costs and mitigating harmful effects on ecosystems [91].

During the adsorption process, sorbate molecules aggregate on the surface of the sorbent without infiltrating its structure [92]. Ideally, effective adsorbent materials exhibit distinct characteristics, including high hydrophobicity, oleophilicity, and rapid uptake capacity [93]. Normally, efficient water/oil separation takes place separating oil from water followed or along with oil adsorption. Sedimentation, hydrocyclone, and centrifugation are the conventional oil/water separation methods that are based on variations in gravity or density.

Biomass-based sorbents such as cellulose, biocarbon, chitosan, and their derivatives are potential candidates for selective sorption of oil spills [94]. The adsorption of oils onto nanocellulose surfaces is influenced by the surface

Table 7.1 Nanocellulose-based sorbents for oil spill cleanup.

Sr#	Materials	Method	Preparation method	Sorption ($g\,g^{-1}$)	Year	Reference
1.	CNC/PVA/MTMS	Acid hydrolysis	Freeze drying	32.7	2019	[95]
2.	CNF/PVA/MTMS/RGO	Acid hydrolysis	Freeze drying	51.7	2024	[96]
3.	BNC	Sorbitol	Membrane	147	2025	[97]
4.	$Ti_3C_2T_x$/TOCNF	Tempo oxidized	Freeze drying	98.92	2025	[98]

characteristics, including surface functionalization and morphology (roughness). Cellulosic sorbents, comprising cellulose, hemicellulose, and lignin, exhibit varying degrees of hydrophobicity. Although lignin contributes to hydrophobic properties, cellulose and hemicellulose are naturally hydrophilic. To improve their performance in water treatment, researchers have explored physical, chemical, and thermal modifications to enhance the hydrophobicity and oleophilicity of cellulose-based biopolymers [24].

Nanocellulose-based sorbents recently reported are summarized in Table 7.1.

7.3.2 Anions Removal

CNF-based adsorbents can be applied not only for metal cations adsorption but also for removal of anion pollutants that include NO_3^-, PO_4^{3-}, SO_4^{2-}, I^-, and F^-. Introducing positively charged functional groups onto the surface of CNF can amplify their affinity for negatively charged anions, facilitating enhanced adsorption through a combination of electrostatic, ion exchange, and complexation mechanisms [99].

7.3.3 Heavy Metal Ions Removal

The term "heavy metal" encompasses a group of dense metals (density > $5.0\,g\,cm^{-3}$) with high atomic weights (>65) that exhibit toxic properties even at minute concentrations (parts per million or billion). Examples of heavy metals include mercury (Hg), cadmium (Cd), lead (Pb), hexavalent chromium (Cr), copper (Cu), and arsenic (As), among others [100, 101]. Although certain heavy metals (e.g., Cu, Se, and Zn) are important trace elements required for human metabolic processes, higher concentrations can result in toxicity. Wastewater discharged by chemical industries often contains toxic elements like arsenic, which can cause chronic poisoning leading to skin keratinization and cancer. The maximum adsorption capacity for arsenic has been observed to reach up to $350\ mg\ g^{-1}$, varying with cellulose-based adsorbent systems. Effective pH ranges for As(V) removal efficiency of 95% were found to be between 3.0 and 8.8, while for As(III) removal, the effective pH range was between 6 and 9 [102].

A variety of technologies, including adsorption, chemical precipitation, solvent extraction, reverse osmosis, and electrochemical treatment, have been employed to remove heavy metals and radioactive ions from wastewater, ensuring effective treatment and environmental protection. Adsorption is one of the appropriate methods owing to its high adsorption capacity, cost effectiveness, and facile operation. Diverse nanosorbents including carbon nanotubes (CNTs), graphene oxide (GO), polymers, metal and metal oxides, zeolites, and their composites have been employed for heavy metal removal. Nanocellulose aerogels have demonstrated high efficiency for heavy metal ion adsorption [103]. The removal

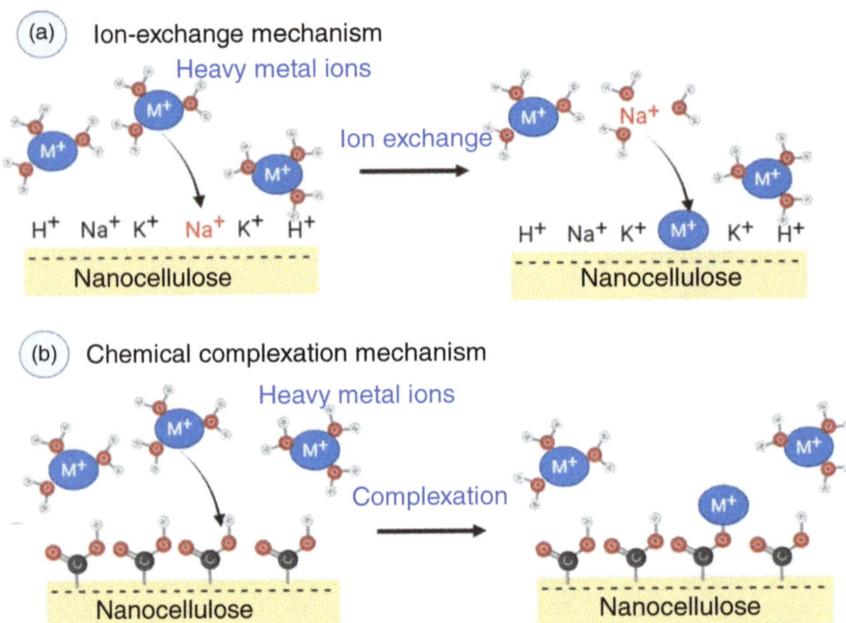

Figure 7.6 Heavy metal removal mechanism from water system using nanocelluloses: (a) Ion exchange mechanism which involves the adsorption of hazardous metal ions (Mn^{+}) takes the place of other ions (K^{+}, Na^{+}, H^{+}) already associated with the nanocellulose surface; (b) chemical complexation mechanism, in which the carboxyl (−COO−) and hydroxyl (−OH) groups of the nanocelluloses have specific site interactions with particular hazardous metal ions. *Source:* Reprinted with permission from [104] Copyright 2021 under Creative Commons Attribution (CC BY) license.

of heavy metals by nanocellulose-based adsorbents is primarily governed by two key mechanisms: ion exchange and chemical complexation, which facilitate the efficient capture and retention of metal ions (see Figure 7.6) [104].

- Ion exchange mechanism: This process entails metal ions adsorption (Mn^+) by nanocellulose surfaces via preassociated ions such as K^+, Na^+, and H^+ (see Figure 7.2a).
- Chemical complexation: This mechanism involves specific interactions between surface functional groups of nanocellulose, such as carboxyl (—COO—) and hydroxyl (—OH) groups, with hazardous metal ions (Mn^+) (see Figure 7.2b).

The adsorption capacity of nanocelluloses is capped by several key factors, including surface area, surface chemistry, and stoichiometric constraints, which generally restrict the maximum uptake to less than half of the available surface ionic sites. Therefore, enhancing the surface area through surface functionalization can increase complexing sites for metal ion adsorption. Cellulose nanofibrils (CNF) have been more commonly utilized as nanocellulose adsorbents compared to CNCs and BNCs.

7.3.4 Adsorption of Dyes

The textile industry's massive consumption of approximately 700,000 tons of dyes annually poses significant environmental and health concerns, as these dyes are largely nonbiodegradable. Various methods, including adsorption, membrane filtration, coagulation, flocculation, ozonation, and biological treatment, have been explored for dye removal. Among these, adsorption stands out as the most practical, straightforward, and cost-effective approach. Ongoing research focuses on developing natural, recyclable adsorbents with high surface areas and tailored functional groups to effectively capture hazardous dyes [105].

Dyes can be classified as:

- Direct dyes: The anionic nature and water solubility of these dyes enable them to form strong bonds with nanocellulose materials.
- Acid dyes: Although these anionic dyes are water-soluble, they display a relatively low affinity for adsorption onto nanocellulose surfaces.
- Basic dyes: These cationic dyes are water-soluble.
- Sulfur dyes (—S—, —S—S—, —Sn—) and azoic (—N=N—), vat dyes: These dyes are water-insoluble and are primarily known for their high affinity for adsorption onto nanocelluloses.
- Reactive dyes: The chemical structures of these water-soluble synthetic chromophores feature chlorotriazine and ethylsulfonyl functional groups, which enable the formation of covalent bonds with cellulose surfaces.

Table 7.2 Nanocellulose-based adsorbents for dyes.

Sr#	Adsorbent	Dye	Adsorption capacity ($mg\,g^{-1}$)	pH	Year	Reference
1.	Pd-mPDA-BNC membrane: BNC embedded with mesoporous polydopamine (mPDA) and decorated with palladium (Pd) nanoparticles	MB	99% and water flux 137.4 $L\,m^{-2}\,h^{-1}$	—	2020	[109]
2.	Melanized-C-CNF foams	Crystal violet (cationic dye)	93% first cycle (removal efficiency)	—	2021	[110]
3.	TOCN/PDMDAAC: TEMPO-oxidized cellulose nanofibrils (TOCNs)/diallyl dimethyl ammonium chloride (PDMDAAC)	Methyl orange (MO, an anionic dye)	404.86	7	2024	[111]
4.		MB, a cationic dye	520.83			
5.	CNCs	Indigo carmine blue (ICB) dye	746.24	4	2024	[112]
6.	1.55% guar gum/1.46% NC/0.84% borax	MB	24.05	8	2024	[113]

The first two categories of dyes interact with nanocellulose through mechanisms involving hydrogen bonding, dipolar forces, and nonspecific hydrophobic interactions. Cationic surface modifications of nanocellulose have been shown to enhance the uptake of direct and acid dyes. The enhancement of adsorption and filtration capacities in nanocellulose-based adsorbents can be achieved by employing porous films or nanocellulose aerogels with tunable porosity.

In a recent study, hydrogel beads (NCC—CH) were synthesized by immobilizing nanocellulose onto chitosan. The resulting hydrogel beads demonstrated a specific surface area (SSA) of $25.77\,m^2\,g^{-1}$ and a zeta potential of +50.6 mV. Analysis of functional groups further confirmed the existence of suitable functional groups on the adsorbent for the adsorption of textile dyes. The adsorption of crystal violet (CV) was 94.75%, $60\,mg\,L^{-1}$ (0.5 g NCC—CH) for 1 hour at pH 9 and for methylene blue (MB), 95.88% removal, $30\,mg\,L^{-1}$ (0.13 g NCC—CH) for 1 hour at pH 9 [106].

In another study, a reusable composite film consisting of poly (vinyl alcohol), chitin, nanocellulose, and maleic acid demonstrated exceptional adsorption capabilities, with a maximum methylene blue (MB) uptake of $467.5\,mg\,g^{-1}$. Furthermore, the film's reusability was confirmed, maintaining an average adsorption efficiency of 83.67% after five cycles [107].

A research study assessed the effectiveness of a novel adsorbent material, comprising nanocellulose and chitosan microbeads (CCMB), in removing direct blue 78 dye (DB78) from aqueous solutions. The CCMB, characterized by a specific surface area of $10.4\,m^2\,g^{-1}$ and a positive net surface charge, achieved maximum removal efficiencies of 91.5% and 88.4% at initial dye concentrations of 50 and $100\,mg\,L^{-1}$, respectively. These optimal removal efficiencies were obtained under acidic conditions (pH 3.5) and a mixing time of 120 minutes. The equilibrium data were accurately modeled using the Langmuir isothermal equation (R2 = 0.99) [108].

There are several reports for removal of dyes via nanocellulose-based adsorbents, few are summarized in Table 7.2.

7.3.5 Elimination of Drugs/Pharmaceuticals

Phenolics are habitually found in wastewater discharged from pharmaceutical, oil refining, petrochemical, and coal processing industries. The most common of these phenolic compounds are phenol, 2,4,6-trichlorophenol, cresol, and 2,4-dinitrophenol. Typically, the physicochemical wastewater treatment methods such as filtration, coagulation, flotation, sedimentation, and lime softening are used for their removal [114].

Laccase-like nanozymes are commonly applied in oxidizing phenols catalytically. But they are not stable in harsh environmental conditions. Recently,

BNC/Cu laccase-like nanozyme was prepared via in situ reduction of copper on the BNC. The BNC/Cu bionanozyme showed excellent specificity toward 2,4-dichlorophenol (2,4-DP) with K_m =0.187 mM. The BNC/Cu bionanozyme demonstrated remarkable thermal stability, exhibiting optimal activity at 85 °C. Notably, it retained 38% of its laccase activity after 11 days and 46.77% after four cycles of use. Furthermore, this bionanozyme efficiently oxidized over 70% of 1,4-dichlorophenol and phenol within a 5-hour timeframe [115].

Nanocellulose are also effective for removal of hazardous drugs such as diclofenac, ciprofloxacin, ceftriaxone, and metronidazole from industrial wastewater. Cyclodextrin-modified CNC nanocomposites with paramagnetic Fe_3O_4/SiO_2 nanorods separated the anesthetic drug, procaine hydrochloride, and the antidepressant drug imipramine hydrochloride effectively [116].

Similarly, a novel nanocellulose-based adsorbent, modified with the block copolymer Jeffamine ED 600 (NC-Jeffamine), demonstrated effective removal of sulfamethoxazole, acetaminophen, and *N*,*N*-diethyl-meta-toluamide (DEET) from aqueous solutions through electrostatic interactions [117].

7.3.6 Removal of Pesticides

Pesticides and chemical fertilizers in agricultural activities have contaminated water with their residues and metabolites, causing a high risk to human health and the climate. The removal of pesticides from contaminated water is essential to reduce the ill effects of pesticides on human beings. Current research has been focused on biocompatible nanocellulose-based derivatives (e.g., nanocomposites with magnetic or photoactive nanoparticles, clay, and activated carbon) in removing pesticides. The percent removal or adsorption capacity of pesticides depends on various parameters, for example, adsorbent dosage, initial pesticide concentration, pH, time of contact, and specifically functionalization of nanocellulose. Furthermore, the type of loaded nanoparticles on nanocellulose also affects the efficiency of the nanocomposites [118].

On the other hand, ingested pesticides can alter the bioavailability and toxicity of food compounds within the human body. A recent study by Ho et al. explored the potential of bi-functional modified CNFs to reduce fat digestion and mitigate pesticide (boscalid) bioavailability in the gastrointestinal tract (GIT) through the formation of inclusion complexes and interactions with surface hydroxyl groups (Figure 7.7). In this study, (2-hydroxypropyl)-β-cyclodextrin (HPBCD) was used to functionalize CNFs through the cross-linker (citric acid) via esterification. Functionalized CNFs showed enhanced performance compared to CNFs in both applications [119].

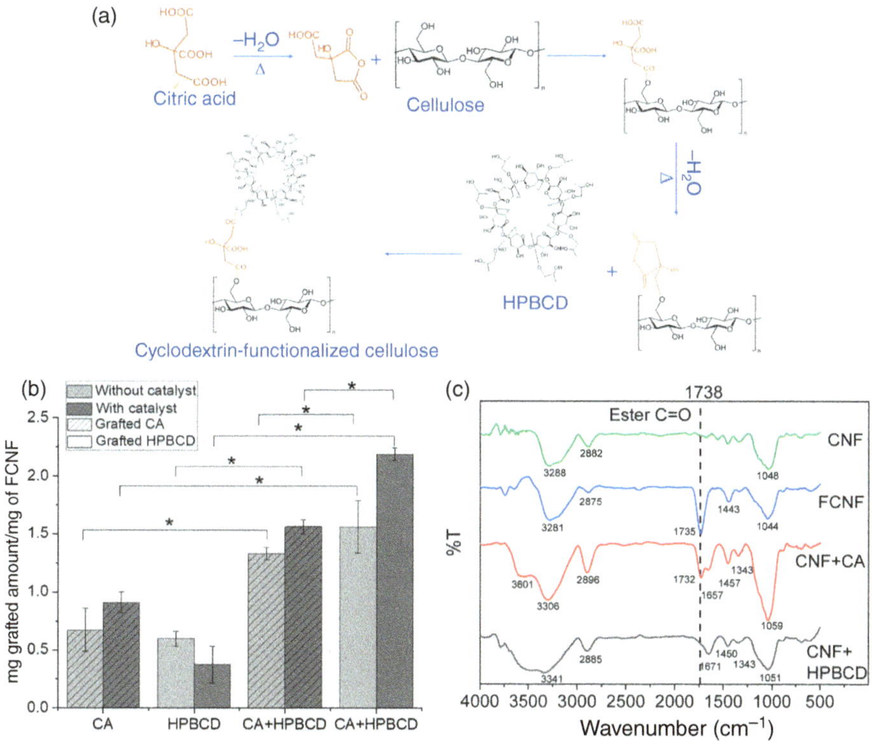

Figure 7.7 Cyclodextrin-functionalized CNF (FCNF) formation. (a) Proposed mechanism of functionalization of the CNF via esterification. (b) Amount of CA and HPBCD grafted onto the CNF across different conditions. (c) FTIR analysis indicates ester bonds resulting from the functionalization. Data are means ± SD, $n = 3$, one-way ANOVA, Tukey's HSD post hoc test, $^*p < 0.05$, significant between compared groups. *Source:* Reprinted with permission from [119] Copyright 2023 under CC-BY-NC-ND 4.0 license.

7.3.7 Antimicrobial and Antibacterial

Contamination of water sources by microorganisms can lead to the spread of various waterborne illnesses, including cholera and giardiasis, gastroenteritis, and cryptosporidiosis. These outbreaks commonly involve bacteria, e.g., *Escherichia coli*, *Shigella dysenteriae*, *Legionella* sp., *Vibrio cholera*, and *Campylobacter jejuni*. Removal of microbes is executed through filtration, partitioning and fractionation, and chromatography (ion-exchange, affinity, gel permeation). Recently, nanocellulose-based antimicrobial agents have gained attention owing to their tunable surface functionalities. Additionally, this biopolymer also provides support to other antimicrobial agents having instability issues. Nanocellulose composites with organic (e.g., Ag, ZnO) and inorganic (polymer matrices such as chitosan) materials have shown improved performance as compared to pristine antimicrobial agents. The presence of nanocellulose brings the synergistic effects to enhance the antimicrobial effect with improved strength [120].

Existing challenges to nanocellulose-based hybrid antimicrobial agents are:

- Efficacy against bacteria or fungi with no cytotoxicity to human/animal cells
- Leaching of metallic nanoparticles from nanocellulose support that would reduce antimicrobial activity
- Stability, reusability, and safety of hybrids
- Economic and sustainable preparation methods to fulfill large volume remediation demands

The superoleophobic nanocellulose-based membranes are promising environmentally friendly adsorbents for oil/water separation. However, bacterial erosion results in the structural damage of these biopolymer membranes and affects their separation efficiency. Recently, Peng et al. reported the in situ decoration of silver nanoparticles (AgNPs) on TCNCs by hydrothermal reduction using silver nitrate. Nanocomposite membranes fabricated through vacuum-assisted filtration method using AgNPs@TCNC suspension exhibited superhydrophilicity, nanoporous structure alongside underwater superoleophobicity. The AgNPs@TCNC membranes exhibited >324 $L\,m^{-2}\,h^{-1}\,bar^{-1}$ water flux with >99% oil rejection. Also, the nanocomposite membranes showed antibacterial (*E. coli* and *S. aureus*) efficacy indicating their stable performance suitable for longer use and storage [121].

7.4 Challenges and Limitations

Nanocellulose-based composite sorbents have garnered significant attention in recent years, driven by their eco-friendly properties and potential for sustainable applications. However, their widespread adoption faces several challenges and limitations. A major hurdle is the inherent hydrophilicity of nanocellulose obtained by acid hydrolysis, which requires surface functionalization to enhance sorption capacity. However, many conventional surface modification techniques, such as silylation by chemical vapor deposition (CVD) or atomic layer deposition (ALD), are time-consuming and energy-intensive. In addition, the use of toxic chemicals in these processes poses hazards to both social health and the atmosphere, detracting from the green appeal of the material.

Furthermore, while surface modification increases sorption capacity, it also reduces the biodegradability of nanocellulose by introducing surface functional groups that inhibit bacterial proliferation. Balancing this trade-off is critical, especially considering the importance of recyclability and regeneration in sustainable sorbents. Achieving this requires careful consideration of material manufacturing factors.

Despite these challenges, surface modification of nanocellulose remains a promising method for contaminant removal. It offers flexibility in design, allowing tailor-made solutions based on the type of contaminant, level of contamination, purification method, and production cost. By optimizing parameters such as surface area, wettability, mechanical strength, flexibility, and recyclability, the sorption capacity of nanocellulose can be significantly enhanced. However, it's worth noting that surface-modified nanocellulose may have limited selectivity in pollutant adsorption, highlighting the need for the development of multifunctional nanocellulose materials.

Similarly, in the arena of water treatment, polymer nanocomposites offer immense potential but also pose dangers to social health, aquatic life, and the atmosphere [122]. These risks arise from the possible discharge of nanoparticles into the atmosphere, either directly or through the polymer matrix. Despite notable progress, important uncertainties remain, requiring continued research and innovation to effectively address these challenges.

Limitations and Challenges

- Time and energy consumption for isolation or extraction of nanocellulose materials
- Solubility and dispersion issues of nanocellulose
- Environmentally unfriendly surface modification of nanocellulose
- Low production yield of nanocellulose and costly long extraction routes
- Challenging industrial scale production

7.5 Conclusion and Future Perspectives

Cellulose-based adsorbents offer an eco-friendly and cost-effective solution for wastewater treatment. Moreover, nanocellulose, a biodegradable and renewable natural polymer, has shown great promise in environmental remediation due to its biocompatibility, reusability, and sustainability. Few considerations are appealing from the literature review:

1) Nanocellulose-based aerogels are excellent sorbents due to their promising sorption properties.
2) Surface functionalization of nanocellulose is a prerequisite for improved sorption capacity.

3) The complex extraction methods and costly modification processes are the chief obstacles aimed at the commercialization of nanocellulose-based adsorbents. Therefore, the upcoming perspective is to discover the cost-effective and less toxic extraction methods and modifications for nanocellulose.
4) Adsorption techniques using chemical and polymer grafting have been observed to be the most effective practices for environmental remediation and sustainability.
5) Nanocellulose based on natural/biopolymeric compounds (starch, chitosan, pectin, etc.) to support the adsorption process in wastewater treatment.
6) Sustainability, physico-chemical properties, surface functionalities, catalytic activity, chemical stability, and significant specificity for pollutants can be tuned by computational modeling and rational design of nanocellulose-based adsorbents.

In short, nanocellulose-based adsorbents have promising potential as abundant, low-cost, and environmentally friendly alternatives to conventional adsorbent materials.

References

1 Machado, A.C.M., Veról, A.P., Battemarco, B.P., and Miguez, M.G. (2020). Proposal of a complementary tool to assess environmental river quality: the river classification index (RCI). *Journal of Cleaner Production* 254: 120000.

2 Gaafar, M., Mahmoud, S.H., Gan, T.Y., and Davies, E.G.R. (2020). A practical GIS-based hazard assessment framework for water quality in stormwater systems. *Journal of Cleaner Production* 245: 118855.

3 Irby, I.D. and Friedrichs, M.A. (2019). Evaluating confidence in the impact of regulatory nutrient reduction on Chesapeake Bay water quality. *Estuaries and Coasts* 42 (1): 16–32.

4 Chen, B., Wang, M., Duan, M. et al. (2019). In search of key: protecting human health and the ecosystem from water pollution in China. *Journal of Cleaner Production* 228: 101–111.

5 Gupta, V.K., Carrott, P.J.M., Ribeiro Carrott, M.M.L., and Suhas (2009). Low-cost adsorbents: growing approach to wastewater treatment—a review. *Critical Reviews in Environmental Science and Technology* 39 (10): 783–842.

6 Snoeyink, V.L. and Jenkins, D. (1980). *Water Chemistry*. Wiley.

7 Singh, N.B. and Susan, A.B.H. (2018). 21—Polymer nanocomposites for water treatments. In: *Polymer-based Nanocomposites for Energy and Environmental Applications* (ed. M. Jawaid and M.M. Khan), 569–595. Woodhead Publishing.

8 Nemerow, N.L. and Agardy, F.J. (1998). *Strategies of Industrial and Hazardous Waste Management*. Wiley.

9 Helmer, R. and Hespanhol, I. (1997). *Water Pollution Control: A Guide to the Use of Water Quality Management Principles*. Abingdon, UK: CRC Press.

10 Silva, L.F., Silva, L.F.O., Izquierdo, M. et al. (2011). Leaching of potential hazardous elements of coal cleaning rejects. *Environmental Monitoring and Assessment* 175 (1): 109–126.

11 Padhi, B. (2012). Pollution due to synthetic dyes toxicity & carcinogenicity studies and remediation. *International Journal of Environmental Sciences* 3 (3): 940.

12 Anku, W.W., Mamo, M.A., and Govender, P.P. (2017). Phenolic compounds in water: sources, reactivity, toxicity and treatment methods. *Phenolic Compounds-natural Sources, Importance and Applications* 12: 419–443.

13 Faust, S.D. and Aly, O.M. (1964). Water pollution by organic pesticides. *Journal-American Water Works Association* 56 (3): 267–279.

14 Martínez, R.C., Gonzalo, E.R., Laespada, M.E., and San Roman, F.J. (2000). Evaluation of surface-and ground-water pollution due to herbicides in agricultural areas of Zamora and Salamanca (Spain). *Journal of Chromatography A* 869 (1–2): 471–480.

15 Freitas, L.D.A.A. and Radis-Baptista, G. (2021). Pharmaceutical pollution and disposal of expired, unused, and unwanted medicines in the brazilian context. *Journal of Xenobiotics* 11 (2): 61–76.

16 Francke, I. and Castro, J. (2013). Carbon and water footprint analysis of a soap bar produced in Brazil by Natura Cosmetics. *Water Resources and Industry* 1: 37–48.

17 Chowdhury, S. and Balasubramanian, R. (2014). Recent advances in the use of graphene-family nanoadsorbents for removal of toxic pollutants from wastewater. *Advances in Colloid and Interface Science* 204: 35–56.

18 Järup, L. (2003). Hazards of heavy metal contamination. *British Medical Bulletin* 68 (1): 167–182.

19 Choudhury, R.R., Sahoo, S.K., and Gohil, J.M. (2020). Potential of bioinspired cellulose nanomaterials and nanocomposite membranes thereof for water treatment and fuel cell applications. *Cellulose* 27 (12): 6719–6746.

20 Zhou, C. and Wang, Y. (2020). Recent progress in the conversion of biomass wastes into functional materials for value-added applications. *Science and Technology of Advanced Materials* 21 (1): 787–804.

21 Das, R., Lindstrom, T., Sharma, P.R. et al. (2022). Nanocellulose for sustainable water purification. *Chemical Reviews* 122 (9): 8936–9031.

22 Brown, R.M. Jr. (1996). The biosynthesis of cellulose. *Journal of Macromolecular Science, Part A: Pure and Applied Chemistry* 33 (10): 1345–1373.

23 Ali, S.D., Zoia, L., and Orlandi, M.E. (2018). Lignocellulosic biomass: a way to bio-based economy. Biofillers from renewable lignocellulosic feedstock for elastomers. p. 1.

24 Ifelebuegu, A.O. and Johnson, A. (2017). Nonconventional low-cost cellulose- and keratin-based biopolymeric sorbents for oil/water separation and spill cleanup: a review. *Critical Reviews in Environmental Science and Technology* 47 (11): 964–1001.

25 Spagnuolo, L., D'Orsi, R., and Operamolla, A. (2022). Nanocellulose for paper and textile coating: the importance of surface chemistry. *ChemPlusChem* 87 (8): e202200204.

26 Nasra, A.E.S., Hasanien, Y.A., and El Mogy, S.A. The potential of nanocellulose composites for sustainable food packaging applications. *Polymer Composites* 46 (6): 5081.

27 Sreedharan, M., Vijayamma, R., Liyaskina, E. et al. (2024). Nanocellulose-based hybrid scaffolds for skin and bone tissue engineering: a 10-year overview. *Biomacromolecules* 25 (4): 2136–2155.

28 Kassie, B.B., Getahun, M.J., Azanaw, A. et al. (2024). Surface modification of cellulose nanocrystals for biomedical and personal hygiene applications. *International Journal of Biological Macromolecules* 282: 136949.

29 Combariza, M.Y., Martínez-Ramírez, A.P., and Blanco-Tirado, C. (2021). Perspectives in nanocellulose for crude oil recovery: a minireview. *Energy & Fuels* 35 (19): 15381–15397.

30 Araki, J., Wada, M., and Kuga, S. (2001). Steric stabilization of a cellulose microcrystal suspension by poly (ethylene glycol) grafting. *Langmuir* 17 (1): 21–27.

31 Saba, N., Mohammad, F., Pervaiz, M. et al. (2017). Mechanical, morphological and structural properties of cellulose nanofibers reinforced epoxy composites. *International Journal of Biological Macromolecules* 97: 190–200.

32 Rostamabadi, H., Bist, Y., Kumar, Y. et al. (2024). Cellulose nanofibers, nanocrystals, and bacterial nanocellulose: fabrication, characterization, and their most recent applications. *Future Postharvest and Food* 1 (1): 5–33.

33 Hu, M., Lv, X., Wang, Y. et al. (2024). Guideline for the extraction of nanocellulose from lignocellulosic feedstocks. *Food Biomacromolecules* 1 (1): 9–17.

34 John, R., Mauricio, B., and Yhors, C. (2015). *Current Trends in the Production of Cellulose Nanoparticles and Nanocomposites for Biomedical Applications* (ed. P.M. Cellulose and J.H.L. Ornaghi), 8. IntechOpen: Rijeka.

35 Trache, D., Tarchoun, A.F., Derradji, M. et al. (2020). Nanocellulose: from fundamentals to advanced applications. *Frontiers in Chemistry* 8: 392.

36 Lin, N. (2014). *Cellulose Nanocrystals: Surface Modification and Advanced Materials*. Université de Grenoble.

37 Ghasemlou, M., Daver, F., Ivanova, E.P. et al. (2021). Surface modifications of nanocellulose: from synthesis to high-performance nanocomposites. *Progress in Polymer Science* 119: 101418.

38 Zhang, Z., Ahmed, A.I.S., Malik, M.Z. et al. (2023). Cellulose/inorganic nanoparticles-based nano-biocomposite for abatement of water and wastewater pollutants. *Chemosphere* 313: 137483.

39 Cassidy, H.G. (1951). Adsorption and chromatography. *Soil Science* 72: 81.

40 Robens, E. (1994). Some intriguing items in the history of adsorption. In: *Studies in Surface Science and Catalysis* (ed. J. Rouquerol et al.), 109–118. Elsevier.

41 Bergaya, F. and Lagaly, G. (2013). *Handbook of Clay Science*. Newnes.

42 Churchman, G.J., Gates, W.P., Theng, B.K., and Yuan, G. (2006). Clays and clay minerals for pollution control. *Developments in Clay Science* 1: 625–675.

43 Chiou, C.T. (2003). *Partition and Adsorption of Organic Contaminants in Environmental Systems*. Wiley.

44 O'Day, P.A. and Vlassopoulos, D. (2010). Mineral-based amendments for remediation. *Elements* 6 (6): 375–381.

45 Malamis, S. and Katsou, E. (2013). A review on zinc and nickel adsorption on natural and modified zeolite, bentonite and vermiculite: examination of process parameters, kinetics and isotherms. *Journal of Hazardous Materials* 252–253: 428–461.

46 Bandosz, T.J. (2006). *Activated Carbon Surfaces in Environmental Remediation*. Elsevier.

47 Al-Majed, A.A., Adebayo, A.R., and Hossain, M.E. (2012). A sustainable approach to controlling oil spills. *Journal of Environmental Management* 113: 213–227.

48 Ge, J., Zhao, H.Y., Zhu, H.W. et al. (2016). Advanced sorbents for oil-spill cleanup: recent advances and future perspectives. *Advanced Materials* 28 (47): 10459–10490.

49 Linsen, B.G. (1970). Physical and chemical aspects of adsorbents and catalysts.

50 Dave, D. and Ghaly, A.E. (2011). Remediation technologies for marine oil spills: a critical review and comparative analysis. *American Journal of Environmental Sciences* 7 (5): 423–440.

51 Faiz Norrrahim, M.N., Mohd Kasim, N.A., Knight, V.F. et al. (2021). Nanocellulose: a bioadsorbent for chemical contaminant remediation. *RSC Advances* 11 (13): 7347–7368.

52 Niinivaara, E. and Cranston, E.D. (2020). Bottom-up assembly of nanocellulose structures. *Carbohydrate Polymers* 247: 116664.

53 Mahfoudhi, N. and Boufi, S. (2017). Nanocellulose as a novel nanostructured adsorbent for environmental remediation: a review. *Cellulose* 24 (3): 1171–1197.

54 Thakur, V., Guleria, A., Kumar, S. et al. (2021). Recent advances in nanocellulose processing, functionalization and applications: a review. *Materials Advances* 2 (6): 1872–1895.

55 Abdelhamid, H.N. (2024). Nanocellulose-based materials for water pollutant removal: a review. *International Journal of Molecular Sciences* 25 (15): 8529.

56 Zhang, H., Jung, J., and Zhao, Y. (2017). Preparation and characterization of cellulose nanocrystals films incorporated with essential oil loaded β-chitosan beads. *Food Hydrocolloids* 69: 164–172.

57 Alves, J., dos Reis, K.C., Menezes, E.G.T. et al. (2015). Effect of cellulose nanocrystals and gelatin in corn starch plasticized films. *Carbohydrate Polymers* 115: 215–222.

58 Rescignano, N., Fortunati, E., Armentano, I. et al. (2015). Use of alginate, chitosan and cellulose nanocrystals as emulsion stabilizers in the synthesis of biodegradable polymeric nanoparticles. *Journal of Colloid and Interface Science* 445: 31–39.

59 Corsello, F.A., Bolla, P.A., Anbinder, P.S. et al. (2017). Morphology and properties of neutralized chitosan-cellulose nanocrystals biocomposite films. *Carbohydrate Polymers* 156: 452–459.

60 El Achaby, M., El Miri, N., Aboulkas, A. et al. (2017). Processing and properties of eco-friendly bio-nanocomposite films filled with cellulose nanocrystals from sugarcane bagasse. *International Journal of Biological Macromolecules* 96: 340–352.

61 Hambardzumyan, A., Foulon, L., Bercu, N.B. et al. (2015). Organosolv lignin as natural grafting additive to improve the water resistance of films using cellulose nanocrystals. *Chemical Engineering Journal* 264: 780–788.

62 Neto, W.P.F., Silvério, H.A., Dantas, N.O. et al. (2013). Extraction and characterization of cellulose nanocrystals from agro-industrial residue–Soy hulls. *Industrial Crops and Products* 42: 480–488.

63 Yu, Z., Sun, L., Wang, W. et al. (2018). Soy protein-based films incorporated with cellulose nanocrystals and pine needle extract for active packaging. *Industrial Crops and Products* 112: 412–419.

64 Ma, L., Zhang, Y., Meng, Y. et al. (2015). Preparing cellulose nanocrystal/acrylonitrile-butadiene-styrene nanocomposites using the master-batch method. *Carbohydrate Polymers* 125: 352–359.

65 Ning, N., Wang, Z., Yao, Y. et al. (2015). Enhanced electromechanical performance of bio-based gelatin/glycerin dielectric elastomer by cellulose nanocrystals. *Carbohydrate Polymers* 130: 262–267.

66 Fortunati, E., Luzi, F., Jiménez, A. et al. (2016). Revalorization of sunflower stalks as novel sources of cellulose nanofibrils and nanocrystals and their effect on wheat gluten bionanocomposite properties. *Carbohydrate Polymers* 149: 357–368.

67 Chen, Y., Zhang, Y., Xu, C., and Cao, X. (2015). Cellulose nanocrystals reinforced foamed nitrile rubber nanocomposites. *Carbohydrate Polymers* 130: 149–154.

68 Mariano, M., El Kissi, N., and Dufresne, A. (2016). Cellulose nanocrystal reinforced oxidized natural rubber nanocomposites. *Carbohydrate Polymers* 137: 174–183.

69 Cao, X., Xu, C., Wang, Y. et al. (2013). New nanocomposite materials reinforced with cellulose nanocrystals in nitrile rubber. *Polymer Testing* 32 (5): 819–826.

70 Kumar, A., Lee, Y., Kim, D. et al. (2017). Effect of crosslinking functionality on microstructure, mechanical properties, and in vitro cytocompatibility of cellulose nanocrystals reinforced poly (vinyl alcohol)/sodium alginate hybrid scaffolds. *International Journal of Biological Macromolecules* 95: 962–973.

71 Zhang, S., Xia, C., Dong, Y. et al. (2016). Soy protein isolate-based films reinforced by surface modified cellulose nanocrystal. *Industrial Crops and Products* 80: 207–213.

72 Oun, A.A. and Rhim, J.-W. (2016). Isolation of cellulose nanocrystals from grain straws and their use for the preparation of carboxymethyl cellulose-based nanocomposite films. *Carbohydrate Polymers* 150: 187–200.

73 de Mesquita, J.P., Donnici, C.L., Teixeira, I.F., and Pereira, F.V. (2012). Bio-based nanocomposites obtained through covalent linkage between chitosan and cellulose nanocrystals. *Carbohydrate Polymers* 90 (1): 210–217.

74 Wu, Y., Lin, Z.Y.(W.), Wenger, A.C. et al. (2018). 3D bioprinting of liver-mimetic construct with alginate/cellulose nanocrystal hybrid bioink. *Bioprinting* 9: 1–6.

75 Roshanghias, S. and Madadlou, A. (2018). Functional and gel properties of whey protein nanofibrils as influenced by partial substitution with cellulose nanocrystal and alginate. *International Dairy Journal* 81: 53–61.

76 Ma, X., Li, R., Zhao, X. et al. (2017). Biopolymer composite fibres composed of calcium alginate reinforced with nanocrystalline cellulose. *Composites Part A: Applied Science and Manufacturing* 96: 155–163.

77 Singla, R., Soni, S., Padwad, Y.S. et al. (2017). Sustained delivery of BSA/HSA from biocompatible plant cellulose nanocrystals for in vitro cholesterol release from endothelial cells. *International Journal of Biological Macromolecules* 104: 748–757.

78 Sarkar, A., Li, H., Cray, D., and Boxall, S. (2018). Composite whey protein–cellulose nanocrystals at oil-water interface: towards delaying lipid digestion. *Food Hydrocolloids* 77: 436–444.

79 Trigueiro, J.P.C., Silva, G.G., Pereira, F.V., and Lavall, R.L. (2014). Layer-by-layer assembled films of multi-walled carbon nanotubes with chitosan and cellulose nanocrystals. *Journal of Colloid and Interface Science* 432: 214–220.

80 Naseri, N., Mathew, A.P., and Oksman, K. (2016). Electrospinnability of bionanocomposites with high nanocrystal loadings: the effect of nanocrystal surface characteristics. *Carbohydrate Polymers* 147: 464–472.

81 Karim, Z., Mathew, A.P., Grahn, M. et al. (2014). Nanoporous membranes with cellulose nanocrystals as functional entity in chitosan: removal of dyes from water. *Carbohydrate Polymers* 112: 668–676.

82 Zhang, X., Marchand, C., Sullivan, B. et al. (2018). Naloxone access for emergency medical technicians: an evaluation of a training program in rural communities. *Addictive Behaviors* 86: 79–85.

83 García, J., García-Galán, M.J., Day, J.W. et al. (2020). A review of emerging organic contaminants (EOCs), antibiotic resistant bacteria (ARB), and antibiotic resistance genes (ARGs) in the environment: increasing removal with wetlands and reducing environmental impacts. *Bioresource Technology* 307: 123228.

84 Mbisana, M., Keroletswe, N., Nareetsile, F. et al. (2024). Nanocellulose composites: synthesis, properties, and applications to wastewater treatment. *Cellulose* 31 (18): 10651–10678.

85 Mautner, A. (2020). Nanocellulose water treatment membranes and filters: a review. *Polymer International* 69 (9): 741–751.

86 Bidgoli, H., Mortazavi, Y., and Khodadadi, A.A. (2019). A functionalized nano-structured cellulosic sorbent aerogel for oil spill cleanup: synthesis and characterization. *Journal of Hazardous Materials* 366: 229–239.

87 Yin, T., Zhang, X., Liu, X., and Wang, C. (2017). Resource recovery of Eichhornia crassipes as oil superabsorbent. *Marine Pollution Bulletin* 118 (1–2): 267–274.

88 Parmar, K.R., Dora, D.T.K., Pant, K.K., and Roy, S. (2019). An ultra-light flexible aerogel-based on methane derived CNTs as a reinforcing agent in silica-CMC matrix for efficient oil adsorption. *Journal of Hazardous Materials* 375: 206–215.

89 Lu, Y. and Yuan, W. (2018). Superhydrophobic three-dimensional porous ethyl cellulose absorbent with micro/nano-scale hierarchical structures for highly efficient removal of oily contaminants from water. *Carbohydrate Polymers* 191: 86–94.

90 Purohit, B.K., Tewari, S., Prasad, K.S.N.V. et al. (2024). Marine oil spill clean-up: a review on technologies with recent trends and challenges. *Regional Studies in Marine Science* 80: 103876.

91 Bi, H., Mulligan, C.N., Lee, K. et al. (2025). Nanotechnology for oil spill response and cleanup in coastal regions. *Environmental Science: Nano* 12 (1): 41–47.

92 Zhou, L. and Xu, Z. (2020). Ultralight, highly compressible, hydrophobic and anisotropic lamellar carbon aerogels from graphene/polyvinyl alcohol/

cellulose nanofiber aerogel as oil removing absorbents. *Journal of Hazardous Materials* 388: 121804.

93 Sayyad Amin, J., Vared Abkenar, M., and Zendehboudi, S. (2015). Natural sorbent for oil spill cleanup from water surface: environmental implication. *Industrial & Engineering Chemistry Research* 54 (43): 10615–10621.

94 Ouyang, D., Lei, X., and Zheng, H. (2023). Recent advances in biomass-based materials for oil spill cleanup. *Nanomaterials* 13 (3): 620.

95 Gong, X., Wang, Y., Zeng, H. et al. (2019). Highly porous, hydrophobic, and compressible cellulose nanocrystals/poly(vinyl alcohol) aerogels as recyclable absorbents for oil–water separation. *ACS Sustainable Chemistry & Engineering* 7 (13): 11118–11128.

96 Lu, J., Feng, Q., Wang, J. et al. (2024). Efficient solar-driven crude oil cleanup via graphene/cellulose aerogel with radial and centrosymmetric design. *Journal of Hazardous Materials* 477: 135418.

97 Jagtap, A., Krushnatray, A., and Dastager, S.G. (2025). Bacterial nanocellulose (BNC) produced from sorbitol as a sustainable nano-filter for oil-water separation. *International Journal of Biological Macromolecules* 284: 137953.

98 Guo, Y., Zhang, J., Wang, S. et al. (2025). Hydrophobic Ti3C2Tx/TEMPO oxidized cellulose nanofibers composite aerogel for efficient oil-water separation. *Polymers* 17 (3): 273.

99 Prasad, C., Jeong, S.G., Won, J.S. et al. (2024). Review on recent advances in cellulose nanofibril based hybrid aerogels: synthesis, properties and their applications. *International Journal of Biological Macromolecules* 261: 129460.

100 Kaur, J., Sengupta, P., and Mukhopadhyay, S. (2022). Critical review of bioadsorption on modified cellulose and removal of divalent heavy metals (Cd, Pb, and Cu). *Industrial & Engineering Chemistry Research* 61 (5): 1921–1954.

101 Abouzeid, R.E., Khiari, R., el-Wakil, N., and Dufresne, A. (2019). Current state and new trends in the use of cellulose nanomaterials for wastewater treatment. *Biomacromolecules* 20 (2): 573–597.

102 Motloung, M.T., Magagula, S.I., Kaleni, A. et al. (2023). Recent advances on chemically functionalized cellulose-based materials for arsenic removal in wastewater: a review. *Water* 15: https://doi.org/10.3390/w15040793.

103 Yu, K., Yang, L., Zhang, S., and Zhang, N. (2025). Nanocellulose-based aerogels for the adsorption and removal of heavy-metal ions from wastewater: a review. *Materials Today Communications* 43: 111744.

104 Salama, A., Abouzeid, R., Leong, W.S. et al. (2021). Nanocellulose-based materials for water treatment: adsorption, photocatalytic degradation, disinfection, antifouling, and nanofiltration. *Nanomaterials* 11 (11): 3008.

105 Qiao, A., Cui, M., Huang, R. et al. (2021). Advances in nanocellulose-based materials as adsorbents of heavy metals and dyes. *Carbohydrate Polymers* 272: 118471.

106 Poornachandhra, C., Jayabalakrishnan, R.M., Prasanthrajan, M. et al. (2023). Cellulose-based hydrogel for adsorptive removal of cationic dyes from aqueous solution: isotherms and kinetics. *RSC Advances* 13 (7): 4757–4774.

107 Mok, C.F., Ching, Y.C., Osman, N.A.A. et al. (2020). Adsorbents for removal of cationic dye: nanocellulose reinforced biopolymer composites. *Journal of Polymer Research* 27 (12): 373.

108 Bassyouni, M., Zoromba, M.S., Abdel-Aziz, M.H., and Mosly, I. (2022). Extraction of nanocellulose for eco-friendly biocomposite adsorbent for wastewater treatment. *Polymers* 14: https://doi.org/10.3390/polym14091852.

109 Gholami Derami, H., Gupta, P., Gupta, R. et al. (2020). Palladium nanoparticle-decorated mesoporous polydopamine/bacterial nanocellulose as a catalytically active universal dye removal ultrafiltration membrane. *ACS Applied Nano Materials* 3 (6): 5437–5448.

110 Tran-Ly, A.N., de France, K.J., Rupper, P. et al. (2021). Melanized-cationic cellulose nanofiber foams for bioinspired removal of cationic dyes. *Biomacromolecules* 22 (11): 4681–4690.

111 Song, R., Zhu, E., Hou, R. et al. (2024). Nanocellulose-based composite beads as the amphoteric adsorbent for effective adsorption of cationic and anionic dyes. *Cellulose* 31 (8): 5077–5095.

112 El-Fattah, W.A., Guesmi, A., Hamadi, N.B. et al. (2024). A green synthesis of cellulose nanocrystals biosorbent for remediation of wastewater containing industrial dye. *Colloids and Surfaces A: Physicochemical and Engineering Aspects* 681: 132729.

113 Rana, H., Anamika, Sareen, D., and Goswami, S. (2024). Nanocellulose-based ecofriendly nanocomposite for effective wastewater remediation: a study on its process optimization, improved swelling, adsorption, and thermal and mechanical behavior. *ACS Omega* 9 (8): 8904–8922.

114 Patel, M., Kumar, R., Kishor, K. et al. (2019). Pharmaceuticals of emerging concern in aquatic systems: chemistry, occurrence, effects, and removal methods. *Chemical Reviews* 119 (6): 3510–3673.

115 Achamyeleh, A.A., Ankala, B.A., Workie, Y.A. et al. (2023). Bacterial nanocellulose/copper as a robust laccase-mimicking bionanozyme for catalytic oxidation of phenolic pollutants. *ACS Omega* 8 (45): 43178–43187.

116 Chen, L., Berry, R.M., and Tam, K.C. (2014). Synthesis of β-cyclodextrin-modified cellulose nanocrystals (CNCs)@Fe_3O_4@SiO_2 superparamagnetic nanorods. *ACS Sustainable Chemistry & Engineering* 2 (4): 951–958.

117 Herrera-Morales, J., Morales, K., Ramos, D. et al. (2017). Examining the use of nanocellulose composites for the sorption of contaminants of emerging concern: an experimental and computational study. *ACS Omega* 2 (11): 7714–7722.

118 Rana, A.K., Mishra, Y.K., Gupta, V.K., and Thakur, V.K. (2021). Sustainable materials in the removal of pesticides from contaminated water: perspective on macro to nanoscale cellulose. *Science of the Total Environment* 797: 149129.

119 Ho, C.G., Setyawati, M.I., DeLoid, G.M. et al. (2023). Cellulose nanofiber platform for pesticide sequestration in the gastrointestinal tract. *ACS Omega* 8 (18): 16106–16118.

120 Hemraz, U.D., Lam, E., and Sunasee, R. (2023). Recent advances in cellulose nanocrystals-based antimicrobial agents. *Carbohydrate Polymers* 315: 120987.

121 Peng, K., Huang, Y., Peng, N., and Chang, C. (2022). Antibacterial nanocellulose membranes coated with silver nanoparticles for oil/water emulsions separation. *Carbohydrate Polymers* 278: 118929.

122 Mittal, V. (2012). *Characterization Techniques for Polymer Nanocomposites*. Wiley.

8

Graphene Bionanocomposites: Preparation and Applications in Wastewater Remediation

P. K. Praseetha[1], R. Shalini[2], K. Ravichandran[2], M. Varshini[3], A. Viji[4], and A. R. Umayal Sundari[5]

[1]*Department of Nanotechnology, Noorul Islam Centre for Higher Education, Kumaracoil, Tamil Nadu, India*
[2]*PG & Research Department of Physics, Thanthai Hans Roever College (Autonomous) (Affiliated to Bharathidasan University, Tiruchirappalli), Perambalur, Tamil Nadu, India*
[3]*PG & Research Department of Physics, AVVM Sri Pushpam College (Affiliated to Bharathidasan University, Tiruchirappalli), Thanjavur, Tamil Nadu, India*
[4]*Department of Physics, Kongunadu College of Engineering and Technology (Affiliated to Anna University), Thottiyam, Tamil Nadu, India*
[5]*Department of Physics, Periyar Maniammai Institute of Science and Technology (Deemed to be University), Thanjavur, Tamil Nadu, India*

8.1 Introduction

Graphene stands tall among the carbon nanomaterials (CNMs) as it has supreme properties like lightweight, high surface area, with prominent thermal and mechanical properties. The graphene or graphene oxides added to biomolecules can amend their electrical, thermomechanical, and barrier performance that makes them perfect for broad applications in industrial, agricultural, and medical applications. Polysaccharides, proteins, and biopolymers could be used for the purpose of synthesizing such materials. Recent literature has revealed the development of drug delivery vehicles, electrode materials, biosensors, energy storage devices, and alternate antibiotics from these bionanocomposite substrates of graphene [1].

Graphene nanocomposites are broadly classified based on their nature of synthesis, the bio-molecules they conjugate with, and the mode of application they are used for. Varied methodologies could be used for the synthesis of graphene at the laboratory scale such as chemical vapor deposition, arc discharge, microwave synthesis, chemical synthesis, molecular beam epitaxy, solvothermal techniques, pyrolysis, colloidal suspension, epitaxial growth, scotch tape method, electrically assisted synthesis, carbon nanotube separation, and exfoliation. Stable rigid foams of bionanocomposites were synthesized by emulsion template, acid precipitation, or by free drying methods [2].

The general characteristics of the bionanocomposite materials prepared were estimated for their optical, morphometric, and temperature tolerance by using UV-visible spectroscopy, Fourier Transform Infrared Spectroscopy (FTIR), photoluminescent spectrum, near infrared spectrum, Raman spectroscopy, Scanning Electron Microscopy (SEM), Tunneling Electron Microscopy (TEM), particle size analysis, and Thermogravimetric/Differential Thermal Analysis (TG/DTA), respectively. Specific characterization of the material for its hydrophobicity, sensing, bioadsorption, photocatalysis, antibacterial, toxicity, permeation, temperature, and pH sensitivity were all estimated based on the criterion for their circumstantial application.

Graphene is present as sheets with sp2 hybridization of carbon atoms that are perfectly arranged in a hexagonal lattice structure. The chemical interactions that hold the graphene are van der Waals, electrostatic (pi-pi interactions), and chemical bonding forming a lamellar structure [3]. The carbon atoms of every layer are bond to three different carbons in the neighborhood through one pi and three sigma bonds. The pi electron of each carbon atom is de-localized during the bonding.

The properties also could be disturbed by poor dispersibility. Graphene oxides have an advantage of processing as they could be exfoliated in water and hence bionanocomposites could be fabricated by one-step method such as water-soluble polymers or through water-linked latex technique. Certain common processes to include graphene and graphene derivative fillers are available: melt intercalation, solution intercalation, wet spinning, intercalative polymerization, drop casting, and electro spinning [4].

Stable fabrication of bionanocomposites of graphene/graphene derivatives requires a good/uniform distribution of material in polymeric matrices that influences the thermal, mechanical, permeation, and gas barrier properties. The most essential challenge in this domain is agglomeration, improper exfoliation, and restacking of graphene or graphene derivatives while incorporation of biomolecules.

8.1.1 Significance of Bionanocomposites

The biomolecules used for the production of nanocomposites, especially the biopolymers, can fill up the pores in the coating pattern, thus preventing the penetration of molecules like digestive enzymes, bile salts, hydrogen ions. Hence, they are the best fit for usage in biomedical applications like formulation of effective pharmaceutical agents and smart drugs. Such nanocomposites hence improve the hardness, compressibility, and adhesion of micro-capsules. Water resistance and super-hydrophobicity are also enhanced as these biofillers reduce the permeation of water molecules and vapor, widening the application prospects.

The graphene material used for the composite formation is required to exhibit porous property as there is scope for adhesion, growth, and transport of cells and metabolic waste materials. The neuritic branches proliferate and spread themselves to expand faster so that they also exhibit better expression levels of proteins and secondary metabolites. The growth rate of neurons also was double when excited with electrical signals in the presence of bionanocomposite scaffolds than with pristine graphene or conventional scaffolds [5].

8.1.2 Classification of Bionanocomposites

Graphene is available for composite formation in various forms listed as pristine graphene, graphene oxide, reduced graphene oxide, graphene nanopores, and graphene quantum dots (GQDs). Pristine graphene is further classified into very few layer graphene with 1–3 layers, few layer graphene with 2–5 layers, multilayered graphene with 2–10 layers, and graphene platelets that contain stacks of graphene sheets [2].

Biopolymers that make a perfect bionanocomposite of graphene for multifaceted applications include

- Polysaccharides (e.g., Starch, Lignin, Cellulose, Chitosan, Alginate, Pullulans)
- Proteins (e.g., Gelatin, Collagen, Zein)
- Polyesters (e.g., Polylactic Acid, Polycaprolactone, PHBV)
- Polynucleotides (e.g., DNA, RNA)

Based on the modes of dispersion, graphene nanocomposites could be classified as

- Phase separated micro-composites
- Intercalated nanocomposites
- Exfoliated nanocomposites

8.1.3 Properties of Bionanocomposites

Graphene, in addition to its superior properties, is biocompatible and less toxic when compared to other nanomaterials. Hence, it is used for specific applications involving biological interventions. This includes photocatalysis, infection care, probiotics, theranostics, scaffold production, biosensing, bioimaging, tissue engineering for skin and neuronal repair, analysis of DNA and RNA, and gene or drug delivery. Graphene oxide-chitosan-based bionanocomposite has proved to be an efficient catalytic material that helps easy separation, promotes high yield, and reveals short reaction time [6]. Another study had opted for the use of graphene cellulose nanobiocomposite for use in batteries as electrodes. Cellulose incorporation provides uniform porosity in the nanocomposite network for transferring electrons and thus increases the electrochemical properties [7]. Bionanocomposites of graphene oxide developed with silk fibroin can play a significant role in increasing surface roughness, thus demonstrating a loss in insulation properties and elevated conducting behavior [8]. Microbial methods of producing graphene oxide (GO) and reduced graphene oxide (rGO) are available in the literature, e.g., *Bacillus sphaericus*, that has an advantage of low toxicity and biocompatibility. They are found to possess unique biomolecules of medical significance on their surface, making them applicable as biomedical graphene nanocomposites [9]. The biomedical applications of these nanomaterials are unique in that they have immense influence on scaffold development, apoptosis, regeneration, cell adhesion, and differentiation. Stem cell research and human dental pulp regeneration are also areas of concern [10].

[G]Polyaniline, hydrogels, polyurethane, activated carbon, metal and metal oxides, and nanospheres are available for energy storage applications with graphene sheets [11].

Osteogenic differentiation, cell adhesion, and proliferation were found to improve in bone marrow stromal cells when graphene oxide and silk fibroin grafts were used as scaffolds. Metal-based nanocomposites suffer corrosion when utilized for bioimplants, and the released ions cause a rise in pH, inflammation, and foreign body reaction. Elastic modulus of titanium-based alloys or stainless steel is comparatively higher, and hence the biomedical implants usually suffer stress shielding. Hence lightweight scaffold materials constructed with graphene derivatives enforced with hydroxyapatite/collagen minimize degradation [12]. Pristine graphene and its derivatives in their bionanocomposite formulations have given room for sensing and sequencing of nucleotides/genes. Graphene interacts with DNA and RNA through pi-pi bonding that could be easily influenced by the polarizability of the oligonucleotide bases. Detection of environmental pollutants like endocrinal disrupters and microplastics with the peculiar properties of graphene nanocomposites has been investigated. Sensitive determination could be made with matrix assemblies of the composites that adhere to hydrophobic and lipophilic microplastics available in drinking water systems [13].

Nanopores precisely engineered in graphene stacks have served as a platform for genetic studies related to detection and sequencing. The external voltage, physical and chemical properties of the pore are all found to interact with the conformation of the DNA strands. The electron mobility that favors easy sequencing is larger in graphene nanoribbons and hence is simple to carryout measurements at higher bandwidths [2]. Graphene nanocomposites with rGO derivatives have been recently used for packaging food materials as they exhibit antioxidant properties that enhance the shelf life of food stuffs. Biodegradable polymers like polysaccharides, proteins and polyesters are employed for the production of such materials.

Another breakthrough introduced by graphene derivatives is their application in pulsed electric field technology for the development of flexible packaging components. Significant enhancement of mechanical properties was exhibited when chitosan, PHBV, alginate, cellulose, starch, and PLA were reinforced even at very low concentrations for such applications. Tensile strength, Young's modulus, and elasticity increased manifold on incorporation of such fillers. Further, the thermal stability was found to improve as the temperature of degradation was delayed to 5–20 °C in comparison to control material that did not possess the filler materials [3].

Graphene when used solely has exhibited few toxicity issues at higher doses of cellular applications. Cellular apoptosis, autophagy, organelle and protein dysfunction, necrosis, oxidative stress, cytoskeletal disorders, and physical destruction are some of the anomalies caused. Recent research suggests that the toxicity diminishes when graphene is reduced or conjugated by polymers to form composites. Hence forth, research on graphene nanocomposites is oriented more toward mitigating certain toxicities at high dosage levels by restricting permeability and enhancing hydrophobicity. Further research to optimize the preparation of stable nanocomposites from organic sources and sustainable materials that help in the production of nontoxic substrates for novel applications should be encouraged.

8.2 Preparation Methods

As graphene bionanocomposites represent a cutting-edge class of materials that integrate the exceptional properties of graphene such as mechanical strength, electrical conductivity, and large surface area with the biocompatibility and functionality of biological molecules or biopolymers, they have shown great promise in different applications that include drug delivery, tissue engineering, biosensing, biomedical implants, and so forth. As different applications need different properties, researchers attempt to customize the properties of these materials to suit for a variety of applications (see Table 8.1). The properties and performance of graphene bionanocomposites are intricately linked to the preparation methods employed during their fabrication. Several methods are employed to prepare graphene-based bionanocomposites (Figure 8.1), each tailored to achieve specific properties in order to meet the requirements of different applications. The primary preparation methods include:

- mechanical mixing,
- solution mixing,
- in situ synthesis,
- layer by layer assembly,

Table 8.1 Preparation procedures and applications of graphene bionanocomposites.

S. no.	Methods	Procedures	Applications
1.	Mechanical mixing	1. Graphene preparation and dispersion in solvent 2. Biopolymer solution prepared via mechanical mixing and evaporation of solvent 3. Posttreatment—drying and curing	1. Drug delivery 2. Tissue engineering 3. Biosensing 4. Biomedical implants
2.	Solution mixing	1. Graphene preparation and dispersion in solvent 2. Biopolymer solution prepared via solution mixing 3. Removal of solvent and composite formation by casting, freeze-drying, and evaporation 4. Posttreatment—drying and curing	1. Drug delivery 2. Tissue engineering 3. Biosensing 4. Biomedical implants 5. Conductive films
3.	In situ synthesis	1. Graphene preparation and dispersion in monomer solution 2. Initiation and formation of polymerization 3. Removal of solvent and composite formation by casting, freeze-drying, and evaporation 4. Posttreatment—drying and curing	1. Drug delivery 2. Tissue engineering 3. Biosensing 4. Biomedical implants 5. Conductive films
4.	Layer-by-layer assembly	1. Substrate preparation 2. Graphene preparation and dispersion in biopolymer solution 3. Layer deposition and drying 4. Repetition of deposition cycles by thickness control 5. Posttreatment—cross-linking and thermal treatment	1. Drug delivery 2. Tissue engineering 3. Biosensing 4. Biomedical coatings 5. Electronics
5.	Wet spinning	1. Graphene preparation and dispersion in solvent 2. Polymer dissolution mixing with graphene dispersion 3. Wet spinning process through extrusion, coagulation bath, and fiber formation 4. Post-spinning treatment—washing, drying, and drawing	1. Biomedical devices 2. Textiles 3. Electronics
6.	Melt mixing	1. Graphene preparation and polymer selection 2. Mixing process by melting the polymer and adding graphene 3. Extrusion and molding 4. Cooling and solidification	1. Biomedical devices 2. Automotive components 3. Packaging materials 4. Consumer goods 5. Electronics

7.	Electrospinning	1. Graphene preparation and dispersion in solvent 2. Polymer dissolution mixing with graphene dispersion 3. Electrospinning process by using electric field and fiber formation 4. Collection and posttreatment	1. Biomedical applications 2. Energy storage 3. Sensors 4. Filtration 5. Textiles
8.	Chemical vapor deposition	1. Preparation of the substances—selection of substance, cleaning, and annealing 2. Graphene growth—decomposition, deposition, and cooling 3. Transfer of graphene—coating, etching, and transfer to target substrate 4. Integration into bionanocomposite	1. Biomedical devices 2. Energy storage 3. Filtration systems 4. Advanced coatings 5. Flexible electronics
9.	Solvent evaporation process	1. Graphene preparation and dispersion in solvent 2. Adding matrix materials 3. Casting or coating 4. Solvent evaporation 5. Posttreatment—annealing and cross-linking	1. Biomedical devices 2. Energy storage 3. Packaging materials 4. Coatings 5. Flexible electronics
10.	Solvent casting	1. Graphene preparation and dispersion in solvent 2. Preparation of polymer solution 3. Mixing graphene and polymerization by homogenization 4. Casting the solution—mold preparation and pouring 5. Solvent evaporation 6. Posttreatment—thermal annealing and cross-linking	1. Biomedical devices 2. Energy storage 3. Packaging materials 4. Protective coatings 5. Flexible electronics
11.	Freeze-drying	1. Graphene preparation and dispersion in solvent 2. Adding biopolymer 3. Freezing and mold preparation 4. Sublimation—primary and secondary drying	1. Biomedical 2. Energy storage 3. Water purification and air filtration 4. Catalysis 5. Sensors
12.	Hydrothermal/ solvothermal	1. Graphene preparation and dispersion in solvent 2. Mixing and dispersion by homogenization 3. Hydrothermal reaction—autoclave loading, temperature-pressure control, and reaction time 4. Cooling and recovery 5. Posttreatment—thermal annealing and cross-linking	1. Biomedical 2. Energy storage 3. Environmental remediation 4. Catalysis 5. Flexible electronics

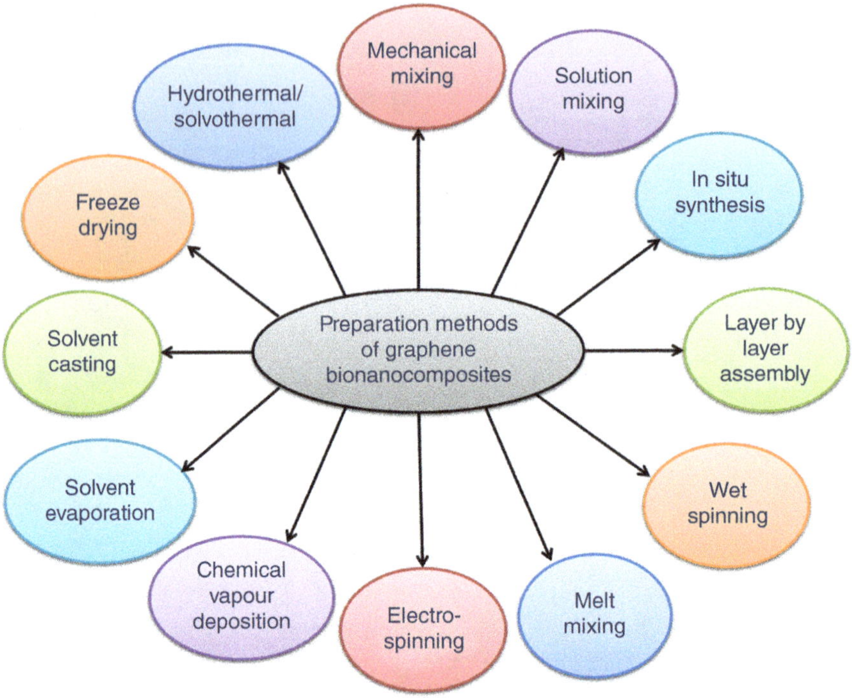

Figure 8.1 Various preparation methods for graphene bionanocomposites.

- wet spinning,
- melt mixing,
- electrospinning,
- chemical vapor deposition,
- solvent evaporation,
- solvent casting,
- freeze-drying, and
- hydrothermal/solvothermal.

The methods in general and the process parameters in particular can greatly influence the properties of bionanocomposite. The method of preparation and the process parameters play crucial roles in determining the structure, composition, dispersion, distribution, morphology, and interfacial interactions between graphene and the biological components. Hence, by appropriately controlling the process parameters, bionanocomposite materials with desired characteristics can be achieved.

8.2.1 Mechanical Mixing

Mechanical mixing is one of the fundamental methods for preparing graphene bionanocomposites using mechanical forces. This method is a very simple and widely used technique for incorporating graphene into the polymer matrix to form bionanocomposites. In order to produce graphene dispersion and distribution inside the polymer matrix, this method includes the technique of physical agitation of both graphene and the biopolymer or bionanomaterial.

Advantages

- The mechanical mixing process is a straightforward method which does not require complex or specialized equipment, and it is easy to implement.
- Compared to other fabrication techniques, it is a cost-effective method that is particularly beneficial for large-scale production.
- The process can be easily scaled up from laboratory to industrial scale, enabling mass production of graphene bionanocomposites.
- Mechanical mixing allows for the incorporation of various types and concentrations of graphene and biopolymers, providing the flexibility to tailor the composite properties in order to make it suitable for specific applications.

Hailan Kang reported that graphene oxide (GE) was reduced in order to develop GE/NR (natural rubber) nanocomposites through straightforward mechanical mixing. Analytical techniques like FTIR, XPS, and XRD were used to analyze the characteristics of GE. The addition of GE improved the mechanical characteristics of GE/NR nanocomposites. In contrast to graphite composites and carbon black composites, GE composites demonstrated reduced internal friction loss and higher strength at low contents [14].

8.2.2 Solution Mixing

Solution mixing is—also known as solution blending—a versatile method for preparing graphene-based bionanocomposites. It involves dispersing graphene in a solvent followed by mixing with a biopolymer solution. This technique ensures a uniform distribution of graphene within the polymer matrix.

Advantages

- The process is simple and does not require specialized equipment, and hence is accessible for most laboratories.
- Solution mixing is a low-cost method compared to other fabrication techniques.
- Solution mixing allows for the incorporation of various concentrations of graphene and biopolymers, enabling the customization of composite properties.

According to Chetan Sharma, using melt and solution-mix techniques, bionanocomposite films of starch/polyvinyl alcohol (PVOH) with different concentrations of laponite RD can be effectively created, with a thickness that varied from 200 to 400 μm. Because of the potential for interaction between the OH groups and clay, the water absorption, water solubility, and water vapor uptake ratio is generally reduced as the clay content increased. Additionally, a coat containing 10% laponite RD and a 25:75 graphene:lipid ratio was shown to have the least amount of water absorption for better barrier qualities. Therefore, starch/PVOH/laponite RD nanocomposite films with enhanced mechanical and water vapor barrier qualities along with regulated water resistance (solubility) qualities may find application as eco-friendly food packaging materials to increase shelf life [15].

8.2.3 *In situ* Synthesis

In situ synthesis is an effective method for fabricating graphene bionanocomposites which offers control over the composition, structure, and properties of the resulting materials. This approach involves the simultaneous formation of graphene and the polymer or biocomponent within the same reaction system, leading to the high degree of integration of the graphene into the polymer matrix at the molecular level, and it allows for strong interfacial bonding and uniform distribution. In situ synthesis allows tailored interactions between graphene and the polymer, resulting in enhanced properties of the bionanocomposites.

Advantages

- The polymerization process ensures that polymer chains form directly around the graphene sheets, leading to strong interfacial bonding and improved mechanical properties.
- The dispersion of graphene in the monomer solution before polymerization helps achieve a uniform distribution of graphene within the polymer matrix.
- The resulting composites often exhibit superior mechanical strength, electrical conductivity, and thermal stability compared to other methods.
- This method can be used with a wide range of monomers and biopolymers allowing for the customization of composite properties to suit specific applications.

Parushuram Nilogal stated in the literature review that this work effectively created highly conductive bacterial nanocelluloses (BNCs) and realized graphene oxide to decreased graphene oxide nanosheets in the silk fibroin biopolymer. The reduced graphene oxide nanosheets are created by the biopolymer that is utilized as a reduction agent. The X-ray diffraction analysis demonstrated the formation of graphene oxide from graphite powder and its subsequent reduction to reduced graphene oxide within the host matrix. The SEM images verified that the decreased GO nanosheets that were generated in situ were evenly distributed throughout the SF matrix. The surface roughness of the BNCs increased as the amount of rGO concentration in the SF increased. It was noted from the XPS analysis that the functional groups with oxygen had been partially eliminated. The increased number of sp2 domains generated during the in situ reduction process may be the cause of the intensity increase observed in the SF-rGO bionanocomposite samples, according to Raman spectroscopy. The current-voltage (I-V) characteristics demonstrated that when the concentration of rGO nanosheets rose, the electrical conductivity increases steadily. One possible explanation for the high conductivity could be the enormous surface area of the rGO sheets that were generated in situ in the SF matrix. The creation of innovative conducting SF-rGO bionanocomposite will open up promising prospects in tissue engineering, biosensors, and many other fields [8].

8.2.4 Layer-by-Layer (LbL) Assembly

Layer-by-layer (LbL) assembly is a bottom-up fabrication technique that constructs thin films or coatings through the alternate deposition of positively and negatively charged species. In the context of graphene-based bionanocomposites, LbL assembly involves the use of graphene or GO and biopolymers, which can be natural or synthetic. This process entails the adsorption of polyelectrolytes and graphene layers alternately onto a substrate. It allows precise control over film properties such as thickness, composition, and morphology. LbL assembly can be used to fabricate coatings, membranes, or biomedical devices with tailored properties.

Advantages

- The process allows for precise control over the thickness, composition, and structure of the composite material.
- A wide range of materials can be used in the LbL assembly process including various types of graphene, biopolymers, and other functional materials.
- The sequential deposition ensures uniform layer formation and consistent properties across the composite film.
- The properties of the composite can be tailored by adjusting the number of layers, the type of materials used, and the deposition parameters.

According to Eungjin Ahn et al., a field guide for the LbL assembly of multilayer nanocomposites based on graphene has been developed. It draws on the expertise and understanding that have been acquired throughout our earlier research. Graphene, polymer, and metal NP have been successfully used to create graphene-based LbL nanocomposites. If the components and techniques for assembling graphene-based nanocomposites are properly planned, then the LbL assembly process can be applied to a wide range of possible counterpart materials. The LbL approach offers a way to modify the quantity of bilayers and internal structures in a functional composite material, which can be used to maximize device performance. Moreover, it provides structural insight into nanoscale composite engineering and a basic understanding of the interfaces between conjugate nanomaterials and graphene. We hope that the comprehensive procedures described in this paper will help scientists create new graphene-based nanocomposites with desired and controllable features [16].

8.2.5 Wet Spinning

Wet spinning method is a fiber production process used for polymers that do not melt and can only be processed from solution. It involves dispersing graphene in a polymer solution by using a spinneret, which is then extruded into a coagulation bath where the fibers are formed. Graphene can be added to the polymer matrix using an adaptation of this technique, and this method is particularly effective for creating high-performance graphene bionanocomposite fibers with enhanced mechanical and functional properties.

Advantages

- Wet spinning enables the continuous production of fibers, making it suitable for large-scale manufacturing.
- The mixing of graphene with the polymer solution ensures a uniform dispersion of graphene within the fibers, leading to consistent properties.
- The resulting composite fibers exhibit enhanced mechanical properties.
- A wide range of polymers and solvents can be used in wet spinning, allowing for the customization of composite properties to suit various applications.

According to Xiaozhen Hu et al., his group created a simple, quick, and effective wet spinning assembly method for creating continuous GO-SA (sodium alginate) nacre-mimetic films. Generally, it took 4 minutes to obtain 6-m long films with an organized multilayer structure. The GO-SA liquid crystals' regular arrangement, the GO sheets' fluid-induced alignment, and the preservation of homogeneous structures throughout the coagulation process gave the composite films, their flawless B&M layered structures and superior mechanical performance. Wet spinning was also used to create the ultra-strong large flakes of GO (LGO)-SA nacre-mimic fibers. Nacre-mimic films with hierarchical nanostructures may be readily made with this technique, providing a means of producing robust and multipurpose biomimetic materials on a wide scale [17].

8.2.6 Melt Mixing

Melt mixing, also known as melt compounding, is a prominent method for the preparation of polymer-based composites, including graphene bionanocomposites. In this method, graphene is dispersed directly into the polymer matrix during the melting process, leading to the formation of a homogeneous mixture. This method offers several advantages: simplicity, high scalability, and versatility, which make it suitable for both laboratory-scale research and industrial production of graphene bionanocomposites.

Carlos A. Avila-Orta et al. reported that to prepare BNCs, the rGO was first pre-exfoliated using ultrasound in a 1 wt.% aqueous solution, and then it was melt combined with thermoplastic waxy corn starch (TWCS) and glycerol. Because the two phases interact so well, the exfoliated GO in TWCS prevents the retrogradation of waxy starch. With a 17.48 °C temperature increase, the thermal stability of BNCs was higher than that of TWCS. Good interfacial interactions between GO and the waxy starch matrix led to increase in the tensile strength and Young's modulus of 140% and 230%, respectively. Because of the good dispersion and exfoliation of the GO, which increases the interfacial contacts between components, the addition of rGO pre-exfoliation and the melt mixing preparation improve the thermal and mechanical properties of TWCS. These findings suggest that the BNCs of GO and starch made by melt mixing may represent a significant substitute for its application in several sectors, including the food packaging, medical, and automotive industries [18].

8.2.7 Electrospinning

Electrospinning method is a highly effective technique for the fabrication of nanofibrous materials. This method leverages an electrostatic field to draw a polymer solution into ultrafine fibers, which can be enhanced by incorporating graphene. A polymer solution containing dispersed graphene is subjected to a high electric field, resulting in nanofibers with graphene incorporated throughout their structure. Electrospun bionanocomposites exhibit high surface area and porosity suited for tissue engineering and filtration applications.

Advantages

- The nanofibers produced have a high surface area to volume ratio, which is beneficial for applications requiring high reactivity or surface interactions.
- The process allows for a uniform dispersion of graphene within the polymer matrix, ensuring consistent properties throughout the nanofibers.
- Electrospinning can be scaled up for industrial production while maintaining control over fiber diameter and morphology.

Roberto Scaffaro and Francesco Lopresti reported that in order to investigate the properties-morphology correlations of the polylactic acid/graphene nanoplatelets (PLA/GnPs) nanocomposites, GnPs were added to aligned and randomly oriented electrospun PLA mats. The morphological analysis demonstrated how the high-speed collecting rotary drum allowed the fibers to align effectively. When comparing PCL/GnPs suspensions to neat PLA mats, the average fiber diameter increases due to the rise in complicated viscosity. Tensile testing demonstrated that adding GnPs to PLA significantly enhances the materials' elastic qualities. Specifically, for every system under investigation, the elastic modulus rose as the GnPs wt.% increased. The reinforcing effect of GnP is more prominent in aligned fibers. Additionally, the PLA/GnPs aligned systems exhibited an increase in elongation, which is consistent with the presence of residual solvent as shown by DTA and DSC measurements. According to GnPs' capacity to slow down the pace at which the solvent mixture evaporates, it was discovered that the amount of residual solvent in the polymer matrix increased as GnPs was added. When compared to clean PLA, TGA studies demonstrated that the addition of GnPs to PLA electrospun mats improves the thermal stability of the composites; nevertheless, aligned and randomly oriented mats exhibited identical behavior [19].

8.2.8 Chemical Vapor Deposition

Chemical vapor deposition (CVD) is a widely adopted technique for producing high-quality graphene films on various substrates which can be subsequently incorporated into bionanocomposites. For graphene production, CVD involves the deposition of carbon atoms onto a substrate, typically copper or nickel, resulting in the formation of a graphene layer. This method is renowned for producing graphene with high crystallinity and large area uniformity, making it ideal for high-performance applications. Strong interfacial contacts between graphene and the substrate are made possible by this method, which results in improved characteristics.

Advantages

- CVD produces graphene with high crystallinity, minimal defects, and large-area uniformity.
- The process can be scaled for large-area graphene production for industrial applications.
- The process allows precise control over the number of graphene layers.
- CVD-grown graphene can be transferred to various substrates and integrated into different polymer matrices.

8.2.9 Solvent Evaporation Process

The solvent evaporation method is a straightforward and effective technique for preparing bionanocomposites, where a polymer matrix is combined with nanomaterials to form a composite material. This method involves dissolving the polymer and dispersing the nanomaterial in a common solvent, mixing them to create a homogeneous solution, and then controlled evaporating the solvent to form the composite material.

Advantages

- The process is simple and can be adopted to various graphene precursors, matrix materials, and fabrication techniques.
- The solvent evaporation process allows for precise control over the dispersion of graphene within the matrix material, leading to uniform composites with tailored properties.
- The process can be scaled up for industrial production, making it suitable for large-scale manufacturing.
- Solvent-based methods are generally cost-effective and require minimal equipment, making them accessible to researchers and industry alike.

Films based on a ST/L biopolymer blend reinforced by GO can be successfully made using the casting/solvent evaporation process, according to Meryem Aqlil et al. The hydrogen bond interactions allowed the lignin to be well blended into the starch matrix, resulting in the creation of a homogenous biocompatible mix matrix. The noninteracting hydroxyl groups of starch and lignin were found to interact further after the addition of GO, leading to the creation of an interconnected network. These interactions led to considerable improvements in the generated bionanocomposites' mechanical, thermal, and water-swelling properties as well as moisture absorption, hydrolytic degradation, and water vapor permeability. The efficacy of the method for generating biodegradable films with improved characteristics, potentially making it a viable option for food packaging uses [20].

8.2.10 Solvent Casting

The solvent casting method is an efficient technique used to produce composite films or membranes by casting a polymer solution onto the substrate and allowing the solvent to evaporate, leaving behind a solid film. This method is widely used in various fields such as materials science, biotechnology, and electronics for creating thin, uniform films with controlled thickness and properties.

Advantages

- This process is straightforward and does not require complex equipment, making it accessible and cost-effective.
- The thickness of the composite film can be precisely controlled by adjusting the amount of solution and the spreading technique.
- Proper mixing and controlled evaporation result in a uniform dispersion of graphene within the polymer matrix.
- The process is easily scalable for industrial production suitable for producing large-area films.

8.2.11 Freeze-Drying

Freeze-drying is also known as lyophilization—a sophisticated technique for fabricating graphene-based bionanocomposites. This method involves freezing a dispersion of graphene and biopolymers, followed by sublimation of the solvent under reduced pressure. Freeze-drying preserves the perishable materials' structural integrity and produces highly porous composites with enhanced mechanical, electrical, and thermal properties. This technique is particularly useful for creating composites with high porosity and uniform graphene distribution.

Applications

- The process produces composites with high porosity, which is advantageous for applications requiring high surface area and permeability.
- The method preserved the uniform distribution of graphene within the biopolymer matrix.
- Since the process involves low temperatures, thermal degradation of the biopolymer and graphene is minimized.
- The resulting composites exhibit enhanced mechanical strength, thermal stability, and electrical conductivity.

According to F. Mohandesa and M. Salavati-Niasari, by using a straightforward freeze-drying technique, a composite of chitosan (CT), graphene oxide (GO), and hydroxyapatite (HAP) nanoparticles was created [21].

8.2.12 Hydrothermal/Solvothermal Method

Hydrothermal synthesis leverages high-temperature and high-pressure conditions to facilitate chemical reactions and crystal growth in an aqueous solution. This method is particularly useful for synthesizing nanomaterials, including graphene-based bionanocomposites, as it promotes uniform dispersion and strong interaction between graphene and the polymer matrix.

Advantages

Hydrothermal synthesis offers several significant advantages for fabricating graphene-based bionanocomposites.

- The method produces composites with high purity and crystallinity which are essential for many advanced applications.
- The high-pressure and high-temperature conditions promote uniform dispersion of graphene within the polymer matrix.
- The process enhances the interaction between graphene and the polymer, resulting in composites with superior mechanical and thermal properties.
- Hydrothermal synthesis is scalable and can be adopted for large-scale production of bionanocomposites.

8.3 Graphene Bionanocomposites Applications in Wastewater Treatment

Wastewater pollutants are broadly classified into three main categories: (i) inorganic pollutants, (ii) organic pollutants, and (iii) microorganisms/virus contaminations. Inorganic pollutants include heavy metals, nitrates/nitrites, phosphates, and oils removals, and organic pollutants include dyes, phenols, BOD, COD, and TOC removals (Figure 8.2).

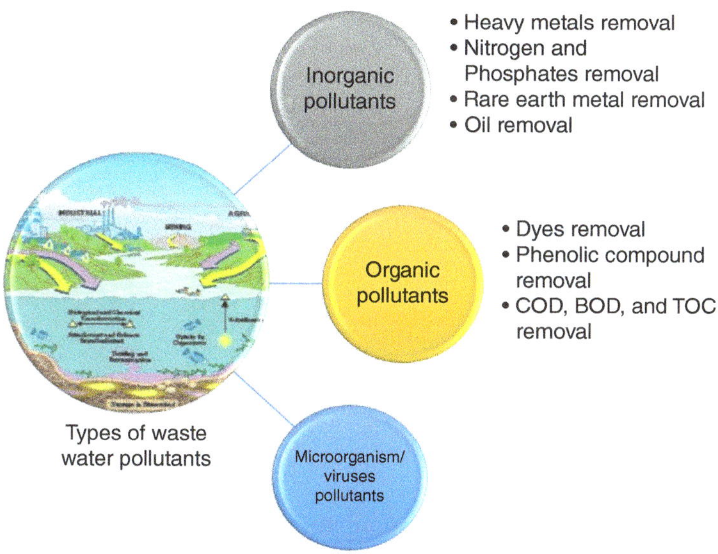

Figure 8.2 Types of wastewater pollutants.

8.3.1 Inorganic Pollutants

8.3.1.1 Heavy Metal Removal

It is an obvious fact that the contamination of heavy metals even in trace amount can cause an adverse effect on the environment as it has a tendency to accumulate in biological systems. The main driven sources for the production of heavy metal pollution are the chemical industrial sector, mining, agricultural activity, municipal, and domestic waste. The conventional methods to remove heavy metals from water comprise physical, chemical, and biological techniques. Good removal efficiencies for eliminating heavy metal ions can be achieved by adopting expensive membrane filtration or coagulation/flocculation techniques. Even though the above-mentioned techniques show good results, they lead to the formation of huge particles and highly concentrated sludge. Oxidation and electrochemical treatments serve as quick and practical processes for heavy metal removal but consume large amounts of energy. Biological treatment also lacks efficacy.

Therefore, researchers have developed a new novel cost-effective and green-routed material to manage heavy metal ion contamination, which has paved the way for the utilization of graphene bionanocomposites. It offers a high specific surface area along with enormous amounts of functional groups, which make it very suitable for binding the heavy metals by adsorption without leaving any toxic by-products. Josphat Phiri et al. reported that GO's porous structure and large surface area make it an effective adsorbent for detecting and eliminating metal ions when incorporated with CMC biopolymers [22].

8.3.1.2 Nitrogen and Phosphates Removal

Perhaps the most common pollutants in drinking water and wastewater are compounds with phosphate and nitrogen bases. These ions cause eutrophication, which depletes the dissolved oxygen in the water and kills aquatic life. They are discharged into the environment from a variety of sources, including chemical fertilizers, farming wastes, and so forth. Moreover, they directly endanger human health when present in large amounts because they can cause illnesses and malignancies.

Physicochemical procedures and biological denitrification are the two most widely used conventional ways for removing nitrate from water. These methods are costly and necessitate posttreatment for residual wastes including nitrite, nitrate, and ammonia. They also require high energy and time investment. Therefore, graphene-based bionanocomposites were extensively researched for the removal of nitrate and nitrite from contaminated water due to their ecologically safe character.

8.3.1.3 Rare Earth Metal Removal

In contrary to the name suggests, rare earth elements are not that much rare. They are natural elements which are abundantly available in the earth and are normally found agglomerated in certain varieties of minerals and rocks. It consists of 17 elements viz., yttrium, scandium, and 15 lanthanides (Neodymium, Lanthanum, Praseodymium, Cerium, Gadolinium, Promethium, Europium, Thulium, Terbium, Samarium, Holmium, Erbium, Lutecium, Ytterbium, and Dysprosium). Nowadays, there is an increasing demand for the separation and the extraction of rare earth elements thanks to their strategic potentials in various fields such as metallurgy, magnetics, catalysts, superconductors, and rechargeable batteries. Graphene bionanocomposites have proven their versatility and suitability for the application of rare earth elements separation and extraction.

8.3.1.4 Oil Removal

Oils are another kind of obstinate wastewater pollutant. The free-floating oils are the major by-products generated by many industrial sectors such as food processing, textiles, petrochemicals, oil, and metalworking industries. It is found in the forms, viz., grease, fats, lubricants, and light and heavy hydrocarbons. When oil emulsions are discharged into the environment, they will impede the oxygen absorption and sunlight penetration by the water resources. This may lead to the chemical oxygen demand and biological oxygen demand and affect the natural environment due to the inadequate separation technique. Despite having well-known membrane technology for separating the oils, researchers have designed graphene bionanocomposites as a new alternative to remove the oils from the water. Zhang et al. produced a poly-ionic liquid-modified GO-coated cotton fabric with switchable wettability, efficiently segregating water and oil [23].

8.3.2 Organic Pollutants

8.3.2.1 Dye Removal

Dyes are colored chemical substances utilized widely in paper, textiles, drugs, leather, rubber, cosmetics, and food industries. A discharge of huge quantities of harmful synthetic dyes into the water bodies can cause severe health problems to humans and other living beings in the environment. Most of the dyes are highly toxic which can cause carcinogenic and mutagenic effects. Consequently, it's critical to cleanse dye-containing water before releasing it into the environment in order to reduce toxicity, prevent water pollution, and safeguard the ecosystem. Moreover, dyes are not biologically biodegradable and are highly toxic, even carcinogenic, to both human populations and microbes. They are also resistant to aerobic digestion. They could stay in the water system for a very long period.

These dyes frequently cause respiratory issues, skin irritation, allergies, watery eyes, coughing, wheezing, and immune system disruption.

Dyes can be removed from industrial wastewaters using a variety of procedures, including adsorption, flotation, oxidation, biological treatment, coagulation, and electrochemical approaches. Though research on graphene-based BNCs in water remediation is still in its early stages, these materials show great promise as environmentally friendly and long-lasting adsorbents for the safe and clean removal of a wide range of water contaminants. As green membranes, it combines the advantages of inorganic and organic materials with those of natural polymers.

The self-assembly of graphene oxide and chitosan was used to design and create graphene oxide-based composite hydrogels. The results of the catalytic capacity tests show that, in good agreement with the pseudo-second-order model, the synthesized composite can remove the dye molecules, MB and Rhodamine B, from polluted water [24]. GO was wrapped in calcium alginate and created macro-porous alginic beads to adsorb acridine orange dye. Incorporating GO into the CA matrix improved adsorption kinetics and pH range compared to using CA beads alone [25]. Li et al. studied the adsorption characteristics of GO/CA composites, synthesized using the sol-gel technique for methylene blue elimination. The eco-friendly composite demonstrated a high adsorption capability of $181.81\,mg\,g^{-1}$. It has a wide pH range and may be easily isolated from water solutions [26].

8.3.2.2 Phenolic Compound Removal

One class of extremely harmful and resistant organic pollutants is phenol (hydroxy-benzene) and its derivatives. These chemicals are typically released into the environment after being used as raw materials in a variety of chemical companies, or from agricultural resources, water chlorination in wastewater treatment plants, or other processes. Their low biodegradability, carcinogenicity, and mutagenicity are attributed to their high solubility in water, oils, and various organic solvents. Because of this, wastewaters containing phenolic compounds should be carefully handled to prevent environmental release of these chemicals. Biological processes, membrane extraction, ozonation, adsorption, electrochemical techniques, and liquid-liquid extraction using organic solvents are some of the techniques for phenolic compound absorption.

8.3.2.3 COD, BOD, and TOC Removal

Identifying organic matter (over 40 million molecules) is challenging, costly, and time-consuming due to its diversity. To effectively classify organic contamination in wastewater analysis, three common metrics are used: the chemical oxygen demand (COD), biochemical oxygen demand (BOD), and total organic carbon (TOC). Higher COD, BOD, or TOC levels indicate more pollution in water. Graphene bionanocomposites are employed to remove organic materials, including COD, BOD, and TOC, from contaminated water.

8.3.2.4 Pathogenic Microorganism Inactivation

There are many different types of water pollution, including microbial contamination and both organic and inorganic pollution. The former comes on by the existence of parasites, viruses, and other harmful microbes. The existence of microorganisms in water is the cause of numerous infectious diseases that affect human health. Even more hazardous is the tendency to use chlorination to manage bacteria disinfection, as this process produces known carcinogenic metabolites that are harmful to human health.

References

1 Azizi-Lalabadi, M. and Jafari, S.M. (2021). Bio-nanocomposites of graphene with biopolymers; fabrication, properties, and applications. *Advances in Colloid and Interface Science* 292: 102416. https://doi.org/10.1016/j.cis.2021.102416.

2 Wu, X., Mu, F., Wang, Y., and Zhao, H. (2018). Graphene and graphene-based nanomaterials for DNA detection: a review. *Molecules* 23: 2050. https://doi.org/10.3390/molecules23082050.

3 Barra, A., Santos, J.D.C., Silva, M.R.F. et al. (2020). Graphene derivatives in biopolymer-based composites for food packaging applications. *Nanomaterials* 10: 2077. https://doi.org/10.3390/nano10102077.

4 Jiménez-López, B.A., Martín-Sampedro, R., Leyva-Ramos, R. et al. (2024). Nanoarchitectonics of lignin-sepiolite bionanocomposite foams for application in environmental remediation. *Sustainable Chemistry for the Environment* 6: 100096. https://doi.org/10.1016/j.scenv.2024.100096.

5 Lazăr, A.-I., Aghasoleimani, K., Semertsidou, A. et al. (2023). Graphene-related nanomaterials for biomedical applications. *Nanomaterials* 13: 1092. https://doi.org/10.3390/nano13061092.

6 Maleki, A. and Paydar, R. (2015). Graphene oxide–chitosan bionanocomposite: a highly efficient nanocatalyst for the one-pot three-component synthesis of trisubstituted imidazoles under solvent-free conditions. *RSC Advances* 5: 33177–33184. https://doi.org/10.1039/C5RA03355A.

7 Das, A.K., Islam, M.N., Ghosh, R.K., and Maryana, R. (2023). Cellulose-based bionanocomposites in energy storage applications—a review. *Heliyon* 9: e13028. https://doi.org/10.1016/j.heliyon.2023.e13028.

8 Nilogal, P., Uppine, G.B., Rayaraddi, R. et al. (2021). Conductive *in situ* reduced graphene oxide–silk fibroin bionanocomposites. *ACS Omega* 6: 12995–13007. https://doi.org/10.1021/acsomega.1c00013.

9 Xu, Q., Lin, X., Gan, L. et al. (2022). Green reduction of graphene oxide using Bacillus sphaericus. *Journal of Colloid and Interface Science* 605: 881–887. https://doi.org/10.1016/j.jcis.2021.07.102.

10 López-García, S., Aznar-Cervantes, S.D., Pagán, A. et al. (2024). 3D graphene/silk fibroin scaffolds enhance dental pulp stem cell osteo/odontogenic differentiation. *Dental Materials* 40: 431–440. https://doi.org/10.1016/j.dental.2023.12.009.

11 Ikram, R., Mohamed Jan, B., Abdul Qadir, M. et al. (2021). Recent advances in chitin and chitosan/graphene-based bio-nanocomposites for energetic applications. *Polymers (Basel)* 13: 3266. https://doi.org/10.3390/polym13193266.

12 Li, Y., Liao, C., and Tjong, S.C. (2019). Synthetic biodegradable aliphatic polyester nanocomposites reinforced with nanohydroxyapatite and/or graphene oxide for bone tissue engineering applications. *Nanomaterials* 9: 590. https://doi.org/10.3390/nano9040590.

13 Baumgarten, L.G., Freitas, A.A., Santana, E.R. et al. (2023). Graphene and gold nanoparticle-based bionanocomposite for the voltammetric determination of bisphenol A in (micro)plastics. *Chemosphere* 334: 139016. https://doi.org/10.1016/j.chemosphere.2023.139016.

14 Kang, H., Tang, Y., Yao, L. et al. (2017). Fabrication of graphene/natural rubber nanocomposites with high dynamic properties through convenient mechanical mixing. *Composites Part B: Engineering* 112: 1–7. https://doi.org/10.1016/j.compositesb.2016.12.035.

15 Sharma, C., Manepalli, P.H., Thatte, A. et al. (2017). Biodegradable starch/PVOH/laponite RD-based bionanocomposite films coated with graphene oxide: preparation and performance characterization for food packaging applications. *Colloid & Polymer Science* 295: 1695–1708. https://doi.org/10.1007/s00396-017-4114-9.

16 Ahn, E., Lee, T., Gu, M. et al. (2017). Layer-by-layer assembly for graphene-based multilayer nanocomposites: the field manual. *Chemistry of Materials* 29: 69–79. https://doi.org/10.1021/acs.chemmater.6b02688.

17 Hu, X., Rajendran, S., Yao, Y. et al. (2016). A novel wet-spinning method of manufacturing continuous bio-inspired composites based on graphene oxide and sodium alginate. *Nano Research* 9: 735–744. https://doi.org/10.1007/s12274-015-0952-2.

18 Ávila-Orta, C.A., Soriano Corral, F., Fonseca-Florido, H.A. et al. (2018). Starch-graphene oxide bionanocomposites prepared through melt mixing. *Journal of Applied Polymer Science* 135: https://doi.org/10.1002/app.46037.

19 Scaffaro, R. and Lopresti, F. (2018). Properties-morphology relationships in electrospun mats based on polylactic acid and graphene nanoplatelets. *Composites. Part A, Applied Science and Manufacturing* 108: 23–29. https://doi.org/10.1016/j.compositesa.2018.02.026.

20 Aqlil, M., MoussembaNzenguet, A., Essamlali, Y. et al. (2017). Graphene oxide filled lignin/starch polymer bionanocomposite: structural, physical, and mechanical studies. *Journal of Agricultural and Food Chemistry* 65: 10571–10581. https://doi.org/10.1021/acs.jafc.7b04155.

21 Mohandes, F. and Salavati-Niasari, M. (2014). Freeze-drying synthesis, characterization and in vitro bioactivity of chitosan/graphene oxide/hydroxyapatite nanocomposite. *RSC Advances* 4: 25993. https://doi.org/10.1039/c4ra03534h.

22 Phiri, J., Gane, P., and Maloney, T.C. (2017). General overview of graphene: production, properties and application in polymer composites. *Materials Science and Engineering B* 215: 9–28. https://doi.org/10.1016/j.mseb.2016.10.004.

23 Zhang, X., Chen, Q., Wei, R. et al. (2020). Design of poly ionic liquids modified cotton fabric with ion species-triggered bidirectional oil-water separation performance. *Journal of Hazardous Materials* 400: 123163. https://doi.org/10.1016/j.jhazmat.2020.123163.

24 Jiao, T., Zhao, H., Zhou, J. et al. (2015). Self-assembly reduced graphene oxide nanosheet hydrogel fabrication by anchorage of chitosan/silver and its potential efficient application toward dye degradation for wastewater treatments. *ACS Sustainable Chemistry & Engineering* 3: 3130–3139. https://doi.org/10.1021/acssuschemeng.5b00695.

25 Sun, L. and Fugetsu, B. (2014). Graphene oxide captured for green use: influence on the structures of calcium alginate and macroporous alginic beads and their application to aqueous removal of acridine orange. *Chemical Engineering Journal* 240: 565–573. https://doi.org/10.1016/j.cej.2013.10.083.

26 Li, Y., Du, Q., Liu, T. et al. (2013). Methylene blue adsorption on graphene oxide/calcium alginate composites. *Carbohydrate Polymers* 95: 501–507. https://doi.org/10.1016/j.carbpol.2013.01.094.

9

Lignocellulosic Biocomposite Adsorbents for Removal of Pollutants from Urban Waste

Fanqing Meng, Yinglong Wang, Hongwei Xu, Xiaxia Chen, Chao Liu, Ranran Wei, Xiaoying Zhang, and Xuewen Wei

College of Chemical Engineering, Qingdao University of Science and Technology, Qingdao, China

9.1 Introduction

Urban wastewater refers to the wastewater generated in urban daily life, industrial production, and commercial activities. With the acceleration of urbanization and population growth, the discharge of urban wastewater has sharply increased, bringing enormous pressure and challenges to the water supply environment, leading to serious wastewater pollution problems. Urban wastewater contains a large amount of pollutants, including organic matter, inorganic matter, heavy metals, nutrients, etc. If these pollutants are not effectively treated, they will cause serious harm to the water environment and ecosystem. Firstly, organic matter is one of the most common pollutants in urban wastewater. Domestic sewage contains a large amount of organic matter, such as food residues, oils, detergents, etc. After these organic substances enter the water through the urban drainage system, they consume dissolved oxygen in the water, leading to oxygen deficiency in the water and making it difficult for aquatic organisms to survive. In addition, organic matter in water can undergo microbial decomposition to produce greenhouse gases such as carbon dioxide and methane, exacerbating the problem of climate change. Secondly, inorganic substances are also one of the important pollutants in urban wastewater. Inorganic substances include heavy metals, nitrogen, phosphorus, etc. Industrial wastewater often contains heavy metal ions, such as mercury, lead, chromium, etc., which are highly toxic to aquatic organisms. Long-term accumulation can disrupt the balance of aquatic ecosystems and pose a threat to biodiversity. In addition, urban wastewater also contains a large amount of nutrients such as nitrogen and phosphorus. Excessive discharge into water bodies can lead to eutrophication, excessive growth of algae, and damage to water quality. Once again, the treatment of urban wastewater is urgent, mainly based on the following reasons. Firstly, wastewater treatment can improve the quality of the water environment and protect the living environment of aquatic organisms. By removing pollutants and harmful substances from waste, water bodies can be purified, water quality can be improved, and the health of water ecosystems can be protected. Secondly, wastewater treatment helps to save water resources. The treated wastewater can be reused, such as for agricultural irrigation and landscape water filling, to achieve the recycling of water resources. In addition, wastewater treatment can also achieve resource recycling and utilization, such as extracting valuable substances and energy from wastewater, further promoting the development of circular economy. With the acceleration of urbanization, the amount of urban wastewater discharge is showing a rapid growth trend, leading to serious wastewater pollution problems. The pollutants in urban wastewater cause serious harm to the aquatic ecosystem, such as water degradation, biological death, and soil quality degradation. Therefore, the treatment of urban wastewater is urgent. Effective wastewater treatment technologies can purify the water environment, protect ecosystems, save water resources, achieve resource recycling, and promote the achievement of sustainable development goals.

Firstly, let's understand the main sources of urban wastewater. Urban wastewater mainly includes the following categories: Domestic sewage: wastewater generated from household, commercial institutions, schools, and other domestic water sources, including bath water, laundry and tableware water, toilet flushing water, etc.; Industrial wastewater: wastewater generated during industrial production, including production wastewater from various factories, such as chemical factories, textile factories, electroplating factories, etc.; Rainwater and surface runoff: runoff formed by rainwater and snowmelt, including urban rainwater discharge, rainwater on roads and buildings, and water containing pollutants in soil; Agricultural wastewater: wastewater generated during agricultural production, including agricultural irrigation wastewater, aquaculture wastewater, etc. These wastewaters contain a variety of pollutants, such as organic matter, heavy metals, nitrogen, phosphorus, etc. If not properly treated, it will seriously endanger the environment and people's health. Therefore, the treatment of urban wastewater is very important. Among numerous wastewater treatment methods, biomaterial treatment methods are widely used. This method uses biodegradation as the main mechanism, utilizing the enzyme system of biomaterials to convert harmful substances into nontoxic or low-toxic substances. The use of biomaterials to treat urban wastewater has several important reasons: firstly, biomaterial treatment methods have lower energy consumption and operation and maintenance costs. Compared to traditional physical and chemical methods, biomaterial treatment methods do not require a large amount of chemical agents and high energy consumption equipment, thereby reducing the energy consumption and economic costs required for the treatment process. This helps to ensure the economic feasibility of wastewater treatment, especially in cases of large urban scale and huge wastewater discharge. Secondly, the treatment method of biomaterials is an environmentally friendly technology. In the process of biodegradation, biomaterials convert harmful substances in wastewater into environmentally friendly substances through their own metabolic activities, reducing the risk of secondary pollution to water bodies and ecosystems. In addition, biomaterials can self-repair and renew through growth and reproduction, which is sustainable and helps to treat wastewater stably in the long term. Most importantly, biomaterial treatment methods have a wide range of applicability. Different types of wastewater can be effectively treated using biomaterial treatment methods, including household wastewater, industrial wastewater, and farmland drainage. By optimizing the selection of biomaterials and cultivation conditions, efficient degradation and removal of different pollutants can be achieved. This makes biomaterial treatment methods a flexible and adjustable technological choice that can adapt to the needs of different environments and wastewater characteristics. In summary, as an efficient, environmentally friendly, and economical wastewater treatment technology, biomaterial treatment methods have the advantages of low energy consumption, environmental friendliness, and wide applicability. Therefore, they are widely used in urban wastewater treatment and are an important means to solve the problem of urban wastewater pollution (Figure 9.1).

The following are the steps for treating wastewater with general biomaterials, each step aimed at achieving the best wastewater treatment effect: (i) Primary treatment: Wastewater undergoes primary treatment, including filtration and solid-liquid separation processes. By filtering, large particle impurities and floating sediment in wastewater can be removed, reducing damage to subsequent treatment equipment and improving treatment efficiency. (ii) Bioreactor: The treated wastewater enters the bioreactor. In the bioreactor, add an appropriate amount of biological material, such as microbial strains, bacteria, algae, etc. These biomaterials have the ability to decompose and degrade organic matter. They use organic matter as energy and decompose it into smaller molecules through metabolism and enzyme action, making it less likely to cause harm to the environment. (iii) Aerobic and anaerobic processes: Depending on the characteristics of the wastewater, aerobic or anaerobic treatment can be chosen. The aerobic process requires oxygen supply and is suitable for treating wastewater that can be completely oxidized by oxygen, such as organic matter, ammonia nitrogen, etc. The anaerobic process does not require oxygen supply and is suitable for treating organic substances that are difficult to be oxidized by oxygen, such as sulfides, nitrates, etc. The combination of aerobic and anaerobic processes can more

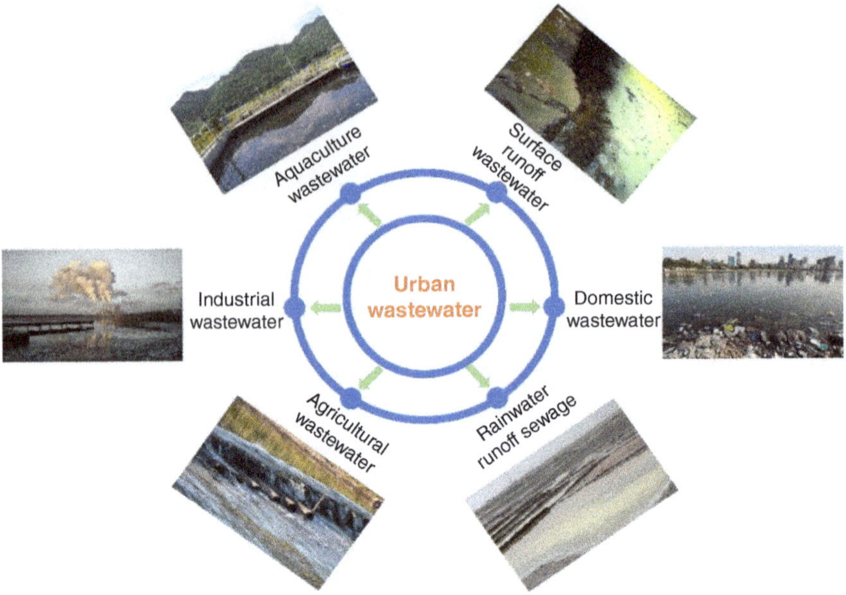

Figure 9.1 Sources of urban wastewater pollutants and wastewater. *Source:* jwvein / Pixabay / CC-0 / https://pixabay.com/photos/monolithic-part-of-the-waters-nature-3137978/ / last accessed on Oct 07, 2025.

comprehensively degrade pollutants in wastewater. (iv) Advanced treatment: The wastewater that passes through the bioreactor may still contain some organic matter and organisms and requires advanced treatment. Common advanced treatment methods include activated carbon adsorption, ozone oxidation, etc. By using activated carbon adsorption, residual organic matter in wastewater can be further removed and the purification degree of water quality can be improved. Ozone oxidation can degrade organic matter in wastewater and kill remaining microorganisms. (v) Clarification and disinfection: After advanced treatment, wastewater can undergo clarification treatment, such as sedimentation clarification or membrane filtration, to further remove suspended solids and turbidity, and ensure that the wastewater meets the discharge standard. At the same time, disinfection treatment can also be carried out, such as ultraviolet radiation or chemical disinfection, to ensure the complete removal of microorganisms in the wastewater and prevent potential harm to the environment and humans. Biomaterial treatment of wastewater is a complex process that involves multiple steps and technological applications. In practical operation, professional technical personnel are required to monitor and manage to ensure that the wastewater treatment effect meets the expected requirements [1]. At the same time, it is necessary to continuously optimize the wastewater treatment process based on the characteristics of wastewater, the selection of treatment equipment, and the adjustment of operating parameters, in order to achieve efficient, economic, and sustainable wastewater treatment.

Biomaterials for wastewater treatment are of great significance and importance in protecting the water environment, ensuring health, resource recovery, and promoting sustainable development. By applying biomaterial treatment technology, we can achieve effective degradation and removal of wastewater, improve water environment quality, protect ecosystem health, and have a positive impact on sustainable development and human well-being. Firstly, biomaterials used to treat wastewater can effectively convert harmful substances into harmless ones. Microorganisms and enzymes in biomaterials can decompose and degrade organic matter and pollutants in wastewater, transforming them into harmless substances, thereby reducing pollution to the water environment. This is crucial for maintaining the health of water bodies and protecting the living environment of aquatic organisms. Secondly, using biomaterials to treat wastewater can improve the quality of the water environment. The organic matter and pollutants in wastewater, if not effectively treated, can lead to water quality deterioration and pose serious risks to the ecological environment and human health. The use of biological materials for treatment can effectively remove organic substances from wastewater, improve water quality, and protect the sustainable utilization of the source area. In addition, biomaterials used for wastewater treatment can also achieve resource recovery and utilization. Wastewater contains many valuable substances and energy sources, such as organic matter, nitrogen, phosphorus, etc. Through the treatment of biomaterials, these valuable substances and energy can be extracted and recycled, achieving the resource utilization of waste, reducing resource waste, and promoting the development of circular economy. Finally, the concept of sustainable development in the treatment of wastewater using biomaterials. Biomaterials are environmentally friendly and can replace traditional chemical treatment methods, reduce energy consumption and chemical agent usage, reduce carbon dioxide emissions and waste generation, and meet the requirements of low-carbon environmental protection. By promoting and applying biomaterial processing technologies, the goal of sustainable development can be promoted and the future of humanity and the natural environment can be protected [2].

In summary, biomaterials for wastewater treatment are of great significance and importance in protecting the water environment, ensuring health, resource recovery, and promoting sustainable development [3]. By applying biomaterial treatment technology, we can achieve effective degradation and removal of wastewater, improve water environment quality, protect ecosystem health, and have a positive impact on sustainable development and human well-being.

9.2 Application and Mechanism Study of Composite Materials in Wastewater Treatment

9.2.1 Preparation and Wastewater Treatment Performance of Lignin Reinforced Cement Composite Materials

Lignin reinforced cement composite material, abbreviated as wood cement composite material, refers to a material obtained by mixing, blending, or composite processing of lignin with inorganic materials such as cement. Lignin is a natural high molecular organic compound present in plant cell walls, characterized by high strength, high stiffness, durability, and corrosion resistance [4]. Lignin reinforced cement composite material is a new type of material that combines lignin with cement. Lignin is a natural high molecular weight organic compound found in plant substrates such as wood and plant fibers, which has properties such as pressure resistance, acid, and alkali resistance. The preparation process of

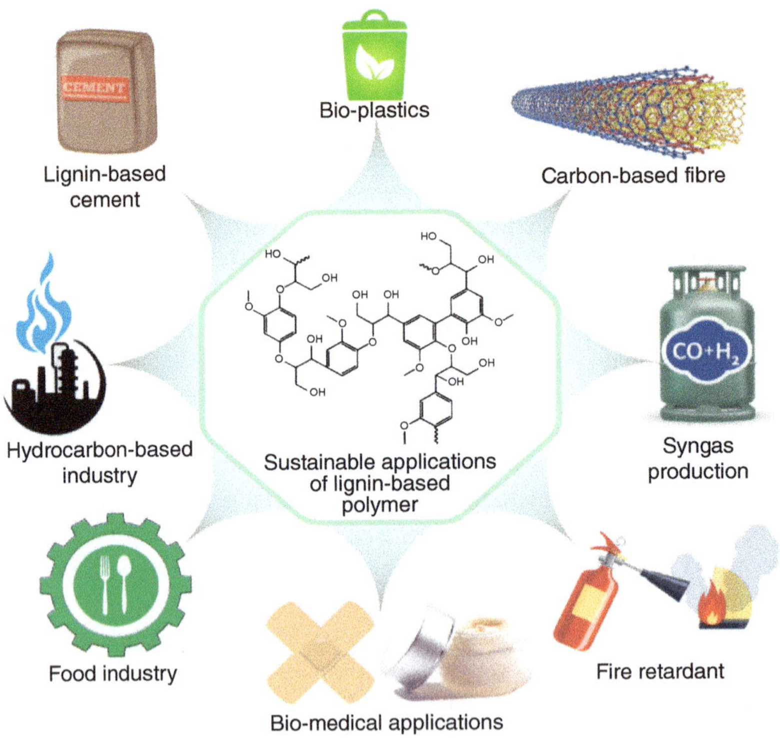

Figure 9.2 Sustainable applications of lignin in different industrial and biomedical sectors [5] / with permission of Elsevier.

lignin reinforced cement composite materials includes steps such as raw material treatment, lignin extraction, lignin modification, and cement composite material preparation. As shown in Figure 9.2, the lignin reinforced cement composite material prepared has the characteristics of increased strength, environmental friendliness, and excellent fire resistance. It has extensive applications in fields such as construction, roads, and water conservancy engineering. In the future, the development trend of lignin reinforced cement composite materials includes the application of new materials, process improvement, research and development of multifunctional materials, and the improvement of construction engineering standards.

The preparation of lignin reinforced cement composite materials usually involves mixing lignin and cement in a certain proportion, and then undergoing appropriate processing techniques (such as extrusion, compression, pouring, etc.) to form a composite material. The performance indicators are shown in Table 9.1. Lignin as a reinforcing agent can effectively improve the mechanical properties of cement matrix, such as compressive strength, flexural strength, and crack resistance. At the same time, the addition of lignin can also improve the durability of cement materials, improve water resistance, chemical corrosion resistance, and

Table 9.1 Key properties of lignin reinforced cement composites.

Performance index	Description	Value/unit
Compressive strength	The ability of a material to retain its shape and strength under certain pressure	30–60 MPa
Tensile strength	The bearing capacity of a material under tensile force	5–10 MPa
Toughness	The ability of a material to resist cracking or fracture under impact or vibration	High
Density	The ratio of mass to volume of a material	1.8–2.3 g cm^{-3}
Corrosion resistance	Durability of materials in chemical environments	Optimal
Environmental performance	The ability to use renewable resources and reduce industrial waste	High
Weight	The total weight of the material	Light
Service life	The life of the material under normal conditions of use	50 years

wear resistance. The preparation process of lignin reinforced cement composite materials can be divided into the following steps:

Raw material preparation: Prepare the necessary raw materials such as lignin and cement. Lignin can be selected from natural lignin or lignin extracts obtained through chemical treatment.

Raw material mixing: Mix lignin and cement in a certain proportion. Mechanical mixing equipment, such as a mixer or concrete mixer truck, can be used to thoroughly and evenly mix the two raw materials.

Adding appropriate admixtures: Suitable admixtures such as fine fly ash, silica fume, slag, etc., can be added during the mixing process as needed to further improve the performance of the material.

Adjusting cement slurry: According to specific needs, the water cement ratio of the cement slurry can be adjusted to achieve ideal fluidity and viscosity.

Mold type: Pour the mixed lignin cement slurry into a pre-prepared mold, which can be a board mold, a shaped mold, or other shaped mold.

Processing: According to the requirements, the slurry can be processed into the desired shape of lignin reinforced cement composite materials using processes such as compression, extrusion, and pouring.

Curing and curing: After initial molding, the lignin reinforced cement composite needs to be cured and protected. Usually, wet curing is adopted, which is carried out under one humidity and temperature condition to gradually harden the material and achieve the expected performance.

Inspection and testing: The prepared lignin reinforced cement composite material needs to be inspected and tested to evaluate and verify its mechanical properties, durability, and other properties.

It should be noted that the preparation process of lignin reinforced cement composite materials can be adjusted and improved according to specific material requirements and application fields. The above is only an overview of the general preparation process, and the specific steps and parameters should be adjusted and optimized according to the actual situation.

Lignin reinforced cement composite materials have broad application prospects by enhancing the mechanical properties, durability, and environmental friendliness of the material by enhancing the cement matrix. Lignin reinforced cement composite materials have the following advantages [4]:

Strength and stiffness improvement: Lignin, as a reinforcing agent, can effectively improve the mechanical properties of cement matrix, such as compressive strength, flexural strength, and stiffness. It can increase the load-bearing capacity of materials, making composite materials more sturdy and durable [6].

Durability improvement: The addition of lignin can improve the durability of cement materials. Lignin has characteristics such as corrosion resistance, water resistance, and chemical corrosion resistance, which can improve the water resistance, weather resistance, and chemical corrosion resistance of materials.

Environmentally friendly: Lignin is a natural organic compound with biodegradability and renewability. Lignin reinforced cement composite materials are more environmentally friendly than traditional cement materials, reducing the consumption of natural resources and environmental loads.

Improvement of processing performance: The addition of lignin can improve the processing and processing performance of cement materials. Lignin can increase the viscosity and fluidity of materials, making them easier to process and shape, and improving construction efficiency [7].

Basically no shrinkage: Lignin reinforced cement composite materials have almost no shrinkage during the solidification process, which can avoid cracks and damage caused by the solidification shrinkage of conventional cement materials.

Plasticity: By using admixtures and adjusting the cement water cement ratio and lignin ratio, the plasticity of lignin reinforced cement composites can be adjusted to meet different application needs.

Lignin reinforced cement composite materials have broad application prospects in sewage treatment. Lignin is a natural organic polymer with high mechanical strength and chemical stability. It can form composite materials by combining with materials such as cement, which can effectively improve the mechanical properties and durability of the material.

In sewage treatment, lignin reinforced cement composite materials can be applied in the following aspects:

Sewage pipeline repair: Lignin reinforced cement composite materials can be used to repair aged, cracked, or damaged sewage pipelines. Their high strength and corrosion resistance can effectively extend the service life of pipelines and reduce the problems of water and sewage leakage.

Components of sewage treatment equipment: Lignin reinforced cement composite materials can be used to manufacture components in sewage treatment equipment, such as sedimentation tanks, reactors, etc. Its high strength and chemical corrosion resistance can ensure the normal operation of the equipment and reduce the frequency of maintenance and replacement.

Coating for sewage treatment equipment: Lignin reinforced cement composite material can be applied as a coating on the surface of sewage treatment equipment, providing an additional protective layer. This coating can enhance the corrosion resistance of the equipment, reduce the accumulation of dirt, and improve the efficiency and lifespan of the equipment.

Building materials for sewage treatment plants: Lignin reinforced cement composite materials can be used as building materials in the construction structure of sewage treatment plants. Its high strength and durability can ensure the stability and safety of buildings, while reducing maintenance and repair costs.

In summary, lignin reinforced cement composite materials have broad application prospects in the field of sewage treatment, which can improve the performance and lifespan of sewage treatment equipment and pipelines, while reducing operation and maintenance costs and environmental risks [7, 8].

Lignin reinforced cement composite materials are commonly used in fields such as building structures, concrete repair, cement pipelines, etc., and have good mechanical properties, durability, and environmental friendliness. They can improve the strength, stability, and durability of engineering structures, extend their service life, and reduce environmental loads. Lignin reinforced cement composite material is a multifunctional material with broad application prospects. It has a wide range of application needs in fields such as architecture, roads, and water conservancy engineering, which can solve the problems existing in traditional cement materials and meet people's requirements for high strength, durability, and environmental friendliness of building materials. With the further development and promotion of technology, lignin reinforced cement composite materials will have a greater market share and contribute to the development of the construction industry [8].

The application of new materials: With the advancement of science and technology, more new materials will be applied in the preparation of lignin reinforced cement composite materials. For example, nanomaterials, fiber reinforced materials, etc., can further improve the performance of composite materials, such as strength, toughness, and wear resistance.

Process improvement: With the research and improvement of the preparation process of lignin reinforced cement composite materials, the automation, refinement, and efficiency of the material preparation process will be achieved. For example, adopting new mixing equipment and improved preparation processes to improve production efficiency and quality stability.

The development and application of multifunctional materials: With the continuous improvement of material performance requirements, there will be more research and application of multifunctional materials. For example, adding waterproofing agents, preservatives, insulation materials, etc., to lignin reinforced cement composite materials to achieve integration of multiple functional requirements.

Green and sustainable development: In the process of material preparation, more attention will be paid to environmental protection and sustainable development. For example, research on optimizing the use and recycling of raw materials, reducing energy consumption and carbon dioxide emissions, etc., to mitigate negative impacts on the environment.

The improvement of construction engineering standards: With the increasing requirements for the performance of lignin reinforced cement composite materials, corresponding construction engineering standards and specifications will be formulated and improved. This is conducive to standardizing and promoting the application of lignin reinforced cement composite materials in the construction field, ensuring that the composite materials meet relevant requirements and safety performance.

Structural design optimization: With in-depth research and understanding of the properties of lignin reinforced cement composite materials, the optimization of material structure design will be achieved. By optimizing the ratio, microstructure, and macroscopic structure of composite materials, the mechanical properties and durability of the materials can be further improved.

9.2.2 Study on the Properties and Mechanism of Carbon Nanotubes/Metal Oxide Composite Adsorbents for Heavy Metal Ion Adsorption

Lignin-derived nanocomposites are a kind of composite materials that combine lignin derivatives and nanomaterials. This material is designed to take full advantage of the structural and mechanical properties of lignin while enabling new functional and performance-superior applications by introducing the properties of nanomaterials. Due to their abundant surface functional groups, lignin biopolymers promote and improve the stabilization and immobilization/modification of metal or metal oxide nanoparticles (NPs) and are very promising nanomaterials and effective carriers of (photocatalysts). Lignin is usually derived from plant cell walls and can be extracted and derived by chemical treatment, biotechnology, or other methods. The choice of derivatives depends on the desired properties and application [4].

Some typical lignin derivatives include cellulose, lignin alcohols, lignin ketones, etc. [5, 6]. Nanomaterials, such as metal oxides, carbon nanotubes, graphene, etc., are introduced based on lignin. Among them, nanostructures include nanofibers and nanofilms. By introducing electrically conductive nanoparticles, such as carbon nanotubes or metal oxides, the lignin composite can be imbibed with conductive properties. This conductivity has potential applications in electronic devices, sensors, and other fields [8]. Improvements in optical performance can be achieved by introducing nanoparticles with fluorescent or other optical properties for bioimaging or other optical applications.

Urban wastewater pollution refers to the phenomenon that the wastewater produced by urban life and industrial activities contains various pollutants, which cause adverse effects on the water environment. These pollutants include organic matter, inorganic substances, heavy metals, nutrients, etc., and their sources involve many aspects such as family life, industrial production, and agricultural emissions. Urban wastewater pollution is a potential threat to water quality, aquatic organisms, and human health. First of all, the sewage discharged by household life mainly includes domestic sewage and stormwater sewage. Domestic sewage contains a lot of organic matter, such as grease, food residue, etc., as well as chemical substances, such as detergents, cleaners, etc. After the discharge of these substances, if not effectively treated, they will enter the water body, causing problems such as water eutrophication and algae reproduction, and destroying the water's ecological balance. Secondly, industrial wastewater discharge is another major source of urban wastewater pollution. Chemical substances and heavy metals used in industrial production, if not strictly treated, may enter water bodies and have serious adverse effects on water quality. Some industrial wastewater contains toxic and harmful substances, which may cause direct harm to aquatic organisms, and also pass through the food chain, posing a threat to human health.

Some conventional purification and treatment processes are used to remove harmful/toxic substances from water; research topics include reverse osmosis, membrane ultrafiltration/separation, ion exchange, (electric) dialysis, coagulation/flocculation, precipitation, chemical/electrochemical precipitation, etc. However, traditional wastewater pretreatment methods are generally difficult to operate, high cost, poor adaptability, and easy to pollute [7, 9]. In addition, such systems are not effective in repairing difficult-to-degrade organic pollutants. There are other sewage treatment technologies, that is, oxidation, odor, adsorption, or advanced oxidation. Adsorption method is an important means to control water pollution and eliminate organic/inorganic pollutants because of its characteristics of high efficiency, easy treatment, economy, simplicity, good repeatability, and sludge reduction. At present, scholars at home and abroad are committed to the development of a "green," cheap, inexpensive, practical biological adsorption material. It was found that lignin showed good adsorption capacity to dyes, heavy metals, and other harmful substances in water, and it was a potential biological adsorbent. Lignin can achieve [10, 11], reduction/degradation and other functions, and can effectively remove azo-containing dyes, dyes, nitrobenzene, and other toxic organic pollutants and heavy metals.

Lignin-based nanomaterials exhibit excellent properties in the field of wastewater treatment because of their high specific surface area and porous structure. Lignin-based nanomaterials have a high specific surface area and abundant porous structure, which enables them to provide a large number of active adsorption sites. In wastewater, these nanomaterials can fully expose adsorption sites, increase the contact area with pollutants in wastewater, and improve adsorption efficiency [12]. The surface of lignin-based nanomaterials is usually modified with various functional groups, such as hydroxyl, carbonyl, etc. These functional groups give the material specific affinity (as shown in Table 9.2, which summarizes the different types of functionalized lignin adsorbents). These functional groups adsorb organic matter and heavy metals in wastewater and realize

Table 9.2 Summary of different kinds of functionalized lignin-based adsorbents for the removal of dyes and heavy metal ions from water.

Functionalized lignin-based adsorbents	Preparation method, particle size (nm)	Pollutant	Adsorption capacity ($mg\ g^{-1}$)/removal efficiency (%)	Reaction/process conditions	Reference
Lignosulfonate/N-methylaniline composite	Cross-linking method	Cr(VI)	1264.8	Adsorbent (50 mg), 45 °C, pH 2.0, 10 min	[13]
Triethylenetetramine activated lignin	Mannich reaction, 50–100	As(V)	62.5	Adsorbent (10 mg), 60 min	[14]
5-Sulfosalicylic acid modified lignin	Chemical method	MB, Pb^{2+}	83.2 (MB), 39.3 (Pb^{2+})	Adsorbent (5.0 $g\ L^{-1}$), 45 °C, 12 h	[15]
Polyethylenimine (PEI)-modified lignin	Coating and cross-linking process	Cr(VI)	657.9	Adsorbent (0.1 g), 24 h	[16]

efficient adsorption of pollutants in wastewater through electrostatic attraction and hydrogen bonding. The introduction of photocatalysts or nanometal particles into lignin-based nanomaterials makes them photocatalytic. In the process of wastewater treatment, through light, these materials can produce reactive oxygen species, such as hydroxyl radicals and superoxide anions, which can promote the degradation of the organic pollutants in the wastewater and enhance the purification efficiency of the water quality. At the same time, lignin-based nanomaterials also show good electrochemical activity in wastewater treatment. Through electrochemical reactions, these materials can catalyze the reduction or oxidation of substances in wastewater, such as the reduction of heavy metal ions to metals or the oxidation of organic matter to harmless products, thus completing the wastewater treatment process. The nanomaterial surface is usually resistant to microbial attachment, reducing the formation of biofilms. This helps prevent the breeding of microorganisms and the contamination of wastewater treatment equipment while ensuring the long-term stable operation of nanomaterials. In summary, lignin-based nanomaterials achieve efficient removal and degradation of various pollutants in wastewater treatment through the synergistic effects of multiple mechanisms such as high specific surface area, porous structure, functional groups, photocatalytic activity, and electrochemical activity, providing a potential green solution for environmental protection [17].

Lignin-derived nanocomposites play an important role in the environmental field, such as for the preparation of materials with efficient adsorption and catalytic degradation of pollutants, for water treatment, air purification, and so on. Lignin and its derived nanomaterials have the characteristics of nontoxicity or low toxicity, good biodegradability, good stability, and high-cost effectiveness, and are easy to apply to environmental remediation, especially for the treatment of water/wastewater pollutants. Nanostructured materials with high surface area, excellent chemical reactivity, low power consumption, significant mechanical properties, and well-defined and controllable shape/size can be effectively used as adsorbents for water remediation/treatment [13]. Figure 9.3 outlines the environmental remediation applications of various techniques/extraction lignin and lignin-derived (nano) materials [6].

Lignin-derived nanocomposite adsorbents have wide application potential in the treatment of municipal wastewater pollution. By combining the natural adsorption properties of lignin with the high surface area and versatility of the nanomaterials, this composite material can effectively remove different pollutants from wastewater. Its main applications include, but not limited to, the efficient adsorption of heavy metal ions, organic pollutants, and oils, the removal of pharmaceutical and personal care product residues, and the reduction of ammonia nitrogen concentrations in wastewater. In addition, lignin-derived nanocomposite adsorbents may also exhibit antimicrobial properties against microorganisms

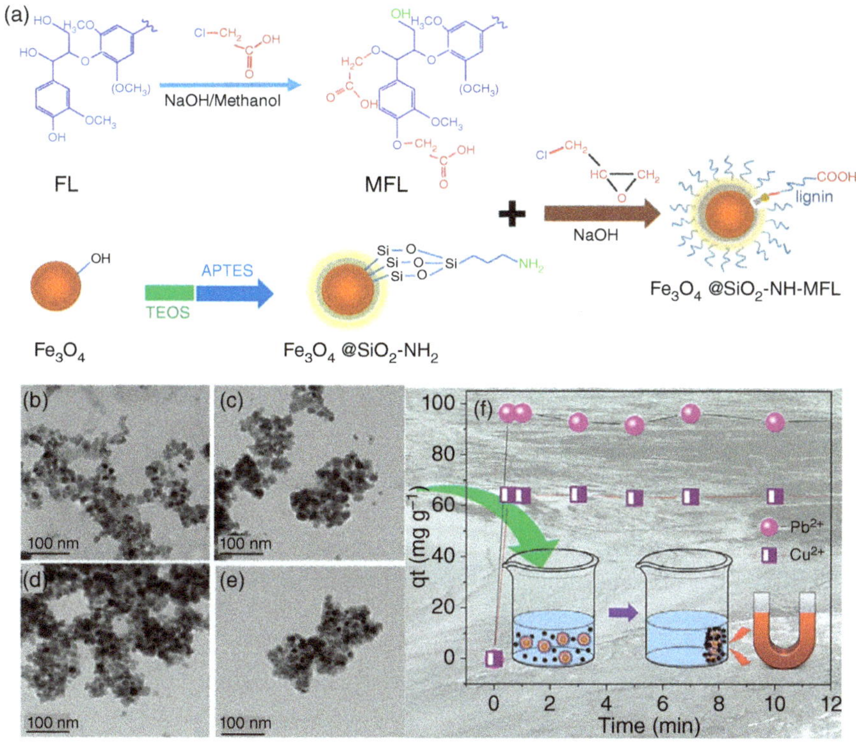

Figure 9.3 Preparation characterization and performance research of lignin-supported impurity magnetic (a) preparation process; representative TEM analysis of (b) Fe_3O_4, (c) Fe_3O_4/SiO_2, (d) Fe_3O_4/SiO_2-NH_2, (e) Fe_3O_4/SiO_2-NH-MFL, and (f) influence of contact time on the ability of hybrid magnetic adsorption of Cu^{2+} and Pb^{2+} [18] / MDPI / CC BY 4.0.

and bacteria, contributing to improving the biosafety of wastewater. These characteristics make it a comprehensive material choice to deal with the diversified pollution of urban wastewater and provide an environmentally friendly and efficient solution for wastewater treatment. However, to achieve the best results, in-depth research on the preparation process, structural design, and application conditions of this composite material is required to meet the specific wastewater contaminant components and treatment requirements. Ponomarev et al. [14] reported that granular L-L-Hydroxide nanocomposites were highly effective in the removal of Pb^{2+}, Ni^{2+}, and Cd^{2+} in single-and multi-component solutions. The results show that the synergic effect of hydrolyzing lignin waste and nano-Mg $(OH)_2$ (~9.94 nm) is good. Moser et al. [18] prepared effective and environmentally friendly lignin, stearic acid, and alumina biocomposite to remove bright green (BG) dyes from water media. In this paper, the separation of lignin from guava stem was studied. The results show that the adsorption data agree well with the Freundlich isotherm and the pseudo-second-order (PSO) dynamics model. A new kind of chitosan/lignin NPs was reported by Sohni et al. [19]. It can be used to remove the dye contamination in water. Chitosan-based lignin nanocomposites (chit/LNP) were prepared using these lignin nanocomposites (LNP) to effectively remove MB from aqueous media.

Furthermore, the lignin-based hydrogels have been widely used in the removal of various organic and inorganic pollutants from water due to their typical structure, excellent biocompatibility, and low cost. Li et al. synthesized a highly porous LS-modified graphene hydrogel by an environmentally friendly one-step hydrothermal method, and studied the adsorption activity of Pb^{2+} ions in aqueous solution. The synergy between LS and graphene can effectively improve and provide sufficient active sites for the rapid adsorption of 308 and 1210 mg^{-1} toxic Pb^{2+} in 40 and 240 minutes, respectively. Hydrogels with different pore sizes (nm to μm) or micropores or mesoporous obtain a high surface area (459.3 m^2 g^{-1}), while hydrogels with large pores guarantee the accessibility of this surface. The three-dimensional porous structure can be formed by the overlap/coalescence of hydrophilic oxidation functional groups and graphene sheets, while the noncovalent interaction between LS and rGO sheets can be explained by hydrogen bonding and π–π conjugation. Dominguez-Robles et al. [20] cross-linked different lignins (organic solvent lignin, straw alkali lignin, and cork lignin) with PMVE-MA (polymethyl vinyl ether-complete acid) in ammonium salts and sodium hydroxide aqueous solutions through a cost-effective thermal process. A series of highly porous lignin-based hydrogels (OSL/PMVE-MA, SSL/PMVE-MA, and SKL/PMVE-MA) with excellent expansion capacity (130 $g\,g^{-1}$) were prepared and adsorbated MB from aqueous solutions. The findings demonstrate that the two primary factors influencing the adsorbent's performance are the lignin's structure and the alkali (sodium hydroxide or ammonium hydroxide) used during manufacture. Compared to hydrogels made with NaOH, those made with SSL and NH_4OH were more effective. As a result, SSL-NH_4OH/PMVE-MA functions as the best hydrogel since it can absorb and release significant volumes of MB and water. Adsorption experiments indicate that at normal temperature, the clearance rate of MB varies between 440 and 840 $mg\,g^{-1}$. Magnetic cellulose/biopolymer hydrogel microspheres were successfully prepared via a sol-gel method using 1-ethyl-3-methylimidazole acetate [Emim] [Ac] as an oil-in-oil emulsifier [21]. As a biodegradable support material, cellulose is used due to its excellent mechanical and chemical properties. The surface properties of hydrogel microbeads were modified by blending cellulose with chitosan, lignin, starch, and carrageenan. Compared with cellulose hydrogel microbeads, the absorption capacity of cellulose/alkali lignin (AL) microbeads and cellulose/organic solvent lignin microbeads for crystal violet (CV) cationic dyes increased by 1.3 and

1.2 times, respectively. Jiang et al. [1] studied the adsorption/removal effect of sulfuric acid/lysine (CLS/Lys) hydrogel as an excellent biosorbent on heavy metal ions (Co^{2+} and Cu^{2+}) in aqueous solution. The main adsorption mechanism of CLS/Lys and its excellent adsorption performance and analytical reuse efficiency are mainly attributed to chemical adsorption through electrostatic attraction (Figure 9.4a). Through FTIR spectrum analysis, it was confirmed that the participation of functional groups such as amine group, sulfonic acid group, and phenolic group (-OH) in the adsorption process of Co^{2+} and Cu^{2+} (Figure 9.4b). Using PSO dynamics and Langmuir isotherms, the adsorption dynamics and isotherms were fitted. In another study, the modified bentonite/Na LS grafted maleic anhydride and acrylamide (AM) hydrogel was prepared by solution graft copolymerization technology, which was used to remove Pb^{2+} from polluted water source/wastewater. The prepared hydrogel showed excellent adsorption capacity and efficiency for Pb^{2+} ions in a quaternary metal solution containing cadmium, copper, and zinc ions. The results showed that the synthesized hydrogel mainly adsorbed lead ions through chelation or ion exchange mechanism. At 25 °C and pH 5.0, the saturated adsorption capacity of Pb^{2+} is 314.8 $mg\,g^{-1}$.

Water is well known to be a valuable resource for the environment and nature, and its quality and availability are crucial for life on earth. The increasing global pollution problem is putting huge pressure on available water resources as a result of the alarming rise in water scarcity and water pollution in developing and industrialized/advanced countries. Nanostructured (nano) materials, including (nano) composites, mixtures, hydrogels, functionalized/modified, porous and carbon-based materials, due to their specific structural and physico-chemical characteristics, such as workability, ease of modification, good chemical/mechanical stability, diversity of forms, improved potential, significant biodegradability/biocompatibility, cost effectiveness, low toxicity, and noncontamination, have attracted wide attention. Lignin and lignin-derived (nano) materials can effectively absorb, adsorb, and/or eliminate toxic heavy metal ions and organic/azodyes by chelation, ion exchange, π–π stacking, hydrogen bonding, and/or electrostatic interactions. Nanomaterials exhibit superior selectivity, sustainability/stability, and enhanced adsorption capacity compared to alternative compounds in environmental pollution remediation.

The water remediation industry is promoting simplified and cost-effective (nano) systems to develop various technologies for water/wastewater treatment, contributing to sustainable development. The application of lignin-derived (nano) materials in dye contaminants and water metal ion remediation/treatment is broadly encouraging, but some critical issues regarding their toxicity, biosafety concerns, and mechanisms should be fully evaluated. The search for low-cost, green, high adsorption capacity, and recyclable (nano) materials still needs to be further studied.

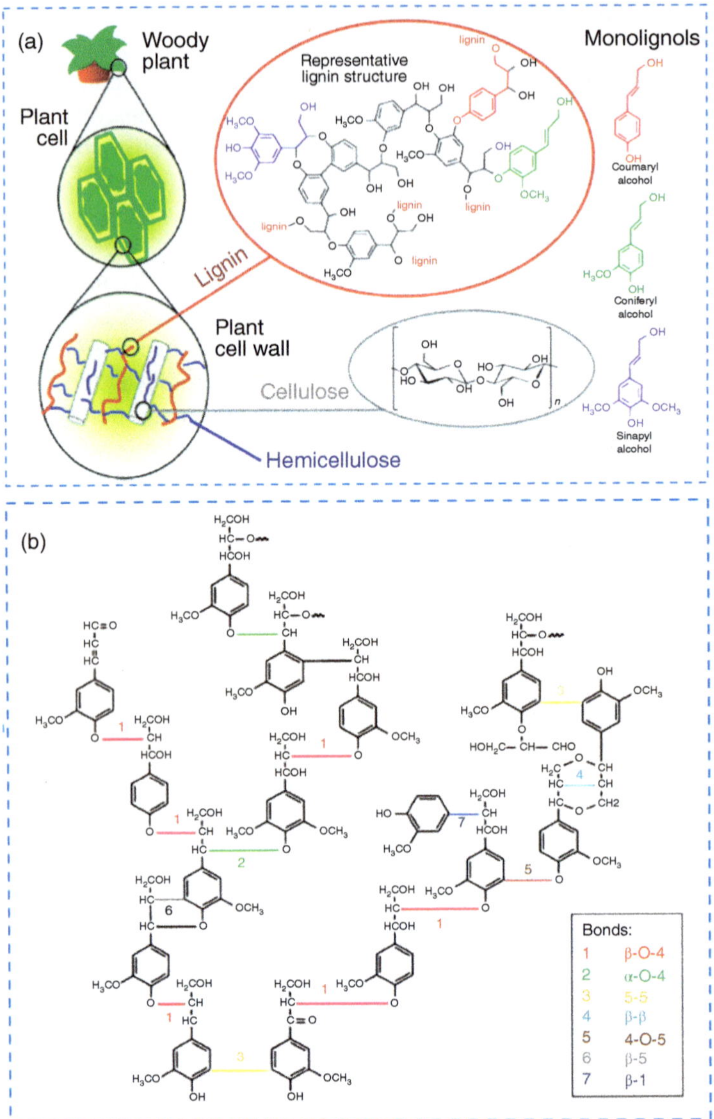

Figure 9.4 (a) The lignin structure and three main lignin precursors [30] / Science and Education Publishing / CC BY 4.0, (b) typical interunit linkages present in softwood lignin [31] / Springer Nature.

9.2.3 Study on Mechanism of Lignin Acid Alcohols Composites in Wastewater Treatment

Lignin-polymer composites are a kind of new materials with excellent properties by combining lignin, a natural aromatic polymer, with other polymers. Lignin-polymer composites can be used in various fields, such as elastomers, foams, adhesives, coatings, etc. Among them, the application of lignin-polymer composite materials in urban wastewater pollution is mainly as adsorption materials, using the active groups in lignin molecules, such as phenol hydroxyl, alcohol hydroxyl, etc., to chemically or physically interact with organic matter or heavy metal ions in wastewater, so as to achieve the purpose of removing pollutants. Ran Haiyan et al. introduced a composite material made of lignin and polylactic acid, which has good biodegradability and adsorbability, and can effectively adsorb dyes and heavy metal ions in wastewater [24]. Another paper reported a composite hydrogel prepared with lignin and polyacrylic acid, which has high strength, high toughness, and high water absorption, and can be used to adsorb copper ions and methylene blue dyes in wastewater [25]. A composite material prepared by lignin and polyvinyl alcohol was studied. After sulfonation modification, this material can be used to adsorb lead ions and phenol in wastewater [26]. As shown in Table 9.3, lignin is described in various ways as a filler in lignin-based biocomposites.

The composite material prepared by lignin and polylactic acid is a new material with excellent properties by combining the natural aromatic polymer of lignin with the biodegradable synthetic polymer of polylactic acid. The composite materials prepared by lignin and polylactic acid can be used in various fields, such as elastomers, foams, adhesives, coatings, etc. Among them, the application of composite materials prepared by lignin and polylactic acid in urban wastewater pollution is mainly as adsorption materials, using the active groups in lignin molecules, such as phenol hydroxyl, alcohol hydroxyl, etc., to chemically or physically interact with organic matter or heavy metal ions in wastewater, so as to achieve the purpose of removing pollutants. Here are some specific practical applications.

The utility model relates to a composite material prepared by lignin and polylactic acid, which has good biodegradability and adsorbability and can effectively adsorb dyes and heavy metal ions [3] in wastewater. The preparation method of the composite material is to melt and extrude the lignin and polylactic acid after mixing, and then obtain the composite material with the desired shape by hot pressing or injection molding. The adsorption performance of the composite is related to the content, molecular weight, and structure of lignin. Generally speaking, the higher the lignin content, the better the adsorption performance, but it will also affect the mechanical properties and thermal stability of the composite. The adsorption mechanism of the composite material is to form

Table 9.3 Description of lignin as filler in lignin-based biocomposites.

Polymer matrices	A common limitation in fabrication	Possible application of lignin-based composites
PLA	The need for compatibilizer	Packaging, food packaging
PVA	The hydrophilic properties of PVA	Packaging, drug delivery system, biomedical wound healing application, agricultural application
Cellulose	The need for compatibilizer	Seed coating
Starch	The hydrophilic properties of starch	Packaging, food packaging
Epoxy	The need for compatibilizer due to the incompatibility resulted in hydrophilic properties	Coating
Natural rubber	The need for compatibilizer due to the incompatibility resulted in hydrophilic properties	Tire industry, automotive devices
Polypropylene	The need for compatibilizer due to the incompatibility; the reduction in elongation; the susceptibility to microorganism	Automotive devices
Commercial adhesives	Difficulties in particles distribution	Filler in adhesives system

complexes with dyes and heavy metal ions through electrostatic interaction, hydrogen bonding, van der Waals force, and π–π stacking.

The utility model relates to a composite material prepared by grafting lignin and polylactic acid to maleic anhydride. The material has good compatibility and adsorbability and can be used to adsorb copper ions and methylene blue dye in wastewater. The preparation method of the composite material is to mix polylactic acid grafted with maleic anhydride and lignin, melt and blend, and then obtain the composite material with the required shape by injection molding. The adsorption performance of the composite is related to the content of polylactic acid grafted maleic anhydride, the type and amount of lignin, pH value and temperature, etc. Generally speaking, the higher the content of polylactic acid grafted maleic anhydride, the better the adsorption performance, but it will also affect the mechanical properties and thermal stability of the composite.

The adsorption mechanism of the composite is to form a complex with copper ions and methylene blue dyes through ion exchange, coordination, electrostatic interaction, and hydrogen bonding.

The utility model relates to a composite material prepared by grafting glycidyl methacrylate with lignin and polylactic acid/vinyl butyl acrylate. The material has good compatibility and absorbability and can be used to adsorb oil-water mixture [5] in wastewater. The preparation method of the composite is to mix the lignin with polylactic acid/vinyl butyl acrylate grafted with glycidyl methacrylate, melt and blend, and then lyophilize or vacuum dry to obtain the porous composite. The adsorption properties of the composite are related to the content of lignin, the content of glycidyl methacrylate grafted by ethylene butyl acrylate, pH value and temperature, etc. Generally speaking, the higher the lignin content, the better the adsorption properties, but it will also affect the water absorption and mechanical properties of the composite. The adsorption mechanism of the composite material is that it can effectively separate the oil-water mixture through the characteristics of super hydrophilic, low toxicity, and ecological friendliness, while maintaining durable oil resistance and anti-fouling properties, and the material is easy to recycle.

The composite material prepared by lignin and polyacrylic acid is a new material with excellent properties by combining lignin, a natural aromatic polymer, with polyacrylic acid, a biodegradable synthetic polymer. The composite materials prepared by lignin and polyacrylic acid can be used in various fields, such as elastomers, foams, adhesives, coatings, etc. Among them, the practical application of lignin and polyacrylic acid composite materials in urban wastewater is mainly as adsorption materials, using the active groups in lignin molecules, such as phenol hydroxyl, alcohol hydroxyl, etc., to react chemically or physically with organic matter or heavy metal ions in wastewater, so as to achieve the purpose of removing pollutants. Here are some specific practical applications:

The utility model relates to a composite hydrogel prepared by using lignin and polyacrylic acid. The hydrogel has high strength, high toughness, and high water absorption, and can be used to adsorb copper ions and methylene blue dye in wastewater [4]. The preparation method of the composite hydrogel is to mix lignin and polyacrylic acid solution for cross-linking reaction and then get porous composite hydrogel by freeze-drying or vacuum drying. The adsorption performance of the composite hydrogel is related to the content of lignin, the type and dosage of cross-linking agent, pH value and temperature, etc. Generally speaking, the higher the lignin content, the better the adsorption performance, but it will also affect the water absorption and mechanical properties of the composite hydrogel. The adsorption mechanism of the composite hydrogel is to form a complex with copper ions and methylene blue dyes through ion exchange, coordination, electrostatic interaction, and hydrogen bonding.

The utility model relates to a composite material prepared by grafting lignin and polyacrylic acid to maleic anhydride. The material has good compatibility and adsorbability and can be used to adsorb dyes and heavy metal ions [2] in wastewater. The preparation method of the composite material is to mix the polyacrylic acid grafted with maleic anhydride and lignin, melt and blend, and then obtain the composite material with the required shape by injection molding. The adsorption performance of the composite is related to the content of maleic anhydride grafted by polyacrylic acid, the type and amount of lignin, pH value and temperature, etc. Generally speaking, the higher the content of maleic anhydride grafted by polyacrylic acid, the better the adsorption performance, but it will also affect the mechanical properties and thermal stability of the composite. The adsorption mechanism of the composite material is to form complexes with dyes and heavy metal ions through ion exchange, coordination, electrostatic interaction, and hydrogen bonding.

The invention relates to a composite material prepared by grafting glycidyl methacrylate with lignin and polyacrylic acid/vinyl butyl acrylate. The material has good compatibility and absorbability and can be used to adsorb oil-water mixture in wastewater. The preparation method of the composite is to mix the lignin and polyacrylic acid/vinyl butyl acrylate grafted with glycidyl methacrylate, melt and blend, and then lyophilize or vacuum dry to obtain the porous composite. The adsorption properties of the composite are related to the content of lignin, the content of glycidyl methacrylate grafted by ethylene butyl acrylate, pH value and temperature, etc. Generally speaking, the higher the lignin content, the better the adsorption properties, but it will also affect the water absorption and mechanical properties of the composite. The adsorption mechanism of the composite material is that it can effectively separate the oil-water mixture through the characteristics of super hydrophilic, low toxicity, and ecological friendliness, while maintaining durable oil resistance and anti-fouling properties, and the material is easy to recycle.

In summary, the composite material prepared by lignin and polyacrylic acid is a new material with broad application prospects, which can not only use lignin as a low-cost and renewable biomass resource but also improve the performance of polyacrylic acid and provide an effective solution for the treatment of urban wastewater pollution.

The composite material prepared by lignin and polyvinyl alcohol is a new material with excellent properties by combining the natural aromatic polymer of lignin with the biodegradable synthetic polymer of polyvinyl alcohol. The composite materials prepared by lignin and polyvinyl alcohol (PVA) can be used in various fields, such as elastomers, foams, adhesives, coatings, etc. Among them, the practical application of the composite material prepared by lignin and polyvinyl alcohol in urban wastewater is mainly as an adsorption material, using the active groups in lignin molecules, such as phenol hydroxyl, alcohol hydroxyl, etc., to chemically or physically interact with organic matter or heavy metal ions in wastewater, so as to achieve the purpose of removing pollutants. Here are some specific practical applications:

The utility model relates to a fully degradable composite material prepared by lignin and polyvinyl alcohol, which has good biodegradability and adsorbability and can effectively adsorb dyes and heavy metal ions [3] in wastewater. The preparation method of the composite is to mix lignin and polyvinyl alcohol solution, freeze dry, and then, through formaldehyde cross-linking reaction, get lignin-polyvinyl alcohol composite. The adsorption performance of the composite is related to the content of lignin, the amount of formaldehyde, pH value and temperature, etc. Generally speaking, the higher the lignin content, the better the adsorption performance, but it will also affect the mechanical properties and water absorption of the composite. The adsorption mechanism of the composite material is to form complexes with dyes and heavy metal ions through electrostatic interaction, hydrogen bonding, van der Waals force, and π–π stacking.

The utility model relates to a composite hydrogel prepared by using lignin and polyvinyl alcohol. The hydrogel has high strength, high toughness, and high water absorption, and can be used to adsorb copper ions and methylene blue dye in wastewater. The preparation method of the composite hydrogel is to mix lignin and polyvinyl alcohol solution for cross-linking reaction, and then get porous composite hydrogel by freeze-drying or vacuum drying. The adsorption performance of the composite hydrogel is related to the content of lignin, the type and dosage of cross-linking agent, pH value and temperature, etc. Generally speaking, the higher the lignin content, the better the adsorption performance, but it will also affect the water absorption and mechanical properties of the composite hydrogel. The adsorption mechanism of the composite hydrogel is to form a complex with copper ions and methylene blue dyes through ion exchange, coordination, electrostatic interaction, and hydrogen bonding.

A composite material prepared by lignin and polyvinyl alcohol can be used to adsorb lead ions and phenol in wastewater [5] after sulfonation modification. The preparation method of the composite is to mix the lignin and PVA solution, freeze dry, and then obtain the lignosulfonate-PVA composite through sulfonation reaction. The adsorption properties of the composite are related to factors such as the degree of sulfonation, pH value and temperature, etc. Generally speaking, the higher the degree of sulfonation, the better the adsorption properties, but it will also affect the mechanical properties and thermal stability of the composite. The adsorption mechanism of the composite material is to form a complex with lead ion and phenol through ion exchange, coordination, electrostatic interaction, and hydrogen bonding.

In summary, lignin-polymer composite is a new material with broad application prospects, which can not only use lignin as a low-cost and renewable biomass resource but also improve the performance of the polymer and provide an effective solution for the treatment of urban wastewater pollution.

9.2.4 Study on the Adsorption Characteristics and Mechanism of Heavy Metal Ions by *Aspergillus niger*–Rice Straw Composite Adsorbent

Since the beginning of the 20th century, people have been enjoying the fruits of science and technology, but they are also facing great dangers, such as environmental pollution caused by the development of science and technology, which has seriously damaged people's health. From the 2015 environmental monitoring analysis report, it can be seen that the world's water resources are seriously polluted.

Metals with a specific gravity greater than 5 are considered heavy metals, such as lead, cadmium, copper, mercury, chromium, and about 45 species. The source of heavy metals is generally mostly the industrial wastewater discharged from large-scale heavy industries, such as mining, metallurgy, electroplating, pharmaceuticals, and even the coating industry, and the industrial wastewater discharged from these industries is directly discharged into the natural water system without relevant treatment, causing great harm to the ecological environment, and the pollution of heavy metals is also increasing day by day [27].

Heavy metals in water bodies are generally hidden, long-term, and not easy to degrade. Heavy metals such as copper ions in the order of ppb can catalyze the metabolism of substances, once they reach the order of ppm, they will be enriched through the biological chain, and finally accumulate in the human body step by step. Once the limit of the human body is exceeded, it will cause various diseases. Heavy metals can also be combined with other harmful substances existing in the natural environment, and generate highly toxic harmful substances. No matter what chemical form of heavy metal pollution sources exist, once they enter the ecological environment, it is difficult to be decomposed and eliminated, and will continue to accumulate and transfer, causing harm to the ecological environment and human beings [28].

The adsorption method is the most effective way to solve the above problems. The adsorption method has many advantages, such as high separation efficiency, economy, and simple operation. The adsorption method has also been highly valued by scholars in the field of environmental protection. The biological adsorption method is a relatively special treatment method that has emerged recently. The principle of biological adsorption is to adsorb certain specific heavy metal ions in industrial wastewater through the characteristics of the organism itself, and then remove the heavy metal ions in the waste liquid through solid–liquid separation.

There are many types of biological adsorbents, including bacteria, fungi, actinomycetes, algae, and lignocellulosic biomass. Traditional processes often use a single type of biomass adsorbent. Fungi and lignocellulosic biomass are combined using solid-phase fermentation technology to produce inexpensive and green composite adsorbents. At the same time, Cu^{2+} is used as a probe to systematically explore and study the biological adsorption mechanism.

Biological agents have many advantages in treating wastewater, such as low cost and high efficiency, as well as outstanding environmental performance, which has attracted widespread attention. So far, polysaccharides extracted from bacteria, yeast [29], genetically engineered bacteria, filamentous fungal organisms, algae, carbon fibers, etc., have been used to study the adsorption of heavy metal ions. The research results indicate that the adsorption degree of biological agents on different metal ions also varies, mainly manifested in two aspects: specific and nonspecific adsorption. Different types of microorganisms have varying degrees of tolerance to different metals, with fungi being the most tolerant, followed by bacteria, and actinomycetes being relatively less tolerant. In addition, adsorption materials can be obtained from agricultural and forestry waste, which have been widely studied due to their economic and practical characteristics. Table 9.4 lists the adsorption of heavy metals by biomass materials.

Compared with the traditional methods mentioned earlier, the advantages of biological adsorption are as follows: (i) low cost and easy availability of raw materials, with a wide variety of varieties; (ii) green, no secondary pollution [30]; (iii) simple equipment and low operating costs; (iv) can selectively remove metals from

Table 9.4 The removal performance of biosorbents for heavy metal ion.

Microorganism	Metal ion	Adsorption capacity
Biomatric	Cu^{2+}	19.3
Sargassum horneri	Cr^{6+}	62.0
Trichoderma	Cr^{6+}	45.0
Bagasse	Pb^{2+}	41.3
Pseudomonitroides	Cu^{2+}	25.6
HCl treatment of wheat stalk	Cu^{2+}	5
NaOH treats black mold	Cu^{2+}	28.7
Shaddock peel	Ni^{+}	46.1
Wheat straw	Cd^{2+}	12.2
Bagasse	Ni^{+}	2
Straw stalk	Cd^{2+}	19.3

low concentration wastewater; (v) wide range of pH value and temperature; and (vi) can effectively recover precious metals.

Due to the small size of bacteria, they are commonly present in the environment and can adapt well to the natural environment, making them suitable as adsorbent materials for organisms. Wood cellulose biomass resources are abundant and diverse, such as bark, straw, rice straw, corn cobs, etc. [31]. The treatment of heavy metal ions with wood fiber raw materials can not only achieve the expected treatment effect, but also reasonably save resources and reduce costs. There are two reasons why it serves as a heavy metal adsorbent: firstly, lignocellulosic materials have a relatively loose physical structure, high specific surface area and pore volume, and can react with a large number of metal ions to achieve adsorption effect; secondly, it contains a large number of active substances, such as hydroxyl and amino groups, which can undergo displacement reactions with heavy metal ions and more effectively adsorb metal ions. The porous structure and multiple active groups inherent in wood fiber materials make their research and utilization prospects very promising.

Wood cellulose materials have certain advantages as heavy metal adsorbents, but there are also disadvantages, mainly manifested in the limited active substances contained in them and the low adsorption rate for heavy metals. Therefore, some scholars use acid-base or organic solvents to modify it to produce easily cross-linked active groups, or expose more adsorption sites to enhance its adsorption of heavy metal ions. There are generally two methods for modification: physical modification and chemical modification. After modification, although the adsorption capacity of lignocellulosic materials has been improved, there are also some shortcomings that cannot be ignored. One reason is that physical modification is relatively expensive: although biochar prepared from agricultural and forestry waste is environmentally friendly and has good adsorption effects, it requires continuous high temperature during the preparation process to ensure the amount of carbon generated, resulting in higher costs. Secondly, chemical modification can easily cause secondary pollution: During the chemical modification process, chemicals can cause the modified lignocellulose material to have a lower or higher pH, and after adsorption, new wastewater will be generated, which cannot achieve the goal of purifying water quality [32]. Therefore, if cellulase is used for biological modification, it will lead to an increase in active sites related to adsorption on its surface, an increase in adsorption capacity, and at the same time, biological modification is more economical and environmentally friendly than chemical modification. In Peng's [33] study, corn stover was modified with cellulase to break cellulose polysaccharides into glucose molecules, while increasing the fiber's looseness and causing it to expand, making it easy for heavy metal ions to enter the fiber structure and be adsorbed. *Aspergillus niger* can secrete a large amount of cellulase.

The traditional preparation process of *A. niger* adsorbent has problems such as cumbersome preparation process, expensive liquid culture medium, and unsatisfactory adsorption effect. At present, the adsorbents for *A. niger* are all produced by liquid-phase fermentation [34] through the following steps: shaking the bottle to cultivate the mycelium balls, washing, filtering, drying, grinding, and sieving. At the same time, reading literature, it was found that the adsorption capacity of the adsorbent produced by liquid shaking bottle cultivation technology in traditional processes is relatively low. When the dosage is 1–5 $g\,L^{-1}$, the initial concentration is 0–50 $mg\,L^{-1}$, and the pH is appropriate, the adsorption capacity for the target pollutant copper ions is around 10 $mg\,g^{-1}$ [35]. In addition, the exploration of biological adsorption mechanisms varies depending on different types of biomass materials, and there are even fewer reports on the efficiency and mechanism of spore adsorption of heavy metals. Such as a low-cost magnetic adsorbent derived from straw removes elemental mercury from flue gas (Figure 9.5).

Therefore, how to simplify the preparation steps, increase adsorption capacity, and explore the biological adsorption mechanism has become a problem that current technology should solve.

In this study, *A. niger*–straw composite adsorbent was prepared by solid-phase fermentation technology. The effects of pH value, contact time, adsorbent dosage, initial solution concentration, and interference ions on the removal of heavy metals in the solution were systematically investigated. Several common kinetic models, isothermal adsorption models, and thermodynamic models were used to

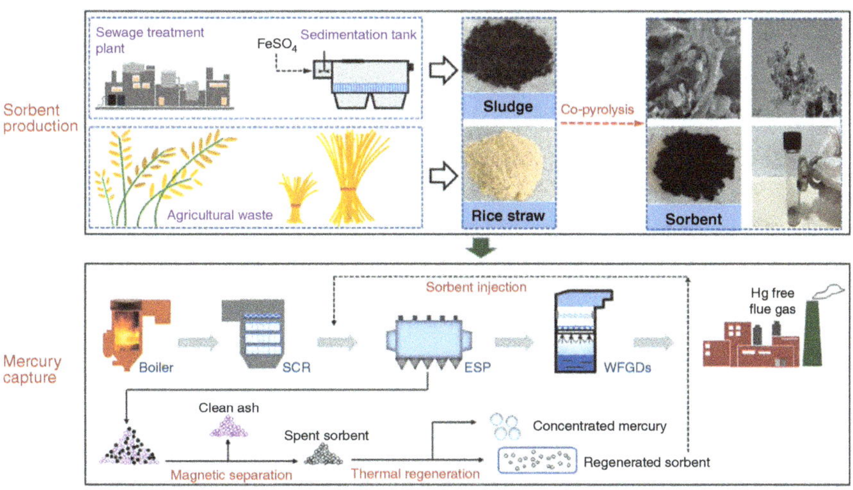

Figure 9.5 A low-cost magnetic adsorbent derived from straw removes elemental mercury from flue gas.

fit and analyze the adsorption process, analyzing and inferring the mechanism of the reaction process based on the characterization characteristics of the adsorption material and the changes in adsorption performance. The main conclusions are summarized as follows:

1) *Aspergillus niger* by solid-phase fermentation, as an alternative technology to the traditional liquid-phase fermentation to prepare *A. niger* adsorbent, has been greatly improved, while its preparation method has been simplified. The iodine value of the adsorbent of *A. niger* solid-phase fermentation product and liquid-phase fermentation product were 1239.8 and 593.7 $mg\,g^{-1}$, respectively. When the initial concentration was 30 $mg\,L^{-1}$, the dosage of adsorbent was 0.1 g, the pH was not adjusted, the temperature was 30 °C, and the rotation speed was 100 rpm, the adsorption capacity of solid-phase fermentation of *A. niger* gas hypha adsorbent for Cu^{2+} was 20.3 $mg\,g^{-1}$, at this time, the adsorption capacity of *A. niger* liquid phase fermentation product was only 8.1 $mg\,g^{-1}$. The optimum pH for *A. niger* adsorption by solid-phase fermentation was 4.0–6.0. FTIR showed that functional groups (—COOH, —NH_2, —PO_4^{3-}, —OH) were involved in the adsorption mechanism of Cu^{2+}, and the second-order kinetic equation and freundlich model could well fit the value of the whole adsorption process, which meant that the adsorption of Cu^{2+} by *A. niger* aerial hypha was a multi-layer adsorption process dependent on chemical adsorption.
2) The solid-phase fermentation process parameters of rice straw were optimized by response surface methodology as follows: inoculum size 1 × 10%, culture temperature 30 °C, and solid-liquid ratio 1:1. The adsorption capacities of *A. niger*–straw composite adsorbent and untreated straw for Cu^{2+} ranged from 4.8 to 1.8 $mg\,g^{-1}$. Meanwhile, the iodine value (348.92 $mg\,g^{-1}$) and specific surface area (6.3139 $m^2\,g^{-1}$) of *A. niger*–straw composite adsorbent were higher than those of straw (233.68 $mg\,g^{-1}$) and specific surface area (1.2287 $m^2\,g^{-1}$). In addition, when *A. niger*–straw composite adsorbent was 0.1 g, the initial concentration was 30 $mg\,L^{-1}$, the pH was not adjusted, the temperature was 30 °C, and the rotating speed was 100 rpm, *A. niger*–straw adsorbent could adsorb Pb^{2+}, Hg^{2+}, and the adsorption capacities of Cd^{2+} were 7.02, 8.07, and 5.28 $mg\,g^{-1}$. The second-order kinetic equation and freundlich model can well fit the whole adsorption process. FTIR showed that —OH, C=O, —C=O, and C=C were involved in the reaction, and the whole process conformed to the second-order kinetics and freundlich equation. The optimal pH of *A. niger*–straw composite adsorbent was not adjusted, and the adsorption equilibrium was achieved within 30 minutes.
3) Since the solid-phase and liquid-phase fermentation products have roughly the same structure and composition, it can be seen that the strong adsorption

capacity of solid-phase fermentation product agent and *A. niger*–straw composite adsorbent is due to the presence of spores. When *A. niger* spore dosage was 0.1 g, initial concentration was 30 mg L^{-1}, pH was not adjusted, temperature was 30°C, and rotation speed was 100 rpm, the adsorption capacities of Pb^{2+}, Hg^{2+}, Cd^{2+}, and Cu^{2+} were 24.8, 26.8, 22.1, and 219 mg g^{-1}, respectively, while the adsorption capacities of commercially available activated carbon were 25.3, 27.9, 23.4, and 22.4 mg g^{-1}, respectively. There was no significant difference between them ($P > 0.05$). When *A. niger* spores were added to 0.5, 0.4, 0.6, and 0.3 g, respectively, the concentrations of Hg^{2+}, Pb^{2+}, Cd^{2+}, and Cu^{2+} reached the national comprehensive wastewater discharge standard. XPS showed that *A. niger* spores adsorbed Pb^{2+}, Hg^{2+}, and Cd^{2+} without valence change, and the adsorption sites were not completely consistent. The adsorption of *A. niger* adsorbent was slow, and the saturated adsorption capacity was reached within 60 minutes when the optimal pH was not adjusted.

4) *Aspergillus niger* spore is a small ball with a diameter of 2 μm, and its surface has grooves and gullies. The experimental results of adsorbing copper ion showed that the maximum capacity of the adsorbent was 23.1 mg g^{-1} under the conditions of 0.1 g dosage of adsorbent, 30 mg L^{-1} initial concentration, 30 °C temperature, and 100 rpm rotating speed. The effect of cations in the concentration range of 100–300 mg L^{-1} on the adsorption capacity of *A. niger* spores for Cu^{2+} was $Mg^{2+} > Ca^{2+} > Na^{+}$, while the presence of anions SO_4^{2-} and PO_4^{3-} promoted the adsorption of Cu^{2+} by *A. niger* spores, which inversely proved the existence of the complexation mechanism. The model fitting results of kinetics, isotherm, and thermodynamics showed that the adsorption process was chemical, multilayer, spontaneous, and endothermic. In addition, the mechanism of Cu^{2+} adsorption is electrostatic attraction, ion exchange (P, K, Ca, Mg), and complexation reaction (C—O, C=O, C—N, C—H). The results of field emission transmission microscopy (FETEM) during the reaction process show the formation of Cu—O, which is confirmed by XPS. There is no change in the valence of Cu^{2+}, and the coordination complexation is the main adsorption mechanism.

9.3 Conclusion

The preparation process of lignocellulosic biocomposite adsorbent is relatively simple, the cost is low, and the raw material source is wide, which is suitable for large-scale application. The preparation process mainly includes cellulose extraction, composite material synthesis, and posttreatment. In order to improve the adsorption performance, cellulose can be pretreated by chemical modification or physical activation, such as the introduction of carboxyl, amino, and other functional groups to enhance its selective adsorption capacity for specific pollutants. In conclusion, lignocellulosic biocomposite adsorbents have broad application prospects in the removal of pollutants from municipal refuse. Future studies can further optimize its structure and performance, explore its applicability in the treatment of different pollutants, and promote its practical application in the field of environmental protection. At the same time, the study of adsorption mechanism is strengthened to reveal the microscopic adsorption process, which provides theoretical support for the development of more efficient adsorption materials. Through this series of efforts, lignocellulosic biocomposite adsorbents are expected to become an important means to solve the problem of urban waste pollution and help sustainable urban development.

References

1 Zishuai, J., Shih-Hsin, H., Xin, W. et al. (2021). Application of biodegradable cellulose-based biomass materials in wastewater treatment. *Environmental Pollution* 290 (1): 118087.

2 Anshuman, R., Vandana, S., Megha, M. et al. (2022). Sustainable treatment of dye wastewater by recycling microalgal and diatom biogenic materials: biorefinery perspectives. *Chemosphere* 305: 135371.

3 Huiying, L., Wen, H., Bin, Q. et al. (2022). Effective removal of proteins and polysaccharides from biotreated wastewater by polyaniline composites. *Advanced Composites and Hybrid Materials* 5: 888–1898.

4 Nair, N., Haque, M.I., Siddique, S. et al. (2024). Role of delignified and lignin-containing cellulose nanofibers in enhancing durability performances of portland cement composites. *Cement and Concrete Composites* 145: 105316.

5 Rai, A., Sirotiya, V., Mourya, M. et al. (2022). Sustainable treatment of dye wastewater by recycling microalgal and diatom biogenic materials: biorefinery perspectives. *Chemosphere* 305: 135371.

6 Wang, L., Zhang, J., Wang, F. et al. (2023). Investigation on the effects of polyaniline/lignin composites on the performance of waterborne polyurethane coating for protecting cement-based materials. *Journal of Building Engineering* 64: 105665.

7 Klapiszewska, I., Parus, A., Ławniczak, Ł. et al. (2021). Production of antibacterial cement composites containing ZnO/lignin and ZnO–SiO_2/lignin hybrid admixtures. *Cement and Concrete Composites* 124: 104250.

8 Tuğluca, M.S., Özdoğru, E., İlcan, H. et al. (2023). Characterization of chemically treated waste wood fiber and its potential application in cementitious composites. *Cement and Concrete Composites* 137: 104938.

9 Liu, J., Nero, M., Jansson, K. et al. (2023). Photonic crystals with rainbow colors by centrifugation-assisted assembly of colloidal lignin nanoparticles. *Nature Communications* 14 (1): 3099.

10 Li, C., An, X., Ren, Q. et al. (2023). Nanogrinding/ethanol activation facilitating lignin fractionation for preparation of monodispersed lignin nanoparticles. *International Journal of Biological Macromolecules* 227: 608–618.

11 Zhu, B., Xu, Y., and Xu, H. (2022). Preparation and application of lignin nanoparticles: a review. *Nano Futures* 6 (3): 032004.

12 Hu, W., Zhang, J., Liu, B. et al. (2021). Synergism between lignin, functionalized carbon nanotubes and Fe_3O_4 nanoparticles for electromagnetic shielding effectiveness of tough lignin-based polyurethane. *Composites Communications* 24: 100616.

13 Ren, M., Kong, F., Zhou, C. et al. (2023). Green, one-pot biomass hierarchical utilization strategy for lignin-containing cellulose nanofibrils and fractionated lignin preparation. *Industrial Crops and Products* 203: 117193.

14 Zhou, H., Shi, X., Wu, W. et al. (2020). Facile preparation of lignosulfonate/N-methylaniline composite and its application in efficient removal of Cr(VI) from aqueous solutions. *International Journal of Biological Macromolecules* 154: 1194–1204.

15 Huang, C., Shi, X., Wang, C. et al. (2019). Boosted selectivity and enhanced capacity of As(V) removal from polluted water by triethylenetetramine activated lignin-based adsorbents. *International Journal of Biological Macromolecules* 140: 1167–1174.

16 Jin, Y., Zeng, C., Lü, Q.-F., and Yu, Y. (2019). Efficient adsorption of methylene blue and lead ions in aqueous solutions by 5-sulfosalicylic acid modified lignin. *International Journal of Biological Macromolecules* 123: 50–58.

17 Makri, S.P., Xanthopoulou, E., Valera, M.A. et al. (2023). Poly (lactic acid) composites with lignin and nanolignin synthesized by in situ reactive processing. *Polymers* 15 (10): 2386.

18 Kwak, H.W., Lee, H., and Lee, K.H. (2020). Surface-modified spherical lignin particles with superior Cr(VI) removal efficiency. *Chemosphere* 239: 124733.

19 Sajjadi, M., Ahmadpoor, F., Nasrollahzadeh, M., and Ghafuri, H. (2021). Lignin-derived (nano)materials for environmental pollution remediation: current challenges and future perspectives. *International Journal of Biological Macromolecules* 178: 394–423.

20 Domínguez-Robles, J., Peresin, M.S., Tamminen, T. et al. (2018). Lignin-based hydrogels with "super-swelling" capacities for dye removal. *International Journal of Biological Macromolecules* 115: 1249–1259.

21 Park, S., Oh, Y., Yun, J. et al. (2020). Cellulose/biopolymer/Fe_3O_4 hydrogel microbeads for dye and protein adsorption. *Cellulose* 27 (5): 2757–2773.

22 Rauber, D., Conrad, M., Huwer, J. et al. (2017). Demonstrating sustainable biomass utilization and processing using ionic liquids—an introduction to undergraduate chemistry laboratories. *World Journal of Chemical Education* 5: 158–163.

23 Guadix-Montero, S. and Sankar, M. (2018). Review on catalytic cleavage of C-C ter-unit linkages in lignin model compounds: towards lignin depolymerisation. *Topics in Catalysis* 61: 183.

24 Ran, H., Zeng, Y., Luo, C. et al. (2021). Research progress of polylactic acid/ lignin composites. *Chemical Engineering and Equipment* 5.

25 Li, P., Caiwen, W., Liu, C. et al. (2022). Research progress on properties and applications of lignin composite hydrogels. *Chinese Journal of Bioengineering* 38: 2489–2498.

26 Zhang, L., Wang, H., Li, Q., and Yang, D. (2022). Research progress of lignin-derived adsorbent materials and their applications in wastewater treatment. *Advances in Chemical Industry* 41: 3731–3744.

27 Nadeem, R., Manzoor, Q., Iqbal, M. et al. (2016). Biosorption of Pb(II) onto immobilized and native mangifera indica waste biomass. *Journal of Industrial and Engineering Chemistry* 35: 185–194.

28 Abdolali, A., Ngo, H.H., Guo, W. et al. (2016). A breakthrough biosorbent in removing heavy metals: equilibrium, kinetic, thermodynamic and mechanism analyses in a lab-scale study. *Science of the Total Environment* 542 (Part A): 603–611.

29 Di Caprio, F., Altimari, P., Uccelletti, D. et al. (2014). Mechanistic modelling of copper biosorption by wild type and engineered Saccharomyces cerevisiae biomasses. *Chemical Engineering Journal* 244: 561–568.

30 Won, S.W., Kotte, P., Wei, W. et al. (2014). Biosorbents for recovery of precious metals. *Bioresource Technology* 160: 203–212.

31 Abdolali, A., Guo, W.S., Ngo, H.H. et al. (2014). Typical lignocellulosic wastes and by-products for biosorption process in water and wastewater treatment: a critical review. *Bioresource Technology* 1 (60): 57–66.

32 Chen, S.H., Yue, Q., Gao, B. et al. (2012). Adsorption of hexavalent chromium from aqueous solution by modified corn stalk: a fixed-bed column study. *Bioresource Technology* 113: 114–120.

33 Terry, P.A. and Stone, W. (2002). Biosorption of cadmium and copper contaminated water by scenedesnmus abundans. *Chemosphere* 47 (3): 249–255.

34 Kapoor, A. and Viraraghavan, T. (1997). Heavy metal biosorption sites in aspergillus niger. *Bioresource Technology* 61 (3): 221–227.

35 Buyukgungor, H. (2013). Equilibrium and kinetic studies on the biosorption of 2-chlorophenol and 4-chlorophenol by live aspergillus niger. *Ekoloji* 22 (88): 1–12.

10

Magnetic Biocomposites for Wastewater Remediation

Feihu Li and Mengyu Ma*

Collaborative Innovation Center of Atmospheric Environment and Equipment Technology, Jiangsu Key Laboratory of Atmospheric Environment Monitoring and Pollution Control, School of Environmental Science and Engineering, Nanjing University of Information Science and Technology, Nanjing 210044, China

10.1 Introduction

Meeting the growing demand for clean water is a critical challenge for the world in the context of global climate change [1]. As the global population continues to grow, so does the volume of wastewater generated through a variety of pathways, with a predicted annual production of 574 billion m^3 by 2050 [2]. These wastewaters, if not treated effectively, no doubt pose a significant threat to human health and the environment since they can spread diseases, pollute waterways, and harmfully affect aquatic life [3]. The growing burden of wastewater treatment is further exacerbating this issue, as conventional treatment methods often rely on energy-intensive processes and chemical usage, and generate large quantities of sludge as by-products [4, 5]. These limitations underscore the urgent need for sustainable and efficient wastewater treatment technologies to meet the United Nations' Sustainable Development Goal (SGD) 2030 of universal access to clean water and sanitation.

In the past two decades, magnetic biocomposites (MBCs) have emerged as a promising and sustainable candidate for wastewater remediation through adsorption processes [6–8]. These materials are typically composed of a mixture of natural biological components, such as chitosan (CS), cellulose, or other biopolymers, along with magnetic nanoparticles, integrating the advantages of magnetic nanoparticles and biological materials thereof [9, 10], and enabling potential applications in biomedical devices, tissue engineering, drug delivery, and

*Author as the corresponding author with Email address at fhli@nuist.edu.cn, and ORCID link at https://orcid.org/0000-0002-2969-8276.

soft robotics [11–14]. Magnetic nanoparticles (MNPs, e.g., Fe_3O_4) provide high surface area and easy separation using external magnetic fields, while biological components offer high selectivity and specificity for target pollutants. Such composites tend to have high stability, surface area, and many surface functionalities, which prove to be very beneficial for the removal of pollutants from wastewater [15]. Benefiting from their unique properties, such as biocompatibility, tailorability of functional groups, reusability, simplicity of operation, cost-effectiveness, and environmental friendliness, MBCs are effective and attractive in removing a wide range of contaminants from wastewater, including heavy metals, dyes, pharmaceuticals, and pesticides [16–20], offering a sustainable and environmentally friendly approach for improving our water quality.

This chapter aims to provide a comprehensive overview of magnetic biocomposites for wastewater remediation. We first introduce the synthesis strategies and characterization techniques of magnetic biocomposites, with consideration of particular properties required for wastewater remediation applications. In this context, we then discuss the various applications of magnetic biocomposites for adsorptive removal of pollutants from wastewater and evaluate the potential benefits and limitations of using them in wastewater treatment. The most recent advances in wastewater remediation using magnetic biocomposites are presented, with emphasis on the adsorptive removal of inorganic and organic pollutants, and recovery of valuable substances from wastewater. Finally, a perspective on future research directions of MBCs is provided to advance the goal of clean water and sustainability.

10.2 Magnetic Biocomposite: Synthesis Strategy and Characterization Methodology

In this section, some fundamental concepts of MBCs are briefly introduced, with emphasis on the general synthesis strategies of MBCs and relevant characterization methodologies.

10.2.1 Selection of Magnetic Nanoparticles and Biological Matrixes

The successful design of MBCs relies on the judicious selection of MNPs and biological matrixes [21], both of which can offer unique properties that govern the overall functionality and performance of the resulting composites when practicalized for wastewater remediation purposes (Figure 10.1). A crucial aspect of MNP selection is their magnetic behavior. Ideally, superparamagnetic nanoparticles are preferred. These materials exhibit zero remanence magnetization, implying they lose magnetization upon removal of the external magnetic field [22]. Additionally, the size and shape of MNPs play a critical role. Smaller nanoparticles generally

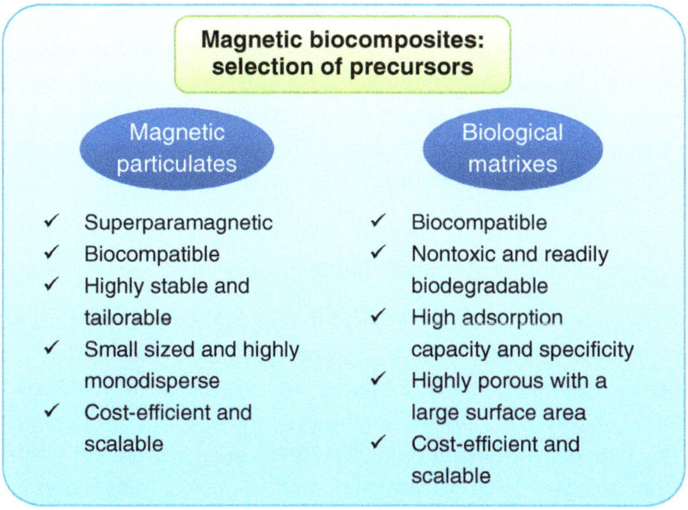

Figure 10.1 The ideal properties of magnetic particulates and biological matrices required for MBCs for wastewater remediation applications.

exhibit enhanced responsiveness to external magnetic fields but may suffer from increased aggregation tendencies [23]. Conversely, larger particles offer better stability but potentially lower magnetic responsiveness. Besides, the surface chemistry of MNPs is another critical consideration. Surface functionalization with biocompatible polymers like polyethylene glycol (PEG) can enhance biocompatibility and dispersibility, and prevent unwanted aggregation [21]. Furthermore, surface modifications can be tailored to capture specific contaminants or elements from wastewater, enabling targeted removal or recovery from the waste stream [19, 24]. Iron oxide nanoparticles, particularly magnetite (Fe_3O_4) and maghemite (γ-Fe_2O_3), are widely employed due to their biocompatibility, tunable magnetic properties, and facile synthesis methods [13]. Other compositions, such as cobalt, nickel, and their alloys, offer distinct magnetic characteristics but might raise concerns regarding cytotoxicity and biocompatibility [23].

Likewise, an ideal biological matrix should not only offer a high surface area and specific functional groups for pollutant binding but also facilitate the incorporation of MNPs. Common biological materials used in MBCs include chitosan, alginate, collagen, fungi, and biomass-based materials, each offering distinct advantages [25–27]. Fungi may possess specific binding sites for targeted contaminants, while algae offer high surface area and potential for in situ remediation [28]. When selecting a biological matrix, factors like the target contaminant type and desired removal efficiency must be considered (Figure 10.1). For example, MBCs with groups of metallophilicity have shown promise in heavy metal removal [7, 29, 30], while algae-based composites may be better suited for organic pollutant adsorption [31, 32]. Additionally, the biocompatibility of the matrix and its potential for regeneration after use are important aspects to consider. Abd Malek et al. [9] demonstrated the effectiveness of a chitosan-based magnetic biocomposite for removing organic dyes from wastewater. Ultimately, selecting the right biological matrix is essential for designing efficient and sustainable MBCs for wastewater remediation.

10.2.2 Synthetic Approaches of Magnetic Biocomposites

Once the magnetic component and the biological matrix are defined separately, the next step is to prepare the MBCs with a proper approach. Depending on whether the magnetic component is pre-prepared, the synthesis of MBCs can be classified as *in situ* (or one-pot, one-step) synthesis and *ex situ* (or two-step) synthesis (Figure 10.2). In the former method, the MNPs are synthesized within the biological matrix, which offers advantages such as better dispersion of nanoparticles within the matrix and potentially enhanced interaction between the magnetic and biological components [25, 33, 34]. However, it may require additional steps to ensure the uniformity and stability of the nanoparticles within the matrix [35]. In the latter, the magnetic component (such as Fe_3O_4 nanoparticles) is pre-prepared separately by coprecipitation, sol-gel, hydrothermal,

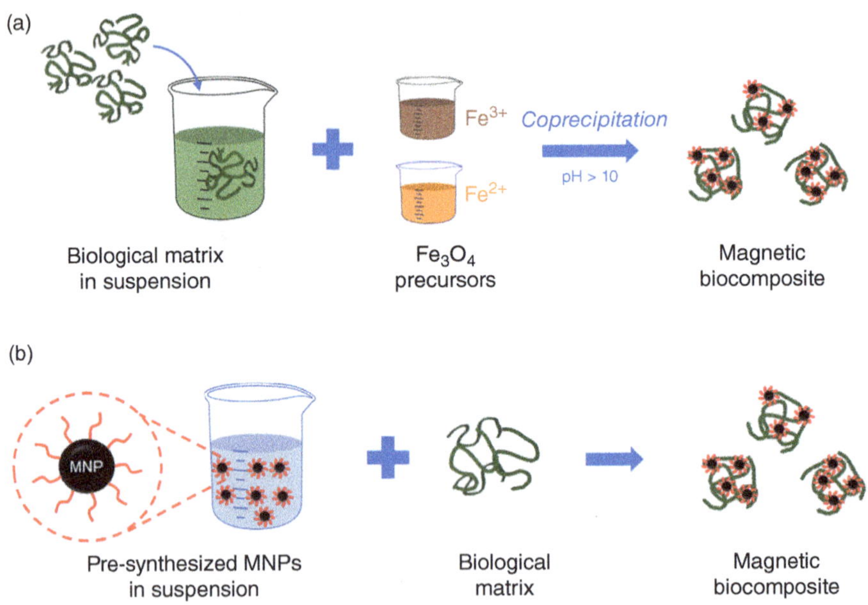

Figure 10.2 Synthesis of MBCs via (a) *in situ* and (b) *ex situ* approaches.

microwave-assisted, and microemulsion methods [23] and then combined with the biological matrix. Based on the bonding nature, the magnetic and biological components can be combined through electrostatic interactions, covalent bonding, or physical encapsulating [36–38]. These approaches allow for more precise control over the magnetic properties and size distribution of the nanoparticles before incorporation into the biocomposite [22, 37]. Besides, layer-by-layer (LbL) assembly, biomimetic synthesis using biomolecules (e.g., peptides, proteins) as templates or scaffolds, and hybrid methods that involve the combination of two or more synthetic methods are also widely applied to create MBCs with enhanced properties [39, 40].

The selection of the appropriate synthetic method for magnetic biocomposite depends on the specific requirements of the application, such as the type of target pollutant, the desired magnetic properties, the stability of the biocomposite, and the cost-effectiveness of the preparation process.

10.2.3 Characterization Techniques

Generally, the physicochemical properties of the synthetic MBCs are fully evaluated before the performance assessment for wastewater remediation in order (i) to establish a convincing relationship between the structural parameters and their activity, and (ii) to understand the underlying mechanisms of pollutant abatement over the surface of the MBCs. To this end, a variety of characterization techniques have been widely used for collecting and analyzing the morphological, crystallographic, structural, chemical, magnetometric, and molecular spectroscopic information of MBCs. Specific techniques include scanning electron microscopy (SEM), transmission electron microscopy (TEM), atomic force microscopy (AFM), X-ray diffraction spectroscopy (XRD), X-ray photoelectron spectroscopy (XPS), X-ray fluorescence spectroscopy (XRF), energy dispersive X-ray spectroscopy (EDX), Fourier transform infrared spectroscopy (FTIR), Raman spectroscopy, nitrogen absorption-desorption isotherm porosimetry, and vibrating sample magnetometry (VSM).

10.2.3.1 Morphological Analysis by SEM, TEM, and AFM

Due to their unique advantages of providing high-resolution images for detailed morphological analysis and inspection, in particular the size and distribution of magnetic particles within the biocomposite matrix, both SEM and TEM are widely used in the morphological characterization of magnetic biocomposite materials [41, 42]. Although both microscopy techniques are very powerful, there are limitations to consider. The sample preparation for SEM analysis would be very time-intensive due to magnetic properties, and a conductive coating on nonconductive samples may also in part affect the morphology. Moreover,

SEM provides only surface imaging, and it may not fully capture the internal structure of the composite. This issue can be complemented by TEM, which allows for detailed visualization of the internal structure of the biocomposite at the nanoscale, especially for thin samples [42]. In the case of thick and soft biocomposites, the TEM sample preparation can be challenging. Additionally, TEM is typically limited to two-dimensional (2D) imaging, which may not fully depict the three-dimensional (3D) structure of the biocomposite [43].

Likewise, AFM offers high-resolution imaging and surface profiling capabilities, allowing for the visualization of the topography and mechanical properties of the biocomposite at the nanoscale [44]. AFM can also be used to map magnetic properties, such as magnetic force and magnetic domain distribution, providing valuable insights into the behavior of the magnetic particles within the composite [45]. While versatile and powerful, AFM may struggle with imaging soft or nonconductive samples and can be limited by the scan size and speed [45]. In short, all the above microscopy offers powerful capabilities for characterizing MBCs, providing detailed insights into their structure and properties at the nanoscale. However, researchers should be mindful of the sample preparation requirements and limitations of each technique when choosing a proper morphological characterization approach.

10.2.3.2 Crystallographic and Compositional Analysis by XRD, XPS, XRF, and EDX

XRD analysis identifies long-range ordering (i.e., crystallographic structure) in crystalline materials and short-range ordering in amorphous materials. XRD technique is often used to verify the successful combination of MNPs with biological matrix [17] and to identify the crystallographic or amorphous nature of the biological matrix and other components [46]. It's sensitive to crystalline phases but may be challenging in sample preparation, especially for soft biocomposites.

XRF is a rapid and nondestructive elemental analysis technique that is of high sensitivity for trace elements. Therefore, it's often employed for chemical composition analysis of biocomposite components [47], and biocomposite before and after applying to wastewater remediation [48]. Likewise, XPS analysis can provide surface-sensitive elemental and chemical information (e.g., high-resolution information on chemical states, and quantitative data of elemental composition). It is widely used for an *ex situ* mechanism investigation based on the surface compositional analysis of MBCs, in particular before and after application to contaminant removal [28, 49]. Similarly, EDX analysis offers elemental composition at high spatial resolution and is often integrated with SEM for qualitative mechanism studies according to the surface elemental data of MBCs [10, 49]. However, EDX is less sensitive to light elements, and a combination with other sensitive techniques may be necessary for a more comprehensive characterization of MBCs targeting pollutants with light elements.

10.2.3.3 Molecular Spectroscopic Analysis by FTIR and Raman Spectroscopy

Both FTIR and Raman spectroscopy can probe the vibrational modes of molecules in biocomposite materials, thereby providing valuable information about the functional groups, bonding configurations, and chemical composition of the magnetic biomaterials based on their interactions with infrared (FTIR) or laser light (Raman) [18, 50]. Nevertheless, FTIR spectroscopy has limited sensitivity to inorganic components [9], and may suffer from spectral interferences, while Raman may also encounter fluorescence interference [51].

Collectively, FTIR and Raman spectroscopy are valuable tools for characterizing MBCs. By utilizing both techniques in conjunction, researchers can obtain comprehensive insights into the surface functional groups and properties of these complex materials, thereby facilitating the mechanism understanding of organic pollutant (e.g., dyes) sequestration [52].

10.2.3.4 Porous Structure Analysis by N_2 Adsorption–Desorption Isotherms

N_2 adsorption–desorption isotherm analysis is a powerful technique usually employed to delve into the intricate porosity of biocomposite materials [53, 54]. By measuring the amount of nitrogen gas adsorbed and desorbed at various pressures, this method provides insights into the surface area, pore size distribution, and pore volume of MBCs [9, 41, 55]. Brunauer-Emmett-Teller (BET) and Barret-Joyner-Halenda (BJH) methods are generally used for the determination of surface area and pore size distribution within the MBCs [49, 56].

However, this technique is only applicable for biocomposites with micropore (<2 nm) and mesopore (2 ~ 50 nm), and limited for biocomposites of macropore (>50 nm). To gain a more in-depth understanding of the porosity of biocomposites, it would be necessary to complement N_2 adsorption–desorption analysis with techniques like SEM or TEM with careful interpretation of the data.

10.2.3.5 Magnetometric Analysis by VSM

VSM analysis is a basic technique for investigating the magnetometric properties of magnetic biocomposite materials [46, 55]. This method measures the magnetization of a sample as it is exposed to a varying magnetic field, providing valuable insights into the material's magnetic behavior. VSM is of high sensitivity and versatility, and offers precise quantification of magnetic parameters, providing a comprehensive understanding of the material's magnetic response. But the sample preparation may be challenging, particularly for soft biocomposites as it often requires grinding and packing to ensure accurate measurement. To obtain a more comprehensive understanding of the magnetic properties and their correlation with the biocomposites' microstructure, integrating VSM with microscopic techniques like TEM or SEM is a routine operation [37, 57].

10.3 Approaches to Evaluate the Performance of Magnetic Biocomposites

In this section, some essential approaches for the performance evaluation of MBCs are presented, with a special focus on the adsorption performance and efficiency of magnetic biocomposite materials in wastewater remediation.

10.3.1 Batch Adsorption Studies

Batch adsorption studies are fundamental for understanding and optimizing the adsorption process, providing a cost-effective and versatile method for evaluating the performance of a magnetic biocomposite adsorbent in removing and/or recovering target substances from wastewater [29, 55]. This technique involves immersing or dispersing a given weight of biocomposite adsorbent in a solution containing the target contaminant or substance of interest under controlled conditions. By measuring the change in concentration of contaminants before and after adsorption, the adsorption capacity and efficiency of the biocomposite can be determined [58], allowing for the investigation of various parameters that influence adsorption performance, such as pH, background ionic strength, adsorbent dosage, contact time, temperature, and competitive ions, etc. [59].

Specifically, batch adsorption studies are often conducted for the following purposes:

(I) pH optimization: The pH of the solution can significantly impact the adsorption process. Both the biocomposites' surface charge and the contaminant's speciation are influenced by pH, affecting their interactions. By conducting experiments at different pH values, researchers can determine the optimal pH range for maximum contaminant uptake, leading to efficient removal.

(II) Dosage optimization: The adsorbent dosage plays a critical role in the adsorption process. Increasing the dosage generally leads to higher contaminant removal, but beyond a certain point, further additions may have minimal impact. Finding the ideal adsorbent dosage ensures efficient contaminant removal without unnecessary material expenditure.

(III) Contact time optimization: This parameter explores the time required for the biocomposite adsorbent to reach equilibrium with the target contaminant. The initial adsorption rate is typically high, followed by a gradual decrease until equilibrium is reached. Determining the optimal contact time ensures maximum contaminant removal within a reasonable timeframe. Besides, fitting experimental data to various kinetic models such as pseudo-first-order or pseudo-second-order kinetics can offer mechanistic insights into the adsorption mechanism and the rate-controlling steps, which is essential for optimizing the adsorption process.

(IV) Understanding the effects of background ionic strength and competitive ions: The presence of other ions, especially those with similar affinity for the magnetic biocomposite adsorbent, in the solution can influence the adsorption process. Understanding the effects of background ionic strength and competitive ions is crucial for assessing the selectivity of the biocomposite and predicting its adsorption behavior in real-world scenarios.

(V) Isotherm and thermodynamic analysis: Isotherm studies involve plotting the equilibrium adsorption data obtained from experiments onto various adsorption isotherm models such as Langmuir or Freundlich equations. These models provide insights into the adsorption capacity and affinity of the magnetic biocomposite for specific contaminants in wastewater. Thermodynamic studies involve analyzing the effect of temperature on adsorption to determine the spontaneity, feasibility, and nature of the adsorption process. Parameters like Gibbs free energy ($\Delta \boldsymbol{G}$), enthalpy ($\Delta \boldsymbol{H}$), and entropy ($\Delta \boldsymbol{S}$) changes can be calculated to understand the thermodynamic feasibility of the adsorption process.

(VI) Understanding adsorption mechanisms: Through analysis of batch experimental data, researchers can gain insights into the interactions between the adsorbent and the contaminant, providing a better understanding of the underlying adsorption mechanism.

(VII) Developing predictive models: The experimental data obtained from batch adsorption studies can be used to develop mathematical models that predict adsorption behavior under various conditions (e.g., surface complexation models), allowing for informed decision-making and optimized process design for specific applications.

By harnessing the power of magnetism (i.e., a magnet), magnetic biocomposite adsorbents provide a sustainable and efficient solution for separating spent adsorbents from adsorption suspensions [59], revolutionizing the field of batch adsorption studies and paving the way for faster, more efficient, and environmentally friendly separation processes.

10.3.2 Fixed-Bed Column Studies

To evaluate the performance of MBCs under dynamic, continuous flow conditions, fixed-bed column studies are often performed to simulate real-world applications, providing crucial insights into their practical applicability [60, 61]. In a typical fixed-bed column, the biocomposite material is packed into a vertical column, and a solution containing the target contaminant is passed through the fixed-bed where the pollutants (e.g., Cu^{2+} and Pb^{2+}) are captured continuously [62]. This setup allows for the investigation of various factors influencing adsorption performance, including the adsorbent's properties (e.g., particle size, rigidity, and porosity), and operational parameters (e.g., flow rate, inlet concentration, fixed-bed height, and volume to area ratio of the bed).

Specifically, fixed-bed column studies enable the study of the following key parameters:

(I) Particle size: The size and uniformity of particles impact flow dynamics and mass transfer within the bed, influencing the adsorption process.
(II) Rigidity: The mechanical strength and structural integrity of the adsorbent influence its long-term stability and performance under continuous flow conditions.
(III) Porosity: The pore structure and surface area of the adsorbent govern its adsorption capacity and the diffusion of contaminants within the material.
(IV) Flow rate: The rate at which the feed solution passes through the column affects the residence time of the contaminant, influencing the extent of adsorption.
(V) Inlet concentration: The concentration of the contaminant in the feed solution directly impacts the adsorption capacity and breakthrough behavior of the bed.
(VI) Fixed-bed height: The height of the adsorbent bed impacts the overall adsorption capacity and the time required for a breakthrough.
(VII) Volume to area ratio of the bed: This ratio, also known as the bed aspect ratio, significantly influences the flow distribution and mass transfer within the column.

To further understand and predict adsorption behavior in fixed-bed column studies, various mathematical models are employed [61], including Thomas Model (THM, to describe the breakthrough curve and predict the breakthrough time based on the adsorption rate constant and other parameters), Bed Depth Service Time Model (BDST, to relate the bed height to the service time—time until breakthrough, and predict the performance of adsorbents under different conditions), and Yoon-Nelson Model (YNM, to provide a simplified representation of the breakthrough curve and estimate of adsorption capacity and breakthrough time). By integrating the fixed-bed column data with appropriate mathematical models, we can effectively evaluate the performance of biocomposite adsorbents under dynamic conditions [63], gain insights into the impact of various factors, and optimize the design and operation of adsorption processes for removal of specific pollutants from wastewater.

10.3.3 Regeneration and Reusability Assessment

Whether in batch adsorption or fixed-bed column adsorption, the magnetic biocomposite adsorbent will inevitably reach saturation after a period of adsorption. Therefore, how to effectively restore its adsorption capacity and extend its reusability is a crucial step forward. Besides, assessing the regeneration and reusability of the spent MBCs is essential to optimize its long-term performance, reduce costs, and promote sustainability [25, 59].

Regeneration aims to restore the adsorption capacity of spent adsorbents, making them reusable for multiple cycles. Based on the underlying mechanism of action, regeneration can be generally categorized into chemical regeneration, thermal regeneration, biological regeneration, and physical regeneration. Chemical regeneration is usually achieved by employing various chemical agents to desorb adsorbed contaminants. Agents like sodium hydroxide (NaOH), hydrogen chloride (HCl), and ethylenediamine tetraacetic acid (EDTA) are often used to desorb heavy metals from MBCs [7, 25, 29]. Sodium chloride (NaCl), dilute nitric acid (HNO_3), and organic solvents like acetone, methanol, and ethanol are effective in desorbing metal species and organic pollutants, respectively [29, 64]. Thermal regeneration involves heating the spent adsorbent to release adsorbed contaminants, which is effective for removing organic impurities from magnetic biochar adsorbents, facilitating the restoration of its adsorption capacity [59, 61].

As mentioned earlier, one of the inherent merits of MBCs is the easy and rapid separation from suspension by external magnets, which facilitates the following regeneration units. Biological techniques using specific microbes to break down adsorbed contaminants over MBCs are time-consuming and, thus, do not match so well to the preceding magnetic separation on the same time scale. Likewise, methods like ultrasonication, agitation, and centrifugation can be employed to remove adsorbed contaminants and restore the adsorbent's surface [65, 66]. However, such physical techniques are seldom used for restoring magnetic biocomposite adsorbents. For a specific MBC adsorbent, cycle life testing via repeated regeneration and adsorption cycling with various regeneration techniques is crucial to evaluate the long-term stability and reusability of the magnetic biocomposite adsorbent and identify the most effective and sustainable methods, paving the way for their widespread application in wastewater remediation [29, 59].

10.3.4 Environmental Risk and Safety Considerations

MBCs, while offering a promising solution for wastewater remediation, also present potential environmental risks that cannot be ignored, particularly for biocomposites involving pyrolysis and those after long-term operation in the removal of toxic compounds [28, 67]. For instance, the preparation of magnetic biochar is likely to release toxic compounds into the environment, particularly if improper methods are used [68]. Additionally, the application of biocomposites can lead to unintended consequences, such as the leaching of toxic compounds from the biocomposite material itself, especially after long-term adsorbing pollutants with weak affinity [67]. If not properly managed, the release of accumulated contaminants (e.g., toxic heavy metals and metalloids) from the biocomposite material into the environment during disposal or breakdown could cause significant harm, effectively shifting the pollution burden rather than eliminating them [59]. Therefore, potential hazards and environmental risks should be considered in the preparation and application of magnetic biocomposite materials to minimize harm to the environment. Short-term simulation studies cannot guarantee that long-term application to water treatment is risk-free. Therefore, long-term water tests and risk assessments are essential for ensuring the sustainable and safe utilization of MBCs in water treatment.

10.4 Applications in Wastewater Remediation

As mentioned earlier, combining MNPs and biological matrices to provide high stability, surface area, and many surface functionalities makes the MBCs particularly interesting in the adsorption of pollutants, recovery of valuable elements, and other state-of-the-art applications for wastewater remediation. The most recent advances will be covered in this section, categorized in the removal of inorganics, organics, recovery of valuable elements, and other applications.

10.4.1 Adsorptive Removal of Inorganic Pollutants

Inorganic pollutants including heavy metals, metalloids, and nutrients in wastewater pose a significant environmental and health risk, yet their removal remains a challenging task in wastewater treatment. Conventional methods, such as chemical precipitation and ion exchange, often struggle to effectively remove such contaminants due to limited adsorption capacities, high operational costs, and the generation of secondary pollutants [24]. Additionally, traditional adsorbents, in particular those on the nanosized scale, can be difficult to separate from the treated water, leading to further environmental contamination. By integrating the high adsorption capacities of biomaterials with the ease of magnetic separation, MBC adsorbents offer a promising solution to these challenges [59, 69].

Over the past two decades, a variety of MBC adsorbents with varying biological components have been developed for the adsorptive removal of heavy metals and other inorganics of concern from wastewater [70, 71]. A part of these novel MBC adsorbents invented over the last 5 years are summarized in Table 10.1. A fuller and more comprehensive list beyond this time scale can be found in the earlier review papers. Heavy metals like Cd(II), Cr(VI), Cu(II), Pb(II), Hg(II), Ni(II), etc., have shown high affinity to these MBC adsorbents under environmentally relevant conditions. For instance, Staron et al. [73] reported an MBC consisting of hydrochar obtained by hydrothermal carbonization of raffia fibers, *Lindnera jadinii* yeast, and MNPs, labeled as Fe_3O_4@hydrochar/*Lindnera jadinii*, exhibits a maximum adsorption capacity of 16.3 mg g^{-1} for Cd(II) at pH 6. Under similar conditions, another fungal-based MBC adsorbent, Fe_3O_4@*P. chrysosporium*, demonstrated a maximum capacity of 18.2 mg g^{-1} toward Cd(II) [72]. Touihri et al. [10] synthesized a novel biocomposite adsorbent—magnetic pine cone gel beads (MPCB) for the removal of Cu(II) and Cr(VI) from aqueous solution—and found that the maximum adsorption capacity for Cu(II) and Cr(VI) were 68.6 and 212.2 mg g^{-1}, respectively. Omer et al. [74] integrated ZIF-67, Fe_3O_4 nanoparticles, and aminated chitosan matrix, yielding core-dual shell Fe_3O_4/ZIF-67@aminated chitosan MBC beads with a maximum capacity of 119.1 mg g^{-1} for Cr(VI). Such hierarchical core-dual shell structure seems to be beneficial for improving their adsorption capacity toward Cr(VI) (Table 10.1), with maximum Cr(VI) capacities of 282.0 and 294.1 mg g^{-1} for Fe_3O_4/CS@aminothiazole and Fe_3O_4/CS@*Azolla pinnata*, respectively [75, 76]. Chitosan (CS) has a large contribution to these large affinities for Cr(VI) since it is an alkaline polysaccharide containing a large number of amino and hydroxyl groups that can chelate with various metal ions [97].

Likewise, as shown in Table 10.1, MBC adsorbents with a hierarchical core/shell/shell structure also demonstrated significantly high affinity toward Cu(II) [18, 77], Pb(II) [79, 81], and Hg(II) [82], probably due to more binding sites are provided from these varying components during the preparation process [77, 82]. The abundance of hydroxyl groups on the surface of nanocellulose (NC) provides an excellent platform for considerable surface modifications, permitting the grafting of various functionalities onto the cellulosic portion, thus allowing easy immobilization of pollutants. Therefore, the MBC adsorbent $CaCO_3$/Fe_3O_4@cellulose nanofiber demonstrated a maximum adsorption capacity of 1010.0 mg g^{-1} toward Pb(II) [81]. Compared to the simple core/shell structured MBC adsorbent, viz. Fe_3O_4/CS [83], the core/shell/shell structured MBC—Fe_3O_4@CS/PSSMA [78], showed higher maximum capacity for Ni(II) under similar conditions (113.1 versus 30.0 mg g^{-1}).

In addition to heavy metals, MBC adsorbents were also widely used for the removal of metalloids and nutrients from wastewater. For example, our group has developed a magnetic biochar composite (i.e., Fe_3O_4@biochar/Mg/Al oxides) with a maximum capacity of 34.5 for As(V) at pH 5 [58]. Luo et al. [57]

Table 10.1 Magnetic biocomposites for the adsorptive removal of pollutants from wastewater reported in the literature over the last 5 years.

MBCs		Target pollutants	Adsorption conditions				Q_{max} (mg g^{-1})	Reference
			pH	T (°C)	$t_{eq.}$ (min)	C_0 (mg L^{-1})		
Fe_3O_4@*P. chrysosporium*	Heavy metals	Cd(II)	6.0	25	60	100	18.2	[72]
Fe_3O_4@hydrochar/*Lindnera jadinii*		Cd(II)	6.0	25	1440	150	16.3	[73]
Fe_3O_4/ZIF-67@aminated chitosan		Cr(VI)	2.0	25	120	50	119.1	[74]
Fe_3O_4/CS@aminothiazole		Cr(VI)	2.0	22	2880	10.4 ~ 499.2	282.0	[75]
Fe_3O_4/CS@*Azolla pinnata*		Cr(VI)	4.0	25	15	50	294.1	[76]
Fe_3O_4/citric acid@chitosan		Cu(II)	6.5	25	66	50	232.6	[77]
Fe_3O_4@CS/PSSMA		Cu(II)	5.0	25	720	5 ~ 200	141.7	[78]
Fe_3O_4@starch-g-PAA/cellulose nanofibers		Cu(II)	5.0	25	540	600	555.6	[18]
Fe_3O_4@TCA@chitosan		Pb(II)	5.5	25	180	20 ~ 250	204.9	[79]
Fe_3O_4/CS@polymer-grafted silica		Pb(II)	6.0	25	480	50 ~ 400	111.1	[80]
$CaCO_3$/Fe_3O_4@cellulose nanofiber		Pb(II)	n/a	25	1440	50 ~ 2000	1010.0	[81]
Fe_3O_4@CS@polypeptide		Hg(II)	5.0	45	120	25 ~ 600	218.8	[82]
Fe_3O_4@CS/PSSMA		Ni(II)	5.0	25	720	5 ~ 200	113.1	[78]
Fe_3O_4/CS		Ni(II)	6.0	25	120	20 ~ 300	30.0	[83]
Fe_3O_4/CS	Metalloids and nutrients	As(V)	3.0	25	1440	5 ~ 500	15.2	[84]
Fe_3O_4@biochar/Mg/Al oxides		As(V)	5.0	10	1440	2 ~ 30	34.5	[58]
Fe_3O_4@camphor seeds husk		As(III)	7.2	n/a	480	10 ~ 100	39.7	[57]
Fe_3O_4@Mg/herbal medicine residues		PO_4^{3-}	8.0	25	180	10 ~ 500	635.4	[85]

Fe_3O_4@peanut husks@NH_2-MIL-101(Fe)		PO_4^{3-}	5.6	25	240	10 ~ 70	20.4	[86]
Fe_3O_4@polyethyleneimine lignin		PO_4^{3-}	6.5	25	1080	10 ~ 100	43.0	[87]
Fe_3O_4@Mg/herbal medicine residues		NH_4^+	8.0	25	180	10 ~ 500	615.6	[85]
Fe_3O_4/AC@*Yarrowia lipolytica* cell		MB	>5.8	25	1440	50 ~ 250	32.1	[19]
Fe_3O_4/CS@salicylaldehyde/nanoclay		AR88	4.0	25	550	20 ~ 200	173.5	[88]
Fe_3O_4/CS-cross-linked-PAM		SY	5.0	25	120	5 ~ 1500	360.2	[89]
Fe_3O_4/corn cob		MB	7.7	55	300	75	13.2	[90]
Fe_3O_4/peanut husk	Dyes	MB	5.7	30	240	30 ~ 240	245.0	[91]
		RhB	6.8			30 ~ 240	57.3	
Fe_3O_4/papaya seed powder		CR	5.0	30	180	100 ~ 1000	216.9	[56]
Fe_3O_4/CS/GO		CR	≤8.0	35	1440	200	395.8	[92]
Fe_3O_4@SiO_2/SA-PEI		X-3B	7.0	25	1440	50 ~ 500	370.9	[93]
Fe_3O_4/alginate/bentonite@CS		CPX	6.0	23	120	100 ~ 600	264.8	[16]
Fe_3O_4-eggshell@ZIF-8		NOR	5.0	25	240	10	53.5	[46]
Fe_3O_4@SiO_2/SA-PEI		DS	5.0	25	1440	25 ~ 200	241.5	[93]
$CuFe_2O_4$/AG-g-PAO	Others	OPP	6.0	25	15	50 ~ 300	769.2	[94]
Fe_3O_4@CS/*chlorella vulgaris*		TC	6~8	25	240	10 ~ 200	834.0	[95]
		CIP					394.9	
		AMX					150.8	
Fe_3O_4/$HCP_{Bn\text{-}glu}$		nitrophenol	6.0	25	30	2 ~ 150	83.3	[96]

AC – activated carbon; AG – Arabic gum; AMX – amoxicillin; AR – Acid Red; $\boldsymbol{C_0}$ – initial concentration; CIP – ciprofloxacin; CPX – ciprofloxacin; CR – Congo Red; CS – chitosan; DS – diclofenac sodium; GO – graphene oxide; HCP – hyper-cross-linked polymer; MB – Methylene Blue; NOR – norfloxacin; OPP – organophosphorus pesticide; PAA – poly(acrylic acid); PAM – polyacrylamide; PAO – polyamidoxime; PEI – poly (ethylene imine); PSSMA – poly(4-styrenesulfonic acid-*co*-maleic acid); $\boldsymbol{Q_{max}}$ – maximum adsorption capacity; RhB – Rhodamine B; SA – sodium alginate; SY – Sunset Yellow; $\boldsymbol{T}$ – temperature; TC – tetracycline; TCA – tetracarboxylic acid; $\boldsymbol{t_{eq.}}$ – equilibrium time; X-3B – reactive red X-3B.

reported recently a magnetically modified biomass camphor seeds husk (i.e., Fe_3O_4@camphor seeds husk) delivering a maximum adsorption capacity of 39.7 mg g^{-1} based on the Langmuir model at pH 7.2, which is much higher than its counterpart without magnetic property. Moreover, nutrient species like phosphate (PO_4^{3-}) and ammonium (NH_4^+) can also be effectively captured by MBC adsorbents, which are likely to be further reused as slow-released fertilizers if handled properly. For instance, Wang et al. [85] developed an MBC adsorbent based on loading magnesium (Mg) to Chinese herbal medicine residues (i.e., Fe_3O_4@Mg/herbal medicine residues). It exhibited remarkable performance for simultaneously removing and recovering phosphate and ammonium from

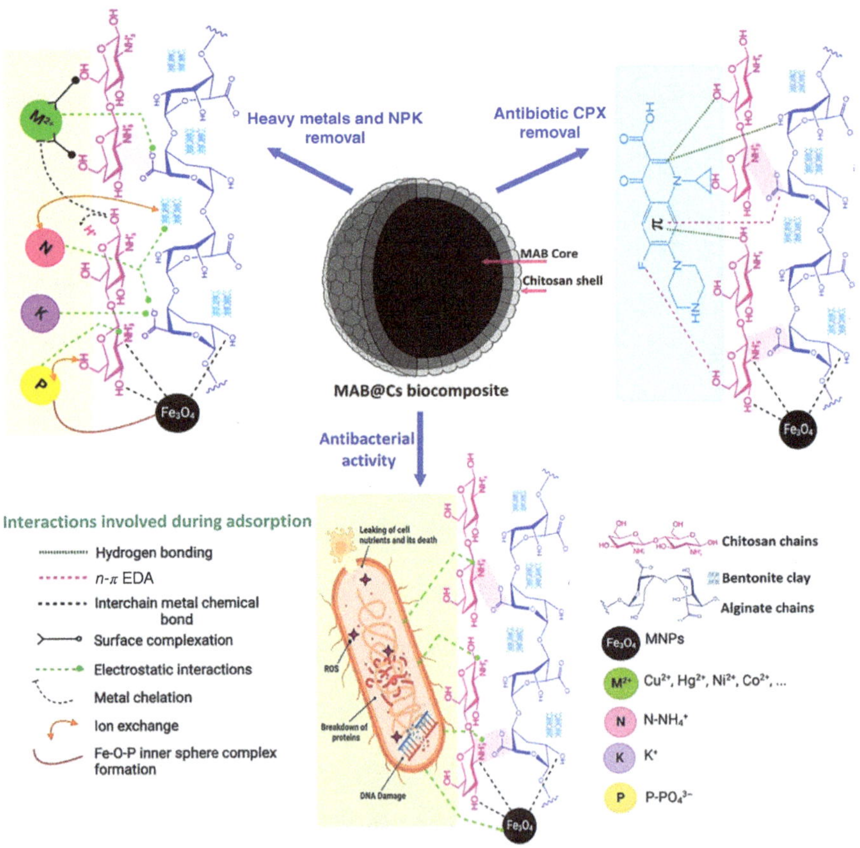

Figure 10.3 Proposed adsorption mechanism of organic and inorganic pollutants by one typical magnetic biocomposite (i.e., Fe_3O_4/alginate/bentonite@CS, labeled as mAB@Cs) and its antibacterial activity. *Source:* Reprinted with permission from [16]. Copyright 2024 Elsevier B. V.

wastewater, with maximum adsorption capacities of 635.35 and 615.57 mg g^{-1}, respectively. Pot tests have verified the feasibility of these nutrient-loaded MBCs as a fertilizer substitute for the growth of a typical forage grass, alfalfa (*Medicago sativa* L.). As depicted in Figure 10.3, the binding mechanisms for the above inorganic pollutants over MBCs usually involved (i) electrostatic interaction, (ii) ion-exchange, (iii) surface complexation, (iv) interchain metal chemical bond, and (v) metal chelation [16].

10.4.2 Adsorptive Removal of Organic Contaminants

The removal of organic pollutants from wastewater is often hindered by their complex properties, such as low biodegradability and high solubility, which can render traditional treatment methods ineffective [19, 98]. Moreover, the high energy intensity and costs associated with advanced treatment processes, such as advanced oxidation and membrane bioreactors, can limit their accessibility, particularly in resource-constrained communities [99, 100]. As a crucial class of organic contaminants, organic dyes are persistent and inherently toxic, which has raised a major environmental challenge and health concern, especially in low-income and middle-income countries [99]. However, recent advancements in MBC adsorbents have demonstrated great potential in overcoming the challenges of synthetic dyes [47, 71].

For instance, research has shown that Fe_3O_4/corn cob can effectively capture methylene blue (MB), with a maximum capacity of 13.2 mg g^{-1} at 55 °C [90], while the yeast-based MBC, viz. Fe_3O_4/AC@*Yarrowia lipolytica* cell has been demonstrated to possess a maximum adsorption capacity of 32.1 mg g^{-1} of MB at 25 °C [19]. Recently, Aryee et al. [91] prepared an MBC based on peanut shells using the coprecipitation method for efficient sequestration of alkaline dyes using MB and rhodamine B (RhB) as model dyes, which showed maximum adsorption capacities of 245.0 and 57.3 mg g^{-1}, respectively, at near-neutral pH (Table 10.1). Nascimento et al. [101] reported a nanosized MBC using yeast waste and MNPs as parent materials by coprecipitation method to remove three vinyl sulfonic acid reactive dyes (i.e., blue 19 (B-19), red 198 (R-198), and yellow 15 (Y-15)) from effluents of industrial-like dyeing processes. However, the adsorption capacities for all dyes were not as high as expected since the saturation of such MBC adsorbents was not reached in these suspensions. Similarly, by incorporating CS inside the MBC adsorbents, for example, three hierarchical core/shell/shell structured MBCs, viz. Fe_3O_4/CS@salicylaldehyde/nanoclay, Fe_3O_4/CS-cross-linked-PAM, and Fe_3O_4/CS/GO have been shown greatly improved capacities toward acid red (AR) 88, sunset yellow (SY), and Congo red (CR), respectively [88, 89, 92]. Munagapati et al. [56] observed that MBC adsorbent based on loading papaya (*papaya L.*) seed powder over MNPs (i.e., Fe_3O_4/papaya seed powder) could sequester up to 216.9 mg g^{-1} of CR at pH 5. A recent study by Wu et al. [93] has unveiled a novel core/shell/shell structured MBC based on loading poly (Benzyldimethyl[2-[(2-methyl-1-oxoally)oxy]ethyl] ammonium chloride) over the magnetic sodium alginate (SA) composite, showing a maximum adsorption capacity as high as 370.9 mg g^{-1} toward reactive red X-3B.

It is well-defined that the solution pH determines the charge on the surface of the adsorbent and also controls the degree of ionization of dissolved substances (e.g., dye molecules), thereby controlling the overall adsorption performance [102]. Studies have shown that the removal of anionic dyes increases at lower pH values, while the opposite is true for cationic dyes [103]. Regarding the binding mechanism of synthetic dyes, Jawad et al. [104] claimed that the reactive groups of MBC adsorbent CTS-GL/MgO/Fe_3O_4 play a crucial role in the adsorption process of Reactive Blue 19 (RB-19). Concretely, the positively charged groups including protonated amino groups ($-NH_3^+$), ($-OH_2^+$), and $Mg(OH)^+$ can effectively bind to the negatively charged ($-SO_3^-$) groups of RB-19 dye by electrostatic attraction. Besides, hydrogen bonding and interactions between the electron donors of the MBC and the acceptors inside the RB-19 dye may probably have contributed to the efficient binding of the dye. In addition, a study by Abd Malek et al. [105] has indicated that the amino and hydroxyl-rich CS component within the MBC, viz. m-CS-PVA/FA, is likely to protonate in acidic media and thus displays a strong affinity toward the anionic dye, Reactive Orange 16 (RO16), by strong electrostatic attraction. Dipole-dipole hydrogen bonding and n–π stacking interactions may also play an important role in the adsorption process.

Apart from organic dyes, organic contaminants like antibiotics, anti-inflammatories, and pesticides, also known as emerging pollutants, have shown remarkable affinity over the surface of many MBCs (Table 10.1). For instance, Mirizadeh et al. [95] investigated the adsorption performance of a novel magnetic chitosan/microalgae biocomposite, Fe_3O_4@CS/*chlorella vulgaris* for capturing three antibiotics, i.e., tetracycline (TC), ciprofloxacin (CIP), and amoxicillin (AMX) from wastewater, and unveiled that the maximal adsorption capacities were as high as 834.0, 394.9, and 150.8 mg g^{-1} for TC, CIP, and AMX, respectively. Hassanzadeh-Afruzi et al. [94] reported a novel MBC adsorbent based on Arabic Gum-grafted-poly amidoxime and $CuFe_2O_4$ magnetic nanoparticles ($CuFe_2O_4$/AG-g-PAO) recently, which displayed a remarkable maximum adsorption capacity as high as 769.2 mg g^{-1} for organophosphorus pesticide (OPP) under ambient conditions (pH 6, 25 °C). This core/shell/shell structured MBC also demonstrated fast adsorption kinetics and excellent recyclability. A recent work by Chkirida et al. [16] reported a new hybrid MBC based on an alginate-bentonite core and a chitosan shell layer (Fe_3O_4/alginate/bentonite@CS), where the MNPs were in situ generated within the alginate-bentonite core via a simple nonthermal coprecipitation approach. The biocomposite has a high ability to magnetically separate and remove organic (ciprofloxacin (CPX), Q_{max} = 264.8 mg g^{-1} at pH 6 and 23 °C) and seven heavy metals from wastewater. The interaction between MBC and these organic and inorganic contaminants is complex, and the high antibacterial activity of the biocomposite is believed to be attributed to the chitosan-MNPs antibacterial coating (Figure 10.3). Despite significant advances of the MBC adsorbents in adsorptive removal of inorganic and organic pollutants from wastewater, the implementation of this technique from laboratory scale to pilot scale and even industrial scale applications is still to be accomplished.

10.4.3 Recovery of Valuable Elements from Wastewater

The recovery of valuable elements, such as precious metals (e.g., gold, palladium, rhodium, silver), radiative metals (e.g., uranium, thorium), rare earth elements (e.g., lanthanum), and energy-critical metals (e.g., lithium) from waste streams is crucial for sustainable resource management and reducing the environmental impact of mining and processing these elements. However, the efficient and cost-effective recovery of these elements from complex waste matrices poses significant challenges. MBCs offer a promising solution due to their ability to selectively adsorb and concentrate target elements through tailored surface modifications, enabling efficient magnetic separation. This approach minimizes energy consumption and chemical reagent use compared to traditional methods. For instance, a MBC based on magnetic chitosan adsorbent (CDF-CS) has been successfully employed to recover gold from electronic waste [106], where a maximum adsorption capacity of 478.47 mg g^{-1} and a >95% recovery rate were achieved at pH 7 and 25 °C. A study by Mohamady [107] has shown that magnetic-chitosan nanocomposite was effective in capturing and recovering Th(IV) from aqueous solutions, with a maximum adsorption capacity of 142.3 mg g^{-1} at pH 3.5. In another study, a mesoporous MBC based on chitosan nanocomposite functionalized with cysteine (Cys) was used to adsorb uranium from contaminated water with a high efficiency of 136.9 mg g^{-1} at pH 3.5 ~ 5 and 25 °C [108].

Additionally, researchers have demonstrated the effective recovery of rare earth elements (REEs) from mining residues or waste streams using MBCs [109–111]. By using ion-imprinting technology, Liu and coworkers developed a Fe_3O_4/carboxymethyl cellulose (Fe_3O_4/CMC) MBC and found that Fe_3O_4/CMC exhibited high adsorption affinity and selectivity toward La(III) in aqueous solution, with Q_{max} up to 61.5 mg g^{-1} at 25 °C [109]. Another study by Mohammadi et al. [110] indicated that the naturally magnetic biomass from the magnetotactic bacterium *Magnetospirillum magneticum* AMB-1 can capture 37.2 mg g^{-1} of La(III) on a dry basis of biomass at pH 6, overperforming that reported for some other bacterial species. More recently, Ye et al. [111] developed a new

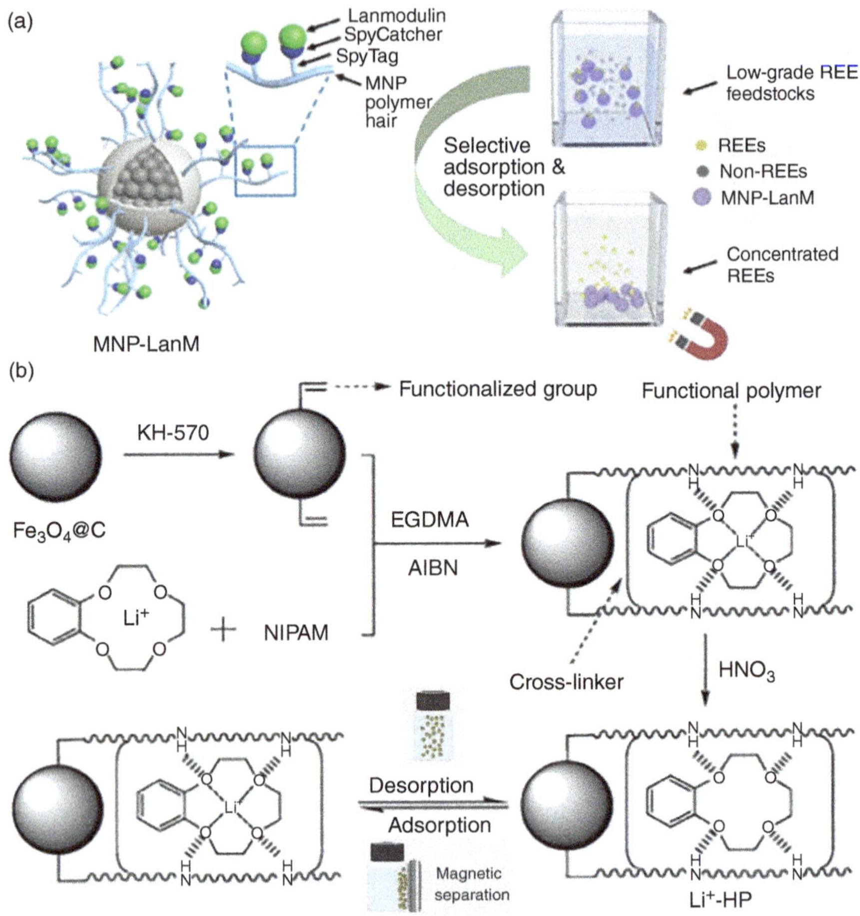

Figure 10.4 Examples of recovery of valuable elements by MBCs from wastewater. (a) Lanmodulin-SpyCatcher-loaded MNPs (MNP-LanM) for selective recovery of rare earth elements (REEs) from waste streams (Reprinted with permission from [111]. Copyright 2023 American Chemical Society). (b) Magnetic ion-imprinted polymers (IIP) for capturing low-concentration lithium ion from wastewater. *Source:* Adapted with permission from [112]. Copyright 2023 Elsevier B. V.

type of MBC (labeled as MNP-LanM) by immobilizing Lanmodulin-SpyCatcher (LanMSpycatcher) on the surface of SpyTag-functionalized MNPs. This MBC had demonstrated an adsorption capacity of $0.95\,mg\,g^{-1}$ for terbium (Tb) with fast adsorption kinetics and high selectivity, providing a promising solution for selective and efficient separation and recovery of REEs from low-grade feedstocks, e.g., the leachate of coal fly ash and geothermal brine (Figure 10.4a). In addition to REE recovery, Liang et al. [112] recently fabricated an ion-imprinted polymer (IIP) over the surface of magnetic carbon nanospheres, which can effectively and selectively capture as much as 23.5 mg g^{-1} of Li(I) from waste streams at 25 °C within 60 minutes. The recognition mechanism of this IIP toward Li(I) is due to the synergism of dehydration effect, electrostatic attraction, and size and imprinting effect (Figure 10.4b), facilitating the adsorption/desorption operations and recovery of Li(I) from aqueous medium. These successes highlight the potential of MBCs for a greener and more sustainable future in resource recovery from waste streams.

10.4.4 Other Applications

Apart from the above applications, MBCs have the potential to effectively break down organic pollutants through catalytic oxidation. Besides, MBCs can be used as adsorbents to capture organic substances from wastewater for the subsequent degradation in advanced oxidation processes. In combination with common advanced oxidation technologies such as photocatalytic oxidation, ozone oxidation, and advanced oxidation processes (e.g., Fenton reaction and photochemical oxidation) [113], MBCs usually play a crucial guiding role in facilitating the oxidant to target the pollutants more accurately in wastewater. For instance, Bornas et al. [114] fabricated a new Mn-Fe_3O_4-SiO_2/amine-glutaraldehyde/chitosan biocomposite (MIOSC-N-et-NH_2@CS-Mn) for the advanced oxidation process. This MBC was combined with an advanced oxidation process for the degradation of methylene orange with a removal efficiency of 98.9%. Ara et al. [115] reported a nanostructured chitosan-modified nickel magnetite biocomposite (CS-NM) recently and found that the MBC can effectively destruct MB and methyl red (MR) under the irradiation of sunlight, with photocatalytic efficiencies of 99% and 96% in 90 minutes for individual MB and MR, respectively. A recent study by Aryee et al. [116] shows that the MBC based on magnetic biomass/MOF composites (MPN@NH_2-MIL-101(Fe)) doped with nitrogen atoms was efficient in activation of peroxydisulfate (PS) for oxidative degradation of tetracycline compounds from wastewater (degradation efficiency ~ 87.6% in 60 minutes). Likewise, it was reported that the MBC based on microwave-assisted method using the methyl cellulose (MC) as a substrate ($CoFe_2O_4$@MC) can act as high-efficient activator of peroxymonosulfate (PMS), yielding a degradation efficiency up to 97% in degradation of reactive blue 19 under optimum conditions [117].

A more recent work by Verma and Moholkar [40] has shown that the Fe_3O_4@Laccase nanocomposite, based on the immobilization of laccase on the surface of amino-functionalized magnetic Fe_3O_4 nanoparticles, was very effective in treating industrial wastewater in the presence of sonication for chemical oxygen demand (COD), total organic carbon, and toxicity reduction. The optimization led to 90.7% COD removal and 70% toxicity reduction due to the synergistic effect of the enzymatic activity and the application of 35 kHz sonication. The Fe_3O_4@Laccase biocatalyst also demonstrated excellent recoverability and 72% retention of enzymatic activity for six consecutive treatment cycles.

10.5 Conclusions

Over the past two decades, magnetic biocomposites (MBCs) have emerged as a promising technology, offering a versatile platform for removing and recovering substances of environmental or valuable concerns from wastewater. Benefiting from their potential biological and ecological compatibility, tunable magnetic behavior, and affinity for various contaminants, these materials offer distinct advantages aimed at developing magnetic separation technologies of interest for water treatment units. These MBC adsorbents offer significant magnetic separation advantages by applying an external magnetic field in a confined space. This aspect is of great importance in waste storage or recycling and the reuse of adsorbents. While this technology holds immense potential, limitations and challenges persist, including the need for greater material scalability, stability, and regeneration methods to optimize their cost-effectiveness and environmental impact. In particular, information on large-scale evaluation and application of MBCs is scarce as most of the reports are laboratory-scale studies. Additionally, further research into tailoring the surface chemistry and structure of MBCs is essential for maximizing their adsorption capacities and selectivity for specific pollutants. To bridge the gap between laboratory results and full-scale application, more research is needed on the optimization of both the MBCs properties and the operation parameters in pilot-scale tests, with consideration of the complexity of real wastewater.

Looking ahead, the future of MBCs in wastewater treatment is bright, particularly with the integration of magnetic biocomposite technology with advanced oxidation processes (AOPs) such as heterogeneous photocatalysis or Fenton reactions for enhanced treatment. This synergistic approach of MBCs-AOPs would harness the unique properties of MBCs for efficient pollutant capture while utilizing AOPs to break down recalcitrant pollutants into less harmful by-products. As research continues to advance, efforts focused on optimizing this integration and developing scalable, cost-effective, and sustainable systems will be critical in transforming wastewater treatment and achieving a cleaner, healthier environment.

References

1 Kummerer, K., Dionysiou, D.D., Olsson, O., and Fatta-Kassinos, D. (2018). A path to clean water. *Science* 361: 222–224.

2 Qadir, M., Drechsel, P., Jiménez Cisneros, B. et al. (2020). Global and regional potential of wastewater as a water, nutrient and energy source. *Natural Resources Forum* 44: 40–51.

3 Vorosmarty, C.J., Mcintyre, P.B., Gessner, M.O. et al. (2010). Global threats to human water security and river biodiversity. *Nature* 467: 555–561.

4 Larsen, T.A., Hoffmann, S., Luthi, C. et al. (2016). Emerging solutions to the water challenges of an urbanizing world. *Science* 352: 928–933.

5 Shannon, M.A., Bohn, P.W., Elimelech, M. et al. (2008). Science and technology for water purification in the coming decades. *Nature* 452: 301–310.

6 Dávila-Jiménez, M.M., Elizalde-González, M.P., Geyer, W. et al. (2003). Adsorption of metal cations from aqueous solution onto a natural and a model biocomposite. *Colloids and Surfaces A: Physicochemical And Engineering Aspects* 219: 243–252.

7 Mahmoodi, N.M., Taghizadeh, M., Taghizadeh, A. et al. (2019). Bio-based magnetic metal-organic framework nanocomposite: Ultrasound-assisted synthesis and pollutant (heavy metal and dye) removal from aqueous media. *Applied Surface Science* 480: 288–299.

8 Safarik, I., Rego, L.F.T., Borovska, M. et al. (2007). New magnetically responsive yeast-based biosorbent for the efficient removal of water-soluble dyes. *Enzyme and Microbial Technology* 40: 1551–1556.

9 Abd Malek, N.N., Jawad, A.H., Abdulhameed, A.S. et al. (2020). New magnetic schiffs base-chitosan-glyoxal/fly ash/Fe_3O_4 biocomposite for the removal of anionic azo dye: An optimized process. *International Journal of Biological Macromolecules* 146: 530–539.

10 Touihri, M., Guesmi, F., Hannachi, C. et al. (2021). Single and simultaneous adsorption of Cr(VI) And Cu (II) on a novel Fe_3O_4/ pine cones gel beads nanocomposite: Experiments, characterization and isotherms modeling. *Chemical Engineering Journal* 416: 129101.

11 Ahmed, N. and Atkinson, G. J. (2022). A review of soft magnetic composite materials and applications. 2022 International Conference On Electrical Machines (Icem), Sept. 5-8 2022 Valencia, Spain. 551-557.

12 Liao, Z.W., Zoumhani, O., and Boutry, C.M. (2023). Recent advances in magnetic polymer composites for biomems: A review. *Materials* 16: 3802.

13 Talaat, A., Suraj, M.V., Byerly, K. et al. (2021). Review on soft magnetic metal and inorganic oxide nanocomposites for power applications. *Journal of Alloys and Compounds* 870: 159500.

14 Wu, S., Hu, W., Ze, Q. et al. (2020). Multifunctional magnetic soft composites: A review. *Multifunctional Materials* 3: 042003.

15 Jain, N., Verma, A., and Singh, V.K. (2019). Dynamic mechanical analysis and creep-recovery behaviour of polyvinyl alcohol based cross-linked biocomposite reinforced with basalt fiber. *Materials Research Express* 6: 105373.

16 Chkirida, S., El Mernissi, N., Zari, N. et al. (2024). Magnetic alginate coated chitosan core@shell beads with excellent performance in simulated and

real wastewater treatment: Behavior, mechanisms, and new perspectives. *International Journal of Biological Macromolecules* 260: 129389.

17 Nata, I.F., Irawan, C., Putra, M.D. et al. (2024). Simultaneous adsorption of Cr(III) ions and contaminants on sugarcane bagasse/rice husk fiber-based amine magnetic biocomposite. *Journal of Hazardous Materials Advances* 13: 100412.

18 Renani, N.M., Etesami, N., and Behzad, T. (2023). Synthesis and characterization of novel magnetic nano-biocomposite hydrogels based on starch-g-poly(acrylic acid) reinforced by cellulose nanofibers for Cu^{2+} ion removal. *ACS Omega* 8: 21929–21940.

19 Staron, P. and Chwastowski, J. (2024). Yeast-based magnetic biocomposite for efficient sorption of organic pollutants. *Applied Sciences* 14: 655.

20 Zhang, Z., Ahmed, A.I.S., Malik, M.Z. et al. (2023). Cellulose/inorganic nanoparticles-based nano-biocomposite for abatement of water and wastewater pollutants. *Chemosphere* 313: 137483.

21 Daniel-Da-Silva, A.L., Silva, N.J.O., Trindade, T., and Gil, A.M. (2011). Biocomposites containing magnetic nanoparticles. In: *Nanocomposite Particles For Bio-Applications: Materials And Bio-Interfaces*, 1e (ed. A.L. Daniel-Da-Silva and T. Trindade). New York, NY: Jenny Stanford Publishing.

22 Stiufiuc, G.F. and Stiufiuc, R.I. (2024). Magnetic nanoparticles: synthesis, characterization, and their use in biomedical field. *Applied Sciences* 14: 1623.

23 Laurent, S., Forge, D., Port, M. et al. (2008). Magnetic iron oxide nanoparticles: Synthesis, stabilization, vectorization, physicochemical characterizations, and biological applications. *Chemical Reviews* 108: 2064–2110.

24 Zang, X., Jiang, R., Zhu, H.Y. et al. (2024). A review on the progress of magnetic chitosan-based materials in water purification and solid-phase extraction of contaminants. *Separation and Purification Technology* 330: 125521.

25 Chen, Y.W. and Wang, J.L. (2011). Preparation and characterization of magnetic chitosan nanoparticles and its application for Cu(II) removal. *Chemical Engineering Journal* 168: 286–292.

26 Kilinç, E., Özdemir, S., Yalçin, M.S., and Soylak, M. (2019). A magnetized fungal solid-phase extractor for the preconcentrations of uranium(VI) and thorium(IV) before their quantitation by ICP-OES. *Microchimica Acta* 186: 355.

27 Rocher, V., Bee, A., Siaugue, J.M., and Cabuil, V. (2010). Dye removal from aqueous solution by magnetic alginate beads crosslinked with epichlorohydrin. *Journal of Hazardous Materials* 178: 434–439.

28 Liu, J.Q., Zhou, R.Y., Yu, J.X. et al. (2022). Simultaneous removal of lead, manganese, and copper released from the copper tailings by a novel magnetic modified biosorbent. *Journal of Environmental Management* 322: 116157.

29 Sheth, Y., Dharaskar, S., Khalid, M., and Sonawane, S. (2021). An environment friendly approach for heavy metal removal from industrial wastewater using chitosan based biosorbent: A review. *Sustainable Energy Technologies and Assessments* 43: 100951.

30 Yavuz, H., Denizli, A., Güngünes, H. et al. (2006). Biosorption of mercury on magnetically modified yeast cells. *Separation and Purification Technology* 52: 253–260.

31 Khorasani, A.C. and Shojaosadati, S.A. (2019). Magnetic pectin-chlorella vulgaris biosorbent for the adsorption of dyes. *Journal of Environmental Chemical Engineering* 7: 103062.

32 Safarikova, M., Pona, B.M.R., Mosiniewicz-Szablewska, E. et al. (2008). Dye adsorption on magnetically modified chlorella vulgaris cells. *Fresenius Environmental Bulletin* 17: 494–500.

33 Chen, X., Huang, Z., Luo, S.Y. et al. (2021). Multi-functional magnetic hydrogels based on millettia speciosa champ residue cellulose and chitosan: Highly efficient and reusable adsorbent for congo red and Cu^{2+} removal. *Chemical Engineering Journal* 423: 130198.

34 Wang, H.T., Xu, X., Ren, Z.F., and Gao, B.Y. (2016). Removal of phosphate and chromium(VI) from liquids by an amine-crosslinked nano-feo biosorbent derived from corn straw. *RSC Advances* 6: 47237–47248.

35 Liao, J. and Huang, H.H. (2019). Magnetic chitin hydrogels prepared from residues with tunable characteristics: A novel biosorbent for Cu^{2+} removal. *Carbohydrate Polymers* 220: 191–201.

36 Copello, G.J., Mebert, A.M., Raineri, M. et al. (2011). Removal of dyes from water using chitosan hydrogel/SiO_2 and chitin hydrogel/SiO_2 hybrid materials obtained by the sol-gel method. *Journal of Hazardous Materials* 186: 932–939.

37 Saber-Samandari, S., Saber-Samandari, S., Joneidi-Yekta, H., and Mohseni, M. (2017). Adsorption of anionic and cationic dyes from aqueous solution using gelatin-based magnetic nanocomposite beads comprising carboxylic acid functionalized carbon nanotube. *Chemical Engineering Journal* 308: 1133–1144.

38 Sharma, S., Hasan, A., Kumar, N., and Pandey, L.M. (2018). Removal of methylene blue dye from aqueous solution using immobilized biomass along with iron oxide nanoparticles as biosorbent. *Environmental Science and Pollution Research* 25: 21605–21615.

39 Chen, X.J., Zhu, J.W., Chen, Z.X. et al. (2011). A novel bienzyme glucose biosensor based on three-layer Au-Fe_3O_4@SiO_2 magnetic nanocomposite. *Sensors and Actuators B: Chemical* 159: 220–228.

40 Verma, K. and Moholkar, V.S. (2024). Investigations in sonoenzymatic treatment of industrial wastewater using Fe_3O_4@laccase nanocomposites. *ACS ES&T Water* 4: 1637–1646.

41 Mahmoud, M.E., Mohamed, A.K., and Salam, M.A. (2021). Self-decoration of N-doped graphene oxide 3-d hydrogel onto magnetic shrimp shell biochar for enhanced removal of hexavalent chromium. *Journal of Hazardous Materials* 408: 124951.

42 Wang, Y., Zhang, Y., Hou, C., and Liu, M.Z. (2016). Mussel-inspired synthesis of magnetic polydopamine-chitosan nanoparticles as biosorbent for dyes and metals removal. *Journal of the Taiwan Institute of Chemical Engineers* 61: 292–298.

43 Algethami, J.S., Alhamami, M.A.M., Alqadami, A.A. et al. (2024). Magnetic hydrochar grafted-chitosan for enhanced efficient adsorption of malachite green dye from aqueous solutions: Modeling, adsorption behavior, and mechanism analysis. *International Journal of Biological Macromolecules* 254: 127767.

44 Thomas, V., Dean, D.R., Jose, M.V. et al. (2007). Nanostructured biocomposite scaffolds based on collagen coelectrospun with nanohydroxyapatite. *Biomacromolecules* 8: 631–637.

45 Alavarse, A.C., Silva, J.B., Ulrich, H., and Petri, D.F.S. (2023). Poly(vinyl alcohol)/sodium alginate/magnetite composites: Magnetic force microscopy for tracking magnetic domains. *Soft Matter* 19: 2612–2622.

46 Zhong, Y.Y., Chen, C., Liu, S. et al. (2021). A new magnetic adsorbent of eggshell-zeolitic imidazolate framework for highly efficient removal of norfloxacin. *Dalton Transactions* 50: 18016–18026.

47 Jawad, A.H., Malek, N.N.A., Abdulhameed, A.S., and Razuan, R. (2020). Synthesis of magnetic chitosan-fly Ash/Fe_3O_4 composite for adsorption of reactive orange 16 dye: Optimization by box-behnken design. *Journal of Polymers and the Environment* 28: 1068–1082.

48 Sudhakaran, S., Abraham, E.V., Mahadevan, H., and Krishnan, K.A. (2021). Crosslinked chitosan-montmorillonite biocomposite with fe intercalation: Enhancing surface chemistry for improved phosphate adsorption. *Surfaces and Interfaces* 27: 101468.

49 He, Y.Y., Wu, P., Xiao, W. et al. (2019). Efficient removal of Pb(II) from aqueous solution by a novel ion imprinted magnetic biosorbent: Adsorption kinetics and mechanisms. *PLoS One* 14: E0213377.

50 Lapo, B., Demey, H., Carchi, T., and Sastre, A.M. (2019). Antimony removal from water by a chitosan-iron(III)[chifer(III)] biocomposite. *Polymers* 11: 351.

51 Elsayed, S.A., Saad, E.M., Butler, I.S., and Mostafa, S.I. (2018). 2-hydroxynaphthaldehyde chitosan schiff-base; new complexes, biosorbent to remove Cadmium(II) ions from aqueous media and aquatic ecotoxicity against green alga. *Journal of Environmental Chemical Engineering* 6: 3451–3468.

52 Roy, C., Chowdhury, D., Sanfui, M.H. et al. (2023). Solid waste collagen-associated fabrication of magnetic hematite nanoparticle@collagen nanobiocomposite for emission-adsorption of dyes. *International Journal of Biological Macromolecules* 242: 124774.

53 Kazemi, J. and Javanbakht, V. (2020). Alginate beads impregnated with magnetic chitosan@zeolite nanocomposite for cationic methylene blue dye removal from aqueous solution. *International Journal of Biological Macromolecules* 154: 1426–1437.

54 Sing, K. (2001). The use of nitrogen adsorption for the characterisation of porous materials. *Colloids And Surfaces A-Physicochemical And Engineering Aspects* 187: 3–9.

55 Asfaram, A., Ghaedi, M., Dashtian, K., and Ghezelbash, G.R. (2018). Preparation and characterization of $Mn_0.4Zn_0.6Fe_2O_4$ nanoparticles supported on dead cells of as a novel and efficient adsorbent/biosorbent composite for the removal of azo food dyes: Central composite design optimization study. *ACS Sustainable Chemistry & Engineering* 6: 4549–4563.

56 Munagapati, V.S., Wen, H.Y., Gollakota, A.R.K. et al. (2022). Magnetic Fe_3O_4 nanoparticles loaded papaya (Papaya L.) seed powder as an effective and recyclable adsorbent material for the separation of anionic azo dye (congo red) from liquid phase: Evaluation of adsorption properties. *Journal of Molecular Liquids* 345: 118255.

57 Luo, Y.W., Wu, Z.L., Guan, Q.H. et al. (2022). Facile synthesis of magnetic porous carbon nanosheets as efficient As(III) adsorbent. *Chemical Papers* 76: 7295–7303.

58 Shen, Z.Y., Jin, J., Fu, J.J. et al. (2021). Anchoring Al- and/or Mg-oxides to magnetic biochars for co-uptake of arsenate and fluoride from water. *Journal of Environmental Management* 293: 112898.

59 Tee, G.T., Gok, X.Y., and Yong, W.F. (2022). Adsorption of pollutants in wastewater via biosorbents, nanoparticles and magnetic biosorbents: A review. *Environmental Research* 212: 113248.

60 Noreen, S., Bhatti, H.N., Iqbal, M. et al. (2020). Chitosan, starch, polyaniline and polypyrrole biocomposite with sugarcane bagasse for the efficient removal of acid black dye. *International Journal of Biological Macromolecules* 147: 439–452.

61 Rosales, E., Meijide, J., Pazos, M., and Sanromán, M.A. (2017). Challenges and recent advances in biochar as low-cost biosorbent: from batch assays to continuous-flow systems. *Bioresource Technology* 246: 176–192.

62 Pan, J.W., Gao, B.Y., Wang, S.Y. et al. (2020). Waste-to-resources: Green preparation of magnetic biogas residues-based biochar for effective heavy metal removals. *Science of the Total Environment* 737: 140283.

63 Oyekanmi, A.A., Alshammari, M.B., Ibrahim, M.N.M. et al. (2022). Highly effective cow bone based biocomposite for the sequestration of organic pollutant parameter from palm oil mill effluent in a fixed bed column adsorption system. *Polymers* 14: 86.

64 Fan, L.L., Luo, C.N., Sun, M. et al. (2013). Synthesis of magnetic B-cyclodextrin-chitosan/graphene oxide as nanoadsorbent and its application in dye adsorption and removal. *Colloids and Surfaces B: Biointerfaces* 103: 601–607.

65 Wang, S.B., Li, H.T., Xie, S.J. et al. (2006). Physical and chemical regeneration of zeolitic adsorbents for dye removal in wastewater treatment. *Chemosphere* 65: 82–87.

66 Yao, Y. (2016). Enhancement of mass transfer by ultrasound: Application to adsorbent regeneration and food drying/dehydration. *Ultrasonics Sonochemistry* 31: 512–531.

67 Amen, R., Bashir, H., Bibi, I. et al. (2020). A critical review on arsenic removal from water using biochar-based sorbents: The significance of modification and redox reactions. *Chemical Engineering Journal* 396: 125195.

68 Fdez-Sanromán, A., Pazos, M., Rosales, E., and Sanromán, M.A. (2020). Unravelling the environmental application of biochar as low-cost biosorbent: A review. *Applied Sciences* 10: 7810.

69 Safarik, I., Baldikova, E., Prochazkova, J. et al. (2018). Magnetically modified agricultural and food waste: Preparation and application. *Journal of Agricultural and Food Chemistry* 66: 2538–2552.

70 Reshmy, R., Philip, E., Madhavan, A. et al. (2022). Nanocellulose as green material for remediation of hazardous heavy metal contaminants. *Journal of Hazardous Materials* 424: 127516.

71 Sharma, A., Mangla, D., and Shehnaz & Chaudhry, S. A. (2022). Recent advances in magnetic composites as adsorbents for wastewater remediation. *Journal of Environmental Management* 306: 114483.

72 Asliyuce, S. (2023). Screening the heavy metal removal capacity of magnetically modified fungal biosorbent. *Chemical Papers* 77: 4331–4344.

73 Staron, P., Kuciakowski, J., and Chwastowski, J. (2023). Biocomposite of hydrochar and with magnetic properties for adsorptive removal of cadmium ions. *Journal of Environmental Chemical Engineering* 11: 110270.

74 Omer, A.M., Abd El-Monaem, E.M., Abd El-Latif, M.M. et al. (2021). Facile fabrication of novel magnetic ZIF-67 MOF@aminated chitosan composite beads for the adsorptive removal of Cr(VI) from aqueous solutions. *Carbohydrate Polymers* 265: 118084.

75 Hamza, M.F., Hamad, D.M., Hamad, N.A. et al. (2022). Functionalization of magnetic chitosan microparticles for high-performance removal of chromate from aqueous solutions and tannery effluent. *Chemical Engineering Journal* 428: 131775.

76 Sarojini, G., Kannan, P., Rajamohan, N., and Rajasimman, M. (2023). Biofabrication of porous magnetic chitosan/Fe_3O_4 nanocomposite using azolla pinnata for removal of chromium - parametric effects, surface characterization and kinetics. *Environmental Research* 218: 114822.

77 Cheraghipour, E. and Pakshir, M. (2021). Environmentally friendly magnetic chitosan nano-biocomposite for Cu(II) ions adsorption and magnetic nanofluid hyperthermia: CCD-RSM design. *Journal of Environmental Chemical Engineering* 9: 104883.

78 Wang, M.X. and You, X.Y. (2023). Efficient adsorption of antibiotics and heavy metals from aqueous solution by structural designed PSSMA-functionalized-chitosan magnetic composite. *Chemical Engineering Journal* 454: 140417.

79 Algamdi, M., Alshahrani, A., and Alsuhybani, M. (2023). Chitosan grafted tetracarboxylic functionalized magnetic nanoparticles for removal of Pb(II) from an aqueous environment. *International Journal of Biological Macromolecules* 225: 1517–1528.

80 Huang, Y.Y., Shen, Y., Zhang, G.Z. et al. (2023). Highly effective and selective removal of lead ions by polymer-grafted silica-coated acid-resistant magnetic chitosan composites. *Separation and Purification Technology* 314: 123561.

81 Zhao, H.X., Li, M.X., Wang, Y. et al. (2024). Preparation of cellulose nanofiber/Fe_3O_4/$CaCO_3$ magnetic biocomposite by air capture and its application on purifying lead-contaminated water: Removal, immobilization and separation. *Applied Surface Science* 646: 158936.

82 Lan, Y.B., Lei, X.M., Zhao, C.Y. et al. (2023). Study on the polypeptide-functionalized magnetic chitosan microspheres that specifically adsorb Hg (II) ions. *Separation and Purification Technology* 323: 124386.

83 Sharifi, M.J., Nouralishahi, A., and Hallajisani, A. (2023). Fe_3O_4-chitosan nanocomposite as a magnetic biosorbent for removal of nickel and cobalt heavy metals from polluted water. *International Journal of Biological Macromolecules* 248: 125984.

84 Kloster, G.A., Valiente, M., Marcovich, N.E., and Mosiewicki, M.A. (2020). Adsorption of arsenic onto films based on chitosan and chitosan/nano-iron oxide. *International Journal of Biological Macromolecules* 165: 1286–1295.

85 Wang, G.Y., Liu, R.Q., Yang, Z.B. et al. (2024). Enhanced simultaneous removal of phosphate and ammonium from swine wastewater using magnetic magnesium-loaded Chinese herbal medicine residues: Performance, mechanism, and resource utilization. *Bioresource Technology* 393: 130103.

86 Aryee, A.A., Gao, C.P., Han, R.P., and Qu, L.B. (2023). Synthesis of a novel magnetic biomass-MOF composite for the efficient removal of phosphates:

adsorption mechanism and characterization study. *Chinese Journal of Chemical Engineering* 62: 202–216.

87 Li, T.T., Tong, Z.H., Zheng, Q.Z. et al. (2021). Fabrication of a lignin-based magnetic nanocomposite adsorbent to recover phosphorus in water for agricultural reuse. *ACS Sustainable Chemistry & Engineering* 9: 10468–10478.

88 Arni, L.A., Hapiz, A., Jawad, A.H. et al. (2023). Fabrication of magnetic chitosan-grafted salicylaldehyde/nanoclay for removal of azo dye: BBD optimization, characterization, and mechanistic study. *International Journal of Biological Macromolecules* 248: 125943.

89 Jiang, R., Shen, T.T., Zhu, H.Y. et al. (2023). Magnetic Fe_3O_4 embedded chitosan-crosslinked-polyacrylamide composites with enhanced removal of food dye: Characterization, adsorption and mechanism. *International Journal of Biological Macromolecules* 227: 1234–1244.

90 Allou, N.B., Tigori, M.A., Koffi, A.A. et al. (2023). Methylene blue magnetic adsorption separation process from aqueous solution using corn cob. *Scientific African* 21: E01828.

91 Aryee, A.A., Gao, C.P., Han, R.P., and Qu, L.B. (2022). Functionalized magnetic biocomposite based on peanut husk for the efficient sequestration of basic dyes in single and binary systems: Adsorption mechanism and antibacterial study. *Journal of Environmental Chemical Engineering* 10: 108205.

92 Ding, C., Xue, M., Zhang, Y. et al. (2023). Novel chitosan/GO@Fe_3O_4 porous microspheres with magnetic separation function for the removal of congo red from aqueous solutions. *Journal of the Taiwan Institute of Chemical Engineers* 149: 105008.

93 Wu, Y.Y., Li, H., An, Y.Y. et al. (2022). Construction of magnetic alginate-based biosorbent and its adsorption performances for anionic organic contaminants. *Separation and Purification Technology* 297: 121566.

94 Hassanzadeh-Afruzi, F., Maleki, A., and Zare, E.N. (2022). Efficient remediation of chlorpyrifos pesticide from contaminated water by superparamagnetic adsorbent based on arabic gum-polyamidoxime. *International Journal of Biological Macromolecules* 203: 445–456.

95 Mirizadeh, S., Solisio, C., Converti, A., and Casazza, A.A. (2024). Efficient removal of tetracycline, ciprofloxacin, and amoxicillin by novel magnetic chitosan/microalgae biocomposites. *Separation and Purification Technology* 329: 125115.

96 Gao, L., Wang, Y., Lu, Y. et al. (2023). Facile synthesis of a magnetic glucose-based hyper-cross-linked polymer by dual-role ferric chloride for nitrophenol removal from sewage. *ACS ES&T Water* 3: 3639–3648.

97 Rosova, E., Smirnova, N., Dresvyanina, E. et al. (2021). Biocomposite materials based on chitosan and lignin: Preparation and characterization. *Cosmetics* 8: 24.

98 Weerasundara, L., Gabriele, B., Figoli, A. et al. (2021). Hydrogels: Novel materials for contaminant removal in water—A review. *Critical Reviews in Environmental Science and Technology* 51: 1970–2014.

99 Lin, J.Y., Ye, W.Y., Xie, M. et al. (2023). Environmental impacts and remediation of dye-containing wastewater. *Nature Reviews Earth and Environment* 4: 785–803.

100 Zhang, S., Zheng, H., and Tratnyek, P.G. (2023). Advanced redox processes for sustainable water treatment. *Nature Water* 1: 666–681.

101 Nascimento, J.R., Bezerra, K.C.H., Martins, T.D. et al. (2021). Textile effluent treatment employing yeast biomass and a new nanomagnetic biocomposite. *Environmental Science and Pollution Research* 28: 27318–27332.

102 Liu, T., Han, X., Wang, Y. et al. (2017). Magnetic chitosan/anaerobic granular sludge composite: Synthesis, characterization and application in heavy metal ions removal. *Journal of Colloid and Interface Science* 508: 405–414.

103 Salleh, M.A.M., Mahmoud, D.K., Karim, W.A.W.A., and Idris, A. (2011). Cationic and anionic dye adsorption by agricultural solid wastes: A comprehensive review. *Desalination* 280: 1–13.

104 Jawad, A.H., Rangabhashiyam, S., Abdulhameed, A.S. et al. (2022). Process optimization and adsorptive mechanism for reactive blue 19 dye by magnetic crosslinked chitosan/MgO/FeO biocomposite. *Journal of Polymers and the Environment* 30: 2759–2773.

105 Abd Malek, N.N., Jawad, A.H., Ismail, K. et al. (2021). Fly ash modified magnetic chitosan-polyvinyl alcohol blend for reactive orange 16 dye removal: Adsorption parametric optimization. *International Journal of Biological Macromolecules* 189: 464–476.

106 Zhao, M., Zhao, J., Huang, Z. et al. (2019). One pot preparation of magnetic chitosan-cystamine composites for selective recovery of Au(III) from the aqueous solution. *International Journal of Biological Macromolecules* 137: 721–731.

107 Mohamady, S.I. (2021). Functionalization of magnetic-chitosan nanocomposite for enhancing Th(IV) ions sorption. *Egyptian Journal of Chemistry* 64: 2095–2111.

108 Al-Ghamdi, A.A., Galhoum, A.A., Alshahrie, A. et al. (2022). Mesoporous magnetic cysteine functionalized chitosan nanocomposite for selective uranyl ions sorption: Experimental, structural characterization, and mechanistic studies. *Polymers* 14: 2568.

109 Liu, L., Chang, S., Wang, Y. et al. (2021). Facile fabrication of ion-imprinted Fe_3O_4/carboxymethyl cellulose magnetic biosorbent: Removal and recovery properties for trivalent la ions. *RSC Advances* 11: 25258–25265.

110 Mohammadi, M., Reinicke, B., and Wawrousek, K. (2022). Biosorption and biomagnetic recovery of La^{3+} by magnetospirillum magneticum AMB-1 biomass. *Separation and Purification Technology* 303: 122140.

111 Ye, Q.H., Jin, X.Y., Zhu, B.T. et al. (2023). Lanmodulin-functionalized magnetic nanoparticles as a highly selective biosorbent for recovery of rare earth elements. *Environmental Science & Technology* 57: 4276–4285.

112 Liang, C., Zhang, X.P., Liu, W.F. et al. (2023). Thermo-responsive ion imprinted polymer on the surface of magnetic carbon nanospheres for recognizing and capturing low-concentration lithium ion. *Minerals Engineering* 201: 108210.

113 Dewil, R., Mantzavinos, D., Poulios, I., and Rodrigo, M.A. (2017). New perspectives for advanced oxidation processes. *Journal of Environmental Management* 195: 93–99.

114 Bornas, B., Faraji, A.R., and Ashouri, F. (2023). Fabrication of a magnetic Mn(II) cross-linked chitosan-amine/glutaraldehyde nanocomposite for the rapid degradation of dyes and aerobic selective oxidation of ethylbenzene. *RSC Advances* 13: 9846–9863.

115 Ara, A., Khattak, R., Khan, M.S. et al. (2022). Synthesis, characterization, and solar photo-activation of chitosan-modified nickel magnetite bio-composite for degradation of recalcitrant organic pollutants in water. *Catalysts* 12: 983.

116 Aryee, A.A., Ma, Y.T., Wang, J.L. et al. (2023). A magnetic biomass/mof composite as a functional material for the oxidative removal of tetracycline: Degradation mechanism and toxicity study. *Journal of Environmental Chemical Engineering* 11: 110663.

117 Nasiri, A., Malakootian, M., Heidari, M.R., and Asadzadeh, S.N. (2021). $CoFe_2O_4$@methylcelloluse as a new magnetic nano biocomposite for sonocatalytic degradation of reactive blue 19. *Journal of Polymers and the Environment* 29: 2660–2675.

11

Alginate Biocomposite Materials: Development and Application in Wastewater Treatment

*Congfei Yao**, *Junkai Zhang**, *Yilei Ding, and Chunguang Chen*

Department of Chemistry, School of Materials and Chemistry, University of Shanghai for Science and Technology, Shanghai, China

11.1 Introduction

Water is a fundamental resource for sustaining life and plays a pivotal role in both economic development and human activities, such as industrial production and everyday living. However, the swift advancement of industrialization and urbanization has significantly contributed to an increase in wastewater discharge. This phenomenon poses severe threats to ecological balance and human well-being. Untreated wastewater often contains harmful substances, including heavy metals, toxic chemicals, and pathogens, all of which can infiltrate natural resources such as water bodies, soil, and the atmosphere, thereby endangering ecosystems and public health. Moreover, the issue of water pollution has escalated into one of the most critical global health challenges. Reports suggest that nearly 6000 children lose their lives each day due to illnesses stemming from inadequate access to safe drinking water, sanitation, and hygiene. Furthermore, hazardous chemicals from polluted water sources can penetrate human systems through contaminated supplies, exacerbating health risks worldwide. Additionally, water pollution intensifies the scarcity of water resources for agricultural use, thereby impacting food production, crop security, and leading to substantial economic losses [1–3]. In summary, the importance and urgency of wastewater management cannot be ignored.

Scholars worldwide have carried out extensive research to identify the most effective wastewater treatment method. The goal is to adopt various technical approaches that can remove organic matter, inorganic salts, microorganisms,

*Authors contributed equally to this work.

suspended solids, and other pollutants from wastewater, enabling it to meet national emission standards or reuse requirements in terms of pollutant quality and quantity [4]. This serves the dual purpose of environmental protection and resource utilization. Although numerous wastewater treatment methods have emerged over the past decades, only a few of them have been successfully implemented and widely adopted by the relevant industries due to the economic limitations and technical difficulties associated with most methods. Typically, wastewater treatment methods can be categorized into three primary types: biological, chemical, and physical treatments [5]. These include specific techniques such as aerobic/anaerobic biological processes, chemical precipitation, chemical oxidation, adsorption, and membrane separation. Among these methods, adsorption is especially preferred due to its straightforward nature, economic feasibility, and high efficiency in eliminating pollutants from wastewater. Numerous adsorbents have been explored for their potential in treating wastewater, such as rice husk biochar [6], garlic peel biochar [7], pyrolyzed petrochemical sediments [8], coconut pulp charcoal [9], activated clay [10], carbon nanotubes [11], and porous carbon materials such as activated charcoal [12]. However, separating water from these carbon-based materials is often challenging, and the loss of adsorbents can lead to environmental pollution in water bodies, which significantly hinders their broader application in environmental protection.

SA is a naturally occurring, water-soluble polysaccharide with a linear structure, commonly extracted from the cell walls of different species of brown algae [13]. The structure of the compound is defined by a repetitive sequence of α-L-guluronic acid (G) block monomer unit and 1,4-linked β-D-mannuronic acid (M) differential isomers. These units feature reactive functional groups, including hydroxyl (—OH) and carboxyl (—COOH) groups, as shown in Figure 11.1 [14]. Due to its remarkable biocompatibility, low toxicity, and hydrophilic properties, SA has found widespread applications across various sectors, such as food processing, cosmetics, pharmaceuticals, paper production, and textile dyeing [15]. Furthermore, alginate, as a natural polysaccharide, is regarded as an ideal carrier for certain nanoadsorbents. It can be employed to encapsulate substances such as activated carbon, carbon nanotubes, graphene oxide, and titanium nanoparticles, among others. This results in the production of alginate-based biocomposite adsorbents, which offer improved performance, ease of separation, and minimized risk of secondary contamination. These substances have shown considerable promise in eliminating contaminants from wastewater, positioning them as a promising avenue for advancing wastewater treatment technologies. For example, Kim et al. [16] used alginate to immobilize powdered activated carbon, obtaining alginate-activated carbon beads with excellent adsorption properties for Cu^{2+}, outperforming both alginate beads and powdered activated carbon. Li et al. [17] created a sustainable adsorbent made from calcium alginate and carbon

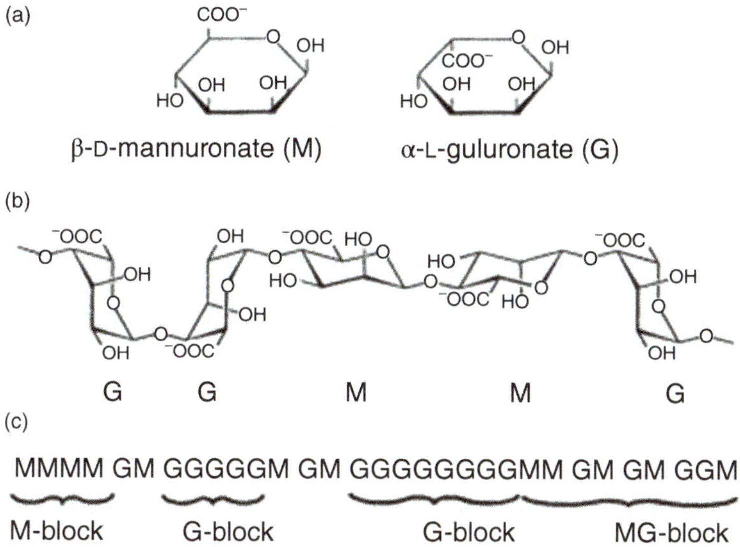

Figure 11.1 Structural formula of alginate: (a) monomer, (b) linear chain, and (c) block distribution. *Source:* Reproduced from ref. [14] copyright (2011), with permission from Food Hydrocolloids.

nanotubes, which exhibited a copper adsorption capacity of $67.9\,mg\,g^{-1}$ at an equilibrium concentration of $5\,mg\,L^{-1}$. Li et al. [18] also developed a composite material combining calcium alginate and graphene oxide, achieving a maximum adsorption capacity of $181.81\,mg\,g^{-1}$ for methylene blue dye. This composite demonstrated both high adsorption performance and removal efficiency over a broad pH range, while being easily separable from the solution and preventing secondary contamination from graphene particles. Mahmoodi et al. [19] synthesized sodium alginate/titanium dioxide nanoparticle composites (SA/n-TiO_2) by immobilizing titanium dioxide nanoparticles onto sodium alginate. These composites exhibited effective adsorption of anionic dyes, including Direct Red 80 and Acid Green 25, in aqueous solution, positioning them as a biocompatible, cost-effective, and high-capacity adsorbent.

This paper presents an in-depth review of the properties and benefits of alginate-based biocomposites in comparison to conventional adsorbent materials. It offers a comprehensive review of the methods and technologies used in the fabrication of various alginate-based biocomposites. And it concludes by exploring the present status and future prospects of these biocomposites in the context of wastewater treatment applications. The aim is to provide innovative ideas that can contribute to the restoration and management of water environments.

11.2 Modification and Preparation of Alginate Biocomposites

Alginate molecules are rich in hydroxyl and carboxyl functional groups, rendering them suitable for a variety of modification schemes. These modification approaches can subsequently lead to the development of diverse alginate biocomposite materials. Alginate modification schemes include physical and chemical strategies. As shown in Figure 11.2, physical modification mainly involves the physical combination of alginate with nanomaterials, polymers, natural macromolecules, etc. In contrast, the chemical modification mainly includes the modification through hydroxyl modification, carboxyl modification, and other reaction pathways. The alginate biocomposites obtained after modification demonstrate significant improvements in physicochemical properties, effectively addressing the limitations of native alginate, such as low limited mechanical strength, inadequate stability, and challenges related to degradation. Moreover, chemical modification can effectively eliminate the oleophobic groups present in alginates, allowing for the incorporation of functional groups or compounds onto the alginate structure. This modification increases the accessibility of active sites, which are essential for specific reactions, such as providing effective adsorption sites for the removal of contaminants like heavy metals, dyes, antibiotics, and radioactive materials from wastewater [20].

11.2.1 Alginate and Mineral Nanomaterials

Dai et al. [21] prepared organo-montmorillonite by employing the cationic surfactant cetyltrimethylammonium bromide. Subsequently, organo-montmorillonite/

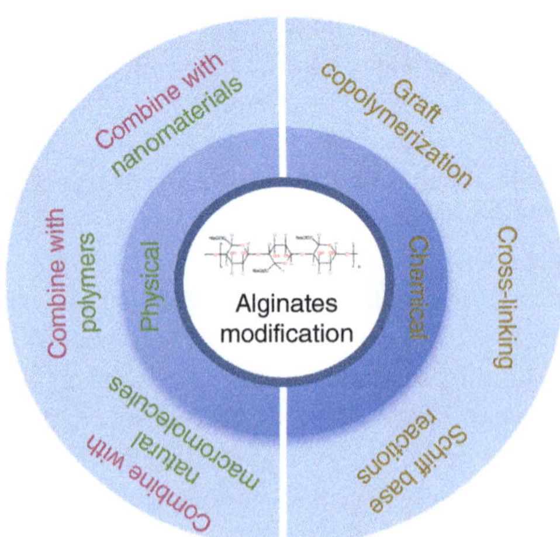

Figure 11.2 Various modification methods for alginate.

SA composites were prepared by coagulating SA and $CaCl_2$ (as cross-linking agents) with the organo-montmorillonite, and encapsulating the organo-montmorillonite into composite microbeads. It has been demonstrated that the composites are capable of effectively adsorbing Polycyclic Aromatic Hydrocarbons (PAHs). Song et al. [22] synthesized SA-loaded nano-$Mg(OH)_2$ composite microbeads, which exhibited remarkable adsorption capabilities. These beads were fabricated using raw materials such as flower-like nano-$Mg(OH)_2$, 1% $CaCl_2$ solution, and sodium alginate powder. The beads are highly effective in extracting and reclaiming rare earth elements from wastewater.

11.2.2 Alginates and Polymers

Sun et al. [23] aimed to enhance the spinnability of sodium alginate by modifying it with acrylonitrile, using potassium persulfate as an initiator. This modification resulted in the synthesis of a sodium alginate-polyacrylonitrile copolymer (SA-g-AN). The researchers then employed the SA-g-AN copolymer in combination with poly(ethylene glycol) (PEG) to create a spinning solution, which was subsequently used to fabricate electrospun nanofibers. In addition, the researchers examined the oil–water separation efficiency and water resistance of the SA-g-AN electrospun fibers. The results demonstrated significant enhancements in both water resistance and thermal stability, while maintaining a stable morphology and consistent performance in oil–water separation. Gutierrez-Gonzalez et al. [24] prepared SA/polyethylene oxide (PEO) composites loaded with curcumin (CU) using the electrospinning method. The incorporation of CU, along with the subsequent cross-linking process, was found to notably enhance the mechanical properties of the materials. By utilizing the biocompatible trifluoroacetic acid (TFA) for cross-linking and taking advantage of the inherent properties of alginate, this alginate-based nanofiber composite proves suitable for applications in water pollution management, particularly in areas like adsorption and filtration.

11.2.3 Alginate and Natural Macromolecules

Chitosan and cellulose are commonly used as natural macromolecular miscible materials for water treatments. Zhao et al. [25] synthesized a composite material by integrating chitosan with electrospun sodium alginate nanofibers, subsequently fabricating chitosan/electrospun sodium alginate nanofiber sponges through freeze-drying in a ternary solvent mixture of acetic acid, water, and dioxane. These composite sponges are abundant in functional groups and feature advanced porous structures along with nanofiber arrangements, all of which contribute to the improvement of their adsorption capacity. It has excellent adsorption performance for systems containing anionic and cationic dyes. In addition, the alginate biocomposite still showed good adsorption performance over a wide pH range and after cyclic adsorption, which has potential application in complex wastewater treatment. Fang et al. [26] developed cellulose-based sodium alginate/iron hydroxide (C/SA/Fe) composite hydrogels, utilizing epichlorohydrin as the cross-linking agent. The adsorption ability of the hydrogel for methylene blue (MB) was subsequently assessed as an adsorbent. The findings revealed that the C/SA/Fe hydrogel demonstrated strong adsorption efficiency, suggesting its promising application in environmental protection areas such as wastewater remediation and resource recovery.

11.2.4 Graft Copolymerization

Graft copolymerization is a frequently utilized technique for modifying sodium alginate, usually initiated by the addition of an initiator, heat, ultraviolet (UV) radiation, or high-energy irradiation. This process targets the hydroxyl groups as the active sites for radical formation. Recent studies have primarily focused on the extraction of contaminants, including dyes and heavy metals, from aqueous solutions by utilizing modified SA materials via graft copolymerization. Shehzad et al. [27] synthesized novel urethane-functionalized calcium alginate hydrogel beads (CA-1) by reacting sodium alginate with 4-phenylcarbamate, followed by cross-linking with Ca^{2+}. In comparison to the unmodified CA hydrogel beads, the CA-1 beads demonstrated enhanced adsorption of U(VI) ions, especially in mildly acidic environments. The CA-1 hydrogel beads exhibited fast adsorption rates, a high capacity for uptake, and a strong affinity for U(VI) ions, underscoring their promising potential for both the removal and recovery of U(VI) ions from wastewater. Zhang et al. [28] employed microwave-assisted copolymerization to prepare carboxymethyl cellulose-iconic acid-sodium alginate (CIS) biocomposites and assessed their flocculation properties against crystalline violet dye (CV). The CIS biocomposites demonstrated a maximum removal efficiency of 92.2% from a $100\,mg\,L^{-1}$ CV-simulated wastewater solution at a CIS concentration of $30\,mg\,L^{-1}$. These findings indicate that the copolymerization technology holds significant promise for furthering the development of innovative alginate-based biocomposites.

11.2.5 Cross-Linking: Hydroxyl or Carboxyl Group

The carboxyl group in SA imparts polyanionic characteristics to the dissolved form of SA. To enhance the reactivity of these carboxyl groups and increase the availability of reaction sites, Kim et al. [29] developed innovative gel beads by cross-linking waste casting ash/SA beads with polyethyleneimine. These beads were utilized to adsorb anionic dyes, including Methyl Red (MR), from wastewater. The interaction during the reaction led to a modification in N-H bonding, resulting in the formation of an amide bond between the carboxyl and amino groups. The cross-linking of the carboxyl groups facilitated electrostatic interactions and strengthened the hydrogen bonding network. In particular, the carboxyl groups of the modified SA materials were essential for the adsorption capacity of organic dyes. Furthermore, the hydroxyl and carboxyl groups present in the glucuronide unit of SA are capable of interacting with cross-linking agents or other polymers, resulting in the formation of a three-dimensional network structure. Hu et al. [30] employed droplet polymerization, using SA as a substrate, and utilized $AlCl_3/CaCl_2$ solution along with glutaraldehyde (GA) as cross-linking agents to fabricate novel alginate-based gel spheres. The results revealed that the GA (SA-Al/Ca) gel spheres displayed exceptional adsorption performance and impressive cyclic stability.

11.2.6 Schiff Base Reaction

The oxidizing agent can attach two aldehyde groups on the carbon-carbon bond of the aldehyde moiety in the SA molecule. The reactivity of the —COO— group is lower than that of the aldehyde group. This structural alteration promotes the Schiff base cross-linking reaction, which in turn enhances the mechanical properties of the resultant material and introduces additional adsorption sites. Capitalizing on this, Wang et al. [31] synthesized a new chitosan derivative by initiating a cross-linking reaction between the chitosan Schiff base and SA. The resultant adsorbent efficiently captured Cd(II) ions, reaching a maximum adsorption capacity of $262.47\,mg\,g^{-1}$. Additionally, after five cycles, the regenerated adsorbent retained 87.15% of its initial adsorption capacity.

In conclusion, the outstanding properties of alginate-based biocomposites discussed earlier suggest that integrating eco-friendly materials into alginate offers a promising approach to advancing the environmental applications of nanomaterials, particularly in wastewater treatment. Sodium alginate, with its unique structural features and polyanionic nature, can be used to form stable, high-performance adsorbents through physicochemical modification strategies such as blending, graft copolymerization, and cross-linking. These alginate-based biocomposite adsorbents demonstrate outstanding efficiency in adsorbing both organic pollutants and heavy metal ions from aqueous environment, along with notable reusability. Alginate-based materials demonstrate excellent adaptability and hold substantial promise for large-scale production, making them strong contenders for practical applications in wastewater treatment. Continued research and innovation in this area present considerable opportunities for the development of sustainable water purification technologies.

11.3 Application of Alginate Biocomposites in Wastewater Treatment

Recently, there has been extensive attention to using modified materials to remediate wastewater [32–34]. Alginate, including sodium alginate (SA) and calcium alginate (CA), stands out as an exemplary material due to its numerous benefits, such as environmental sustainability, excellent biodegradability, widespread availability, and remarkable stability in various solvents [35]. In contrast to other treatment methods like photocatalysis, electrocatalysis [36], flocculation, coagulation, advanced oxidation, Fenton oxidation, and biodegradation [37], modified alginate-based composites have the ability to adsorb pollutants, such as dyes, heavy metals, and antibiotics, from wastewater, while avoiding the formation of harmful by-products [38]. This composite material is considered to cause minimal or no secondary pollution and has been widely applied in treating wastewater [39].

11.3.1 Metal Ion Removal

Wastewater discharges have always been accompanied by heavy metal pollution problems as industry develops [40–42]. The risks posed by heavy metals in wastewater are not only a matter of environmental concern but also have significant implications for human health and safety. Specific examples of such pollutants include chromium (Cr) [43], copper (Cu) [44], silver (Ag), arsenic (As) [45], and lead (Pb) pollution [46–49]. The widely used treatment methods include precipitation, electrochemical processes, biological removal, and adsorption techniques [50, 51]. However, these approaches often come with various limitations, including the potential for secondary contamination, high operational costs in electrochemical methods, and the complexity of biological removal processes. In contrast, adsorption methods can overcome these problems and achieve fast, effective, repeatable, and simple operation [52].

Various adsorbents, including carbonaceous materials, composite membranes, and biocomposite beads have been reported [20]. In particular, spherical biocomposite microbeads have the advantage of being easy to fill and recycle, making them suitable for industrial applications. During the production of microbeads, it is essential to choose an appropriate cross-linking agent to facilitate the cross-linking of alginate, thereby improving the structural integrity and strength of the microbeads. Among them, calcium chloride is the widely used cross-linking agent, which can be used to obtain alginate microbeads by ion-exchanging with sodium ions of the polymer [53, 54]. However, the SA beads have low removal ability for metal ions. For example, the adsorption for Cr(VI) was reported to be 24.2 $mg\,g^{-1}$ [55]. Thus, the suboptimal adsorption performance can be improved by integrating other materials rich in specific functional groups through chelation. Additionally, further investigation into bead materials with rapid adsorption kinetics and high adsorption capacities holds significant practical potential [56].

An et al. designed chemically altered alginate beads aimed at the removal of heavy metals, specifically chromium (Cr), by incorporating polyethyleneimine (PEI) as a modifying agent. This modification successfully generates a high concentration of active sites on the surface via cross-linking reactions, resulting in the formation of alginate@PEI adsorbents capable of efficiently adsorbing Cr(VI) from wastewater. The process for preparing these adsorbents is straightforward and easily scalable, making it suitable for mass production. The adsorbent exhibited a notably superior adsorption capacity in comparison to numerous other analogous materials, achieving a maximum Cr(VI) adsorption capacity of 431.6 mg g^{-1} under optimal experimental conditions. In addition, it exhibits stability, reusability, and ease of separation from wastewater, thereby facilitating its potential for widespread application (Figure 11.3a) [55].

Liang et al. developed a composite material by integrating sodium alginate with tetramethyl phosphonium sulfate (THPS) to introduce -HPO functional groups, which enhance the material's ability to adsorb chromium (Cr). This modification led to a substantial improvement in the adsorption efficiency compared to conventional alginate salts. Their research demonstrated that the composite effectively removed Cr ions from wastewater. The maximum adsorption capacity of SAB (sodium alginate beads) modified with THPS was 145.77 mg g^{-1}, significantly surpassing the capacity of SAB alone, which was only 21.51 mg g^{-1}. Notably, SAB-THPS demonstrated an exceptionally rapid adsorption rate when the initial Cr(VI) concentration was below 100 mg L^{-1}, achieving removal efficiencies exceeding 99.7% within just 20 minutes. The Cr(VI) removal process by SAB-THPS operates through three distinct mechanisms of adsorption. Primarily, the majority of the adsorption is attributed to the chemical reduction of Cr(VI) to Cr(III), while a smaller portion of the adsorption occurs via ion exchange and electrostatic interactions with —OH and —COOH functional groups. These combined mechanisms give SAB-THPS excellent adsorption performance (Figure 11.3b). SAB-THPS is a sustainable and simple-to-manufacture material, with its bead-like structure effectively addressing the issue of secondary pollution commonly associated with the challenging recycling of conventional powders. This attribute makes it a highly promising option for efficient and rapid wastewater treatment [57].

Germanos et al. synthesized calcium alginate-derived adsorbents (MABs and ABs), incorporating both magnetic and nonmagnetic properties, to investigate the impact of Fe_3O_4 nanoparticles (NPs) on their structural characteristics, adsorption kinetics, and Cu(II) uptake mechanisms. Characterization studies demonstrated

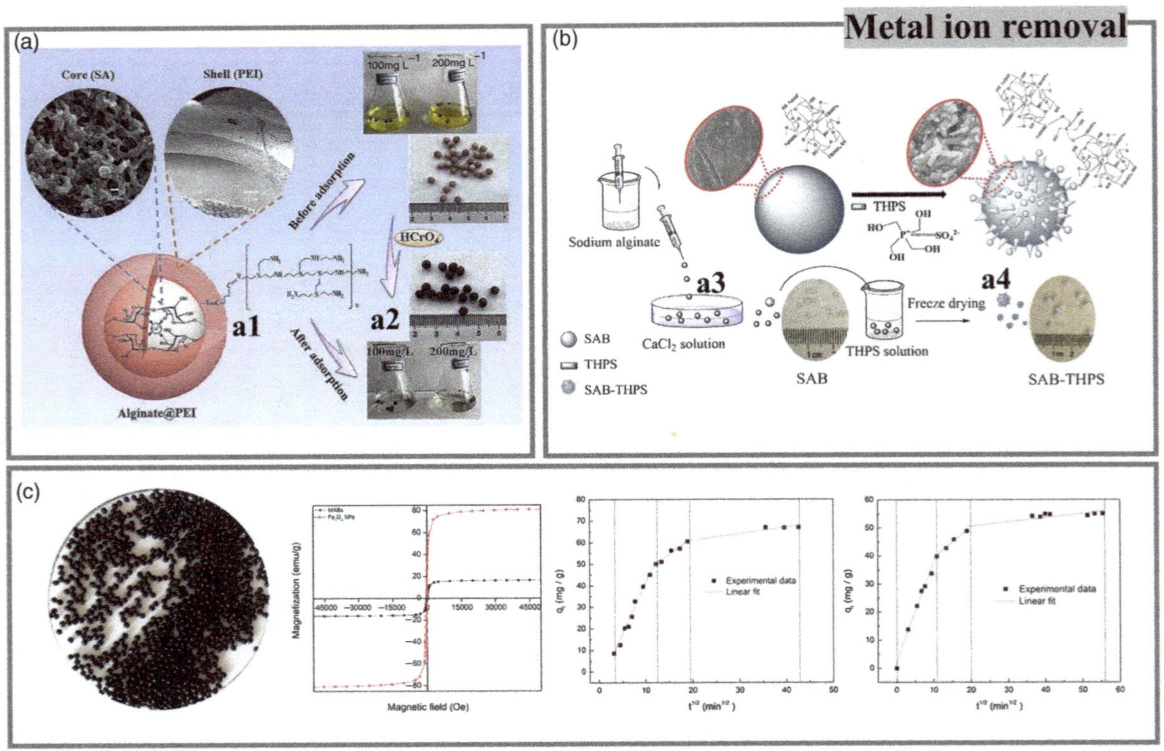

Figure 11.3 Removal of metal ions using (a) Alginate@PEI. *Source:* Reproduced from ref. [55] copyright (2017), with permission from Chemical Engineering Journal. (b) SAB-THPS. *Source:* Reproduced from ref. [57] copyright (2023), with permission from Journal of Environmental Chemical Engineering. (c) MBAs. *Source:* Reproduced from ref. [58] copyright (2020), with permission from Journal of Environmental Chemical Engineering.

that the composite material exhibited a rough surface with numerous protrusions, thereby providing additional active sites for enhanced adsorption. The MABs demonstrated greater thermal stability compared to the ABs. Copper ions adsorption experiments revealed that the MABs and ABs attained adsorption capacities of 54.9 and 66.7 $mg\,g^{-1}$, respectively. The adsorption kinetics conformed to a pseudo-second-order model, with diffusion being a significant factor in regulating the adsorption process (Figure 11.3c) [58].

Chitosan, a polysaccharide derived from crustacean shells, contains amino and hydroxyl groups, which enable it to efficiently adsorb heavy metal ions [59, 60]. Sodium alginate, another polysaccharide obtained from brown algae, has carbonyl groups that can interact with metal ions, facilitating their effective removal from wastewater [61]. Although chitosan offers several advantageous properties, its application is constrained by its low crystallinity, limited surface area, and instability in acidic conditions. In contrast, alginates are characterized by low diffusion rates, restricted porosity, and relatively slow adsorption kinetics [62]. Therefore, chemical modification of these biopolymers to enhance their adsorption performance is crucial. In recent years, a considerable volume of review literature has emphasized that the inclusion of these two biopolymers, or their integration with organic or inorganic precursors, significantly enhances their capacity to adsorb heavy metals. This development seeks to further refine the application of biocomposite materials for the efficient removal of metal ions from wastewater.

Guerrero et al. pioneered a novel composite material by blending chitosan with sodium alginate, with the goal of improving the sorption efficiency of the combined material relative to the use of chitosan or sodium alginate alone as adsorbents. They also optimized various synthesis parameters, including the regeneration method, drying conditions, contact duration, sorbent particle size, and other reaction variables. The materials produced exhibited particle sizes spanning from 150 to 210 μm, 210 to 420 μm, 420 to 840 μm, and above 840 μm. According to their findings, the composite material, synthesized via a one-pot regeneration method followed by freeze-drying, demonstrated the highest adsorption efficiency. Specifically, the CSAR3L sample displayed an adsorption capacity of 288 milligrams of Cu(II) per gram of the composite. The interaction between the components enhanced the composite material's ability to adsorb copper ions from wastewater, leading to improved removal rates. The use of different drying methods results in the development of a crystalline calcium carbonate phase within the material, thereby generating numerous adsorption-active sites. After the adsorption process is completed, a hydroxylated copper sulfate phase is observed, accompanied by a significant decrease in calcium concentration, indicating the occurrence of an ion exchange adsorption mechanism. The observed adsorption kinetics and the characteristics of the adsorption isotherms support this interpretation, suggesting the existence of multiple active sites and the formation of chemisorbed monolayers (Figure 11.4a). In summary, this research highlights the interaction of promoters within the biocomposite material, correlating adsorption rates with its physical and chemical characteristics, and proposes a surface reaction mechanism to explain the enhancement of its performance [63].

Gutierrez et al. developed a range of biocomposite structures and surface characteristics of CSA using various drying techniques, including freeze-drying, methanol with vacuum drying, and a combination of freeze-drying, vacuum drying, and methanol treatments. Furthermore, they synthesized calcium carbonate (CSA-Ca) in situ to facilitate the removal of Cu^{2+} or Pb^{2+} ions from wastewater. Their findings reveal that this advanced composite material offers rapid adsorption and favorable diffusion properties, outperforming individual materials in terms of efficiency. The composite demonstrated an efficient removal of more than 90% of metal ions from wastewater containing concentrations between 10 and 100 mg L^{-1}. Kinetic adsorption analysis showed that the material was capable of adsorbing up to 429 mg L^{-1} of Cu^{2+} and 1742 mg L^{-1} of Pb^{2+}. The morphology of vanadate and calcite within $CaCO_3$, as well as the different drying conditions, significantly influenced both the material's structure and its adsorption capabilities. Moreover, the ability to regenerate and reuse these adsorbents plays a crucial role in mitigating environmental issues while also ensuring their practical and economic feasibility. The research indicated that a 0.10 M EDTA solution was effective in regenerating the material, with removal efficiencies for Cu^{2+} and Pb^{2+} maintained at 75% and 62%, respectively, after three cycles of use (Figure 11.4b). All in all, chitosan and sodium alginate both offer economic, environmental, low-cost, and biodegradable advantages, making them ideal for removing pollutants from wastewater [64].

As mentioned earlier, traditional adsorbents such as chitosan and alginates suffer from drawbacks such as aggregation in heavy metal ion adsorption, difficulty in removal from water, and low utilization rates [66]. Electrospinning is an effective manufacturing technique for producing micro-nano fibers. Fabricating materials with fiber network architectures offers a broad array of adsorption sites and efficient transport pathways, which could facilitate the rapid adsorption of alginates in a short period [20]. Moreover, in contrast to conventional adsorbents, the intertwined fiber network structure helps alleviate volume fluctuations during the adsorption process. In this context, Pan et al. engineered a robust, strengthened alginate composite fiber through electrospinning, integrating graphene oxide as a reinforcement material, with the objective of removing heavy metal ions from wastewater.

The fabricated circular nanofibers exhibited a fairly uniform distribution, with an average diameter of 400 nm. The interwoven fibers created a porous network that promoted effective transfer of heavy metal ions. This material demonstrates a

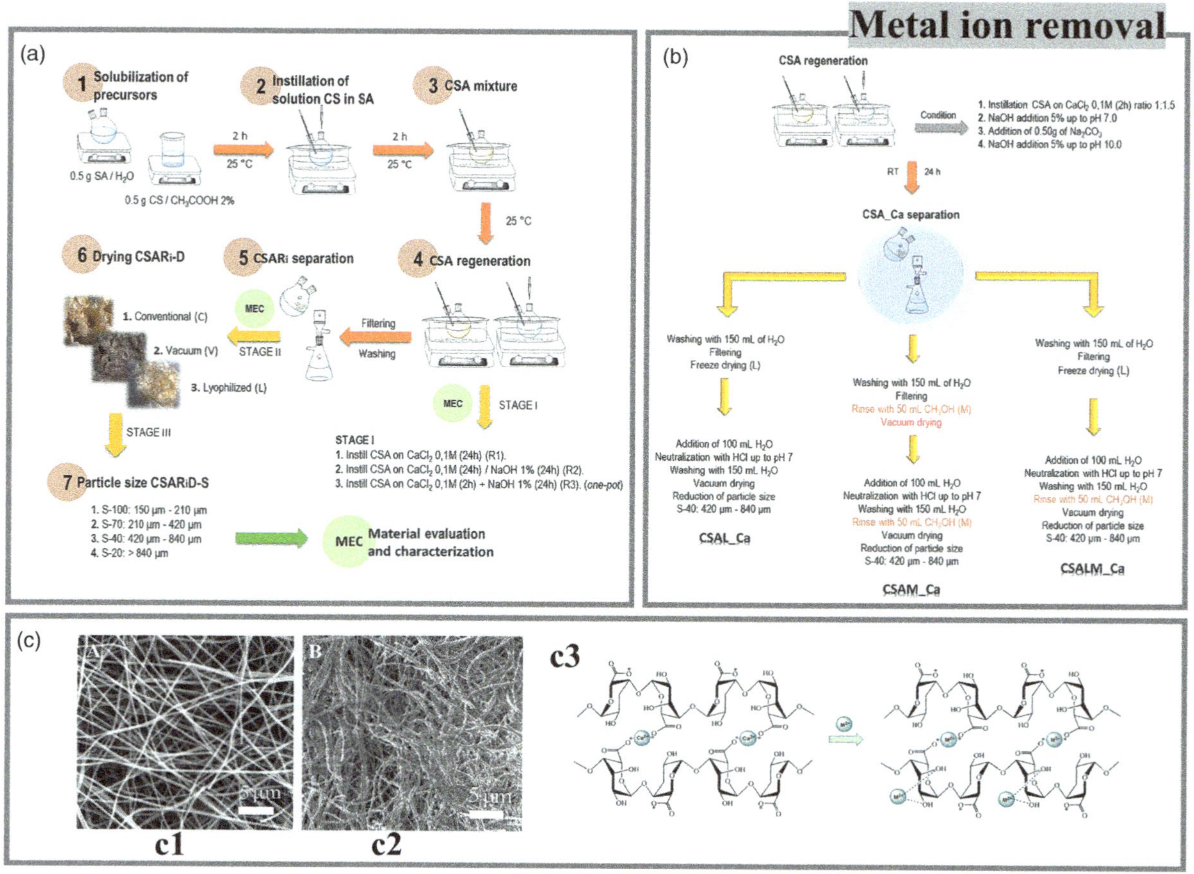

Figure 11.4 Removal of metal ions using (a) chitosan/alginate. *Source:* Reproduced from ref. [63] copyright (2023), with permission from International Journal of Biological Macromolecules. (b) CSA-Ca. *Source:* Reproduced from ref. [64] copyright (2024), with permission from International Journal of Biological Macromolecules. (c) Alginate/graphene oxide. *Source:* Reproduced from ref. [65] copyright (2019), with permission from International Journal of Biological Macromolecules.

notable attraction to Cu^{2+} and Pb^{2+} ions, with its adsorption capacity enhancing as the initial concentration of metal ions in the solution increases. The highest adsorption capacities for Pb^{2+} and Cu^{2+} ions are 386.2 and 102.4 mg g^{-1}, respectively, which substantially outperform the adsorption capabilities of traditional adsorbents found in existing studies. In competitive adsorption tests with Cd^{2+} and Cu^{2+}, the material exhibited a remarkable ability to selectively separate Pb^{2+}, effectively removing 97.2% of Pb^{2+}, 78.6% of Cu^{2+}, and 58.6% of Cd^{2+} from individual metal ion solutions. However, when exposed to mixed Pb^{2+}/Cu^{2+} solutions, the material absorbed 83.6% of Pb^{2+} and 27.4% of Cu^{2+}.

Applying this material to remove Pb^{2+} from wastewater samples from a battery factory (containing 0.08 mM Pb^{2+} and 0.05 mM Cd^{2+}) also yields satisfactory results. An analysis of the adsorption mechanism indicates that the fibers predominantly remove heavy metals via ion exchange and chemical coordination interactions. Owing to the remarkable mechanical strength of graphene oxide, the alginate-based composite fibers demonstrate negligible degradation of their properties during repeated use (Figure 11.4c). In summary, the combination of electrospinning methods with the intrinsic performance advantages considerably enhances the practical effectiveness of alginate composite fibers for wastewater treatment [65].

11.3.2 Dye Removal

Dye-containing waste represents a significant source of water pollution [67]. Dyes are generally organic molecules composed mainly of two essential elements: chromophores (such as —NR, —NHR, —NHCOOH, and —OH groups) and auxochromes (including —N, —NO, and —NO_2 groups). Chromophores are responsible for imparting color to the dyes, serving as the fundamental elements of their color properties [68, 69]. These dyes are extensively utilized across various industries, including plastics, paper production, carpet manufacturing, food processing, printing, cosmetics, and textiles [70].

Globally, the textile industry utilizes around 10,000 tons of dyes each year, with approximately 10–15% of these dyes being discharged into wastewater during the dyeing process [71–73]. The dyes found in wastewater are regarded as hazardous and toxic [74]. Recent studies have highlighted the genetic toxicity and mutagenic effects observed in test organisms exposed to dyes and dye-contaminated effluents [75]. Since the 20th century, chemists have synthesized a wide variety of dyes, which, while providing vibrant colors, contribute significantly to environmental pollution [76].

Alginates, primarily extracted from brown algae, are abundant in all oceans worldwide [15, 77]. They are effective bioadsorbents for removing pollutants from wastewater, such as methyl blue (MB), crystal violet (CV), malachite green (MG), methyl orange (MO), etc. However, their ability to treat dyes needs significant improvement. Based on this, combining alginate with other materials to prepare biocomposites has become a research hotspot.

Sui et al. addressed the challenge of dye removal by using a wet spinning method to composite carbon nanotubes (CNTs) with SA as a biological adsorbent. The structural analysis indicated that the CA fibers contained numerous pores, with the incorporated CNTs well-dispersed within the CA matrix, potentially functioning as microchannels. This configuration likely played a role in enhancing both the surface area and pore volume of the sodium alginate. The composite was then employed to eliminate MB and MO dyes from wastewater. This research explored the dye adsorption performance of the CA/multi-walled carbon nanotube (MWCNT) composite fibers, taking into account the effects of varying MWCNT concentrations and the pH values of the wastewater.

The findings indicated that, in comparison to pristine CA fibers, the addition of CA/MWCNTs composite fibers resulted in a threefold improvement in the adsorption capacity for MO and also accelerated the adsorption rate for MB. Kinetic analysis indicated that the adsorption of both MB and MO onto CA/MWCNT fibers followed a second-order model. The adsorption capacities (q values) were determined to be 190.5 $mg\,g^{-1}$ for MB and 11.4 $mg\,g^{-1}$ for MO, respectively. The highest adsorption capacities observed for MB and MO were 606.1 and 12.5 $mg\,g^{-1}$, respectively. When compared to other adsorbents, the CA/MWCNT fibers demonstrated superior adsorption performance for both MB and MO. For instance, alginate magnetic beads (AC-MAB) had q values of 22.060 $mg\,g^{-1}$ for MB and 0.655 $mg\,g^{-1}$ for MO, while epoxy cross-linked alginate magnetic beads (EpiMAB) showed the q values of 261.73 $mg\,g^{-1}$ for MB and 6.55 $mg\,g^{-1}$ for MO (Rocher, Bee, Siaugue, & Cabuil, 2010). Additionally, GGMWCNT-FeO exhibited a q value of 61.92 $mg\,g^{-1}$ for MB adsorption. Furthermore, desorption experiments indicated desorption rates of 79.7% for MB and 80.2% for MO. These comparisons demonstrate the enormous potential of CA/MWCNTs fibers as dye adsorbents in wastewater treatment (Figure 11.5a) [78].

In recent decades, grass ash, agricultural and industrial by-products, as well as activated carbon, have been employed to eliminate contaminants from wastewater [81]. Nevertheless, these materials have not gained widespread attention due to their limited adsorption capacity, high costs, and challenges associated with regeneration. In contrast, clays offer a more effective, cost-efficient, and environmentally friendly solution for addressing pollution-related issues [82], composites have higher stability and show enhanced physicochemical properties [83]. Abida Kausar et al. introduced a cost-effective adsorbent by combining sodium alginate with acid-treated clay, utilizing both local raw white clay and modified clay as adsorbent materials for the removal of Blue FBN (BFBN) and Rose FRN (RFRN) dyes from wastewater.

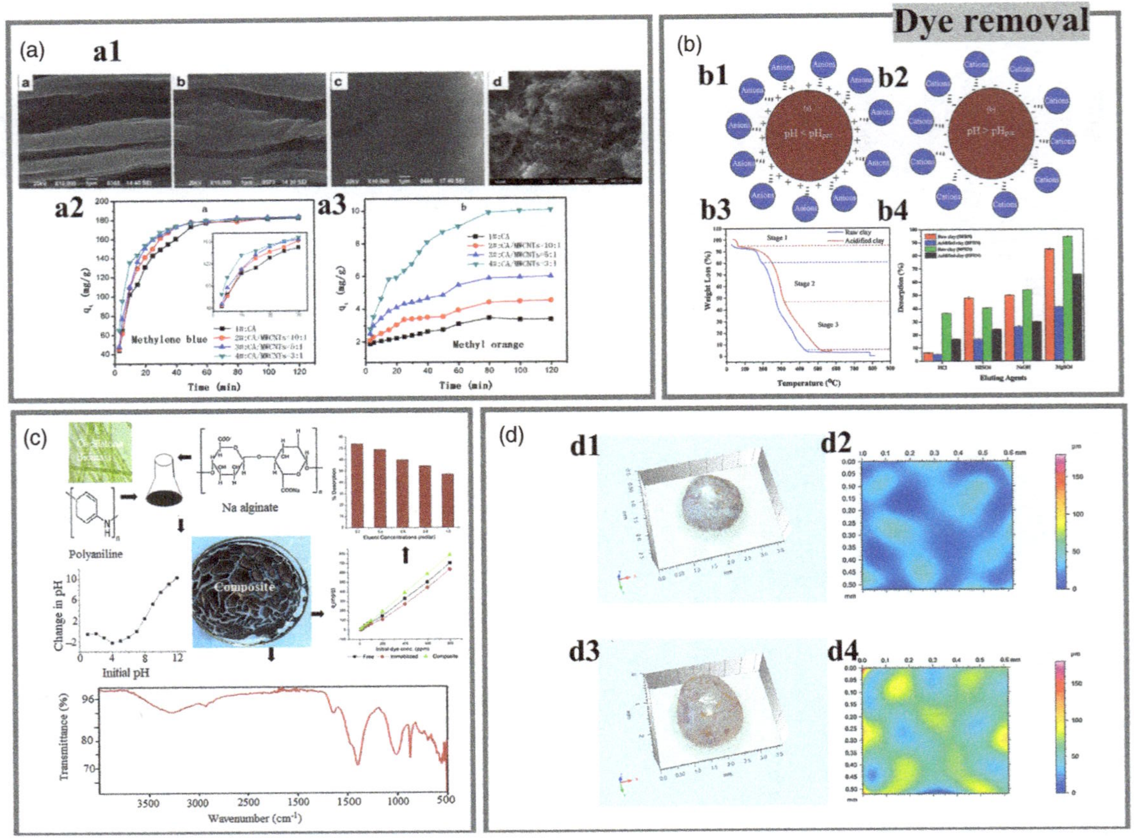

Figure 11.5 Removal of dye using (a) CA/MWCNT. *Source:* Reproduced from ref. [78] copyright (2012), with permission from Carbohydrate Polymers. (b) Alginate/clay. *Source:* Reproduced from ref. [76] copyright (2020), with permission from International Journal of Biological Macromolecules. (c) Na-Alg-ODOB. *Source:* Reproduced from ref. [79] copyright (2020), with permission from Journal of Materials Research and Technology. (d) Lipopeptide biosurfactants/lignocellulosic biocomposites. *Source:* Reproduced from ref. [80] copyright (2015), with permission from Carbohydrate Polymers.

The findings reveal that acidified clay outperforms raw clay in terms of adsorption efficiency, demonstrating effective interactions between the adsorbent and dye ions. Experimental results for two dyes, BFBN and RFRN, revealed that the Freundlich adsorption isotherm provided the best fit for the data. Additionally, the findings suggest that the adsorption of both dyes achieved equilibrium within 80 minutes and followed the pseudo-second-order kinetic model. Under conditions of pH 2.0 and a temperature of 30 °C, the maximum adsorption efficiencies were 76.39% for BFBN and 59.85% for RFRN. The composite material exhibited stability at elevated temperatures and could be regenerated using $MgSO_4$ wash solution. When raw and modified clay were used for the removal of textile effluent color, the removal rates were 50.35% and 54.95%, respectively. The modified clay exhibited excellent performance in adsorbing synthetic BFBN and RFRN dyes from aqueous solutions, positioning it as a highly potential candidate for addressing the treatment of industrial and textile wastewater (Figure 11.5b) [76].

Bhatti et al. developed polyaniline (PANI) and sodium alginate (Na-Alg) biocomposites incorporating Oscillatoria (OS) biomass, employing them for the removal of Basic Blue (BB) dye from wastewater. The results indicated that an increase in the adsorbent amount led to a gradual reduction in adsorption capacity. Among them, the Na-Alg-ODOB biocomposite material had a maximum absorption rate of 634.4 $mg\,g^{-1}$ for BB, reaching adsorption equilibrium within 120 minutes. This could be attributed to enhanced particle interactions leading to aggregation phenomena. The Freundlich model indicated that the adsorption mechanism was primarily physical, while the pseudo-second-order kinetic model closely matched the adsorption data for BB41 by this composite, highlighting its exothermic characteristics. Furthermore, physicochemical analysis of the treated wastewater revealed substantial reductions in chemical oxygen demand (COD), total phosphorus (TP), and total nitrogen (TN), achieving decreases of 93.7%, 80.8%, and 85.5%, respectively (Figure 11.5c). This composite material significantly improved water quality parameters, demonstrating its feasibility and efficiency in wastewater treatment [79].

The application of lipopeptide biosurfactants to lignocellulosic biocomposites to enhance wastewater treatment has been recognized by researchers. Moldes et al. employed a lipopeptide biosurfactant sourced from corn steep liquor to treat lignocellulosic biocomposites, which were encapsulated within calcium alginate beads. This strategy aimed to enhance the effectiveness of removing colored dyes from wastewater. The synthesized biocomposite material exhibited good stability, maintaining stability up to 200 °C, and the incorporation of the biosurfactant significantly reduced its weight loss around 400 °C, making the material more stable and uniform. An analysis of the wastewater color following treatment with various adsorption materials indicated that the biocomposite formulated with biosurfactant resulted in a dye removal efficiency of around 76%, reducing the color index to 74% of its original value. In contrast, the biocomposite without biosurfactant achieved a dye removal rate of approximately 70%, with the color index decreasing to about 69% of its initial value (Figure 11.5d). These results indicate that combining biosurfactants with alginate greatly improves their removal capacity and enhances their application value. Morphological analysis indicates that the enhanced performance is likely due to the denser, more rounded, and rougher structure of the material synthesized with biosurfactant incorporation, which offers an increased number of active sites [80].

11.3.3 Other Applications

As the usage of different antibiotics continues to increase, large amounts of wastewater containing compounds such as tetracycline and ciprofloxacin (CPX) are being discharged into the environment [84]. Ciprofloxacin, a widely utilized antibiotic for the treatment of infections, is particularly challenging to biodegrade and poses serious toxic effects on humans, including leukopenia, vomiting, and oral ulcers. The high concentration of CPX in aquatic environments can also facilitate the growth of antibiotic-resistant bacteria, which in turn threatens the quality of water. Consequently, it is essential to remove CPX from wastewater to acceptable levels prior to its discharge into sewage systems. Among the different adsorbents utilized for CPX removal, such as activated carbon, kaolin, chitosan-metal particles, and coal fly ash, alginate has attracted significant interest owing to its environmentally friendly nature, water affinity, biocompatibility, and a range of other advantageous properties.

Wu et al. introduced an innovative biocomposite fiber by integrating graphene oxide (GO) into CA using a wet spinning technique. The findings demonstrated a significant enhancement in the adsorption capacity of this composite relative to pure CA, with the maximum adsorption capacity rising from 4.4 to 11.8 $mg\,g^{-1}$. This improvement is likely due to the enhanced surface complexation and ion exchange interactions between the CPX molecules and the GO/CA fibers, which stem from the functional groups and microchannels introduced by GO within the fiber structure. The Langmuir model predicted a maximum adsorption capacity of 66.25 $mg\,g^{-1}$, which is lower compared to carbon nanotubes (135 $mg\,g^{-1}$). Nonetheless, the GO/CA fiber offer the advantage of easy separation from the aqueous solution postadsorption, thereby minimizing the risk of secondary contamination that might occur due to the presence of dispersed carbon nanotubes. The wet spinning method was utilized to prepare this novel biocomposite fiber, and the incorporation of GO notably improved the adsorption capacity of ciprofloxacin onto GO/CA fibers (Figure 11.6a) [85].

Muhammad Amir Zeb and colleagues [86] developed a biocomposite material by combining magnetized graphene oxide with the natural polymer sodium

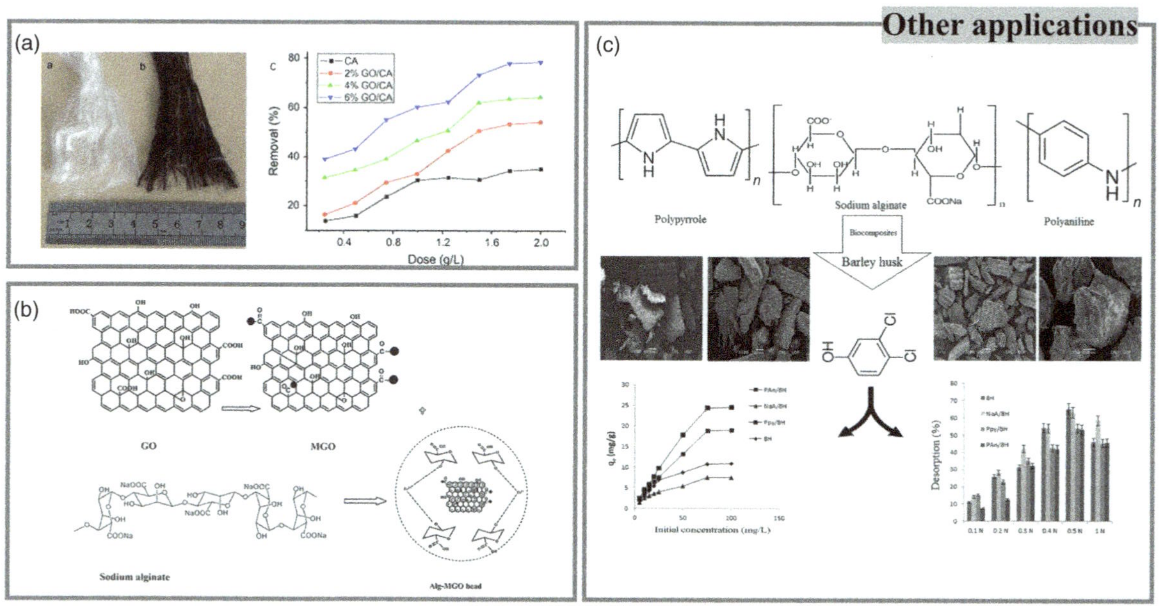

Figure 11.6 Removal of (a) ciprofloxacin. *Source:* Reproduced from ref. [85] copyright (2013), with permission from Chemical Engineering Journal. (b) Aromatic amines. *Source:* Reproduced from ref. [86] copyright (2023), with permission from Environmental Science and Pollution Research. (c) 2,4-Dichlorophenol using alginate biocomposite. *Source:* Reproduced from ref. [87] copyright (2020), with permission from International Journal of Biological Macromolecules.

alginate. This material was utilized for the removal of aromatic amines, such as aniline, para-chloroaniline (PCA), and para-nitroaniline (PNA) from aqueous solutions. This composite effectively mitigates the aggregation of graphene oxide, thereby enhancing its adsorption capacity and facilitating its subsequent separation from the liquid phase. The findings demonstrate that the functional groups present in both graphene oxide and sodium alginate are preserved in the magnetic biocomposite. When tested on water samples, the material efficiently adsorbed aniline, PCA, and PNA. At room temperature and with a solution pH of 4, the maximum adsorption capacities were 18.39 mg g^{-1} for aniline, 17.13 mg g^{-1} for PCA, and 15.24 mg g^{-1} for PNA.

At the same time, it exhibits effective separation capabilities for different real water samples. In various samples, the extraction efficiency of aniline ranges from 93.17% to 98.82%, PCA from 90.22% to 96.65%, and PNA from 89.82% to 93.55% (Figure 11.6b). Additionally, regeneration of the composite material can be effectively achieved using ethanol, demonstrating excellent reusability. This indicates that environmentally friendly alginate-magnetic graphene oxide biocomposite materials can serve as useful adsorbents for removing organic pollutants such as aromatic amines in water treatment processes without causing secondary pollution. The advantages of alginate-magnetic graphene oxide lie in its ease of magnetic separation and simple magnetic sorting, making the composite beads highly promising for applications in wastewater treatment [86].

Phenolic compounds are important chemical raw materials used in the industry for producing insecticides, explosives, dyes, and more. However, they are also toxic substances that can coagulate proteins and have high toxicity, making them common highly toxic and difficult-to-degrade organic compounds in industrial wastewater. Industrial wastewater often contains high concentrations of phenolic compounds. With strict policies in various countries to control these compounds, while maximizing economic benefits, finding a method to significantly reduce the concentration of phenolic compounds in wastewater and potentially recover them has become an important challenge facing the chemical industry.

Bhatti et al. prepared a biocomposite material by integrating Na-Alg with cellulose biomass extracted from barley husk (BH), demonstrating its effectiveness in removing 2,4-dichlorophenol (2,4-DCP) from aqueous solutions. The kinetics and thermodynamic properties of 2,4-DCP removal were investigated. The composite material has a simple synthesis method, is easy to operate and apply, and shows promising removal rates with substantial adsorption capacity. It has broad applicability in wastewater containing varying levels of phenol, with the adsorption capacity increasing progressively as the phenol concentration rises. However, when the phenol concentration exceeds 100 mg L^{-1}, the adsorption process stabilizes, reaching an equilibrium state. This can likely be attributed to the concentration gradient between the solid and liquid phases at higher concentrations, which creates a driving force but also increases the mass transfer resistance between the two phases. In addition, the presence of salts in the aqueous solution has a negligible impact on the adsorption of phenolic compounds. When HCl is used as the eluting agent (ranging from 0.1 to 1 N), the desorption efficiency of phenol can surpass 60% (Figure 11.6c). The findings from these experiments suggest that the sodium alginate-based biocomposite material shows potential for effectively removing 2,4-DCP from industrial wastewater and could also be useful in eliminating other similar pollutants [87].

Algae can absorb nutrients like nitrogen and phosphorus as they grow, making them useful for nutrient removal in wastewater treatment and enabling energy and nutrient recovery [88, 89]. In contrast to alternative approaches such as chemical treatments, utilizing algae eliminates the need for significant chemical inputs and the production of sludge. Researchers have immobilized algae capable of removing nutrients (nitrogen and phosphorus) from wastewater into alginate beads, achieving effective wastewater treatment using cost-effective and safe food-grade alginates. However, considering that algae cells are small, have low concentrations, and are dispersed throughout the water, this method shows limitations such as slow nutrient removal and difficulties in separating algae from treated wastewater [90, 91]. Therefore, immobilizing algae becomes crucial.

Studies have indicated that the immobilization process generally entails the incorporation of algae cells into a sodium alginate solution, which is subsequently introduced into a solution containing multivalent cations, typically calcium (Ca). Since calcium has favorable characteristics for bead formation and has minimal toxicity, it serves as a cross-linking agent, binding the alginate polymer to create solid beads, making this method an efficient approach. However, immobilization can lead to increased production costs, highlighting the need for careful selection of alginates in bead production to improve the cost-effectiveness of algae-based wastewater treatment. Roddick et al. investigated various alginate products for immobilizing Chlorella vulgaris, evaluating factors such as nutrient removal efficiency, biomass productivity, bead stability, and production costs, and compared food-grade, low viscosity, and medium viscosity alginates for their suitability [92].

The experimental findings indicated that low-viscosity alginates (47 $mg\,L^{-1}\,d^{-1}$) resulted in higher algae biomass productivity than medium-viscosity (37 $mg\,L^{-1}\,d^{-1}$) and food-grade (36 $mg\,L^{-1}\,d^{-1}$) alginates, with nutrient removal exhibiting minimal variation across different alginate product batches. As a result, the selection of alginate could prioritize bead stability and cost efficiency. Beads made from medium-viscosity alginates exhibited the least stability, showing the highest strength degradation and releasing more organic carbon during a 4-week exposure to wastewater than the other two types. This result was probably due to the lower L-guluronate content, which hindered the cross-linking process in Ca-alginate esters. In contrast, the enhanced cross-linking of food-grade alginates correlated with higher porosity (ranging from 96.7% to 96.8%) and improved fixation efficiency (95–97%). Despite the reduced algae growth rates, immobilization within food-grade alginates led to comparable nutrient removal performance and minimal loss in stability when exposed to wastewater. This suggests that opting for cost-effective alginates with a sufficiently high G-content could provide an economical alternative for laboratory-scale applications [92].

11.4 Conclusion and Prospects

11.4.1 Summary of Research Content

As scientific and technological advancements accelerate alongside industrialization, a substantial amount of pollution is generated during production processes, highlighting the urgent need to tackle environmental contamination.

Table 11.1 Common alginate materials used for pollutant removal.

Alginate materials	Pollutants	Removal efficiency	Reference
Alginate@PEI	Cr(VI)	$431.6\,mg\,g^{-1}$	[55]
SAB-THPS	Cr(VI)	$145.77\,mg\,g^{-1}$	[57]
MABs	Cu(II)	$54.9\,mg\,g^{-1}$	[58]
ABs		$66.7\,mg\,g^{-1}$	
Chitosan/Alginate	Cu(II)	$288\,mg\,g^{-1}$	[63]
CSA-Ca	Cu(II)	$429\,mg\,L^{-1}$	[64]
	Pb(II)	$1742\,mg\,L^{-1}$	
Alginate/Graphene oxide	Pb(II)	$386.2\,mg\,g^{-1}$	[65]
	Cu(II)	$102.4\,mg\,g^{-1}$	
Alginate/Clay	BFBN	$25.41\,mg\,g^{-1}$	[76]
	RFRN	$18.96\,mg\,g^{-1}$	
CA/MWCNTs	MB	$606.1\,mg\,g^{-1}$	[78]
	MO	$12.5\,mg\,g^{-1}$	
Na-Alg-ODOB	BB	$634.4\,mg\,g^{-1}$	[79]
GO/CA	Ciprofloxacin	$39.06\,mg\,g^{-1}$	[85]
Biocomposite I	Aniline	$18.39\,mg\,g^{-1}$	[86]
	PCA	$17.13\,mg\,g^{-1}$	
	PNA	$15.24\,mg\,g^{-1}$	
Biocomposite II	2,4-Dichlorophenol	$24.57\,mg\,g^{-1}$	[87]

Presently, various persistent pollutants, including pharmaceuticals, heavy metal ions, and dyes, contaminate water bodies. These pollutants often persist in groundwater or surface soils due to improper usage or inadequate wastewater treatment, posing significant threats to both human health and the ecosystem. With increasingly stringent regulations governing industrial wastewater, treatment prior to environmental discharge has become mandatory. In recent years, a variety of wastewater treatment strategies, including physical, chemical, and biological techniques, have been introduced. Nonetheless, each of these approaches offers distinct advantages while also posing specific challenges. Due to the diverse challenges facing wastewater treatment, it is currently difficult to determine a single treatment method that can be universally applicable. However, undoubtedly, in the context of sustainable development and health issues, implementing green and high-performance wastewater treatment methods is crucial. Alginate biocomposites have many advantages in wastewater treatment, such as high efficiency, environmental friendliness, economy, and sustainability, making them a new type of wastewater treatment material with broad application prospects. This article outlines the development and advantages and disadvantages of alginate biocomposites, which will help provide a more complete theoretical guide for the selection and modification of efficient alginate materials and wastewater treatment processes, thereby promoting innovation in the basic theory and technological processes of wastewater treatment. A summary of common alginate materials used for pollutant removal is presented in Table 11.1.

11.4.2 Future Development Trends of Alginate Biocomposites

11.4.2.1 Technological Innovation and Research Hotspots

a) Material design and structural optimization: By designing porous structures and modifying functional groups, the specific surface area and adsorption activity of alginate biocomposites are enhanced to improve their efficient adsorption and degradation capabilities in wastewater treatment.
b) Biological function enhancement: Introducing specific microorganisms or bioactive substances to enhance the material's biodegradation capability and pollution resistance.
c) Process improvement: Optimizing preparation parameters such as solution concentration, temperature, and reaction time to achieve efficient, low-cost large-scale preparation.
d) **Multi-functional applications:** Developing multi-functional composite materials suitable for different types of wastewater treatment to achieve simultaneous removal or conversion of various pollutants.

e) Intelligent and automated applications: Incorporating cutting-edge information technologies, including the Internet of Things and artificial intelligence, to develop smart monitoring and control systems that improve treatment efficiency and guarantee stable operation.
f) Nanoscale and functionalized composite materials: Combining nanomaterials to prepare nanoscale composite materials, designing composite materials with specific functional groups to enhance adsorption activity and wastewater treatment efficiency.
g) Biological enhancement technology: Using biotechnology to improve the microbial composition or function of materials, such as introducing specific bacterial strains or functional genes.

11.4.2.2 Sustainable Development and Environmental Protection

a) Utilization of renewable resources: Alginate, as a renewable natural resource, has a simple preparation process and does not excessively deplete natural resources.
b) Low energy consumption, low pollution: Using physical adsorption and biological degradation methods to reduce energy consumption and chemical agent input, thereby reducing environmental pollution.
c) Efficient pollutant removal: Alginate-based biocomposites offer a notable specific surface area and a rich abundance of active functional groups, allowing them to efficiently adsorb and break down various pollutants in wastewater, such as organic substances, heavy metals, nutrients, and other contaminants.
d) Consideration of environmental friendliness: Emphasizing environmental friendliness in the design and application process by selecting renewable, biodegradable material components to reduce energy consumption and waste emissions.
e) Utilization of waste resources: Materials after wastewater treatment can be utilized as organic fertilizers, soil conditioners, etc., reducing environmental negative impacts.
f) Resource utilization and circular economy: Researching the resource utilization pathways of materials after wastewater treatment, such as soil conditioners, biomass energy, to promote the circular economy.
g) Environmental friendliness assessment: Conducting in-depth assessments of environmental friendliness, considering the impact on water bodies and soils, and application risks to ensure safe and sustainable applications.

References

1 Zamora-Ledezma, C., Negrete-Bolagay, D., Figueroa, F. et al. (2021). Heavy metal water pollution: a fresh look about hazards, novel and conventional remediation methods. *Environmental Technology & Innovation* 22: 101504.

2 Saleem, M.H., Mfarrej, M.F.B., Khan, K.A., and Alharthy, S.A. (2024). Emerging trends in wastewater treatment: addressing microorganic pollutants and environmental impacts. *Science of the Total Environment* 913: 169755.

3 Noreen, U., Ahmed, Z., Khalid, A. et al. (2019). Water pollution and occupational health hazards caused by the marble industries in district Mardan, Pakistan. *Environmental Technology & Innovation* 16: 100470.

4 Alahabadi, A., Hosseini-Bandegharaei, A., Moussavi, G. et al. (2017). Comparing adsorption properties of NH 4 Cl-modified activated carbon towards chlortetracycline antibiotic with those of commercial activated carbon. *Journal of Molecular Liquids* 232: 367–381.

5 Anipsitakis, G.P. and Dionysiou, D.D. (2004). Dionysiou, dionysios D, radical generation by the interaction of transition metals with common oxidants. *Environmental Science & Technology* 38: 3705–3712.

6 Li, Y., du, Q., Liu, T. et al. (2011). Preparation of activated carbon from enteromorpha prolifera and its use on cationic red X-GRL removal. *Applied Surface Science* 257 (24): 10621–10627.

7 Hameed, B.H. and Ahmad, A.A. (2009). Batch adsorption of methylene blue from aqueous solution by garlic peel, an agricultural waste biomass. *Journal of Hazardous Materials* 164 (2-3): 870–875.

8 Aroguz, A.Z., Gulen, J., and Evers, R.H. (2008). Adsorption of methylene blue from aqueous solution on pyrolyzed petrified sediment. *Bioresource Technology* 99 (6): 1503–1508.

9 Kavitha, D. and Namasivayam, C. (2007). Experimental and kinetic studies on methylene blue adsorption by coir pith carbon. *Bioresource Technology* 98 (1): 14–21.

10 Weng, C.-H. and Pan, Y.-F. (2007). Adsorption of a cationic dye (methylene blue) onto spent activated clay. *Journal of Hazardous Materials* 144 (1-2): 355–362.

11 Yao, Y., Xu, F., Chen, M. et al. (2010). Adsorption behavior of methylene blue on carbon nanotubes. *Bioresource Technology* 101 (9): 3040–3046.

12 Karagoz, S., Tay, T., Ucar, S. et al. (2008). Activated carbons from waste biomass by sulfuric acid activation and their use on methylene blue adsorption. *Bioresource Technology* 99 (14): 6214–6222.

13 Firdaus, R.M., Rosli, N.I.M., Ghanbaja, J. et al. (2019). Enhanced adsorption of methylene blue on chemically modified graphene nanoplatelets thanks to favorable interactions. *Journal of Nanoparticle Research* 21 (12): 257.

14 Draget, K.I. and Taylor, C. (2011). Chemical, physical and biological properties of alginates and their biomedical implications. *Food Hydrocolloids* 25 (2): 251–256.

15 Lee, K.Y. and Mooney, D.J. (2012). Alginate: properties and biomedical applications. *Progress in Polymer Science* 37 (1): 106–126.

16 Kim, T.Y., Jin, H.J., Park, S.S. et al. (2008). Adsorption equilibrium of copper ion and phenol by powdered activated carbon, alginate bead and

alginate-activated carbon bead. *Journal of Industrial and Engineering Chemistry* 14 (6): 714–719.

17 Li, Y., Liu, F., Xia, B. et al. (2010). Removal of copper from aqueous solution by carbon nanotube/calcium alginate composites. *Journal of Hazardous Materials* 177 (1-3): 876–880.

18 Li, Y., du, Q., Liu, T. et al. (2013). Methylene blue adsorption on graphene oxide/calcium alginate composites. *Carbohydrate Polymers* 95 (1): 501–507.

19 Mahmoodi, N.M., Hayati, B., Arami, M., and Bahrami, H. (2011). Preparation, characterization and dye adsorption properties of biocompatible composite (alginate/titania nanoparticle). *Desalination* 275 (1–3): 93–101.

20 Guo, H., Qin, Q., Chang, J.S., and Lee, D.J. (2023). Modified alginate materials for wastewater treatment: application prospects. *Bioresource Technology* 387: 129639.

21 Dai, W.-J., Wu, P., Liu, D. et al. (2020). Adsorption of polycyclic aromatic hydrocarbons from aqueous solution by organic montmorillonite sodium alginate nanocomposites. *Chemosphere* 251: 126074.

22 Song, Y., Li, J., Li, Y. et al. (2019). Enrichment and recovery of low-concentration REEs from water using synthesized eco-friendly sodium alginate/nano-Mg(OH)2 composite beads. *Surfaces and Interfaces* 15: 232–238.

23 Sun, F., Guo, J., Liu, Y., and Yu, Y. (2019). Preparation, characterizations and properties of sodium alginate grafted acrylonitrile/polyethylene glycol electrospun nanofibers. *International Journal of Biological Macromolecules* 137: 420–425.

24 Gutierrez-Gonzalez, J., Garcia-Cela, E., Magan, N., and Rahatekar, S.S. (2020). Electrospinning alginate/polyethylene oxide and curcumin composite nanofibers. *Materials Letters* 270: 127662.

25 Zhao, X., Wang, X., and Lou, T. (2022). Simultaneous adsorption for cationic and anionic dyes using chitosan/electrospun sodium alginate nanofiber composite sponges. *Carbohydrate Polymers* 276: 118728.

26 Fang, Y., Liu, Q., and Zhu, S. (2021). Selective biosorption mechanism of methylene blue by a novel and reusable sugar beet pulp cellulose/sodium alginate/iron hydroxide composite hydrogel. *International Journal of Biological Macromolecules* 188: 993–1002.

27 Shehzad, H., Zhou, L., Wang, Y. et al. (2019). Effective biosorption of U(VI) from aqueous solution using calcium alginate hydrogel beads grafted with amino-carbamate moieties. *Journal of Radioanalytical and Nuclear Chemistry* 321 (2): 605–615.

28 Zhang, H., Guan, G., Lou, T., and Wang, X. (2023). High performance, cost-effective and ecofriendly flocculant synthesized by grafting carboxymethyl cellulose and alginate with itaconic acid. *International Journal of Biological Macromolecules* 231: 123305.

29 Kim, H., Purev, O., Myung, E. et al. (2022). Removal of methyl red from aqueous solution using polyethyleneimine crosslinked alginate beads with waste foundry dust as a magnetic material. *International Journal of Environmental Research and Public Health* 19 (15): 9030.

30 Hu, Y., Wei, X., Hu, Y. et al. (2020). Facile preparation of sodium alginate-based gel spheres by droplet polymerization method for removal of levofloxacin from aqueous solution. *Chemical Engineering Journal* 392: 123718.

31 Wang, W., Lu, Q., Ren, X. et al. (2019). Synthesis of novel ECH crosslinked chitosan schiff base-sodium alginate for adsorption of Cd(II) ion from aqueous solution. *Desalination and Water Treatment* 145: 169–178.

32 Wang, Q., Xue, Q., and Li, Y. (2024). An eco-friendly construction of superwetting alginate-based aerogels with self-cleaning performance for multifunctional water treatment. *International Journal of Biological Macromolecules* 261: 129766.

33 Quesada, H.B., de Araújo, T.P., Vareschini, D.T. et al. (2020). Chitosan, alginate and other macromolecules as activated carbon immobilizing agents: a review on composite adsorbents for the removal of water contaminants. *International Journal of Biological Macromolecules* 164: 2535–2549.

34 Holkar, C.R., Jadhav, A.J., Pinjari, D.V. et al. (2016). A critical review on textile wastewater treatments: possible approaches. *Journal of Environmental Management* 182: 351–366.

35 Jiang, X., Zhou, B., and Wang, J. (2023). Super-wetting and self-cleaning polyvinyl alcohol/sodium alginate nanofiber membrane decorated with MIL-88A(Fe) for efficient oil/water emulsion separation and dye degradation. *International Journal of Biological Macromolecules* 253: 127205.

36 Nasrollahzadeh, M., Sajjadi, M., Iravani, S., and Varma, R.S. (2021). Green-synthesized nanocatalysts and nanomaterials for water treatment: current challenges and future perspectives. *Journal of Hazardous Materials* 401: 123401.

37 Jacob, J.M., Karthik, C., Saratale, R.G. et al. (2018). Biological approaches to tackle heavy metal pollution: a survey of literature. *Journal of Environmental Management* 217: 56–70.

38 Ortega-Aguirre, S., Díaz-Nava, M.C., Solache-Ríos, M. et al. (2022). Effect of alginate on the removal of yellow 6 by a biopolymer-ferric zeolite composite. *Separation and Purification Technology* 292: 120971.

39 Udayakumar, G.P., Muthusamy, S., Selvaganesh, B. et al. (2021). Ecofriendly biopolymers and composites: preparation and their applications in water-treatment. *Biotechnology Advances* 52: 107815.

40 Beheshti, H., Irani, M., Hosseini, L. et al. (2016). Removal of Cr (VI) from aqueous solutions using chitosan/MWCNT/Fe_3O_4 composite nanofibers-batch and column studies. *Chemical Engineering Journal* 284: 557–564.

41 Park, H.G. and Chae, M.Y. (2004). Novel type of alginate gel-based adsorbents for heavy metal removal. *Journal of Chemical Technology & Biotechnology* 79 (10): 1080–1083.

42 Vareda, J.P., Valente, A.J.M., and Durães, L. (2019). Assessment of heavy metal pollution from anthropogenic activities and remediation strategies: a review. *Journal of Environmental Management* 246: 101–118.

43 Gopalakannan, V., Periyasamy, S., and Viswanathan, N. (2016). Synthesis of assorted metal ions anchored alginate bentonite biocomposites for Cr(VI) sorption. *Carbohydrate Polymers* 151: 1100–1109.

44 Benali, F., Boukoussa, B., Issam, I. et al. (2023). Assessment of AgNPs@Cu@ alginate composite for efficient water treatment: effect of the content of Cu(II) crosslinking agent. *Journal of Polymers and the Environment* 31 (10): 4170–4183.

45 Shim, J., Kumar, M., Mukherjee, S., and Goswami, R. (2019). Sustainable removal of pernicious arsenic and cadmium by a novel composite of MnO_2 impregnated alginate beads: a cost-effective approach for wastewater treatment. *Journal of Environmental Management* 234: 8–20.

46 Germanos, G., Youssef, S., Abboud, M. et al. (2017). Diffusion and agglomeration of iron oxide nanoparticles in magnetic calcium alginate beads initiated by copper sorption. *Journal of Environmental Chemical Engineering* 5 (4): 3727–3733.

47 Idris, A., Ismail, N.S.M., Hassan, N. et al. (2012). Synthesis of magnetic alginate beads based on maghemite nanoparticles for Pb(II) removal in aqueous solution. *Journal of Industrial and Engineering Chemistry* 18 (5): 1582–1589.

48 Zhu, H., Fu, Y., Jiang, R. et al. (2014). Optimization of copper(II) adsorption onto novel magnetic calcium alginate/maghemite hydrogel beads using response surface methodology. *Industrial & Engineering Chemistry Research* 53 (10): 4059–4066.

49 Nitayaphat, W. and Jintakosol, T. (2015). Removal of silver(I) from aqueous solutions by chitosan/bamboo charcoal composite beads. *Journal of Cleaner Production* 87: 850–855.

50 GracePavithra, K., Jaikumar, V., Kumar, P.S., and SundarRajan, P.S. (2019). A review on cleaner strategies for chromium industrial wastewater: present research and future perspective. *Journal of Cleaner Production* 228: 580–593.

51 Zeraatkar, A.K., Ahmadzadeh, H., Talebi, A.F. et al. (2016). Potential use of algae for heavy metal bioremediation, a critical review. *Journal of Environmental Management* 181: 817–831.

52 Qu, J., Wei, S., Liu, Y. et al. (2022). Effective lead passivation in soil by bone char/CMC-stabilized FeS composite loading with phosphate-solubilizing bacteria. *Journal of Hazardous Materials* 423: 127043.

53 Al-Othman, Z.A., Ali, R., and Naushad, M. (2012). Hexavalent chromium removal from aqueous medium by activated carbon prepared from peanut shell: adsorption kinetics, equilibrium and thermodynamic studies. *Chemical Engineering Journal* 184: 238–247.

54 Saha, B. and Orvig, C. (2010). Biosorbents for hexavalent chromium elimination from industrial and municipal effluents. *Coordination Chemistry Reviews* 254 (23–24): 2959–2972.

55 Yan, Y., An, Q., Xiao, Z. et al. (2017). Flexible core-shell/bead-like alginate@ PEI with exceptional adsorption capacity, recycling performance toward batch and column sorption of Cr(VI). *Chemical Engineering Journal* 313: 475–486.

56 Aziz, F., Achaby, M.E., Lissaneddine, A. et al. (2020). Composites with alginate beads: a novel design of nano-adsorbents impregnation for large-scale continuous flow wastewater treatment pilots. *Saudi Journal of Biological Sciences* 27 (10): 2499–2508.

57 Liang, S., Cai, W., Dang, C. et al. (2023). Synthesis of sodium alginate/ phosphorus tetramethylmethyl sulfate biocomposite beads with exceptional adsorption rate for Cr(VI) removal. *Journal of Environmental Chemical Engineering* 11 (2): 109317.

58 Germanos, G., Youssef, S., Farah, W. et al. (2020). The impact of magnetite nanoparticles on the physicochemical and adsorption properties of magnetic alginate beads. *Journal of Environmental Chemical Engineering* 8 (5): 104223.

59 Gokila, S., Gomathi, T., Sudha, P.N., and Anil, S. (2017). Removal of the heavy metal ion chromiuim(VI) using Chitosan and alginate nanocomposites. *International Journal of Biological Macromolecules* 104: 1459–1468.

60 Azzam, E.M.S., Eshaq, G., Rabie, A.M. et al. (2016). Preparation and characterization of chitosan-clay nanocomposites for the removal of Cu(II) from aqueous solution. *International Journal of Biological Macromolecules* 89: 507–517.

61 Nasrollahzadeh, M., Sajjadi, M., Iravani, S., and Varma, R.S. (2021). Starch, cellulose, pectin, gum, alginate, chitin and chitosan derived (nano)materials for sustainable water treatment: a review. *Carbohydrate Polymers* 251: 116986.

62 Yu, Y., Zhao, C., Liu, X. et al. (2017). Selective flocculation of pollutants in wastewater using pH responsive HM-alginate/chitosan complexes. *Journal of Environmental Chemical Engineering* 5 (6): 5406–5410.

63 Guerrero, J.D., Marchesini, F.A., Ulla, M.A., and Gutierrez, L.B. (2023). Effect of biocomposite production factors on the development of an eco-friendly chitosan/alginate-based adsorbent with enhanced copper removal efficiency. *International Journal of Biological Macromolecules* 253: 126416.

64 Guerrero, J.D., Arias, E.R., and Gutierrez, L.B. (2024). Enhancing copper and lead adsorption in water by in-situ generation of calcium carbonate on alginate/chitosan biocomposite surfaces. *International Journal of Biological Macromolecules* 266: 131110.

65 Pan, L., Wang, Z., Zhao, X., and He, H. (2019). Efficient removal of lead and copper ions from water by enhanced strength-toughness alginate composite fibers. *International Journal of Biological Macromolecules* 134: 223–229.

66 Doyo, A.N., Kumar, R., and Barakat, M.A. (2023). Recent advances in cellulose, chitosan, and alginate based biopolymeric composites for adsorption of heavy metals from wastewater. *Journal of the Taiwan Institute of Chemical Engineers* 151: 105095.

67 Akhtar, K., Khan, S.B., Bakhsh, E.M. et al. (2022). Nanoarchitectured Cu based catalysts supported on alginate/glycyl leucine hybrid beads for tainted water treatment. *International Journal of Biological Macromolecules* 208: 56–69.

68 Bakhsh, E.M., Akhtar, K., Fagieh, T.M. et al. (2021). Development of alginate@tin oxide–cobalt oxide nanocomposite based catalyst for the treatment of wastewater. *International Journal of Biological Macromolecules* 187: 386–398.

69 Tkaczyk, A., Mitrowska, K., and Posyniak, A. (2020). Synthetic organic dyes as contaminants of the aquatic environment and their implications for ecosystems: a review. *Science of the Total Environment* 717: 137222.

70 Ohemeng-Boahen, G., Sewu, D.D., and Woo, S.H. (2019). Preparation and characterization of alginate-kelp biochar composite hydrogel bead for dye removal. *Environmental Science and Pollution Research* 26 (32): 33030–33042.

71 Al-Sakkari, E.G., Abdeldayem, O.M., Genina, E.E. et al. (2020). New alginate-based interpenetrating polymer networks for water treatment: a response surface methodology based optimization study. *International Journal of Biological Macromolecules* 155: 772–785.

72 Bhatti, H.N., Jabeen, A., Iqbal, M. et al. (2017). Adsorptive behavior of rice bran-based composites for malachite green dye: isotherm, kinetic and thermodynamic studies. *Journal of Molecular Liquids* 237: 322–333.

73 Kausar, A., Iqbal, M., Javed, A. et al. (2018). Dyes adsorption using clay and modified clay: a review. *Journal of Molecular Liquids* 256: 395–407.

74 Luo, T., Hao, Y., Wang, C. et al. (2022). Lignin nanoparticles and alginate gel beads: preparation, characterization and removal of methylene blue. *Nanomaterials* 12 (1).

75 Abbas, M., Adil, M., Ehtisham-ul-Haque, S. et al. (2018). Vibrio fischeri bioluminescence inhibition assay for ecotoxicity assessment: a review. *Science of the Total Environment* 626: 1295–1309.

76 Kausar, A., Sher, F., Hazafa, A. et al. (2020). Biocomposite of sodium-alginate with acidified clay for wastewater treatment: kinetic, equilibrium and thermodynamic studies. *International Journal of Biological Macromolecules* 161: 1272–1285.

77 Radoor, S., Karayil, J., Jayakumar, A. et al. (2024). Recent advances in cellulose- and alginate-based hydrogels for water and wastewater treatment: a review. *Carbohydrate Polymers* 323: 121339.

78 Sui, K., Li, Y., Liu, R. et al. (2012). Biocomposite fiber of calcium alginate/multi-walled carbon nanotubes with enhanced adsorption properties for ionic dyes. *Carbohydrate Polymers* 90 (1): 399–406.

79 Maqbool, M., Bhatti, H.N., Sadaf, S. et al. (2020). Biocomposite of polyaniline and sodium alginate with Oscillatoria biomass: a potential adsorbent for the removal of basic blue 41. *Journal of Materials Research and Technology* 9 (6): 14729–14741.

80 Perez-Ameneiro, M., Vecino, X., Cruz, J.M., and Moldes, A.B. (2015). Wastewater treatment enhancement by applying a lipopeptide biosurfactant to a lignocellulosic biocomposite. *Carbohydrate Polymers* 131: 186–196.

81 Harshala, K. and Wagh, N.D. (2022). Use of agricultural waste-based biosorbents for the removal of heavy metals from aqueous solution: a review. *Nature, Environment and Pollution Technology* 21 (3): 1003–1014.

82 Al-a'qarbeh, M.M., Shammout, M.W., and Awwad, A.M. (2020). Nano platelets kaolinite for the adsorption of toxic metal ions in the environment. *Chemistry International* 6 (2): 49–55.

83 García, A., Culebras, M., Collins, M.N., and Leahy, J.J. (2018). Stability and rheological study of sodium carboxymethyl cellulose and alginate suspensions as binders for lithium ion batteries. *Journal of Applied Polymer Science* 135 (17).

84 Kong, Y., Zhuang, Y., and Shi, B. (2020). Tetracycline removal by double-metal-crosslinked alginate/graphene hydrogels through an enhanced Fenton reaction. *Journal of Hazardous Materials* 382: 121060.

85 Wu, S., Zhao, X., Li, Y. et al. (2013). Adsorption of ciprofloxacin onto biocomposite fibers of graphene oxide/calcium alginate. *Chemical Engineering Journal* 230: 389–395.

86 Zeb, M.A., Shah, J., and Jan, M.R. (2023). Alginate magnetic graphene oxide biocomposite synthesis for removal of aromatic amines from aqueous samples. *Environmental Science and Pollution Research* 30 (17): 51099–51112.

87 Bhatti, H.N., Mahmood, Z., Kausar, A. et al. (2020). Biocomposites of polypyrrole, polyaniline and sodium alginate with cellulosic biomass: adsorption-desorption, kinetics and thermodynamic studies for the removal of 2,4-dichlorophenol. *International Journal of Biological Macromolecules* 153: 146–157.

88 Tam, N.F.Y., Lau, P.S., and Wong, Y.S. (1994). Wong Wastewater inorganic N and P removal by immobilized Chlorella vulgaris. *Water Science and Technology* 30: 369–374.

89 Megharaj, M., Pearson, H.W., and Venkateswarlu, K. (1992). Removal of nitrogen and phosphorus by immobilized cells of Chlorella vulgaris and Scenedesmus bijugatus isolated from soil. *Enzyme and Microbial Technology* 14: 656–658.

90 Whitton, R., le Mével, A., Pidou, M. et al. (2016). Influence of microalgal N and P composition on wastewater nutrient remediation. *Water Research* 91: 371–378.

91 Mallick, N. (2002). Biotechnological potential of immobilized algae for wastewater N, P and metal removal: a review. *BioMetals* 15: 377–390.

92 Kube, M., Mohseni, A., Fan, L., and Roddick, F. (2019). Impact of alginate selection for wastewater treatment by immobilised chlorella vulgaris. *Chemical Engineering Journal* 358: 1601–1609.

12

Chitosan/Clay Membranes for Heavy Metals Removal

Ismaila Olalekan Saheed[1,2] *and Faiz Bukhari Mohd Suah*[1]

[1]*Green Analytical Chemistry Laboratory, School of Chemical Sciences, Universiti Sains Malaysia, Minden, Malaysia*
[2]*Department of Chemistry and Industrial Chemistry, Kwara State University, Ilorin, Nigeria*

12.1 Introduction

The prevalent presence of harmful heavy metal contaminants in water systems necessitates sustainable mitigating techniques to guarantee worldwide water security. Heavy metal contamination is one of the primary environmental concerns in which the contamination of the aqueous environment, such as surface water, groundwater, drinking water, and wastewater, has an irreversible negative impact on human health, the environment, and the ecology [1, 2]. In recent times, ecological systems have been severely contaminated by heavy metal ions, including copper (Cu), zinc (Zn), mercury (Hg), cadmium (Cd), manganese (Mn), and lead (Pb) due to natural occurrence and anthropogenic activities. Various remediation procedures such as chemical precipitation, extraction, electrochemical deposition, ultrafiltration, ion exchange, membrane filtration [3]), coagulation, reverse osmosis, bioremediation, and adsorption [1, 4, 5] have been utilized to remove this class of contaminant from various contaminated sources including air, soil, natural water sources, water treatment plants and channels, wastewaters, and industrial effluents. Some of these methods, such as adsorption and membrane filtration, are popular due to their cost effectiveness, heavy metal removal efficiency, and sustainability. Meanwhile, these popular methods require the fabrication of suitable materials to accomplish an effective removal technique related to metal contamination in an aqueous medium. Different organic and inorganic materials of various properties and forms, including polymer-based, carbon-based, clay minerals-based, and metal-organic framework materials, have been applied for heavy metal removal in their original, synthesized or modified form. Additionally, natural biopolymers like chitosan are naturally abundant, inexpensive, biocompatible, biodegradable with low toxicity, high adsorption capability, and better ease of modification, and have been prepared/or modified and systematically applied for the removal of heavy metals from aqueous solution. Moreover, owing to clay's low cost, abundance, and environmental stability, it has also been broadly employed as a significant sorbent in decontaminating heavy metals in polluted water [6].

Membrane adsorbents have recently been made to remove dangerous substances such as heavy metal ions from aqueous medium. Their effectiveness in removing contaminants at trace levels has made membrane adsorbents an appealing and potent technological material [7]. On the other hand, among the most promising hybrid materials for creating ecologically acceptable adsorbents and membranes for treating wastewater containing heavy metal contaminants are polymer composites. Biopolymers like chitosan, derived from shrimp and other crustaceans, are widely available in nature and do not harm living things. As regards developing chitosan-clay composites as adsorbent materials, numerous studies have demonstrated their potential advantages for treating industrial wastewater, such as the ability to recycle the process, good performance, and no need for pH adjustment. The investigations have also focused on gaining a thorough understanding of the impacts of several variables (such as clay and clay mineral type, polymer-to-clay ratio, and polymer characteristics) on the structure of the materials to optimize the efficiency of pollutant removal in a model or real contaminated water. Chitosan-clay composites, being products developed from natural origin, are green materials and have interested researchers as one of the materials envisaged for use to lessen the adverse effects on the environment, including its possibility of being developed into membrane filters for treating heavy metal contaminated water [8]. Integrating chitosan into clay minerals to develop polymer–clay composite membranes is of tremendous environmental relevance [2]. This chapter discusses the chitosan/clay composite membrane for the decontamination of heavy metals from wastewater, featuring their preparation methods, properties, application, and limitations.

12.1.1 Heavy Metals Contamination

Chemical elements with relatively high densities (about five times heavier than water) found naturally in the earth's crust are known as heavy metals. Most of these metals have atomic weights and atomic numbers that are relatively high, with their specific gravities higher than 5.0. Even at low concentrations, these metals are hazardous to humans and could be poisonous. They find their way into the human body through food, drinks, or air and tend to bioaccumulate due to their nondegradable feature, which can lead to significant health disorders [9].

Some industrial and consumer waste discharges, as well as leaching of heavy metals from rocks into rivers, streams, and groundwater, are the primary means by which heavy metals enter the water system. The potential sources of heavy metal ions are not limited to wastewater from industries producing metals, fertilizers, insecticides, paints, pigments, and batteries. Notable of this class of metals (heavy metals) include chromium (Cr), copper (Cu), zinc (Zn), nickel (Ni), silver (Ag), gold (Au), cadmium (Cd), iron (Fe), mercury (Hg), lead (Pb), palladium (Pd), and thallium (Tl) [10]. Heavy metal pollution has attracted much attention due to its high toxicity, significant quantity release, and nonbiodegradability. These heavy metals are toxic to humans; some metals (like Cd and Pb) are widespread in the aquatic environment [11].

However, some heavy metals are indispensable components of mammalian metabolic systems. For example, manganese can support average bone growth and development and maintain normal glucose and fat metabolism. Zinc is crucial for the development of the brain and intellect. Copper is among the components of hemocyanin in humans. Meanwhile, severe harm can result from exposure to heavy metal concentrations above the safe threshold [9]. The maximum permissible limit set as standards for various heavy metal concentrations in drinking water by multiple organizations, including World Health Organization (WHO) [12], United States Environmental Protection Agency (USEPA) [13], and European Union standard (EU Standard) [14], are presented in Table 12.1.

12.1.2 Membrane Technology

Membrane technology offers an intriguing substitute for conventional separation techniques owing to its little energy usage, excellent efficiency, comfortable operating conditions, and simple scaling up [15]. Membrane technological innovation has been considered beneficial in various contemporary ventures as it offers outstanding outcomes, reduced compound utilization, low working costs, and a simple approach with a relatively modest impression. A membrane can also drive molecular separation, which is considered the creation of a high-immaculateness item [16]. In basic terms, a membrane acts as a barrier, allowing some substances to flow through but stopping others from passing across. The streams treated and received via membrane are referred to as permeate, whereas the substances not allowed to pass are known as retentate. Its selectivity and flow rate typically measure the effectiveness of a membrane. Flow rate (or flux) is the fluid volume that passes across a membrane per unit area and time. Conversely, selectivity is defined as the part of a material in a feed stream that needs to be preserved by a membrane. Removing substances through membrane separation is relatively simple. It is often controlled by factors such as size exclusion or steric hindrance, the Donnan exclusion effect (charge–charge repulsion effect), and adsorption efficiency toward particular contaminants [17].

Table 12.1 Heavy metals permissible limits for drinking water and their respective common diseases and effects on human.

Heavy metal	Standard permissible limits in drinking water (in mg L^{-1})			Effect on human
	WHO	USEPA	EU	
Cr	0.050	0.100	0.050	Kidney disease, hypertension, diabetes, tumor, respiratory disease
Mn	0.100	0.050	0.050	—
Fe		0.300	0.200	—
Ni	0.070	—	0.02	Dermatitis, allergy
Cu	2.000	1.300	2.000	Gastroenteritis, hepatocirrhosis
Zn	5.000	5.000	—	
As	0.010	0.010	0.01	Dermal lesions, skin cancer, bladder and lung cancers
Ag	0.100	0.100	—	Argyria
Cd	0.003	0.005	0.005	Bone disease, kidney disease, digestive system
Hg	0.001	0.002	0.001	Hepatopathy, skin disease, kidney disease, digestive and respiratory systems disease
Tl		0.002		Intestine, kidney, or liver problems and hair loss
Pb	0.010	0.015	0.010	Heart disease, anemia, nervous system disease, nephropathy
Sb	0.005	0.006	0.005	Heart disease, skin disease, nervous system disorder, increase in blood cholesterol, hepatopathy, decrease in blood sugar

Membranes have been categorized as organic or inorganic based on their material composition. Organic membranes are usually developed from synthetic organic polymers [16]. Meanwhile, membrane materials can be broadly divided into two groups: polymers and ceramics. Since ceramic membranes are hydrophobic and have a high chemical resistance capacity, they are generally more suitable for treating industrial wastewater than polymers. Nevertheless, the drawbacks to ceramic membranes include their brittle nature and expensive installation and operation costs [18]. Polymer materials like polyvinylidene fluoride, polysulfone, and polypropylene are good candidates for developing commercial membranes because of their chemical resistance.

Membranes function as filters, and some could also perform additional roles as adsorbents depending on the material used for the preparation. As a method of separation, filtration involves removing solid particles from a fluid by passing it through a filter medium while retaining the solid particles on the filter. On the one hand, membrane filtration is a type of separation technique in which a driving force is smeared throughout a membrane to separate substrates. The substrates are separated via membrane filtration according to their sizes, pore sizes of the membrane, concentration of the solution, and applied pressures [18]. Various membrane filtration technologies, including microfiltration, nanofiltration, ultrafiltration, and reverse osmosis, have been applied to directly remove metallic ions without using secondary chemical reagents. Just as reported hitherto, Bhattacharyya et al. examined the viability of employing a charged ultrafiltration membrane for the treatment of electroplating water containing Ni^{2+}, Cu^{2+}, and Zn^{2+} ions by pioneering heavy metals' removal using a membrane-based technique [19]. Among the membranous methods, reverse osmosis is the most efficient membrane filtration technique that is applicable for removing heavy metal ions from water. However, its usage has been limited due to its high capital demand for setting up and its expensive operational costs. Water purified through reverse osmosis is unsuitable for drinking due to its low mineral content [20]. Meanwhile, nanofiltration, as a separation process, lies between reverse osmosis and ultrafiltration, with a nominal pore size of 2 nm. Nanofiltration uses less electrical energy than reverse osmosis because it requires lower operating pressures.

A membrane designed for adsorption application performs dual functions: (i) as membrane filtration and (ii) as membrane adsorption. Adsorption is a process by which molecules (adsorbate) are removed from the liquid medium by transferring onto the surface of a solid material (adsorbent) via a mass transfer process, where binding takes place through chemical and physical interactions [21, 22]. Adsorption relies on these interactions to bind adsorbent materials and contaminants such as heavy metal ions. To effectively remove trace amounts of pollutants like cationic heavy metals, anionic phosphates, and nitrates, the membrane must be developed as an adsorptive membrane [21]. Because of the unique structural characteristics of the adsorptive membranes, their usage for heavy metal recovery/removal from wastewater has proven to be comparatively successful. Based on the source of the materials used for the preparation, different adsorptive membranes, such as polymeric membranes, polymer-ceramic membranes (or polymer-clay membranes), and electrospinning nanofiber membranes, have been developed [22].

Various kinds of thin-film composite membranes have been developed for the treatment of water and wastewater contaminated with substances such as heavy metal ions [17]. These include:

(i) Pressure-driven membrane, (i) low pressure-driven membrane. Examples are microfiltration, ultrafiltration, and distillation [23] (ii) high pressure-driven membrane. Examples are nanofiltration and reverse osmosis [17, 24] (iii) osmotic pressure-driven membrane. Examples are forward osmosis and electrodialysis.
(ii) Liquid membrane.

The main parameters to consider when using membranes in an aqueous medium are the degree of hydrophilicity, pore size, distribution, surface charge, solution flow, and functional groups that facilitate the separation process. The rate at which the filtrate (water) is produced, including the effectiveness of the heavy metals removal, is significantly impacted by the characteristics of the membrane as a whole. Except for microfiltration, membrane processes that are frequently employed to treat water contaminated with heavy metals can be broadly categorized into several major types, such as complexation, enhanced ultrafiltration, micellar-enhanced ultrafiltration, adsorptive ultrafiltration mixed matrix, nanofiltration, reverse osmosis, forward osmosis, liquid membrane, and electrodialysis [17].

Furthermore, potential adsorbents can be infused into polymer membranes to create adsorptive membranes, combining the advantages of membrane filtration and adsorption. When water contaminated with heavy metals penetrates the membranes' cross-section, the adsorbents can effectively trap the heavy metal ions to produce clean permeate. Such a composite membrane primarily overcomes the limitations of polymeric membranes to improve the membrane's permeability and selectivity [25]. Chitosan-clay composite and its derivatives have been found useful in an attempt to create polymer-based composite adsorbents in various forms, including membranes, for removing heavy metals from aqueous medium. The subsequent sections will discuss this in detail.

12.1.2.1 Membrane Separations' Mechanism

By combining sorption, diffusion, and filtration, membranes permit the passage of specific species in a solution. The separation is accomplished via a selective passage of intended components of a stream across the membrane when others are retained on the surface (see Figure 12.1). This process is applied to separate particulate matter from liquid suspensions or gases, dialysis, and separation of ions in electrochemical processes. The retentate is retained on the surface as the permeate is allowed to pass through the membrane. At times, a fluid called "sweep" is used to aid permeate's passage through the pores.

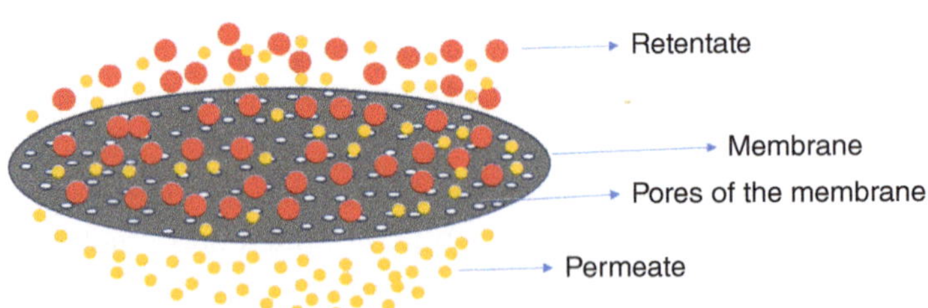

Figure 12.1 Mechanism of separation through adsorptive membrane.

Four distinct mechanisms are typically involved in the separation through membranes [26]: (i) size exclusion brought about by membrane pores or holes that only allow materials smaller than the pore size to pass through; (ii) solution diffusion caused by component molecular diffusion that causes migration across the membrane; (iii) pore flow caused by molecules passing through the pore surface interacting with it, which causes molecules of comparable sizes to be selectively transported more slowly; and (iv) attraction between the components of the solution and the membrane molecules causing separation by sorption (applicable to adsorptive membrane).

Some dynamic forces affect the mass flow across the membranes, which can be described using nonequilibrium thermodynamics as follows:

(i) For schemes that are not in equilibrium, the equilibrium thermodynamic associations are appropriate inasmuch as the gradient is not very large.
(ii) The scheme considers all driving forces, and models each flux as a linear relationship.
(iii) If the force is an odd number and the flux difference is in tonsorial order, coupling of fluxes and forces is not observable (Curie's Postulate).
(iv) In the absence of magnetic fields, the coefficient matrix in flux–force relationships is symmetric (Onsager's reciprocal relations) [27].

12.1.3 Chitosan and Clay

12.1.3.1 Chitosan

Chitosan is a biodegradable biopolymer that is made up of a linear polysaccharide of β-linked D-glucosamine and N-acetyl-D-glucosamine structural units distributed randomly. Its deacetylated structural unit, β-(1 → 4)-2-amino-2-deoxy-D-glucose predominate over the lower percentage acetylated unit β-(1 → 4)-2-aceta-mido-D-glucose (that is often <20%) with both units connected via glycosidic bond (see Figure 12.2) [28, 29]. Chitosan possesses various crystal

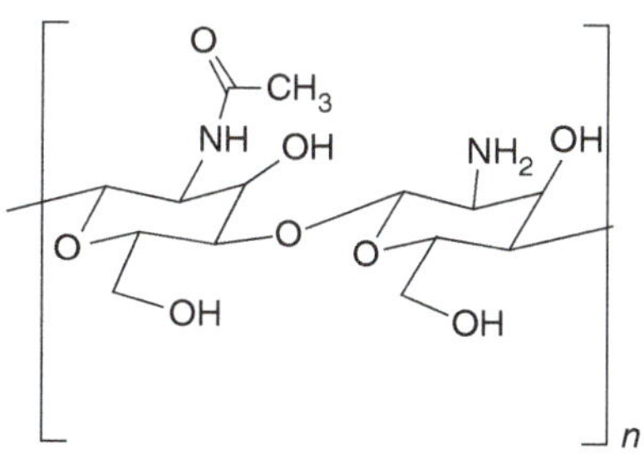

Figure 12.2 Chemical structure of chitosan.

structures: α, β, and γ. α-Chitosan, which is the primary form, is composed of two parallel and inversely oriented polysaccharide chains with a strong intermolecular force and a compact structure. β-Chitosan consists of two polysaccharide chains that are parallel to each other and have weak intermolecular hydrogen bonds between them. Three parallel polysaccharide chains make up γ-Chitosan, with two oriented along the same direction while the other is in the opposite direction [30]. Chitosan is often derived from chitin and also next to cellulose as the most readily available polysaccharide in nature [31, 32]. It can be obtained from various organisms such as crustaceans, insects, arachnids, fungi, algae, mollusks, fishes, among others, through various methods of extraction procedures, featuring chemical methods, enzymatic methods, as well as irradiation applications. Different chitosan crystal structures can be obtained from crustaceans. Crabs and shrimps, squids, and loligos are examples of primary sources of α-chitosan, β-chitosan, and γ-chitosan, respectively [33]. Chemical extraction is the most popular extraction technique among others, and it consists of three fundamental stages, namely: deacetylation, demineralization, and deproteinization [34]. Chitosan's molecular weight and extent of deacetylation influence its chemical and physical properties [35]. Chitosan is advantageous and unique due to its low cost and abundance, as well as its antibacterial property, nontoxicity, macromolecular structure, hydrophilicity, biocompatibility and biodegradability, cationicity, active sites, and relatively high adsorption capacity [36].

Chitosan is obtainable via N-deacetylation of chitin in the presence of hot alkali. The molecular weight and degree of deacetylation in chitosan vary from 5×10^4 Da–2×10^6 Da and 40–98%, respectively [37]. Among other factors, chitosan's molecular weight, degree of deacetylation, solubility, as well as degree of crystallinity all affect how it behaves chemically and physically [38]. Commercial interest in chitosan stems from its higher nitrogen content when compared to synthetic substituted cellulose [29, 39]. In comparison with its common precursor chitin, chitosan is less crystalline, a feature that enhances its accessibility to reagents. But like chitin, it decomposes before melting, which renders it a polymer without a defined melting point range. Chitosan is polymorphic in nature (having various packing forms and various polarities of adjacent chains of the successive sheets), which makes it to display various diffraction patterns of either orthorhombic and monoclinic unit cells with organic-based acids like methanoic and ethanoic acids. The tendon polymorph, which is hydrated, and the annealed polymorph, which is anhydrous, are the primary polymorphs of chitosan [40, 41].

Chitosan dissolves in most acid media, while methanoic and ethanoic acids are frequently used for its solubilization [42]. The protonation of the amino group in its polymeric chain via acid treatment generates a large number of cationic sites which increases its solubility and helps to produce materials with greater polarity and electrostatic repulsion. Chitosan is insoluble in water but soluble in some solvents like p-toluene sulphonic acid and dimethylsulphoxide. Chitosan is also soluble in hydrochloric acid as a result of feasible interaction between its $—NH_2$ group and H^+ of hydrochloric acid, which transforms it into a positively charged polyelectrolyte [30]. The cations of the chitosan's polyelectrolyte destroy the hydrogen bonds within chitosan molecules, which in turn enhance its solubility in water. However, sulfuric acid is often avoided for chitosan dissolution because it reacts with chitosan to produce chitosan sulfate precipitate [37]. The degree of deacetylation and molecular weight of chitosan determine its extent of solubility. The higher the degree of deacetylation, the better the ease of protonation of $—NH_2$ group, and the more readily it dissolves. Whereas, chitosan with higher molecular weight are more difficult to dissolve because they possess polymer chain with higher number of hydrogen bonding [30]. The acetylated portion of chitosan's structure can form hydrogen bonding with other molecules or engage in hydrophobic interactions, both of which significantly enhance the molecule's stability and contribute to a certain degree of rigidity that strengthens its structural characteristics. The heavy metals attachment to chitosan occurs primarily via bonding of metal with the amino group of chitosan's molecule, and the mole ratio of 1:1 is common to divalent heavy metals (M^{2+}) and chitosan's amino group ($—NH_2$) coordination [43]. Additionally, it has been reported that, when compared with some natural polymers derived from seafood wastes and those from natural materials like bark, activated sludge, being utilized for commercial chelating ion-exchange resins, chitosan exhibits the highest chelating ability. For example, chitosan is known to be a more effective Ni^{2+} scavenger from industrial waste than natural sourced materials like polymerized peanut skins, such that binding capacities > 1 mmol g^{-1} for heavy metals (except for chromium) are obtainable. Wastewater with Ni^{2+} concentration of 7 mg L^{-1} can give a concentration <0.1 mg L^{-1} after passing through a chitosan column. Chitosan can be simply regenerated using a buffered ammonium chloride [44].

12.1.3.2 Chitosan Membrane

Chitosan has been developed and processed into a variety of products, including films, threads, sponges, beads, sheets, flakes, membranes, and micro- and nanoparticles for the treatment of wastewater and the removal of heavy metals [45]. These varieties of products are developed to improve the characteristics of chitosan required for effective performance toward various applications including heavy metals removal and wastewater treatment. Despite offering high active surface area and treatment performance, dispersed adsorbents like flakes, beads, and powders still present challenges because they need to be separated from the applied medium after the treatment process. Chitosan-based membranes are advantageous over other chitosan adsorbents due to their ease of separation after processing and lack of need for posttreatment filtration. Additional benefits of membrane-type adsorbents include greater adsorption capacity, rapid kinetics, improved reusability, simplicity in stacking and scaling up, and the potential for nonadsorptive separation mechanisms to improve filtration performance. One of the drawbacks of chitosan during membrane formation is its fragility and difficulty to undergo film-like structure during phase inversion. Inadequate mechanical strength, pH sensitivity, and its susceptibility to swelling are among the critical concerns on the usage of chitosan-based membranes as adsorbents [7, 46].

Unlike cellulose derivatives which are soluble in quite a number of organic solvents, chitosan is soluble in limited solvent and it is commonly dissolved in aqueous acetic acid. This limits the way to create composite membranes to blending or dispersing the filler into a polymer solution. Surfactants can be used to enhance the process, simplifying the dispersion phenomena as well as membrane preparation using any of the currently recognized techniques such as casting or solvent evaporation [47]. The chitosan molecular weight is known to affect the solution's viscosity such that an increase in the molecular weight of chitosan makes the chitosan-acetic acid solution's viscosity increase. The degree of deacetylation, on the other hand, was found to have no effect on the viscosity of chitosan-acetic solutions. The phase inversion approach can be utilized to create chitosan membranes from chitosan solutions of different concentrations and viscosities [48]. Solvent evaporation and casting can be used to create chitosan-based membranes. Casting a chitosan-acetic acid solution onto a plate (such as flat glass or polyethylene film), letting the solvent evaporate, and then coagulating with alkaline solutions or ammonia gas are the general steps in the process of making chitosan membranes [49]. Surfactants play a beneficial role in enhancing the dispersion of chitosan-based composite membranes. It is possible to directly apply chitosan-based membranes onto the surfaces of food or packaging materials. The commonly used coating methods include spread coating, spray coating, dip coating, and immersing coating. Moreover, repeated coating/casting-drying can be utilized to fabricate multilayer membranes. Melt extrusion has found extensive use in producing chitosan-based membranes. The chitosan pellets are extruded into sheets using a twin-screw extruder connected to a flat die or blown into a film through an annular die with a blown film extruder. Various spinning technologies can also be utilized to fabricate chitosan-based membranes composed of nanofibers [50].

Contaminants are obtainable from natural or synthetic sources and have been tracked in the environment up to date. Chitosan-based membranes have shown to be highly promising adsorbents for removing these contaminants. Meanwhile, chemical, thermal, and mechanical resistances are important properties for assessing the stability standpoint of chitosan-based adsorptive membranes. Chemical or thermal cross-linking of the chitosan polymeric matrix can enhance the physical and chemical stability of chitosan-based membranes. Enhancing the chemical and mechanical strength of chitosan-based adsorptive membranes is achievable by mixing chitosan with suitable copolymers, such as polyvinyl alcohol and cellulose derivatives [51]. The free amino groups in chitosan allow it to function as a cationic polyelectrolyte, and as such, chitosan membrane has been employed for chloride-ion conducting due to its ionic characteristics. Similarly, protonation of amino groups may result in hydroxide ions being free in water when a chitosan membrane swells in it; this phenomenon may support ionic conduction in the membrane. But if no electrolyte was introduced to the membrane during production, the ionic conduction it displays is intrinsic [52].

Chitosan membranes exhibit remarkable versatility and capability with respect to the separation of gaseous mixtures, azeotropic mixtures, heavy metal ions, dyes, and direct methanol fuel cell applications [53]. While cross-linking increases the membrane's stability and resistance to plasticization under harsh environments, it however reduces the mobility of the chain, resulting in a more compact structure. Although chitosan membranes are relatively new for gaseous mixture separation, the amino groups present in the polymer chain permit selective passage of CO_2. Heavy metal ions have been removed from aqueous solutions using a variety of chitosan adsorptive membranes; however, these membranes have structural deficiencies such as lack of reactive sites and having limited stability in adsorption systems. Blending, coating, cross-linking, among other methods were effective means of resolving the challenge. Epichlorohydrin and glutaraldehyde are the most widely used cross-linking agents that have been applied to remove heavy metal ions [54]. Qurashi et al. prepared various modified chitosan membranes by blending chitosan with polyvinylpyrrolidone (PVP) in 1 wt.% acetic acid, and their physical, thermal, and mechanical characteristics reveal distinct properties of the blends from pristine chitosan. The blends are seemingly hard and flexible when dry but soft when soaked in water with exceptional water absorption efficiency compared to chitosan. Chitosan exhibits a brittle character, lower percent elongation-at-break but with higher tensile strength compared to the corresponding modified membrane. The blends become more hydrophilic due to the presence of PVP in the chitosan substrate. This makes the blends to swell in three dimensions, and the hydrogen bonds between the chitosan molecules are broken [55].

The synthesis and characterization of hydrophilic polymer membranes based on macromolecular chitosan networks was achieved by Clasen et al. Several modifications made to the membrane's structure during preparation to adjust its qualities, specifically its permeability, surface structure elasticity, and tensile strength were made by addition of flexibilizer, dialdehydes cross-linker, formation of the modified chitosan with polyanion sulfoethyl cellulose, and the addition of synthetic micro- and nanopores filled with silica particles or polyethylene glycol to the chitosan matrix [56].

12.1.3.3 Clay and Clay Mineral

Clays are naturally occurring tiny particles on the earth's surface that are primarily made of silica, alumina, water, and weathered rock. It also refers to materials not only with particles less than 2 μm in size but as well a family of minerals with comparable chemical compositions and common crystal structure properties [57]. Clay's mineralogy consists of phyllosilicates, which are sheets structurally based on tetrahedral sheet ($[SiO_4]^{4-}$) with interlayer sites connected to octahedral sheets ($[AlO_3(OH)_3]^{6-}$) through sharing of apical oxygen atoms [58]. These sheets can be abstracted as $Mg(OH)_2$ (brucite) or $Al(OH)_3$ (gibbsite) layers. Clay minerals may possess a net negatively charged surface due to Al^{3+} or Fe^{3+} replacing Si^{4+}. The most basic of octahedral layer's unit consist of six —OH or oxygen atoms surrounding either Al^{3+} or Mg^{2+}. The Al^{3+} occupies 2/3 of octahedral arrangements to give dioctahedral arrangements, while the divalent cations (Mg^{2+}) of the octahedral sheet form trioctahedral arrangements. Different clay mineral types are produced by combining different tetrahedral and octahedral sheet configurations [59]. For example, the kaolinite group including dickite, nacrite, kaolinite, halloysite, and allophane, with the formula $Al_2Si_2O_5(OH)_4$ (sometimes written in the form of oxides as $Al_2O_3{\cdot}2SiO_2{\cdot}2H_2O$) and a 1:1 clay mineral structure (one tetrahedral sheet of silica linked to one octahedral sheet alumina) with a balanced charge [60, 61]. The 2:1 clay mineral type composed of octahedral sheets sandwiched by two tetrahedral sheets (tetrahedral-octahedral-tetrahedral layers) such that the tetrahedral and octahedral sheets are joined by H-bonding and electrostatic interactions, resulting in a different clay arrangement. Unlike the 1:1 layer type, the 2:1 layers have a net negative charge which are either linked together by separate cations like potassium (for illite type with a nonexpansive 2:1 layer), sodium or calcium (for smectite type with an expandable 2:1 layer) or by octahedral sheets with positive charge (for chlorites type). When SiO_2 from the tetrahedron of the pyrophyllite is substituted for KAl_2, the result is an octahedral structure illite group. The smectite mineral type, which is found in montmorillonite, and the vermiculite group (another 2:1 layer type with a small expandable character) exhibit low charges that permit hydration of the interlayer cations. On the other hand, the chlorites group ($(Mg, Fe, Li)_6\ AlSi_3O_{10}(OH)_8$) has a 2:1:1 layer structure, where an additional brucite octahedral layer intercalates with the 2:1 layer [62].

Many forms of inorganic clay have been added to polymeric networks (such as chitosan backbone) to create organic/inorganic composites, including kaolin, bentonite, sepiolite, mica, vermiculite, and rectorite. Excellent characteristics of clay minerals, including their potential for controlled release, low or zero toxicity, and good biocompatibility, have sparked a constant interest in their applications. The main reasons for these applications are the cationic clays' exceptional swelling property, exceptional hydration ability, high cation exchange capacity, and structural plasticity [63]. Commonly, adsorption technique is widely applied to treat contaminants in aqueous medium, and clay is also among the common, cheap, and abundant adsorbent for adsorption applications. Clay minerals have interesting adsorption properties such as a comparatively large surface area, good porous structure, good chemical and physical stability, and ease of functionalization with either inorganic or organic species. Many researchers have used clay solely or in modified form to remove heavy metals [64]. Conversely, the effectiveness of clays and clay minerals during the adsorption process is influenced by a variety of factors including surface charge, specific surface area, and swelling characteristics. The clay and clay mineral also exhibit cation exchange capacity (the amount of positive ions that can be exchanged for every unit mass of dry clay mineral). The affinity of clay minerals to remove cationic organic and inorganic pollutants, like dyes and heavy metals increases when these cationic counter ions are exchanged for other cations. However, clays and clay minerals have low to medium anion exchange capacities, a drawback that require physical and chemical modification to enhance their surface qualities and structure in order to make suitable for wider water pollutants decontamination [65]. The process of heavy metal adsorption by clay and modified clay involves several intricate adsorption mechanisms, such as surface complexation, ion exchange, and direct bonding of heavy metal cations to clay surfaces. Usually, the modified clays have much higher heavy metal adsorption capacities than natural clays [9].

A modified bentonite mineral was made through various processes including acid activation, aluminum pillaring, and pillaring followed by acid activation [66]. Acid activation resulted in the partial dissolution of the octahedral layer caused by the cations Al^{3+}, Mg^{2+}, and Fe^{3+}, as well as the formation of amorphous silica as a result of the tetrahedral layer being destroyed. The specific surface area increased to $154\,m^2\,g^{-1}$ and the cation exchange capacity decreased to $354\,m^2\,g^{-1}$ following treatment with HCl 0.5 and 3 M, respectively. Chromium was adsorbed onto both the natural and modified clays, and the maximum adsorption capacities vary with the clay treatment applied [66]. Biocomposite bead in various compositions was prepared using kaolinite-rich modified clay and chitosan derived from prawn shells [67]. The beads' potential for lead (II) and methylene blue adsorption measured in standard models of lead (II) and methylene blue solution indicate promising removal efficiency. When applied to real industrial effluents to eliminate Cr(VI), Pb(II), and MB, the composites rich in chitosan removed 50.90% of Cr and 39.50% of Pb from the tannery effluent [67]. Ni(II) and Cd(II) have been adsorbed onto epichlorohydrin cross-linked chitosan–clay composite beads in contaminated solutions containing heavy metals [68]. The obtained results showed conformity of the adsorption of Ni(II) and Cd(II) on epichlorohydrin cross-linked chitosan–clay composite beads to Langmuir and Freundlich models, respectively, while the maximum adsorption capacities for Ni(II) and Cd(II) ions were 32.36 and 72.31 mg g^{-1}, respectively. The recovery of the metal ions and regeneration of the adsorbent carried out on loaded beads using 0.1 M HNO_3 and 0.01 M EDTA show that the metal ions could be recovered more effectively under EDTA conditions due to its high level of metal ions complexing capacity [68].

Pillaring, a method of clay modification, involves the exchange of ions between the cation of the interlayer clay mineral and the cation of the inorganic pillaring specie. In order to create pillared clay, two steps must be completed: first, the pillaring agent must be made; then, the pillaring agent solution must be mixed with the clay suspension. Pillared clays have drawn reasonable attention of researchers as a potential adsorbent due to interesting characteristics including strong surface acidity, high thermal stability, large surface area, pore volume, and extended pore size, as well as catalytic active substrates and metal oxide pillars. The pillared clay shows adsorption selectivity and efficiency, while intercalation of polyoxocations in the interlamellar region of the clay aids increased adsorption efficiency toward heavy metals [69]. The subsequent section focuses on the chitosan/clay membranes and their application for heavy metals removal.

12.2 Chitosan/Clay Membrane

Chitosan is known for its widely accessible, nontoxic, biocompatible, and biodegradable polymer material produced by deacetylating chitin. Similar to this, it has been reported that clay (that usually exists as nanoparticles), which can be produced artificially or naturally, is cheap, abundant, and nontoxic [70]. Chitosan is also known to dissolve readily in acidic media, a feature that has influence on its application as a membrane to treat aggressive solutions. These shortcomings pose significant issues and also restrict its usage as an adsorptive membrane. Therefore, it is imperative to enhance its mechanical, thermal, structural, and physical properties. Various modification approaches including cross-linking and blending to composites have been utilized to enhance the mechanical and thermal characteristics of chitosan membrane. In this way, the reactive functional groups on chitosan's backbone can be utilized as binding/interaction sites for creating a more stable composite membrane [46]. The use of ceramic membranes, particularly those based on clay, is considered being preferred over the sole organic membranes. This is due to its excellent mechanical, thermal, and chemical strength, long lifespan, low fouling tendency, no need for additives, and freedom from temperature restrictions. In contrast to polymeric membranes, reduced fouling in clay-based membranes results from the inherent hydrophilicity that facilitates simple detachment of the cake layer formation on the membrane surface [71].

12.2.1 Preparation Methods of Chitosan/Clay Membrane

The synthesis method adopted for the preparation of chitosan/clay composite membrane has influence on the physical, mechanical, and structural characteristics of the developed membrane, such as morphology, surface area and porosity, chemical and mechanical stability, and physical features, among others. The common preparation methods for chitosan/clay membrane include solvent casting and evaporation process method, self-assembling method, and solution casting and blending method.

12.2.1.1 Solvent Casting and Evaporation Process

Solvent casting and evaporation process involves solubilizing a polymer and plasticizer, and then spreading out the solution over a substrate, the subsequent solvent removal then leads to intercalation of plasticizer molecules as well as molecular orientation of the polymer chains to give film. The sequence of solvent casting for chitosan/clay mineral membrane includes (i) dissolution of chitosan in acid medium (usually acetic acid), (ii) proper mixing of chitosan solution with plasticizer and clay suspension, (iii) casting of the mixture onto surface platform (or mould), (iv) evaporation of the solvent under specified temperature condition, (v) addition of basic solution (usually NaOH solution) to neutralize the acid, and (vi) careful detachment of the membrane from the surface platform [72].

Solvent casting and evaporation process was utilized to create chitosan composite membranes with varying kaolin contents [46]. As revealed through characterization, the kaolin and chitosan in the membranes are strongly bonded, and the binding had a significant impact on the morphology of the membrane by improving its mechanical, chemical, and thermal properties. When compared with chitosan, kaolin improved the water permeability (by increasing porosity) and water resistance of the membrane, making it suitable for treating industrial effluent and purifying water [46]. In another related work done by Bouzid and co-workers, chitosan/kaolin membrane with high flux and hydrophilicity was also developed through solvent casting and evaporation process in the presence of polyethylene glycol. The morphology and hydrophilicity of the as-fabricated membranes were found to be significantly influenced by the combined action of polyethylene glycol as a porogen and kaolin as reinforcing agent. It was possible to create highly porous membranes with a finger-like structure such that as the polyethylene glycol content rose from 0% to 3% wt.%, the pore size and density increased [73]. Solvent casting method was also successful for creating chitosan/kaolin membranes having a cross-linking agent (citric acid) and a porogen (polyethylene glycol) presenting significant combined action [74]. The findings showed that the addition of a hydrophilic porogenic reagent allows for the creation of a more hydrophilic membrane with porous morphology, which in turn increases the permeability of pure water. When compared to pure chitosan films, the cross-linked membranes showed better mechanical, thermal, and water resistance with a pore size of 50 nm that is within the ultrafiltration range [74]. In a separate work by Reddy et al., solvent casting method was described to develop chitosan/clay film. A 2% wt/vol chitosan solution in 1% acetic acid (pH = 4.0), stirred for 30 minutes and then allowed to age was firstly made. Then 1% acetic acid was used to disperse the nanoclays, which were then stirred for 24 hours, ultrasonicated for 30 minutes, combined with a chitosan solution, and stirred some more. The final mixture was placed in glass plates coated with Teflon and left in an oven set at 60 °C for the entire night. For 30 minutes, the resulting composite films were immersed in a 1% NaOH solution to eliminate any excess acidity. It was observed that the capacity to swell was greatly enhanced by having more quantity of chitosan in the chitosan/clay film; the order in which the swelling capacity of the chitosan/clay films is proportional to the reduction in clay content [75]. In fact, chitosan/clay nanocomposite film produced by solvent casting method for use as a covalently linked bromelain carrier in wine-like medium also shows that the inclusion of nanoclay into the chitosan to produce film results in enhanced mechanical properties [76].

12.2.1.2 Layer by Layer Self-Assembly

In this method, the surface is modified to film by mutual attraction and deposition between the alternating polyelectrolytic layers. The hydrogen bonding, electrostatic interactions, hydrophobic interaction, and van der Waals forces existing between charged polymers are examples of the mutual attractions. Chitosan demonstrates both hydrophilic and cationic properties when dissolved in an acidic medium. So, the clay minerals can bind with the chitosan via electrostatic interactions and H-bonding through their oxygen and —OH functional groups. The film deposition could then be accomplished either by submersion or solution spraying onto the substrate. This process is advantageous because multi-component films/membranes can simply be made [72]. Using the self-assembling technique, chitosan/montmorillonite nanocomposites were deposited over the ceramic surface in varying montmorillonite mass ratios. This was made by swelling Na- montmorillonite in distilled water (stirred at 900 rpm, 60 °C, 2 hours) and sonicating prior to adding to chitosan solution made in 1% acetic acid. After further stirring (at 900 rpm, 60 °C) to create the nacre-like bionanocomposite film on the ceramic surface, the nanocomposite solution was then applied to the surface and left in the oven (60 °C) for the entire night [77].

12.2.1.3 Solution Casting/Intercalation

The solution casting/intercalation method entails dissolving clay in a different solvent medium (such as water and chloroform) and then preparing a solution chitosan in a suitable solvent (such as formic acid and acetic acid). The solvent used for the swelling of clay within the interlayer is displaced by the chitosan chains when the two solutions are mixed. The intercalated structure is preserved when the solvent is removed from the mixture, giving the chitosan-clay composite [78–80]. Using chitosan, sodium alendronate, and Na-Mt, new clay mineral-polymer membranes were created by Villaça et al. through solution casting. Low concentrations of Na-Mt were unsuitable for the synthesis of membranes because of their extreme brittleness and integrity loss in water. Nonetheless, using two distinct methods, sonication and stirring, high Na-Mt loads were achieved for the nanocomposite membranes. When compared to polymer-only membranes, the sodium alendronate Na-Mt:CS membrane produced by ultrasonic energy pretreatment exhibited better physical characteristics, particularly in the areas of water absorption capacity, thermal characteristics, tensile strength, and roughness. The addition of hydrophilic and hydrophobic plasticizers produced further advantages [79].

Films of biopolymer-clay nanocomposites were created using chitosan and kaolinite by Baccour et al. The mechanical and physical properties are significantly improved by combining kaolinite and chitosan, and the intercalation of chitosan and clay was well attached as shown by XRD, FTIR spectra, and DSC. All of the films' solubility, thickness, transmittance, transparency, and color were clearly impacted by the addition of clay. As the amount of kaolinite in the chitosan/nanoclay films increased, the values of their tensile strength decreased while the values of their elongation at break decreased [81]. Incorporation of silver nanoparticles/curcumin/clay into chitosan film using an environmentally friendly process has been possible through solution casting technique [82]. The surface or interlayer of the involved clay minerals may experience in situ heterogeneous Ag nanoparticles nucleation upon incorporation of those minerals. While, the chitosan composite films containing exhibited great mechanical properties. The composite films' tensile strength and elongation at break increased by 15.90 MPa and 27.27%, respectively, as in comparison to the chitosan films. As a result, the developed ginger-colored composite film demonstrated exceptional mechanical and water resistance [82].

12.2.1.4 Blending

Blending involves combination of different chemical substances to develop new substances or mixtures with distinct characteristics. The process often involves mixing two or more chemicals together without chemical reaction included. As applicable to chitosan/clay membrane preparation, this method is quite similar to solvent casting and solution casting methods. Ulu et al. blended chitosan, κ-carrageenan, and acid-activated bentonite to develop porous composite membranes. Characterization revealed homogeneous distribution of bentonite in the membrane matrix, with an extensive surface porous structure [83].

12.3 Application of Chitosan/Clay Membranes for Heavy Metals Removal

Contaminants can be trapped by active functional groups on adsorptive membranes when wastewater permeates through the membrane. The unique morphological surface and chemical structural features of adsorptive membranes allow them to efficiently remove heavy metal ions from water, and as well as their capacity for effective regeneration and reuse. Natural or synthetic polymers can have some essential functional groups like amine, carboxyl, and sulfonic acid that permit efficient heavy metal removal through adsorption. Because of the presence of amino and hydroxyl groups on chitosan, chitosan-based adsorptive membranes in particular have been widely used to remove a variety of contaminants from wastewater. The adsorption of metal ions onto chitosan may be related to various mechanisms, such as chelation, electrostatic attraction, ion exchange, etc., which are contingent upon the pH levels, solution composition, and characteristics of the metal ions. Chitosan/clay mineral membranes are majorly used for removing heavy metals from water and wastewater through the water treatment technology.

12.3.1 Binding Oxidation State of Heavy Metal Ions onto Chitosan/Clay Membranes

The chitosan/clay composite membranes behave majorly as adsorptive membranes. The adsorption of heavy metals onto the chitosan/clay membranes does not only depend on the surface properties (like surface area, active functional group, porosity) of the membrane but is also a function of the speciation of the metal ions in solution [84, 85]. Heavy metal ions can be divided into two primary groups: (i) divalent heavy metal ions and (ii) multivalent heavy metal ions, depending on the common oxidation numbers of the metal ions in aqueous solution. Cd^{2+}, Cu^{2+}, Pb^{2+}, Ni^{2+}, Hg^{2+}, and Zn^{2+} are examples of divalent heavy metal ions with an oxidation number of two. Multiple oxidation numbers, or oxidation numbers higher than two, are present in multivalent heavy metals like Cr, As, and Se, which can exist as Cr(III), Cr(VI), As(III), As(V), Se(IV), and Se(VI) ions. The charges of multivalent heavy metal ions are significantly influenced by the pH of the medium, which in turn affects how the metal ions interact with the surface of the adsorptive membrane. For example, compared to As(V), As(III) is more toxic and challenging to treat, and it is mostly found in neutral form in aqueous solutions. Therefore, in order to achieve a higher removal efficiency, As(III) must be preoxidized to As(V) [25]. The removal of these various metal ions onto adsorptive membranes such as chitosan/clay membranes also involves the interactions between the metal ions formed in solution and the amino and hydroxyl functional groups on the chitosan, as shown in Figure 12.3.

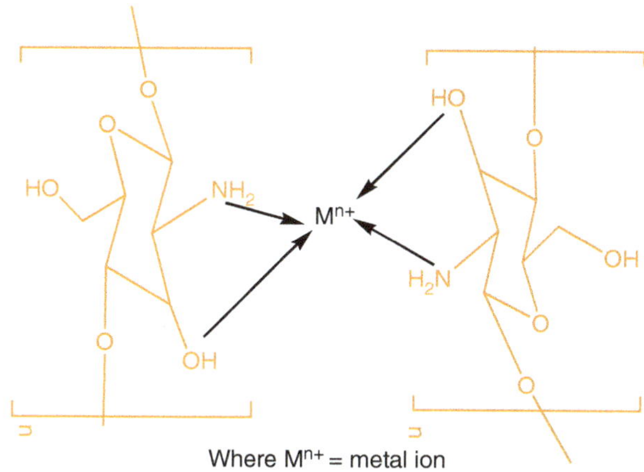

Figure 12.3 Metal ion chelation by chitosan-based adsorptive membranes.

12.3.2 Drawback of Chitosan/Clay Membranes for Heavy Metals Removal

Unlike in sole chitosan membrane (example of polymer membrane) in which its major disadvantages is thermal stability, the chitosan/clay membranes have disadvantages related to fouling, slower and more extreme recovery methods [22]. Membrane fouling is a persistent bottleneck that reduces filtration efficiency by lowering water permeability flux and shortening the membrane's lifespan, both of which result in increased costs. The development of a membrane with appropriate material selection, designed precipitation kinetics, feed pretreatment, operating parameters, and back washing are the primary factors to be taken into consideration to lessen the pervasive fouling problem [45]. The creation of hybrid thin sheet membrane utilizing montmorillonite clay, chitosan/polyvinyl alcohol using a nonsolvent-induced phase inversion technique has been attempted. The hybrid clay-polymeric nanofiltration membrane has a good rejection rate and superior overall performance, including increased hydrophilicity. Through intercalation, clay was uniformly distributed throughout the membrane, resulting in a smaller and suitable filtration pore size. The membrane's pure water flux was found to be 25.72 $l m^{-2} h^{-1}$, and its higher porosity of 81.22% indicated that it was hydrophilic with a mean pore size of 0.4281 nm [45].

12.4 Conclusion

Heavy metals are dangerous, nonbiodegradable with tendency of bioaccumulation in humans causing various diseases, but their existence as contaminants in water systems is highly possible due to natural leaching and wider use for anthropogenic activities. Membrane technological innovation has been considered beneficial in a variety of contemporary applications as it offers outstanding outcomes, reduced compound utilization, low working costs, a simple and straightforward approach, with a relatively modest impression. The adsorptive membrane is relatively effective for heavy metal recovery/removal from wastewater owing to their exclusive structural properties, and the effectiveness is also a function of the materials used for the preparation. Chitosan, as a biodegradable, biocompatible, and abundant biopolymer, is popular for the removal of heavy metals from different types of media due to its active functional group that binds easily with the metals via chelation, and it is simple to modify into various forms such as bead and membrane. Clay, another naturally abundant material, is also popular for eliminating metal contamination from water. Chitosan/clay composite membranes are developed as adsorptive membranes that perform dual functions as filters and adsorbents for removal of heavy metal from aqueous solution. They possess excellent mechanical, thermal, and chemical strength, long lifespan, low fouling tendency, no need for additives, and freedom from temperature restrictions. In contrast to sole polymeric membranes, reduced fouling in clay-based membranes results from their inherent hydrophilicity.

References

1 Noormohammadi, M., Zabihi, M., and Faghihi, M. (2024). Novel chitosan—clay—iron nanocomposites supported on anodic aluminum as an efficient plate-shaped adsorbent for the removal of arsenic and dye from aqueous solutions. *Journal of Physics and Chemistry of Solids* 187: 111874. https://doi.org/10.1016/j.jpcs.2024.111874.

2 Rusmin, R., Sarkar, B., Mukhopadhyay, R. et al. (2022). Facile one pot preparation of magnetic chitosan-palygorskite nanocomposite for efficient removal of lead from water. *Journal of Colloid and Interface Science* 608: 575–587. https://doi.org/10.1016/j.jcis.2021.09.109.

3 Pandey, S. and Mishra, S.B. (2011). Organic—inorganic hybrid of chitosan/organoclay bionanocomposites for hexavalent chromium uptake. *Journal of Colloid and Interface Science* 361: 509–520. https://doi.org/10.1016/j.jcis.2011.05.031.

4 Zeng, H., Wang, L., Zhang, D. et al. (2019). Amido-functionalized carboxymethyl chitosan/montmorillonite composite for highly efficient and cost-effective mercury removal from aqueous solution. *Journal of Colloid and Interface Science* 554: 479–487. https://doi.org/10.1016/j.jcis.2019.07.029.

5 Zia, Q., Tabassum, M., Meng, J. et al. (2021). Polydopamine-assisted grafting of chitosan on porous poly (L-lactic acid) electrospun membranes for adsorption

of heavy metal ions. *International Journal of Biological Macromolecules* 167: 1479–1490. https://doi.org/10.1016/j.ijbiomac.2020.11.101.

6 Foroutan, R., Jamaleddin, S., Mohammadi, R. et al. (2020). Influence of chitosan and magnetic iron nanoparticles on chromium adsorption behavior of natural clay: adaptive neuro-fuzzy inference modeling. *International Journal of Biological Macromolecules* 151: 355–365. https://doi.org/10.1016/j.ijbiomac.2020.02.202.

7 Salehi, E., Daraei, P., and Shamsabadi, A.A. (2016). A review on chitosan-based adsorptive membranes. *Carbohydrate Polymers* 152: 419–432. https://doi.org/10.1016/j.carbpol.2016.07.033.

8 Salvé, J., Grégoire, B., Imbert, L. et al. (2021). Design of hybrid Chitosan-Montmorillonite materials for water treatment: study of the performance and stability. *Chemical Engineering Journal Advances* 6: 100087. https://doi.org/10.1016/j.ceja.2021.100087.

9 Gu, S., Kang, X., Wang, L. et al. (2018). Clay mineral adsorbents for heavy metal removal from wastewater: a review. *Environmental Chemistry Letters* 17. https://doi.org/10.1007/s10311-018-0813-9: 629–654.

10 Vakili, M., Deng, S., Cagnetta, G. et al. (2019). Regeneration of chitosan-based adsorbents used in heavy metal adsorption: a review. *Separation and Purification Technology* 224: 373–387. https://doi.org/10.1016/j.seppur.2019.05.040.

11 Chen, L., Wu, P., Chen, M. et al. (2018). Preparation and characterization of the eco-friendly chitosan/vermiculite biocomposite with excellent removal capacity for cadmium and lead. *Applied Clay Science* 159: 74–82. https://doi.org/10.1016/j.clay.2017.12.050.

12 WHO (2017). *Guidelines for Drinking-water Qaulity. World Health Organisation.*

13 USEPA (n.d.). National primary drinking water regulations.

14 EU (2023). European union (drinking water) regulations.

15 Algieri, C., Chakraborty, S., and Candamano, S. (2021). A way to membrane-based environmental remediation for heavy metal removal. *Environments* 8: 52. https://doi.org/10.3390/environments8060052.

16 Birniwa, A.H., Habibu, S., Abdullahi, S.S. et al. (2024). Membrane technologies for heavy metals removal from water and wastewater: a mini review. *Case Studies in Chemical and Environmental Engineering* 9: 100538. https://doi.org/10.1016/j.cscee.2023.100538.

17 Abdullah, N., Yusof, N., Lau, W.J. et al. (2019). Recent trends of heavy metal removal from water/wastewater by membrane technologies. *Journal of Industrial and Engineering Chemistry* 76: 17–38. https://doi.org/10.1016/j.jiec.2019.03.029.

18 Xiang, H., Min, X., Tang, C.J. et al. (2022). Recent advances in membrane filtration for heavy metal removal from wastewater: a mini review. *Journal of Water Process Engineering* 49: 103023. https://doi.org/10.1016/j.jwpe.2022.103023.

19 Bhattacharyya, D., Moffitt, M., and Grieves, R.B. (1978). Charged membrane ultrafiltration of toxic metal oxyanions and cations from single- and multisalt aqueous solutions. *Separation Science and Technology* 13: 449–463. https://doi.org/10.1080/01496397808058294.

20 Maher, A., Sadeghi, M., and Moheb, A. (2014). Heavy metal elimination from drinking water using nanofiltration membrane technology and process optimization using response surface methodology. *Desalination* 352: 166–173. https://doi.org/10.1016/j.desal.2014.08.023.

21 Khulbe, K.C. and Matsuura, T. (2018). Removal of heavy metals and pollutants by membrane adsorption techniques. *Applied Water Science* 8: 19. https://doi.org/10.1007/s13201-018-0661-6.

22 Vo, T.S., Hossain, M.M., Jeong, H.M., and Kim, K. (2020). Heavy metal removal applications using adsorptive membranes. *Nano Convergence* 7: 36. https://doi.org/10.1186/s40580-020-00245-4.

23 Carmona, B. and Abejón, R. (2023). Innovative membrane technologies for the treatment of wastewater polluted with heavy metals: perspective of the potential of electrodialysis, membrane distillation, and forward osmosis from a bibliometric analysis. *Membranes* 13: 385. https://doi.org/10.3390/membranes13040385.

24 Verma, B., Balomajumder, C., Sabapathy, M., and Gumfekar, S.P. (2021). Pressure-driven membrane process: a review of advanced technique for heavy metals remediation. *Processes* 9: 752. https://doi.org/10.3390/pr9050752.

25 Nasir, A.M., Goh, P.S., Abdullah, M.S. et al. (2019). Adsorptive nanocomposite membranes for heavy metal remediation: recent progresses and challenges. *Chemosphere* 232: 96–112. https://doi.org/10.1016/j.chemosphere.2019.05.174.

26 Darwano, H., Duy, S.V., and Sauvé, S. (2014). A new protocol for the analysis of pharmaceuticals, pesticides, and hormones in sediments and suspended particulate matter from rivers and municipal wastewaters. *Archives of Environmental Contamination and Toxicology* 66: 582–593. https://doi.org/10.1007/s00244-014-0007-9.

27 Buruga, K., Song, H., Shang, J. et al. (2019). A review on functional polymer-clay based nanocomposite membranes for treatment of water. *Journal of Hazardous Materials* 379: 120584. https://doi.org/10.1016/j.jhazmat.2019.04.067.

28 Begum, S., Yuliana, N., and Saleh, N. (2021). Review of chitosan composite as a heavy metal adsorbent: material preparation and properties. *Carbohydrate Polymers* 259: 117613. https://doi.org/10.1016/j.carbpol.2021.117613.

29 Saheed, I.O., Da Oh, W., and Suah, F.M.B. (2021). Chitosan modifications for adsorption of pollutants—a review. *Journal of Hazardous Materials* 408: 124889. https://doi.org/10.1016/j.jhazmat.2020.124889.

30 Fatullayeva, S., Tagiyev, D., Zeynalov, N. et al. (2022). Recent advances of chitosan—based polymers in biomedical applications and environmental protection. *Journal of Polymer Research* https://doi.org/10.1007/s10965-022-03121-3 29.

31 Cao, Y., Zhang, Y., Zhang, Y. et al. (2020). Biodegradable functional chitosan membrane for enhancement of artemisinin purification. *Carbohydrate Polymers* 246: 116590. https://doi.org/10.1016/j.carbpol.2020.116590.

32 Yang, Z., Ye, T., Ma, F. et al. (2022). Preparation of chitosan/clay composites for safe and effective hemorrhage control. *Molecules* 27: 2571.

33 Anitha, A., Sowmya, S., Kumar, P.T.S. et al. (2014). Chitin and chitosan in selected biomedical applications. *Progress in Polymer Science* 39: 1644–1667. https://doi.org/10.1016/j.progpolymsci.2014.02.008.

34 Biswas S, Fatema J, Debnath T, Rashid TU. Chitosan–clay composites for wastewater treatment: a state-of-the-art review. *ACS ES&T Water* 2021 1(5):1055-85. https://doi.org/10.1021/acsestwater.0c00207.

35 Spoiala, A., Ilie, C., Ficai, D., & Ficai, A. (2021). Chitosan-based nanocomposite polymeric membranes for water purification—a review *Materials*, 1–29. https://doi.org/10.3390/ma14092091, 14.

36 Abdulrasool, M.M., Ruaa, K.M., Mays, A.D. et al. (2021). Regeneration of chitosan-based adsorbents used in heavy metal adsorption. *Journal of Life Sciences* 15: 11–19. https://doi.org/10.17265/1934-7391/2021.01.002.

37 Mourya, V.K. and Inamdar, N.N. (2008). Chitosan-modifications and applications: opportunities galore. *Reactive and Functional Polymers* 68 (6): 1013–1051. https://doi.org/10.1016/j.reactfunctpolym.2008.03.002.

38 Hameed, A.Z., Raj, S.A., Kandasamy, J. et al. (2022). Chitosan: a sustainable material for multifarious applications. *Polymers* 14: 2335. https://doi.org/10.3390/polym14122335.

39 Liu, C., Bai, R., and Nan, L. (2004). Sodium tripolyphosphate (TPP) crosslinked chitosan membranes and application in humic acid removal. *AIChE Annual Meeting, Conference Proceedings*, 5561–5574.

40 Baklagina, Y.G., Klechkovskaya, V.V., Kononova, S.V. et al. (2018). Polymorphic modifications of chitosan. *Crystallography Reports* 63 (3): 303–313. https://doi.org/10.1134/S1063774518030033.

41 Lizardi-mendoza, J., Monal, W.M.A., and Valencia, F.M.G. (2016). Chemical characteristics and functional properties of chitosan. In: *Chitosan in the Preservation of Agricultural Commodities* (ed. S. Bautista-baños, G. Romanazzi, and A. Jiménez-aparicio), 11–31. Elsevier Inc. https://doi.org/10.1016/B978-0-12-802735-6/00001-X.

42 Monier, M. (2012). Adsorption of Hg_2, Cu^{2+} and Zn^{2+} ions from aqueous solution using formaldehyde cross-linked modified chitosan—thioglyceraldehyde Schiff' s base. *International Journal of Biological Macromolecules* 50 (3): 773–781. https://doi.org/10.1016/j.ijbiomac.2011.11.026.

43 Wu, F., Tseng, R., and Juang, R. (2010). A review and experimental verification of using chitosan and its derivatives as adsorbents for selected heavy metals. *Journal of Environmental Management* 91 (4): 798–806. https://doi.org/10.1016/j.jenvman.2009.10.018.

44 Varma, A.J., Deshpande, S.V., and Kennedy, J.F. (2004). Metal complexation by chitosan and its derivatives: a review. *Carbohydrate Polymers* 55 (1): 77–93. https://doi.org/10.1016/j.carbpol.2003.08.005.

45 Sangeetha, K., Vinodhini, P.A., Sudha, P.N. et al. (2019). Novel chitosan based thin sheet nanofiltration membrane for rejection of heavy metal chromium. *International Journal of Biological Macromolecules* 132: 939–953. https://doi.org/10.1016/j.ijbiomac.2019.03.244.

46 Bouzid, S., Gassara, S., Bouaziz, J. et al. (2017). Development and characterization of porous membranes based on kaolin/ chitosan composite. *Applied Clay Science* 143: 1–9. https://doi.org/10.1016/j.clay.2017.03.008.

47 Thakur, V.K. and Voicu, S.I. (2016). Recent advances in cellulose and chitosan based membranes for water purification: a concise review. *Carbohydrate Polymers* 146: 148–165. https://doi.org/10.1016/j.carbpol.2016.03.030.

48 Osifo, P.O., Neomagus, H.W.J.P., van der Merwe, H., and Branken, D.J. (2017). Transport properties of chitosan membranes for zinc (II) removal from aqueous systems. *Separation and Purification Technology* 179: 428–437. https://doi.org/10.1016/j.seppur.2017.02.030.

49 Cui, L., Gao, S., Song, X. et al. (2018). Preparation and characterization of chitosan membranes. *RSC Advances* 8: 28433–28439. https://doi.org/10.1039/c8ra05526b.

50 Wang, J. and Zhuang, S. (2022). Chitosan-based materials: preparation, modification and application. *Journal of Cleaner Production* 355: 131825. https://doi.org/10.1016/j.jclepro.2022.131825.

51 Salehi, E., Khajavian, M., Sahebjamee, N. et al. (2022). Advances in nanocomposite and nanostructured chitosan membrane adsorbents for environmental remediation: a review. *Desalination* 527: 115565. https://doi.org/10.1016/j.desal.2022.115565.

52 Wan, Y., Creber, K.A.M., Peppley, B., and Bui, V.T. (2003). Ionic conductivity of chitosan membranes. *Polymer* 44: 1057–1065. https://doi.org/10.1016/S0032-3861(02)00881-9.

53 Xu, D., Hein, S., and Wang, K. (2008). Chitosan membrane in separation applications. *Materials Science and Technology* 24 (9): 1076–1087. https://doi.org/10.1179/174328408X341762.

54 Sanjari, A.J. and Asghari, M. (2016). A review on chitosan utilization in membrane synthesis. *ChemBioEng Reviews* 3 (3): 134–158. https://doi.org/10.1002/cben.201500020.

55 Qurashi, M.T., Blair, H.S., and Allen, S.J. (1992). Studies on modified chitosan membranes. I. Preparation and characterization. *Journal of Applied Polymer Science* 46: 255–261. https://doi.org/10.1002/app.1992.070460206.

56 Clasen, C., Wilhelms, T., and Kulicke, W.-M. (2006). Formation and characterization of chitosan membranes. *Biomacromolecules* 7: 3210–3222. https://doi.org/10.1002/(sici)1097-4628(19980118)67:3<513::aid-app14> 3.3.co;2-5.

57 Uddin, M.K. (2017). A review on the adsorption of heavy metals by clay minerals, with special focus on the past decade. *Chemical Engineering Journal* 308: 438–462. https://doi.org/10.1016/j.cej.2016.09.029.

58 Amari, A., Alzahrani, F.M., Katubi, K.M. et al. (2021). Clay-polymer nanocomposites: preparations and utilization for pollutants removal. *Materials* 14: 1–21.

59 Khan, S., Ajmal, S., Hussain, T., and Rahman, M.U. (2023). Clay-based materials for enhanced water treatment: adsorption mechanisms, challenges, and future directions. *Journal of Umm Al-Qura University for Applied Sciences* https://doi.org/10.1007/s43994-023-00083-0 11: 219–234.

60 Barakan, S. and Aghazadeh, V. (2021). The advantages of clay mineral modification methods for enhancing adsorption efficiency in wastewater treatment: a review. *Environmental Science and Pollution Research* 28: 2572–2599.

61 Khairy, M., Ayoub, H.A., Rashwan, F.A., and Abdel-hafez, H.F. (2018). Chemical modification of commercial kaolin for mitigation of organic pollutants in environment via adsorption and generation of inorganic pesticides. *Applied Clay Science* 153: 124–133. https://doi.org/10.1016/j.clay.2017.12.014.

62 Alomari, A.D.A. (2024). Chemically modified clay for adsorption of contaminants: trends, advantages and limitations—a concise review. *International Journal of Environmental Analytical Chemistry* 105. https://doi.org/10.1080/03067319.2024.2311296: 2302–2325.

63 Nasreen, K. and Sudha, P.N. (2014). Synthesis and characterization of ternary blends of chitosan, vermiculite and montmorillonite clay. *Der Pharmacia Lettre* 6: 79–87.

64 Kinoti, I.K., Karanja, E.M., Nthiga, E.W. et al. (2022). Review of clay-based nanocomposites as adsorbents for the removal of heavy metals. *Journal of Chemistry* 2022 (1): 7504626.

65 Ewis, D., Ba-abbad, M.M., Benamor, A., and El-naas, M.H. (2022). Adsorption of organic water pollutants by clays and clay minerals composites: a comprehensive review. *Applied Clay Science* 229: 106686. https://doi.org/10.1016/j.clay.2022.106686.

66 Arfaoui, S., Frini-srasra, N., and Srasra, E. (2008). Modelling of the adsorption of the chromium ion by modified clays. *Desalination* 222: 474–481. https://doi.org/10.1016/j.desal.2007.03.014.

67 Biswas, S., Rashid, T.U., Debnath, T. et al. (2020). Application of chitosan-clay biocomposite beads for removal of heavy metal and dye from industrial effluent. *Journal of Composites Science* 4 (1): 16.

68 Tirtom, V.N., Dinçer, A., Becerik, S. et al. (2012). Comparative adsorption of Ni(II) and Cd(II) ions on epichlorohydrin crosslinked chitosan–clay composite beads in aqueous solution. *Chemical Engineering Journal* 197: 379–386. https://doi.org/10.1016/j.cej.2012.05.059.

69 Pandey, P. and Saini, V.K. (2019). Pillared interlayered clays: sustainable materials for pollution abatement. *Environmental Chemistry Letters* 17: 721–727. https://doi.org/10.1007/s10311-018-00826-0.

70 Qu, B. and Luo, Y. (2021). A review on the preparation and characterization of chitosan-clay nanocomposite films and coatings for food packaging applications. *Carbohydrate Polymer Technologies and Applications* 2: 100102. https://doi.org/10.1016/j.carpta.2021.100102.

71 Foorginezhad, S., Zerafat, M.M., Mohammadi, Y., and Asadnia, M. (2022). Fabrication of tubular ceramic membranes as low-cost adsorbent using natural clay for heavy metals removal. *Cleaner Engineering and Technology* 10: 100550. https://doi.org/10.1016/j.clet.2022.100550.

72 Parhi, R. and Jena, G.K. (2022). Montmorillonite-chitosan based nano-composites and applications. In: *Advanced Applications of Micro and Nano Clay Biopolymer-based Composites* (ed. A. Al-Ahmed and Inamuddin), 49–86. https://doi.org/10.21741/9781644901915-8.

73 Bouzid, S., Gassara, S., Bouaziz, J. et al. (2019). Enhancing hydrophilicity and permeation flux of chitosan/kaolin composite membranes by using polyethylene glycol as porogen. *Applied Clay Science* 168: 312–323. https://doi.org/10.1016/j.clay.2018.11.029.

74 Rekik, S.B., Gassara, S., Bouaziz, J. et al. (2023). Fabrication, characterization and permeation studies of ionically cross-linked chitosan/kaolin composite membranes. *Periodica Polytechnica, Chemical Engineering* 67 (2): 256–270. hal-04099325. https://hal.umontpellier.fr/hal-04099325.

75 Reddy, B.H.N., Lakshmi, V.V., Mahesh, K.R.V. et al. (2016). Preparation of chitosan/ different organomodified clay polymer nanocomposites: studies on morphological, swelling, thermal stability and anti-bacterial properties. *Nanosystems: Physics, Chemistry, Mathematics* 7 (4): 667–674. https://doi.org/10.17586/2220-8054-2016-7-4-667-674.

76 Benucci, I., Liburdi, K., Cacciotti, I. et al. (2018). Food hydrocolloids chitosan/clay nanocomposite films as supports for enzyme immobilization: an innovative green approach for winemaking applications. *Food Hydrocolloids* 74: 124–131. https://doi.org/10.1016/j.foodhyd.2017.08.005.

77 Yousefi, V., Mohebbi-kalhori, D., and Samimi, A. (2019). Investigation of the self-assembled chitosan/montmorillonite nanocomposite over the ceramic support as a low-cost membrane for microbial fuel cell application. *International Journal of Hydrogen Energy* 45 (7): 4804–4820. https://doi.org/10.1016/j.ijhydene.2019.11.216.

78 Shchipunov, Y. (2012). Bionanocomposites: green sustainable materials for the near future. *Pure and Applied Chemistry* 84: 2579–2607. https://doi.org/10.1351/PAC-CON-12-05-04.

79 Villaça, J.C., Silva, L.C.R.P., de Alexandria, A.K. et al. (2017). Development and characterization of clay-polymer nanocomposite membranes containing sodium alendronate with osteogenic activity. *Applied Clay Science* 146: 475–486. https://doi.org/10.1016/j.clay.2017.06.033.

80 Yadav, T.C., Saxena, P., and Srivastava, A.K. (2020). Potential applications of chitosan nanocomposites: recent trends and challenges. In: *Advanced Functional Textiles and Polymers* (ed. Shahid-ul-Islam and B.S. Butola), 365–403. https://doi.org/10.1002/9781119605843.ch13.

81 Baccour, A., Jridi, M., Kchaou, H. et al. (2020). Preparation, characterization, mechanical and barrier properties investigation of chitosan-kaolinite nanocomposite. *Polymer Testing* 84: 106380. https://doi.org/10.1016/j.polymertesting.2020.106380.

82 Li, S., Mu, B., Zhang, H. et al. (2022). Incorporation of silver nanoparticles/curcumin/clay minerals into chitosan film for enhancing mechanical properties, antioxidant and antibacterial activity. *International Journal of Biological Macromolecules* 223: 779–789. https://doi.org/10.1016/j.ijbiomac.2022.11.046.

83 Ulu, A., Alpaslan, M., Gultek, A., and Ates, B. (2022). Eco-friendly chitosan/κ -carrageenan membranes reinforced with activated bentonite for adsorption of methylene blue. *Materials Chemistry and Physics* 278 (December 2021): 125611. https://doi.org/10.1016/j.matchemphys.2021.125611.

84 Majigsuren, E., Byambasuren, U., Bat-amgalan, M. et al. (2024). Adsorption of chromium (III) and chromium (VI) ions from aqueous solution using chitosan–clay composite materials. *Polymers* 16: 1399.

85 Padmavathy, K.S., Murali, A., Madhu, G., and Sahoo, D.K. (2017). Adsorption of hexavalent chromium (Cr (VI)) from wastewater using novel chitosan/halloysite clay nanocomposite films. *Indian Journal of Chemical Technology* 24: 593–600.

13

Cellulose-Based Composite Membranes for Pollutants Elimination

Abdelhay El Amri[1], Nadia Zari[2], and Rachid Bouhfid[1]

[1] *International Water Research Institute (IWRI), Mohammed VI Polytechnic University, Lot 660 Hay Moulay Rachid, Ben Guerir 43150, Morocco*
[2] *Moroccan Foundation for Advanced Science, Innovation and Research (MAScIR), Mohammed VI Polytechnic University, Lot 660-Hay Moulay Rachid, 43150 Ben Guerir, Morocco*

13.1 Introduction

In the coming years, demand for water for domestic, agricultural, and industrial use will increase, particularly in developing countries. According to the World Water Council, by 2030 more than 3.9 billion people will be living in areas of increasing water scarcity [1]. This problem represents a socio-economic barrier and requires global cooperation to implement effective technical solutions [2]. Several types of industry, including pharmaceutical, food processing, and chemical industries, produce saline wastewater [3, 4]. Water is considered hypersaline if its concentration of inorganic salts is between 1% and 3.6% w/w [5]. Conventional techniques for the treatment of hypersaline wastewater face difficulties such as increased treatment costs and the handling of by-products. For this reason, the research community has explored several sophisticated methods using new materials for efficient water treatment [6]. The use of membranes for desalination in commercial water treatment plants has recently attracted a great deal of interest, particularly in developed countries. This is because these systems outperform chemical treatment plants in terms of efficiency and convenience [7]. The efficiency of a desalination membrane depends on its surface properties, such as hydrophobicity, pore size, reactive functional groups, and surface charge [2].

Recently, nanomaterials have been used to improve the physicochemical properties of membranes. The introduction of nanoparticles into different layers of the membrane by coating, grafting, or implantation can improve its physicochemical properties and performance [8]. Nanocellulose nanocrystals (CNCs), bacterial nanocelluloses (BNCs), and cellulose nanofibrils (CNFs) are among the promising nanomaterials for desalination applications [9]. Nanocellulose (NC) has several advantages including durability, nontoxicity, hydrophilicity, and renewable origin [10, 11]. They have a large surface area, are highly resistant, nontoxic, and ideal for desalination, adsorption, and pollution removal [12–14]. They can contain metals, organic waste, emulsions, organic and inorganic dyes, potentially hazardous microbes, and hypersaline wastewater. The aim of this chapter is to introduce nanocellulose membranes, highlighting their different characteristics, their latest synthesis methods by controlling their size and geometric aspect ratio. It also explains how the surface properties of these membranes can be modified according to their application, i.e., adsorption, filtration, flocculation, ultrafiltration, or nanofiltration for effective removal of contaminants, including toxic metal ions, lubricants, organic dyes, drugs, and pesticides.

13.2 Cellulose: Structure and Surface Chemistry

Cellulose is a naturally occurring polymer that is produced by various organisms such as plants, trees, algae (Valonia), as well as some bacteria (Acetobacter) and marine animals such as tunicate. This biopolymer, which is insoluble in water, is the most abundant on earth, with a biomass production estimated at 1011–1012 tons/year worldwide. Cellulose is completely biodegradable, renewable, biocompatible, and inexpensive, making it a biopolymer of choice for a very wide range of applications in fields ranging from biomedical to energy, thermal insulation, and food packaging.

It is overwhelmingly produced by plants, of which it constitutes the cell wall, thanks to synthetic enzymes called "cellulose synthase." This wall is organized in the form of a nanocomposite: cellulose in the form of microfibers plays the role of reinforcement while hemicellulose and lignin behave as a matrix. The diversity of cellulose sources (plants, trees, bacteria) means that the quality of the raw material varies depending on the species from which it is derived. For example, wood from trees contains between 35% and 50% cellulose, whereas cotton is made up entirely of pure cotton fibers, with around 94% [15]. Also, the variation in the synthesis process leads to a significant disparity in physical properties (crystallinity rate, length and diameter of fibrils, morphology). Cellulose is a linear homopolysaccharide whose repeating unit is D-anhydroglucopyranose ($C_6H_{11}O_5$), each of which carries three hydroxyl groups linked by β-1,4 bonds. The anhydroglucose units (AGUs) adopt a chair conformation, which is the most thermodynamically stable because it minimizes steric and van der Waals interactions between the chains (Figure 13.1a).

The ends of the chains have different properties, one of which is called the "nonreducing end" and has a C1—OH group, while the other is called the "reducing end" and has a C1=O aldehyde function in equilibrium with its

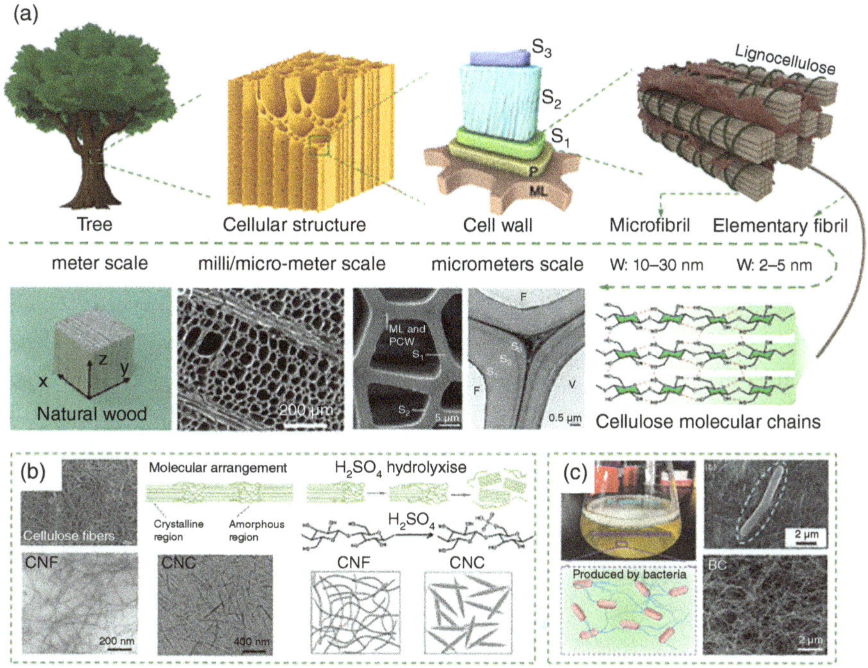

Figure 13.1 (a) Hierarchy of structure in a cellulose fiber from the plant cell wall. (b) The forms and insides of CNF and CNC. Scanning electron micrograph of cellulose fibers. The crystalline and amorphous parts of cellulose's molecular structure are shown here. Hydrolysis using sulfuric acid for the separation of CNC from cellulose. (c) Electron micrograph of CNF [16].

C1—OH hemiacetal form, giving it reducing properties (Figure 13.1b). The degree of polymerization (DP) characterizes the number of AGU units that make up the macromolecular chain and depends on the source of the material, but also on the extraction processes. 300 and 1700, while the cellulose in cotton fibers and other plants has a DP of between 800 and 10,000. After their biological synthesis, cellulose chains link together through inter- and intramolecular interactions. The molecular chains assemble to form a semi-crystalline structure called an "elementary fibril" or "protofibril" with a diameter of 3–5 nm [17].

These microfibrils can have a square cross-section (20 nm × 20 nm) for Valonia, for example, or a rectangular cross-section (6–10 nm × 30–50 nm) in the case of microfibrils produced by Acetobacter, with a length of more than 1 μm. During biosynthesis, crystalline regions alternate with a limited number of disordered amorphous regions along the microfibril (Figure 13.1c). The mass ratio of the crystalline phase to the total mass of the sample is called the crystallinity rate and is strongly dependent on the origin of the cellulose. Various mechanical or chemical treatments (acid hydrolysis, oxidation) can be used to isolate specific structural levels, such as nanofibrils (NFCs) or nanocrystals (CNCs). The resulting materials are commonly referred to as cellulosic nanomaterials or nanocelluloses (CNMs) [18].

13.3 Microfibrillated Cellulose (MFC) and Cellulose Nanocrystals (CNCs)

There are many mechanical processes for developing NFCs, such as homogenization, grinding, refining, ultrasound, or extrusion. These processes disintegrate cellulose fibers to produce highly entangled microfibril networks with high aspect ratios (10–100). The particles formed are longer than 1 μm and have a characteristic width varying from 5 (single fibril, NFC) to 50 nm (multiple fibril bundle, MFC). NFCs can also be obtained by chemical surface treatments, the most

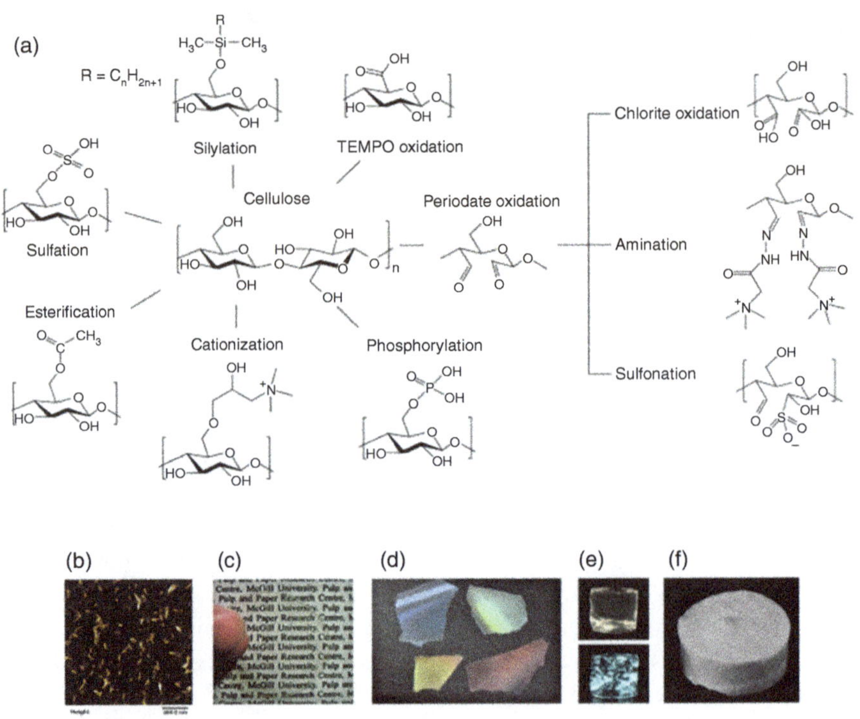

Figure 13.2 (a) Reactions leading to NFC. From top to bottom: oxidation with the TEMPO radical; carboxylation via periodate-chlorite oxidation; sulfonation via periodate-bisulfite treatment; carboxymethylation and quaternization. (b) Hairy nanocellulose as an antibacterial carrier, (c) transparent film, (d) chiral nematic films, (e) hydrogels, and (f) aerogels [19].

common of which is oxidation using the 2,2,6,6-tetramethylpiperidine-1-oxyl radical (TEMPO) in the presence of bromide (NaBr) and sodium hypochlorite (NaClO) at basic pH (Figure 13.2).

This reaction allows the primary hydroxyls located in the C_6 position of the cycle to be oxidized into carboxylic acids with a density of up to 1.52 mmol g^{-1}. The electrostatic repulsions induced by the presence of the negative charges R-COO^- thus allow the delamination of the nanofibrils. Other types of chemical treatments such as enzymatic hydrolysis, carboxylation, carboxymethylation, quaternization, or sulfonation can be used in order to introduce ionic functions on the surface of the nanofibrils and promote their individualization [20]. CNCs, also called whiskers, are obtained by acid hydrolysis of cellulose fibers. During this reaction, the amorphous areas composing the nanofibrils are selectively

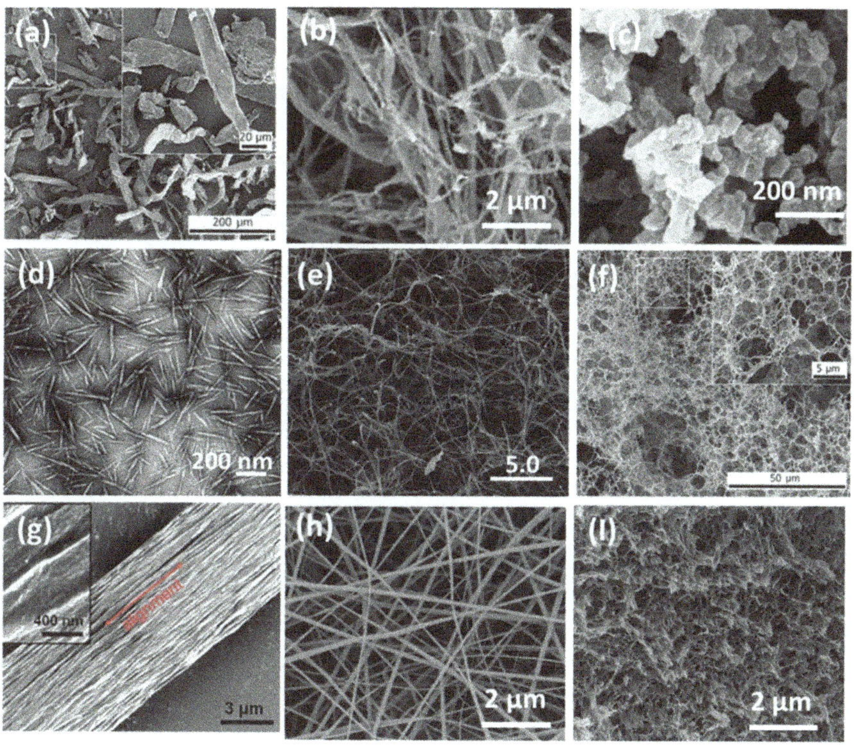

Figure 13.3 Electron microscope micrographs of different types of cellulosic nanomaterials: (a) cellulose microfibers (CMFs), (b) cellulose fibers (CNFs), (c) spherical cellulose nanoparticles (SCNPs), (d) cellulose nanocrystals (CNCs), (e) cellulose nanofibers (CNFs), (f) bacterial cellulose (BC), (g) CNFs assembled into a material stronger than spider silk, (h) electrospun CNFs, and (i) cellulose aerogels [22].

degraded, leaving the crystalline regions intact. Rigid crystalline objects with a needle shape are then produced.

Their sizes as well as their degree of crystallinity (54–84%) vary depending on the sources from which the cellulose comes [21]. CNCs have lower form factors than NFCs with a typical geometry of 50–300 nm in length and 5–10 nm in diameter for CNCs from plants. Nanocrystals resulting from the hydrolysis of bacterial cellulose from Valonia or tunicate have much higher form factors with lengths that can exceed one micrometer. Examples of CNCs are given in Figure 13.3.

CNCs are produced by hydrolysis in concentrated acid medium, the most used being sulfuric acid. Hydrolysis by hydrochloric acid, hydrobromic acid, or phosphoric acid has also been tested. The surface charges of CNCs as well as the appearance of the crystals are determined by the hydrolysis conditions (temperature, reaction time, agitation, concentration, and type of acid). Hydrolysis with phosphoric and sulfuric acid gives rise to negatively charged surfaces due to phosphating or partial sulfation of hydroxyls on the surface of the crystals. The surface charge density is higher with the use of sulfuric acid than with phosphoric acid (30–90 versus 10.8 mmol kg^{-1}, respectively). Studies have shown that it depends on the duration of hydrolysis and that the size of the crystals decreases with increasing charge density. The presence of these charges makes it possible to stabilize aqueous suspensions of CNCs through electrostatic repulsions.

13.4 Chemical Surface Modifications of Nanocellulose

Cellulose has a structure that allows it to be functionalized in different ways. As such, the hydroxyl functions offer many possibilities for chemical modifications, including esterification and etherification. It is also possible to partially oxidize the surface hydroxyls of cellulose to introduce carboxylic acid or aldehyde functions that then allow other modification routes to be used. These chemical modifications can have various objectives, such as making cellulose more hydrophobic, introducing ionic functions to the surface of the fibers to form NFCs or grafting polymers onto them. In addition, the surface of cellulose can interact noncovalently with other chemical species, in particular via hydrogen bonds, electrostatic or van der Waals interactions.

13.4.1 Etherification

Cellulose carboxymethylation is a process developed in 1918 and commonly used to allow better individualization of fibrils in order to prepare NFCs. The first step of the synthesis consists in "activating" the fibers by an alkaline treatment with sodium or potassium hydroxide. The fibers are then put in contact with

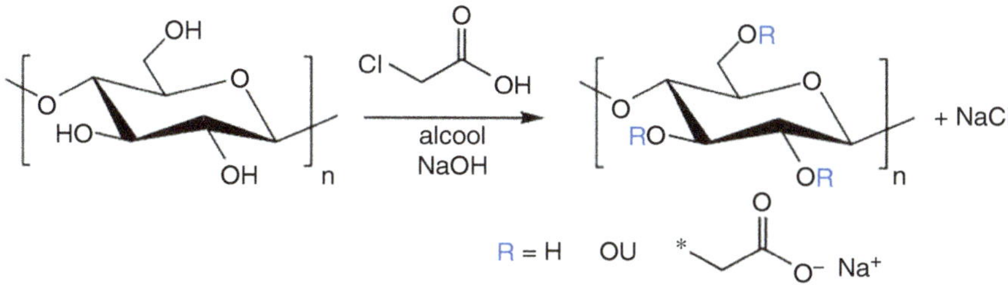

Figure 13.4 Carboxymethylation reaction of cellulose.

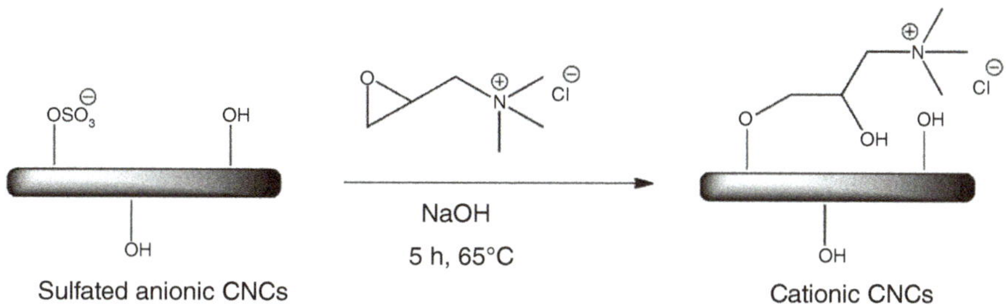

Figure 13.5 Cationization reaction of CNCs by EPTMAC [23].

chloroacetic acid or sodium chloroacetate salt in ethanol or isopropanol to obtain carboxymethyl functions (Figure 13.4).

Nucleophilic substitution occurs preferentially on the O_6 position, then on O_2 and O_3 of the cellulose ring. This process has the disadvantage of using toxic reagents such as sodium chloroacetate or its derivatives. The production of CNF is also permitted via the introduction of positive charges. As such, a classic reaction is the cationization of the surface of NFCs or CNCs by the nucleophilic substitution of 2,3-epoxypropyltrimethylammonium chloride (EPTMAC) (Figure 13.5).

The reaction generally takes place in a basic aqueous medium (NaOH) by heating at 50–65 °C for several hours, NaOH here allowing the conversion of the hydroxyl groups into their reactive alcoholate forms. In the case of a reaction in a dilute medium, the suspension of cationized NFCs obtained has a high viscosity [24]. Values of DS = 0.02 for CNCs from cotton and DS = 0.04 for CNCs from wood are typically obtained for reactions in an alkaline heterogeneous medium. The results obtained by spraying technique showed higher functionalization rates for cotton CNCs (DS = 0.12), the highest values being reached for substrates previously activated by a basic treatment, with DS = 0.25 and 0.29 for mercerized cellulose pulp and dissolving pulp, respectively. This study concludes on the regioselective nature of the substitution by showing that it occurs preferentially on O_6, then on O_2 and finally significantly less on O_3 of the AGU cycle.

13.4.2 Esterification/Acetylation

Esterification of cellulose surface hydroxyls represents a predominant modification pathway because it allows a very wide choice of functionalization. Cellulose esters are generally classified into two groups: inorganic (nitrates, sulfates and cellulose phosphates) and organic. Inorganic esters can be produced by direct esterification of cellulose during hydrolysis reactions with inorganic acids such as sulfuric acid, phosphoric acid, or nitric acid. The synthesis of cellulose sulfates is generally done via the reaction with sulfuric acid, sulfur trioxide, and chloro or fluor sulfonic acid. The reaction with phosphoric acid, phosphorus pentoxide, or phosphoric acid allows the generation of phosphate groups [25].

The introduction of organic cellulose ester functions allows a wide variety of functionalization by the reaction between the hydroxyls of the AGU cycle and a carboxylic acid, an acyl halide, or an acid anhydride. The reaction is generally carried out in a strongly acidic medium due to the low reactivity of the carboxylic acid function. The classic example is the reaction with a mixture of acetic acid and acetic anhydride in the presence of sulfuric acid, leading to the production of cellulose acetate, a plastic commonly used in industry. The addition of a base such as triethylamine or pyridine can activate the carboxylic acid or acyl halide function when it has too limited reactivity (fatty acids and aromatic carboxylic acids) (Figure 13.6a). Activation of carboxylic acids is also possible using p-toluenesulfonyl chloride (TosCl) to form highly

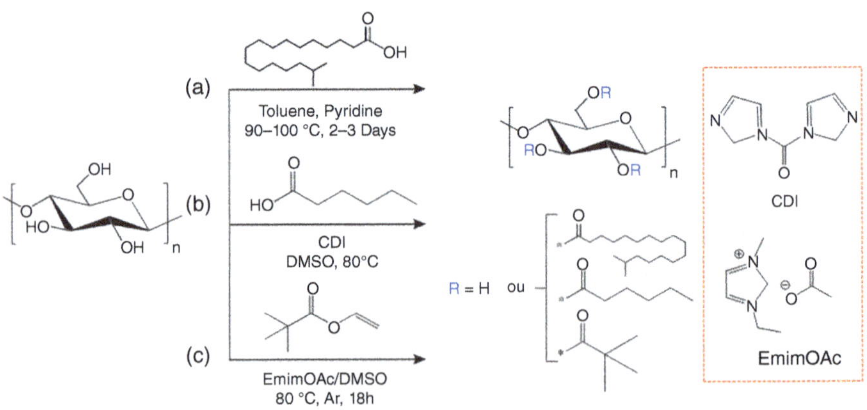

Figure 13.6 Esterification reactions: (a) with fatty acids, using pyridine, (b) with CDI as activator, and (c) by transesterification in the ionic liquid EmimAOc/DMSO.

reactive intermediates and obtain high DSs (0.62–1.42). 4-(4,6-Dimethoxy-1,3,5-triazin-2-yl)-4-methyl-morpholinium chloride (DMTMM) is a less toxic alternative to TosCl, although it leads to lower DSs (0.67). *N.N′*-Carbonyldiimidazole (CDI) is the most widely used nontoxic alternative for in situ activation of cellulose esterification (Figure 13.6b).

In general, esterification can be carried out in heterogeneous media via a modification of the crystal surface or in homogeneous media in solution; the second method requires the use of systems capable of solubilizing cellulose such as *N,N*-dimethylacetamide lithium chloride (DMAC–LiCl) or dimethyl sulfoxide tetra-butylammonium fluoride (DMSO–TBAF). Ionic liquids such as 1-ethyl-3-methylimidazolium acetate (EmimOAc) can also be used to form esters by transesterification with the obtaining of high DS (DS > 2.9) (Figure 13.6c).

13.4.3 Amination on Periodate Oxidized Cellulose

Several studies have examined the amination of cellulose dialdehyde surfaces, by varying the amines used as well as the reaction conditions. The formation of CNCs from wood pulp by a sequenced periodate oxidation reaction followed by a reductive amination. The authors thus grafted aliphatic amines (10 eq/aldehyde) at pH = 4.5, using a-picoline borane as a reducing agent with a conversion of 10–23% of the initial aldehydes into amines. The same team notably used Girard's reagent, which is a hydrazide that does not require the addition of a reducing agent, under similar conditions (Figure 13.7). A conversion of 60–89% of the aldehydes into cationic groups was thus obtained.

Attaching polymers, including quaternary ammonium compounds (QACs), that have antibacterial properties to nanocellulose is another way to make nanocellulose with these properties. The anionic phosphate groups found in the peptidoglycans and phospholipids of Gram-positive and Gram-negative bacteria, respectively, are preferentially bound to by the cationic centers of QACs. Bacterial cell membranes gradually lose their integrity, leading to cell death via rupture.

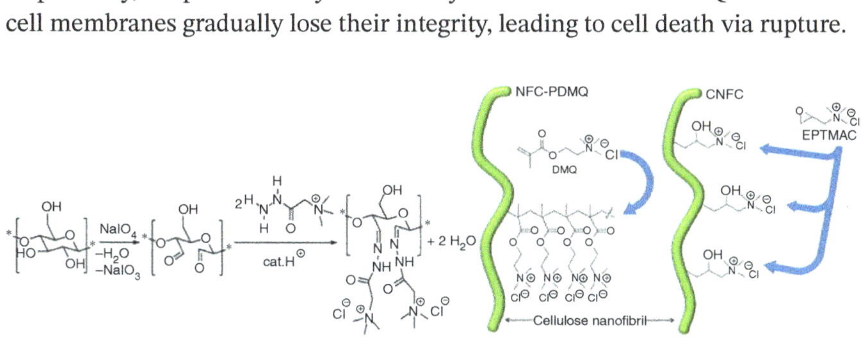

Figure 13.7 Cationization of periodate oxidized cellulose by Girard's T reagent [26, 27].

Various methods can be employed to incorporate QACs onto nanocellulose; a comparison of two methods of cationization of CNFs was conducted. Figure 13.7a shows that two methods were used: epoxypropyl trimethylammonium chloride (EPTMAC) etherification and redox-initiated graft copolymerization using [2-(methacryloyloxy)ethyl] trimethylammonium chloride. Finally, the researchers discovered that both methods produced materials with a wide range of antibacterial activity. However, the process of etherification led to a higher degree of substitution and increased charge density. The process of etherification resulted in the production of CNFs that exhibited more efficacy in eradicating germs, particularly Gram-negative bacteria, compared to copolymerization [9]. The surfaces of aerogels were crucial factors for achieving the most effective antibacterial activity, as seen in Figure 13.7b. Because they couldn't easily reach the aerogel's sorption sites, the quaternary ammonium groups in copolymerized CNFs were less effective at killing bacteria. The goal of treating surfaces with QAC is to achieve electrical neutrality by allowing anionic contaminants to adsorb onto them. In addition, strong evidence suggests that some Quaternary Ammonium Compounds (QACs) lose part of their suppressive effects and bacterial resistance develops with time spent exposed to them. Both of these difficulties provide opportunities for more investigation [27].

13.4.3.1 The Use of Cellulose as a Membrane Material

The potential for sustainable membrane material development of cellulose has attracted significant attention from researchers. High hydrophilicity, excellent porosity, and desired mechanical strength are some of the things researchers are interested in. A number of notable advantages that cellulose provides have contributed to its rising profile. Its nontoxicity, ease of processing, high crystalline structure, natural water-attracting capabilities, and insolubility in various solvents are just some of its many desirable properties. All parts of cellulose contain strong hydrogen bonding, which supports all these benefits. Nanocellulose, which is derived from cellulose-based materials, stands out due to its ample presence of functional groups, including hydroxyl and carboxylate groups. These functional groups provide adsorption sites that are second to none. A wide range of inorganic and organic contaminants can be removed quickly and efficiently. Cellulose also contributes to the development of highly porous, robust membranes. Cellulose is used in multiple ways in water filter membranes. As a filler, it's capable of narrowing the range of pores while boosting mechanical toughness. The membrane can be used as a support layer to improve its overall performance and structural stability, or as a standalone membrane to effectively filter nanoscale particles [28]. Cellulose has a unique mix of properties that make

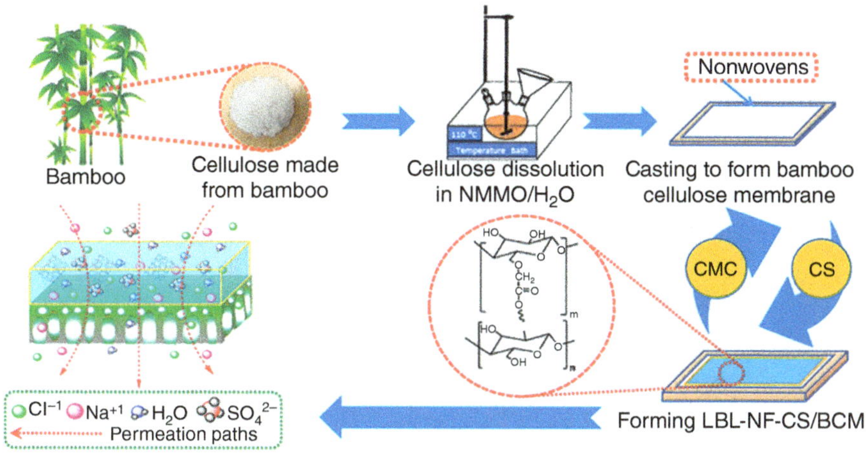

Figure 13.8 The steps involved in developing cellulose membrane technology, from sourcing the material to using it for filtering. A porous substrate made of regenerated cellulose is used to create a nanofiltration film (LBL-NF-CS/BCM) [30].

it an environmentally friendly and very promising product for the development of membrane technology, especially in the development of environmentally friendly water treatment technologies. The LBL-NF-CS/BCM composite membrane has been developed for nanofiltration [29]. The membrane consists of a thin film of alternating layers of carboxymethylcellulose (CMC) and chitosan (CS) deposited by layer-by-layer sputtering. The average pore diameter of the cellulose-based composite membrane is approximately 2.2 nm. The membrane showed a flow rate of $12\,L\,m^{-2}\,h^{-1}$ and a remarkable NaCl rejection rate of 36% under ideal conditions at a differential pressure across the membrane of 0.3 MPa (Figure 13.8).

Figure 13.9 depicts the structural characteristics of membrane materials composed of different cellulose particles. Cellulose fibers are derived from the cells of wood, with their size varying across different species. The basal layer of cellulose fiber paper had pore diameters ranging from 1 to 5 μm, whereas the coating pores had sizes ranging from 50 to 500 nm. The paper has a smooth surface with a coating that has an approximate pore size of 180 nm. The estimated porosity of the material, while its permeability is measured at 0.09 m. The initial paper substrate had pores that were 3–3.5 μm in size, whereas the coating that was applied had pores that were 50–130 nm in size. A porosity range of 32–44% was shown in Figure 13.9, while a range of 20–36% was shown in the coating. It is possible to classify the pigmented coatings of cellulose fiber paper as lower microfiltration or higher ultrafiltration, and the paper itself as a microfiltration membrane [7].

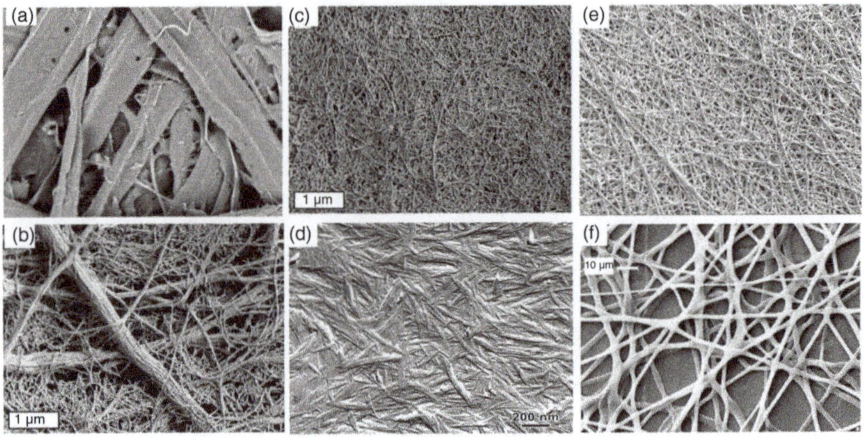

Figure 13.9 Scanning electron microscopy images of the structures of various cellulose-based membranes, such as fibrils, nanofibers, nanocrystals, electrospun nanofibers of cellulose acetate, and rebuilt cellulose (a–f) [7].

13.5 The Use of Nanocellulose as a Membrane Material

Cellulose in its acetate form has long been used for membrane filtration applications. These membranes allow the addition of adsorbents such as MOFs and clays [31] to develop a composite membrane. They also have the ability to remove chemical compounds and dyes. Compared to reverse osmosis membranes, nanofiltration membranes (with pore sizes between 1 and 10 nm) are more popular for water treatment applications due to their high throughput and low energy consumption. Interfacial polymerization of piperazine or trimesoyl chloride onto an ultrafiltration layer is a common technique for producing thin-film composite membranes for nanofiltration. However, these membranes suffer from problems such as biofouling, reduced flow rate, and chlorine degradation. To overcome these problems, nanomaterials such as carbon nanotubes, graphene oxides, and zeolites are used. However, problems arise when these materials are incompatible with both the inorganic components and the polyamide layer, causing them to separate or dissolve [32].

Another approach is to use nanocellulose as an interlayer between the functional thin film and the hydrophobic ultrafiltration membrane. In addition, these interlayers improve substrate permeability and coating efficiency without reducing salt rejection [33].

13.5.1 Membrane Nanocellulose Absorbents in Oil–Water Separation

Many industries, such as mining, oil-bearing wastewater, and the metals and semiconductor industries, contribute to the massive oil spills in the ocean caused by shipping and oil exploration. Every day, 140,000 L of oil-contaminated water are produced by a typical mining operation, which loses more than a fifth of its total oil due to spills, leaks, and overflows. Therefore, absorbents are essential to treat oil-contaminated water [34]. Nanocellulose is an efficient sorbent for oil/water separation because it can be hydrophobically modified to increase hydrophilicity and hydrolysis and produce lightweight aerogels with a large surface area [35, 36].

Nanocellulose-based aerogels may be synthesized using a variety of methods, including chemical vapor deposition, click chemistry, and atomic layer deposition, as mentioned before.

Despite their low compressive modulus, some studies deal with nanocellulose-aerogel composites with excellent absorption capacities.

Foams containing low amounts of CNFs tend to have dense structures with small pores, typically around 50 μm. In contrast, foams with a higher concentration of CNFs exhibit improved interconnectivity, featuring larger cavities (approximately 250 μm) and broader surfaces. For instance, when hydrophobic latex particles are added to CNFs, the hydrophobic properties of the foams are enhanced.

Aerogels based on nanocellulose may be synthesized via a variety of methods, including chemical vapor deposition, click chemistry, and atomic layer deposition, as mentioned previously. Mechanical forces are typically discussed when considering the application of sorbents for oil/water separation to eliminate absorbed oil during absorption cycles. A mechanical pump was used to extract the assimilated oil during intervals of absorption. The initial height could be explained by the application of NaOH to decompose the honeycomb structure (Figure 13.10a) into a spring-like lamellar structure (Figure 13.10b). This additional processing appears to have increased their mechanical strength. Although the lamellar sponge became hydrophobic after being treated with methyltrimethoxysilane, its structure remained intact. The high absorption capacity of organic solvents, silicone oils, and other oils ranges from 16 to 41 times their own weight, as shown in Figure 13.10c. These oils and solvents include hexane, chloroform, and motor oil because of their impressive ability to absorb a variety of oils, organic solvents, and gases, with capacities ranging from 16 to 41 times their own weight (Figure 13.10c).

As illustrated in Figure 13.10b, aerogels' large surface area and interconnected porous structure contribute to their high adsorption capacity, which measures

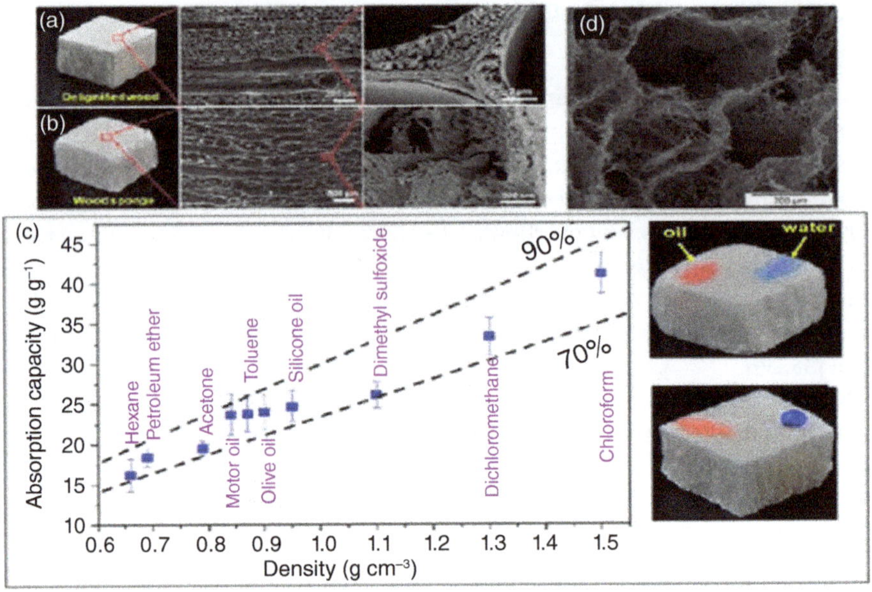

Figure 13.10 (a) Photograph and scanning electron micrographs showing the wood sponge's anisotropic structure: (b) cross-section, (c) silylated wood sponge absorbing oil, and (d) water and oil droplets on untreated and treated wood sponges. Diesel oil and diesel/water exhibit realistic multi-cycle curves. (d) SEM image of an aerogel made with 0.4 weight percent electrospun cellulose fibers in water [37].

373 g g^{-1}. The aerogel achieved a chloroform adsorption rate of about 95%. The aerogels demonstrated an impressive ability to quickly regain their original shape after multiple absorption cycles, achieving a recovery rate of approximately 89% even after experiencing a compressive stress of around 70%. Remarkably, after 50 cycles of compression, the mechanical strength of the cellulose aerogels (PVA) remained relatively stable. These aerogels were created by depositing methyl trichlorosilane on the surface of PVA/CNFs through chemical vapor deposition. They effectively extracted various metal ions from water, including mercury, lead, copper, and silver ions. Despite being hydrophobic, the aerogels maintained a sufficient surface charge for effective electrostatic adsorption, allowing them to attract and retain these ions. Moreover, it has been shown that Cu(II) and Ag(I) ions retained enough surface charge to undergo electrostatic adsorption, even after the hydrophobic treatment [38].

13.5.2 Membrane Nanocellulose-MOF Composites

Metal-organic frameworks (MOFs) are porous crystalline materials formed when inorganic metal ions or groups interact with organic ligands. They offer several advantages, including high thermal stability, a large specific surface area, and an extensive network structure.

MOFs are used in various applications, such as water treatment, where they serve as adsorbents for organic pollutants, dyes, and cations. When water passes through a MOF, it becomes denser, which reduces the flow rate through the adsorption column and decreases the availability of sorption sites. This compaction makes it more challenging to collect, discharge, and reuse contaminants. To address this limitation, the unique properties of nanocellulose can be utilized, particularly the ability of CNFs to function as scaffolds [39]. TEMPO-oxidized CNFs serve not only as scaffolds for ZIF-8 MOFs but also as sites for the nucleation and growth of these MOFs. The electrostatic attraction between Zn^{2+} ions and the carboxyl groups on the CNFs facilitates the formation and enlargement of ZIF-8 crystals. Additionally, ionic interactions and hydrogen bonds between the MOFs and CNFs complicate the removal of MOFs from the aerogel structure. The study concluded that the aerogels outperform individual ZIF-8 MOFs in maximum adsorption capacity, achieving $83.3\,\mathrm{mg\,g^{-1}}$, and demonstrate a higher rhodamine B sorption efficiency of $0.036\,\mathrm{g\,mg^{-1}\,h^{-1}}$. However, an undesirable side effect of this procedure is the leaching of ions, such as Co^{2+}, into the treated water. To address this issue, MOFs can be utilized to produce heterogeneous catalysts. The research also indicated that MOFs containing cobalt effectively degraded p-nitrophenol, achieving a 90% reduction within one hour while minimizing the release of Co^{2+}. Additionally, the study found that a dead-end filtration system exhibited a high flux rate of $84\,\mathrm{L\,m^{-2}\,h^{-1}\,bar^{-1}}$ over 24 hours when TEMPO-oxidized CNFs were combined with zinc-based MOF (ZIF-8). The combination also demonstrated significant adsorption of Janus Green B dye (98.9%) and Methylene Blue dye (93.8%). The data indicates that combining nanocellulose with aerogels effectively reduces catalyst loss and improves flux by minimizing compaction.

The use of wood as a substrate for MOFs has been extensively studied. Wood facilitates the containment of MOFs while minimizing losses and enhancing remediation efficiency by improving the interaction between the absorbent and contaminants. Figure 13.11 [40] illustrates the manufacturing process of a wood-MOF composite, which involves the direct formation of zirconium MOFs within a natural lime tree. Zirconium MOFs are created in the wood channels

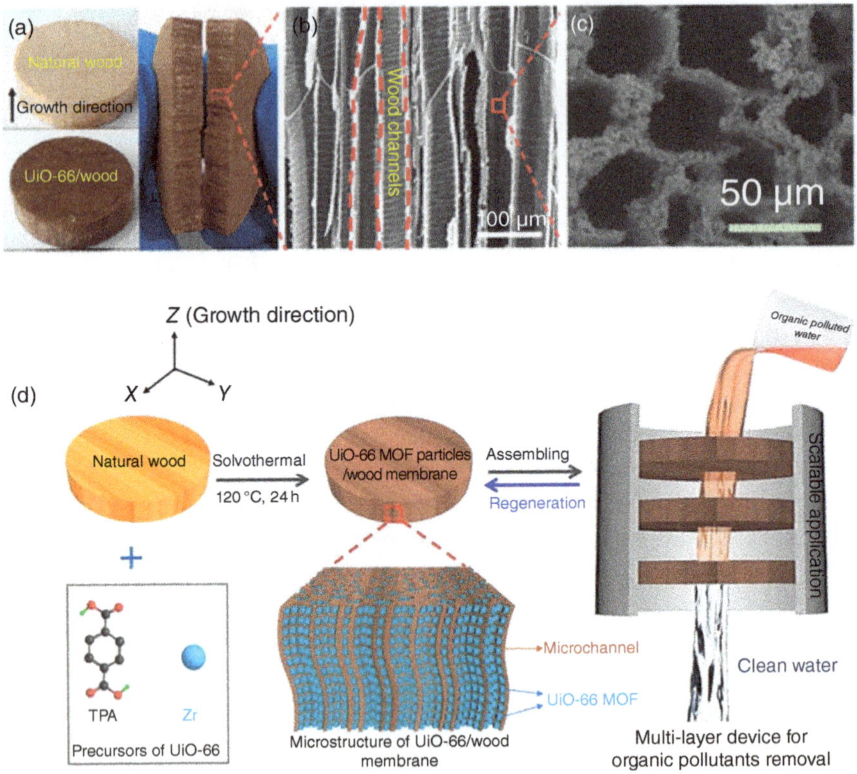

Figure 13.11 (a) Photos of the natural basswood and UiO151 66/wood membrane. (b), (c) SEM image of the UiO-66/wood membrane showing numerous elongated and low-tortuosity microchannels. (d) A schematic depicting the UiO-66 MOFs synthesis process in a 3D wood membrane for organic pollution removal [40].

by adsorbing Zr^{4+} ions, which initiate nucleation on the cellulose hydroxyl chains. Within 5 minutes, the UiO-66/wood membrane absorbed 10 mg L^{-1} of rhodamine B, reducing its concentration to 0.096 mg L^{-1}. Furthermore, after six treatment cycles, the removal efficiency remained consistently above 90%. This membrane demonstrated effective performance compared to previous methods for removing endocrine disruptors such as bisphenol A and bisphenol S, as well as the carcinogen 1-naphthylamine.

The wood-MOF composites were effective in extracting uranyl ions from saltwater using a method that involved producing a zirconium-based MOF with a high loading of polydopamine (UiO-66-NH_2). After treatment, they captured 5.81 mg g^{-1} (71.7 wt.%) of uranyl ions. In another study, a Zn(II)-based magnetic MOF aerogel demonstrated exceptional metal adsorption efficiency, achieving

558.66 mg g^{-1} for lead (Pb). Additionally, an iron-based MOF functionalized with amines showed enhanced performance in reducing hexavalent chromium when exposed to visible light, maintaining a reduction efficiency of over 80% even after 10 tests. If you are working with magnetic adsorbents, particularly bacterial nanocellulose, it's important to note that the properties of the adsorbent may vary depending on the iron precursor used. Combinations of iron (II) sulfate and iron (III) chloride yield composites with better magnetic characteristics than those made with iron (II) acetate and iron (III) chloride. Overall, it has been demonstrated that MOFs can be improved by integrating them with nanocellulose, which serves as a template for MOF structure formation and as an adsorbent support [31].

13.5.3 Nanocellulose-Based Electrolyte Membrane Nanopapers

In the current dynamics of sustainable development, polymer electrolytes based on bio-based materials are of great interest due to their low cost, their abundance, and because they are derived from renewable raw materials. The thermal and chemical stability and their mechanical properties. The term "nanopaper" refers to a membrane made of cellulose nanomaterials (1–100 nm) whose properties depend on several parameters: (i) the type of nanomaterials used (CNCs, NFCs, MFCs), (ii) the initial natural resource (wood, bacterial cellulose, tunicate, algae), and (iii) the preparation method.

The first method involves a gradual and uniform increase in the concentration of CNFs, which enhances interfibrillar interactions to produce a nanopaper. In contrast, vacuum filtration rapidly removes the solvent, causing the fibrils to concentrate quickly at the bottom and form a dense network (see Figure 13.12). Tensile tests on membranes produced by these two methods show that vacuum filtration yields superior mechanical properties. However, a significant drawback

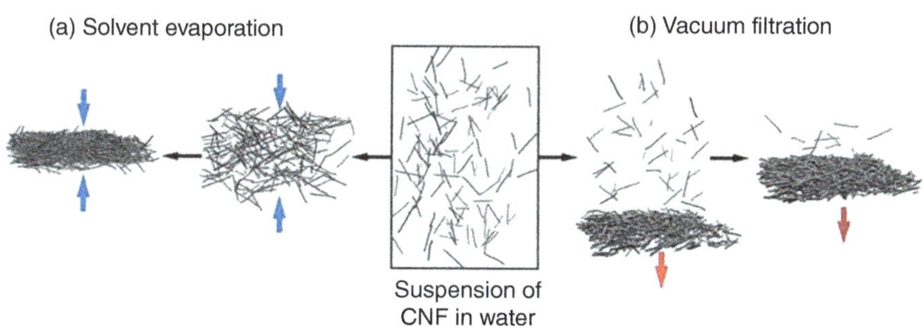

Figure 13.12 Preparation routes for nanopapers: (a) solvent evaporation and (b) vacuum filtration [41].

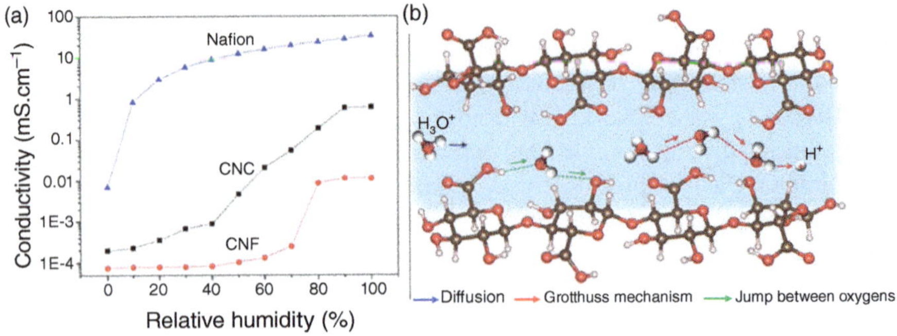

Figure 13.13 Nanocellulose exhibits proton conductivity. (a) Relationship between proton conductivity and humidity for cellulose nanofiber (CNF), cellulose nanocrystal (CNC) paper, and Nafion at a temperature of 30 °C. (b) Diagram illustrating potential processes and routes for proton conduction in nanocellulose [43].

of this technique is the need to optimize the filtration and pressing processes, including controlling pressure, time, and temperature. The solvent evaporation method requires slow evaporation to prevent defects in the membrane, which is a time-consuming process [42]. Cellulose nanopapers have potential applications as proton exchange membranes (PEMs) due to their excellent barrier properties against gases such as hydrogen (H_2) and oxygen (O_2). Generally, cellulose membranes exhibit low conductivity, approximately 0.008 mS cm^{-1} at 40 °C and relative humidity (RH) of 98%. In contrast, membranes made from microfibrillated cellulose (MFC) and CNCs have higher conductivities of around 0.05 mS cm^{-1} at 100 °C and 4.6 mS cm^{-1} at 120 °C, respectively, at 100% RH (as shown in Figure 13.13a). The elevated conductivity observed in CNCs can be attributed to the introduction of sulfuric acid groups during the acid hydrolysis process. This leads to increased hydrophilicity and acidity of the material, as well as a higher number of charge carriers (illustrated in Figure 13.13b).

13.5.4 Dendrimer-Functionalized Cellulose Nanocrystals

In recent years, several biofunctional materials have been developed by grafting dendrons onto cellulose. Each successive generation of dendrons, which are fractal structures, incorporates an increasing number of functional groups. This modification, known as dendronization, significantly enhances the chemical performance of cellulose and opens up possibilities for various bioapplications, including biosensors, drug delivery systems, biocatalysts, and more [44].

Nanoscale branching polymers, known as polyamidoamine (PAMAM) dendrimers, exhibit several advantageous properties, including a large specific surface area, a porous interior, and a controllable symmetrical structure. One of the key strengths of all generations of PAMAM dendrimers is their ability to bind metal ions, thanks to their abundance of nitrogen-containing functional groups such as primary amines, amides, and tertiary amines.

To enhance the ability of nanocrystalline cellulose (NCC) to adsorb copper ions, PAMAM dendrimers can be grafted onto its surface. NCC is not only a natural, biodegradable material with a large surface area, but it also possesses nanometric dimensions and is easy to manipulate, as shown in Figure 13.14.

After the grafting of PAMAM, the adsorption capacity of NCC for copper ions would significantly improve due to the increased number of adsorption sites provided by the functional groups on the PAMAM. Furthermore, the quantity of

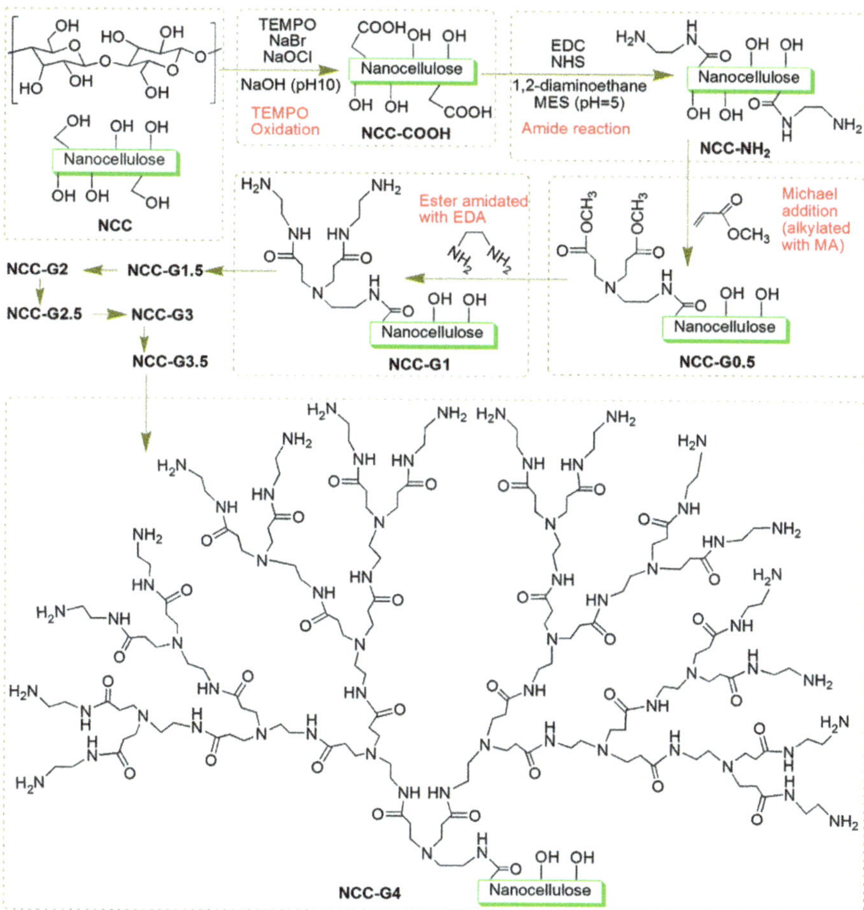

Figure 13.14 Schematic drawing of a bifunctional nanocrystalline cellulose (NCC) surface grafted with dendrons [45].

these functional groups can be adjusted during the synthesis process. To evaluate the effectiveness of these modifications, we measured the equilibrium adsorption capabilities for Cu(II) to determine how well the modified NCC removed copper ions from water-based solutions.

The preferential binding of Cu(II) ions by PAMAM is enhanced by the coordinated action of the tertiary nitrogen and oxygen atoms found in the amide groups. This conclusion is supported by simulations of Cu(II) complexation with CNN-PAMAM dendrimers, as illustrated in Figure 13.14. After 24 hours of adsorption, the system reached equilibrium. At a pH of 3.5, the adsorption capacity was 16.40 $mg\,g^{-1}$. This capacity increased to 19.92 $mg\,g^{-1}$ at pH 4.3 and rose further to 28.84 $mg\,g^{-1}$ at pH 5.4. The increase in pH enhances the adsorption capability of Cu(II) ions on NCC-PAMAM, suggesting that the process involves a complexation or chelation mechanism. At lower pH levels, a higher concentration of protons (H^+) competes with Cu(II) for the amino group's adsorption sites. As the pH rises, the amino groups undergo deprotonation, thereby improving their ability to chelate with Cu(II). This results in an observed increase in adsorption capacity.

13.5.5 Membrane-Based Composites Made from Nanocellulose and Clay

The high cation exchange capacity of clays makes them excellent adsorbents for organic compounds, dyes, metal ions, and pesticides. These low-cost materials feature a high specific surface area and possess strong physical and chemical stability. However, similar to MOFs, their powdered form significantly limits their applications in water treatment. Nevertheless, composites made from nanocellulose and clay have demonstrated synergistic effects. The hydroxyl and carboxylate groups of CNFs not only serve as a structural scaffold but also enhance the sorption of pollutants. When compared to single adsorbents, these CNC-montmorillonite composites showed a greater capacity for adsorbing methylene blue, achieving 183.8 $mg\,g^{-1}$, compared to the 96.6 $mg\,g^{-1}$ of the adsorbents alone. Additionally, as the CNF/clay ratio increased from 4 to 12, the adsorption capacity for Cu^{2+} decreased from 87.1 to 56.5 $mg\,g^{-1}$.

The high cation exchange capacity of clays makes them excellent adsorbents for organic compounds, dyes, metal ions, and pesticides. These low-cost adsorbents have a high specific surface area and high physical and chemical stability. However, as with MOFs, their powder form considerably limits their applications in water treatment. Nevertheless, nanocellulose-clay-based composites showed synergistic effects, indicating that the hydroxyl and carboxylate groups of CNF not only act as a scaffold but also contribute to the sorption of pollutants. Compared with adsorbents alone (140.6 mg g^{-1} versus 96.6 $mg\,g^{-1}$), CNC-montmorillonite composites showed a greater capacity to adsorb methylene blue (183.8 mg g^{-1}) at well-defined proportions. When the CNF/clay ratio increased from 4 to 12, the Cu^{2+} adsorption capacity decreased from 87.1 to 56.5 mg g^{-1}.

The high cation exchange capacity of clays makes them excellent adsorbents for organic compounds, dyes, metal ions, and pesticides. These low-cost adsorbents possess a high specific surface area along with significant physical and chemical stability. However, similar to MOFs, their powdered form limits their applications in water treatment. Despite this limitation, nanocellulose-clay-based composites have demonstrated synergistic effects. The hydroxyl and carboxylate groups in CNFs not only serve as a scaffold but also enhance the sorption of pollutants. When comparing the capacity of these composites to standalone adsorbents

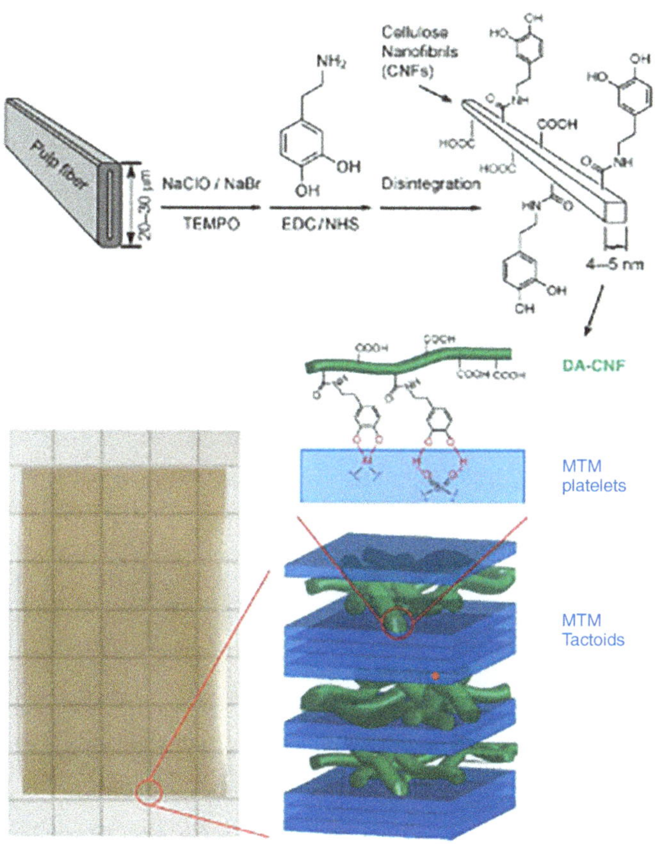

Figure 13.15 Illustration of the process for preparing DA-Conjugated NFC from pulp fibers and subsequently assembling DA-NFC with MTM to create a transparent nanocomposite [46].

(140.6 mg g^{-1} versus 96.6 mg g^{-1}), it was found that CNC-montmorillonite composites had a superior capacity to adsorb methylene blue, reaching 183.8 mg g^{-1} at specific proportions. Furthermore, as the ratio of CNF to clay increased from 4 to 12, the adsorption capacity for Cu^{2+} decreased from 87.1 to 56.5 mg g^{-1}.

This property is advantageous for water treatment applications because it allows pollutants to pass through to the sorption sites located between the clay layers. However, it can also weaken the mechanical integrity of the adsorbents. Mechanical failure and disintegration of clay composites may occur due to fractures that develop during the drying process. The adsorption capacities can be enhanced, for instance, by creating composites of CNFs and montmorillonite [46]. This is achieved by manufacturing layered nanocomposite films through the combination of dopamine-functionalized CNFs with montmorillonite sheets using catalytic chelation, ion-metal bonding, and hydrogen bonding (see Figure 13.15).

13.6 Potential of Nanocellulose-Based Membranes in Large-Scale Water Treatment

Nanocellulose membranes are known to retain water-soluble chemicals [47, 48]. Heavy metals like Ni^{2+}, Hg^{2+}, Cr^{6+}, Cu^{2+}, and Cd^{2+}, along with organic pollutants, anionic and cationic dyes, inorganic anionic contaminants, and cellulose derivatives, are effective at removing these pollutants from the environment.

Figure 13.16 shows different techniques used to recover water-soluble components from wastewater, including adsorption, membrane separation, electrochemical techniques, ion exchange, flotation, coagulation-flocculation, and organic pollutants [50].

As illustrated in Figure 13.16, two techniques—ion exchange and chemical complexation—are employed in nanocellulose adsorption membranes for the removal of heavy metals. Ion exchange involves the adsorption of metal ions, which are then replaced by other ions that are bound to the surface of the adsorbent. On the other hand, chemical complexation refers to the interaction between functional groups on the adsorbent's surface and specific metal ions. To enhance the ability of nanocellulose to absorb metal ions, it is essential to modify its surface to create ionizable complex sites. A common approach to increasing adsorption capacity is by incorporating amino and carboxyl sulfate groups. These functional groups can be added during the nanocellulose production process or introduced by modifying the surface of the nanofibers.

CNFs and CNC have been extensively researched for their potential in heavy metal extraction. In particular, CNCs derived from rice straw have been tested for their ability to adsorb various metal ions, including Pb^{2+}, Ni^{2+}, Cd^{2+}, Ag^{+}, Cu^{2+}, and Fe^{3+}. The measured adsorption capacities for these ions were found to be

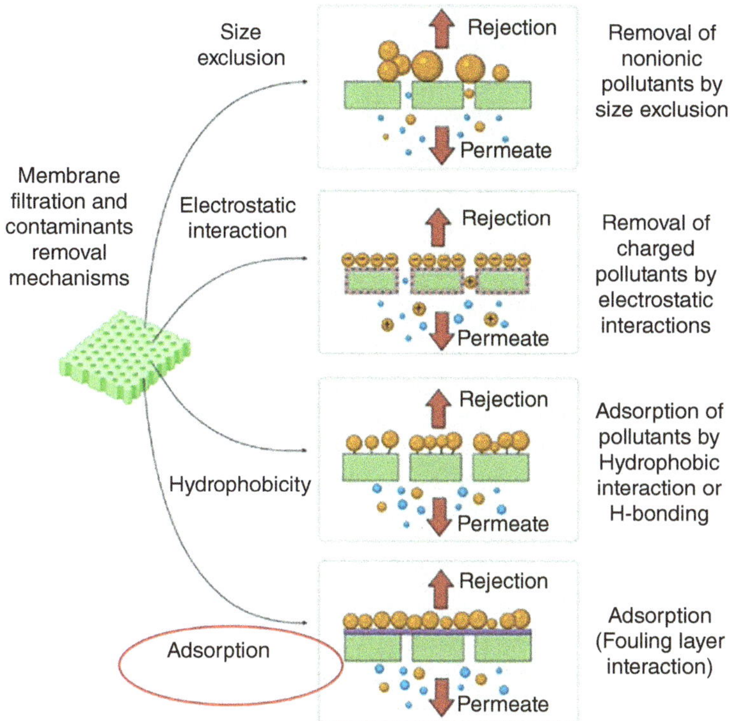

Figure 13.16 Ion exchange and complexation techniques (a), size exclusion (b), hydrophobic interaction (c), electrostatic repulsion (d), and fouling layer interaction (e) [49].

9.7, 9.42, 8.55, 56, 20, and 6.5 $mg\,g^{-1}$, respectively. After undergoing enzymatic treatment, the surface of the CNCs is modified with phosphate groups, which enhances their capacity to adsorb cations. Specifically, the sorption capacities for Cu^{2+}, Ag^{+}, and Fe^{3+} ions are increased to 117, 136, and 115 $mg\,g^{-1}$, respectively. This indicates that the sorption capacity has roughly doubled compared to untreated CNCs [51].

The application of an acid solution or EDTA to remove the adsorbed metal facilitated the restoration of the amino functional groups on the CNFs. In this study, ionic or ionizable groups, such as carboxyl (COOH), sulfate ($SO4^{2-}$), and quaternary ammonium ($—R_3N^+$), are introduced to the surface of CNFs either before or after they are converted into fibrils. A significant portion of these groups undergo chemical treatment before manufacture of CNFs. These groups aid the fibrillation process by generating repulsive forces between the microfibrils, which reduces their cohesion and disrupts their hydrogen bonding [52, 53].

To enhance the binding sites on the surface of nanofibrils, the carboxylation of CNFs was investigated. The modified bacterial cellulose (BC) exhibited more than double the adsorption capacity compared to unmodified BC, primarily due to the introduction of carboxyl groups. Specifically, the adsorption capacity for Cu^{2+} increased from 11 to 20 mg g^{-1}, and for Pb^{2+}, it rose from 25 to 65 mg g^{-1}. The adsorption process is influenced by the interaction between equivalent metal ions and amino groups through complexation and ion exchange. Succinic anhydride-treated CNFs have been utilized to adsorb Cd^{2+} and Pb^{2+} ions from wastewater, with their adsorption capacities being approximately 3 mmol g^{-1} for Cd^{2+} and 12 mmol g^{-1} for Pb^{2+}. In contrast, the adsorption capacity of crude cellulose was only 0.002 mmol g^{-1}.

To enhance the surface of the nanofibrils with binding sites, the carboxylation of CNFs was investigated. Compared to unmodified bacterial cellulose (BC), the adsorption capacity of the modified BC more than doubled due to the introduction of carboxyl groups. Specifically, the adsorption capacity for Cu^{2+} increased from 11 to 20 mg g^{-1}, while the capacity for Pb^{2+} rose from 25 to 65 mg g^{-1}. This adsorption process is influenced by the proportional interaction of metal ions with amino groups through complexation and ion exchange. Succinic anhydride-treated CNFs have been utilized to adsorb Cd^{2+} and Pb^{2+} ions from wastewater, achieving adsorption capacities of approximately 3 mmol g^{-1} for Cd^{2+} and 12 mmol g^{-1} for Pb^{2+} [54]. In contrast, the adsorption capacity of crude cellulose was only 0.002 mmol g^{-1}.

Copper adsorption capacity varies significantly depending on the source of the material. For instance, sugar beet pulp has an adsorption capacity of 0.33 mmol g^{-1}, while aspen wood fibers reach 4 mg g^{-1}, sawdust reaches 1.79 mg g^{-1}, wheat bran has an impressive capacity of 51.5 mg g^{-1}, and bentonite clay adsorbs 44.84 mg g^{-1}. The oxidized nanofibers exhibited comparable or lower copper adsorption capacity than mercerized cellulose treated with succinic anhydride, which has a capacity of 139 mg g^{-1}. Researchers developed a hybrid nanocomposite adsorbent capable of binding chromium ions (Cr^{3+}), lead ions (Pb^{2+}), and manganese ions (Mn^{2+}) by synthesizing magnetic nanocellulose that integrates (BC) spheres with Fe_3O_4 nanoparticles. In terms of adsorption capacity, the spherical Fe_3O_4/CNF nanocomposites showed capacities of approximately 25 mg g^{-1} for Cr^{3+}, 65 mg g^{-1} for Pb^{2+}, and 33 mg g^{-1} for Mn^{2+} [49].

13.7 Challenges and Prospects for Nanocellulose-Based Membrane

The removal of organic dyes and heavy metal ions using cellulose-based composite materials presents a promising technology; however, it is challenging due to difficulties in controlling porosity, permeability, and specific adsorption. Various methods, such as electrospinning, coating, molding, and freeze-drying, are being explored to regulate the morphology of these membranes. Although the manufacturing process for these membranes is relatively straightforward, ensuring their quality and performance presents significant challenges. For example, thicker membranes are often needed to compensate for defects, but this approach can reduce permeability and increase pressure drop. To enhance the structure of high-flux membranes, it is crucial to manage the film creation process and drying procedures effectively. This control is essential for achieving affordable and cost-effective large-scale production of suitable membranes. One promising solution to the issue of nanocellulose agglomeration during film drying is to apply a thin functional layer of nanocellulose on top of existing filters or membrane structures. To enhance selectivity and antifouling properties, modifying the surface of materials is crucial. There are several challenges that must be addressed before developing composite materials that are effective in removing pollutants. Native nanocellulose membranes often exhibit poor performance in terms of photocatalytic degradation and adsorption. Functionalizing the membrane surface with either organic or inorganic materials can promote chemisorption or hydrogen bonding, which are key processes for pollutant adsorption. The presence of functional groups, particularly —COOH and —OH, plays a significant role in the inorganic mineralization and immobilization of metal oxides during the adsorption process, as they interact notably with pollutant molecules [55]. To achieve better control over the aggregation of nanofibers into a web-like structure with regulated porosity and surface characteristics, new processing techniques need to be developed. These techniques should improve the reproducibility of individual nanofiber geometries. Additionally, adjustments to processing methods should likely involve the use of environmentally friendly solvents and carefully controlled fillers applied to the nanofiber surface. This approach aims to eliminate the need for traditional wet chemistry methods [56]. Nanocellulose composite membranes have the potential to enhance the functioning of complex water systems, including mechanical stability, chemical resistance, adsorption specificity, and photodegradation [57, 58]. The use of cellulose nanoparticles as an organic component in nanocomposites is an innovative method that is expected to attract more attention in the future. Further investigation into composites, especially those including cellulose nanoparticles and inorganic substances, is required. Further investigation on cellulose/inorganic hybrids is necessary for the extended treatment of wastewater and natural water sources. However, there are still other aspects of the intricate process involved in the formation of cellulose/inorganic composite materials that remain unknown to us. In situ precipitation is one of the methods that have been employed to synthesize nanocomposite membranes. Immobilizing heavy metals and initiating the development of inorganic nanoadsorbents are both made easier once cellulose is functionalized by oxidation or grafting [59]. In fact, the homoaggregation of composite components can have a detrimental effect on the final composite materials' shape, efficiency, and performance. Soluble cellulose derivatives, cellulose solvents, and inorganic mineralization control [60] are among the methods investigated for this issue. Nevertheless, nanocomposite membrane regeneration and recyclability must be taken into account throughout the design phase.

Nanocellulose composite membranes exhibit significant potential for applications in water treatment due to their mechanical stability, chemical resistance, adsorption specificity, and ability to photodegrade [57, 58]. The incorporation of cellulose nanoparticles as an organic component in nanocomposite materials represents an innovative approach that is likely to gain more attention in the future. However, further research is necessary to explore composites, particularly those that incorporate cellulose nanoparticles alongside inorganic substances. There remain many unknowns regarding the complex processes involved in forming cellulose/inorganic composite materials. One method utilized for synthesizing these nanocomposite membranes is in situ precipitation. The functionalization of cellulose through oxidation or grafting aids in the immobilization of heavy metals and the development of inorganic nanoadsorbents [59]. It is important to note that homoaggregation of composite components can negatively impact the shape, efficiency, and performance of the final composite materials. Methods such as soluble cellulose derivatives, cellulose solvents, and the control of inorganic mineralization [60] are being studied to address this issue. However, the regeneration and recyclability of nanocomposite membranes must be considered during the design phase.

13.8 Conclusion and Outlook

The development of environmentally friendly materials for water treatment with enhanced performance has garnered significant interest from both academia and industry. One of the key challenges facing humanity in the 21st century is ensuring access to uncontaminated water. To address this issue, various methods have been developed. Nanocelluloses, a type of nanomaterial, show great promise for wastewater treatment due to their high performance and efficiency. They are lightweight yet mechanically strong, chemically inert, and possess versatile surface chemistry. Additionally, nanocelluloses are inexpensive to produce and safe to handle, making them a more favorable option compared to inorganic nanomaterials. Recent advances in the synthesis of nanocellulose have led researchers to create nanocellulose-based materials specifically for wastewater treatment. These materials include hydrogels, aerogels, sponges, and membranes. Extensive studies have been conducted on the effectiveness of different forms of nanocellulose in treating water, such as bacterial cellulose, cellulose fibers, and cellulose nanocrystals. These studies have shown potential in a range of applications, particularly in membranes and filters, focusing on the filtration of nanoparticles and the creation of affinity membranes. Nanocellulose has demonstrated significant potential as an adsorbent for a variety of chemicals, including heavy metal ions, dyes, pharmaceuticals, insecticides, and fertilizers. Materials based on nanocellulose exhibit impressive scalability and regeneration capabilities, maintaining consistent performance across multiple adsorption-desorption cycles. A new generation of renewable and sustainable cellulose composites can be produced efficiently and cost-effectively through surface grafting or by mixing nanocellulose with different inorganic elements, polymers, and carbon compounds. The unique properties of nanocellulose, along with its high compatibility with various modifiers, make it suitable for numerous water treatment applications, including adsorption, absorption, disinfection, antifouling, ultrafiltration, nanofiltration, and reverse osmosis.

References

1. Stanislas, T.T., Bilba, K., De Oliveira Santos, R.P. et al. (2022). Nanocellulose-based membrane as a potential material for high performance biodegradable aerosol respirators for SARS-CoV-2 prevention: a review. *Cellulose* 29 (15): 8001–8024. https://doi.org/10.1007/s10570-022-04792-3.
2. Grzybek, P., Dudek, G., and Van Der Bruggen, B. (2024). Cellulose-based films and membranes: a comprehensive review on preparation and applications. *Chemical Engineering Journal* 495: 153500. https://doi.org/10.1016/j.cej.2024.153500.
3. Sapuła, P., Bialik-Wąs, K., and Malarz, K. (2023). Are natural compounds a promising alternative to synthetic cross-linking agents in the preparation of hydrogels? *Pharmaceutics* 15 (1): 253. https://doi.org/10.3390/pharmaceutics15010253.
4. Wang, Z., Han, M., Zhang, J. et al. (2020). Investigating and significantly improving the stability of tannic acid (TA)-aminopropyltriethoxysilane (APTES) coating for enhanced oil-water separation. *Journal of Membrane Science* 593: 117383. https://doi.org/10.1016/j.memsci.2019.117383.
5. Vatanpour, V., Pasaoglu, M.E., Barzegar, H. et al. (2022). Cellulose acetate in fabrication of polymeric membranes: a review. *Chemosphere* 295: 133914. https://doi.org/10.1016/j.chemosphere.2022.133914.

6 Islam, M.D., Uddin, F.J., Rashid, T.U., and Shahruzzaman, M. (2023). Cellulose acetate-based membrane for wastewater treatment—a state-of-the-art review. *Materials Advances* 4 (18): 4054–4102. https://doi.org/10.1039/D3MA00255A.

7 Wang, J., Abbas, S.C., Li, L. et al. (2024). Cellulose membranes: synthesis and applications for water and gas separation and purification. *Membranes* 14 (7): 148. https://doi.org/10.3390/membranes14070148.

8 Mu, M., Li, Y., Yu, H.-Y. et al. (2021). Construction of nanocellulose aerogels with mechanical flexibility and ph-responsive properties via a cross-linker structure design strategy. *ACS Sustainable Chemistry & Engineering* 9 (29): 9951–9960. https://doi.org/10.1021/acssuschemeng.1c03503.

9 Etale, A., Onyianta, A.J., Turner, S.R., and Eichhorn, S.J. (2023). Cellulose: a review of water interactions, applications in composites, and water treatment. *Chemical Reviews* 123 (5): 2016–2048. https://doi.org/10.1021/acs.chemrev.2c00477.

10 Bahsaine, K., Benzeid, H., El Allaoui, B. et al. (2024). Porous polyvinyl fluoride coated cellulose beads for efficient removal of Cd(II) from phosphoric acid. *International Journal of Biological Macromolecules* 254: 127867. https://doi.org/10.1016/j.ijbiomac.2023.127867.

11 Bahsaine, K., el Allaoui, B., Benzeid, H. et al. (2023). Hemp cellulose nanocrystals for functional chitosan/polyvinyl alcohol-based films for food packaging applications. *RSC Advances* 13 (47): 33294–33304. https://doi.org/10.1039/D3RA06586C.

12 Saud, A., Saleem, H., and Zaidi, S.J. (2022). Progress and prospects of nanocellulose-based membranes for desalination and water treatment. *Membranes* 12 (5): 462. https://doi.org/10.3390/membranes12050462.

13 Fatima, A., Ortiz-Albo, P., Neves, L.A. et al. (2023). Biosynthesis and characterization of bacterial cellulose membranes presenting relevant characteristics for air/gas filtration. *Journal of Membrane Science* 674: 121509. https://doi.org/10.1016/j.memsci.2023.121509.

14 Wang, D. et al. (2018). High aspect ratio carboxylated cellulose nanofibers cross-linked to robust aerogels for superabsorption–flocculants: paving way from nanoscale to macroscale. *ACS Applied Materials & Interfaces* 10 (24): 20755–20766. https://doi.org/10.1021/acsami.8b04211.

15 Mishra, S., Prabhakar, B., Kharkar, P.S., and Pethe, A.M. (2023). Banana peel waste: an emerging cellulosic material to extract nanocrystalline cellulose. *ACS Omega* 8 (1): 1140–1145. https://doi.org/10.1021/acsomega.2c06571.

16 Chen, Z., Hu, Y., Shi, G. et al. (2023). Advanced flexible materials from nanocellulose. *Advanced Functional Materials* 33 (18): 2214245. https://doi.org/10.1002/adfm.202214245.

17 Xu, T., Du, H., Liu, H. et al. (2021). Advanced nanocellulose-based composites for flexible functional energy storage devices. *Advanced Materials* 33 (48): 2101368. https://doi.org/10.1002/adma.202101368.

18 Cruz-Tato, P., Ortiz-Quiles, E.O., Vega-Figueroa, K. et al. (2017). Metalized nanocellulose composites as a feasible material for membrane supports: design and applications for water treatment. *Environmental Science & Technology* 51 (8): 4585–4595. https://doi.org/10.1021/acs.est.6b05955.

19 Tavakolian, M., Jafari, S.M., and Van De Ven, T.G.M. (2020). A review on surface-functionalized cellulosic nanostructures as biocompatible antibacterial materials. *Nano-Micro Letters* 12 (1): 73. https://doi.org/10.1007/s40820-020-0408-4.

20 López Durán, V., Hellwig, J., Larsson, P.T. et al. (2018). Effect of chemical functionality on the mechanical and barrier performance of nanocellulose films. *ACS Applied Nano Materials* 1 (4): 1959–1967. https://doi.org/10.1021/acsanm.8b00452.

21 Gharsallah, F., Ben Ayed, E., Ghorbel, N. et al. (2024). Aminated nanocellulose-mediated synthesis of polyaniline/nanocellulose hybrid materials for sustainable conductive and supercapacitors nanopapers. *Materials Today Sustainability* 27: 100811. https://doi.org/10.1016/j.mtsust.2024.100811.

22 Salama, A., Abouzeid, R., Leong, W.S. et al. (2021). Nanocellulose-based materials for water treatment: adsorption, photocatalytic degradation, disinfection, antifouling, and nanofiltration. *Nanomaterials* 11 (11): 3008. https://doi.org/10.3390/nano11113008.

23 Sunasee, R. and Hemraz, U.D. (2018). Synthetic strategies for the fabrication of cationic surface-modified cellulose nanocrystals. *Fibers* 6 (1): 15. https://doi.org/10.3390/fib6010015.

24 Lin, N., Gèze, A., Wouessidjewe, D. et al. (2016). Biocompatible double-membrane hydrogels from cationic cellulose nanocrystals and anionic alginate as complexing drugs codelivery. *ACS Applied Materials & Interfaces* 8 (11): 6880–6889. https://doi.org/10.1021/acsami.6b00555.

25 Lu, Q., Cai, Z., Wang, S. et al. (2017). Controlled construction of nanostructured organic–inorganic hybrid material induced by nanocellulose. *ACS Sustainable Chemistry & Engineering* 5 (9): 8456–8463. https://doi.org/10.1021/acssuschemeng.7b02394.

26 Sirviö, J., Honka, A., Liimatainen, H. et al. (2011). Synthesis of highly cationic water-soluble cellulose derivative and its potential as novel biopolymeric flocculation agent. *Carbohydrate Polymers* 86 (1): 266–270. https://doi.org/10.1016/j.carbpol.2011.04.046.

27 Littunen, K., De Castro, J.S., Samoylenko, A. et al. (2016). Synthesis of cationized nanofibrillated cellulose and its antimicrobial properties. *European Polymer Journal* 75: 116–124. https://doi.org/10.1016/j.eurpolymj.2015.12.008.

28 Li, S., Liu, S., Huang, F. et al. (2018). Preparation and characterization of cellulose-based nanofiltration membranes by interfacial polymerization with piperazine and trimesoyl chloride. *ACS Sustainable Chemistry & Engineering* 6 (10): 13168–13176.

29 Qiu, C., Zhang, X., Zhang, Y. et al. (2021). Bamboo-based biofoam adsorbents for the adsorption of cationic pollutants in wastewater: methylene blue and Cu(II). *ACS Omega* 6 (36): 23447–23459. https://doi.org/10.1021/acsomega.1c03438.

30 Li, S., Wang, D., Xiao, H. et al. (2021). Ultra-low pressure cellulose-based nanofiltration membrane fabricated on layer-by-layer assembly for efficient sodium chloride removal. *Carbohydrate Polymers* 255: 117352. https://doi.org/10.1016/j.carbpol.2020.117352.

31 Raveena, Kumari P. Nanocellulose@ gallic Acid-based MOFs: a novel material for ecofriendly food packaging. *ACS Omega* 2024 9(33):35654-65. https://doi.org/10.1021/acsomega.4c03847.

32 Durairaj, V., Wester, N., Etula, J. et al. (2019). Multiwalled carbon nanotubes/nanofibrillar cellulose/nafion composite-modified tetrahedral amorphous carbon electrodes for selective dopamine detection. *Journal of Physical Chemistry C* 123 (40): 24826–24836. https://doi.org/10.1021/acs.jpcc.9b05537.

33 Han, J., Wang, S., Zhu, S. et al. (2019). Electrospun core–shell nanofibrous membranes with nanocellulose-stabilized carbon nanotubes for use as high-performance flexible supercapacitor electrodes with enhanced water resistance, thermal stability, and mechanical toughness. *ACS Applied Materials & Interfaces* 11 (47): 44624–44635. https://doi.org/10.1021/acsami.9b16458.

34 He, H., Jiang, H., Chen, C. et al. (2019). Efficient oil/water separation by zwitterionic poly(sulfobetaine methacrylate)@$Cu(OH)_2$ nanoneedle array-coated copper meshes with superwetting and antifouling properties. *ACS Sustainable Chemistry & Engineering* 7 (16): 13815–13826. https://doi.org/10.1021/acssuschemeng.9b01774.

35 Kollarigowda, R.H., Abraham, S., and Montemagno, C.D. (2017). Antifouling cellulose hybrid biomembrane for effective oil/water separation. *ACS Applied Materials & Interfaces* 9 (35): 29812–29819. https://doi.org/10.1021/acsami.7b09087.

36 Ao, C., Zhao, J., Li, Q. et al. (2020). Biodegradable all-cellulose composite membranes for simultaneous oil/water separation and dye removal from water. *Carbohydrate Polymers* 250: 116872. https://doi.org/10.1016/j.carbpol.2020.116872.

37 Liu, T., Zhang, X., Gu, A. et al. (2022). In-situ grown bilayer MOF from robust wood aerogel with aligned microchannel arrays toward selective extraction of uranium from seawater. *Chemical Engineering Journal* 433: 134346. https://doi.org/10.1016/j.cej.2021.134346.

38 Zheng, Q., Cai, Z., and Gong, S. (2014). Green synthesis of polyvinyl alcohol (PVA)–cellulose nanofibril (CNF) hybrid aerogels and their use as superabsorbents. *Journal of Materials Chemistry A* 2 (9): 3110–3118. https://doi.org/10.1039/C3TA14642A.

39 Guan, H., Cheng, Z., and Wang, X. (2018). Highly compressible wood sponges with a spring-like lamellar structure as effective and reusable oil absorbents. *ACS Nano* 12 (10): 10365–10373. https://doi.org/10.1021/acsnano.8b05763.

40 Guo, R., Cai, X., Liu, H. et al. (2019). In situ growth of metal–organic frameworks in three-dimensional aligned lumen arrays of wood for rapid and highly efficient organic pollutant removal. *Environmental Science & Technology* 53 (5): 2705–2712. https://doi.org/10.1021/acs.est.8b06564.

41 Benítez, A.J. and Walther, A. (2017). Cellulose nanofibril nanopapers and bioinspired nanocomposites: a review to understand the mechanical property space. *Journal of Materials Chemistry A* 5 (31): 16003–16024. https://doi.org/10.1039/C7TA02006F.

42 Gustafsson, S., Manukyan, L., and Mihranyan, A. (2017). Protein–nanocellulose interactions in paper filters for advanced separation applications. *Langmuir* 33 (19): 4729–4736.

43 Bayer, T., Cunning, B.V., Selyanchyn, R. et al. (2016). High temperature proton conduction in nanocellulose membranes: paper fuel cells. *Chemistry of Materials* 28 (13): 4805–4814. https://doi.org/10.1021/acs.chemmater.6b01990.

44 Montañez, M.I., Hed, Y., Utsel, S. et al. (2011). Bifunctional dendronized cellulose surfaces as biosensors. *Biomacromolecules* 12 (6): 2114–2125. https://doi.org/10.1021/bm200201y.

45 Wang, Y. and Lu, Q. (2020). Dendrimer functionalized nanocrystalline cellulose for Cu(II) removal. *Cellulose* 27 (4): 2173–2187. https://doi.org/10.1007/s10570-019-02919-7.

46 Yao, K., Huang, S., Tang, H. et al. (2017). Bioinspired interface engineering for moisture resistance in nacre-mimetic cellulose nanofibrils/clay nanocomposites. *ACS Applied Materials & Interfaces* 9 (23): 20169–20178. https://doi.org/10.1021/acsami.7b02177.

47 Karim, Z. and Monti, S. (2021). Microscopic hybrid membranes made of cellulose-based materials tuned for removing metal ions from industrial effluents. *ACS Applied Polymer Materials* 3 (8): 3733–3746. https://doi.org/10.1021/acsapm.1c00105.

48 Wang, Q., Liu, S., Liu, J. et al. (2022). Sustainable cellulose nanomaterials for environmental remediation—achieving clean air, water, and energy: a review. *Carbohydrate Polymers* 285: 119251. https://doi.org/10.1016/j.carbpol.2022.119251.

49 Barhoum, A., Deshmukh, K., García-Betancourt, M.L. et al. (2023). Nanocelluloses as sustainable membrane materials for separation and filtration technologies: principles, opportunities, and challenges. *Carbohydrate Polymers* 317: 121057. https://doi.org/10.1016/j.carbpol.2023.121057.

50 Rajinipriya, M., Nagalakshmaiah, M., Robert, M., and Elkoun, S. (2018). Importance of agricultural and industrial waste in the field of nanocellulose and recent industrial developments of wood based nanocellulose: a review. *ACS Sustainable Chemistry & Engineering* 6 (3): 2807–2828. https://doi.org/10.1021/acssuschemeng.7b03437.

51 Liu, P., Borrell, P.F., Božič, M. et al. (2015). Nanocelluloses and their phosphorylated derivatives for selective adsorption of Ag^+, Cu^{2+} and Fe^{3+} from industrial effluents. *Journal of Hazardous Materials* 294: 177–185. https://doi.org/10.1016/j.jhazmat.2015.04.001.

52 Gholami Derami, H., Jiang, Q., Ghim, D. et al. (2019). A robust and scalable polydopamine/bacterial nanocellulose hybrid membrane for efficient wastewater treatment. *ACS Applied Nano Materials* 2 (2): 1092–1101. https://doi.org/10.1021/acsanm.9b00022.

53 Chaker, A. and Boufi, S. (2015). Cationic nanofibrillar cellulose with high antibacterial properties. *Carbohydrate Polymers* 131: 224–232. https://doi.org/10.1016/j.carbpol.2015.06.003.

54 Reshmy, R., Philip, E., Paul, S.A. et al. (2020). Nanocellulose-based products for sustainable applications-recent trends and possibilities. *Reviews in Environmental Science and Biotechnology* 19 (4): 779–806. https://doi.org/10.1007/s11157-020-09551-z.

55 Alomar, T., Qiblawey, H., Almomani, F. et al. (2023). Recent advances on humic acid removal from wastewater using adsorption process. *Journal of Water Process Engineering* 53: 103679. https://doi.org/10.1016/j.jwpe.2023.103679.

56 Li, S., Wang, X., Guo, Y. et al. (2022). Recent advances on cellulose-based nanofiltration membranes and their applications in drinking water purification: a review. *Journal of Cleaner Production* 333: 130171. https://doi.org/10.1016/j.jclepro.2021.130171.

57 Liu, W., Sun, F., Jiang, L. et al. (2019). Surface structure patterning for fabricating non-fluorinated superhydrophobic cellulosic membranes. *ACS Applied Polymer Materials* 1 (5): 1220–1229. https://doi.org/10.1021/acsapm.9b00215.

58 Geng, B., Wang, H., Wu, S. et al. (2017). Surface-tailored nanocellulose aerogels with thiol-functional moieties for highly efficient and selective removal of Hg(II) ions from water. *ACS Sustainable Chemistry & Engineering* 5 (12): 11715–11726. https://doi.org/10.1021/acssuschemeng.7b03188.

59 Shi, L., Ding, P., Wang, Y. et al. (2019). Self-healing polymeric hydrogel formed by metal–ligand coordination assembly: design, fabrication, and biomedical applications. *Macromolecular Rapid Communications* 40 (7): 1800837. https://doi.org/10.1002/marc.201800837.

60 Y. Jin, Q. Zhu, W. Zhong, K. Yan, D. Wang, Superfast, porous, and organic solvent-sensitive actuator based on EVOH nanofibrous membrane and PS microspheres, *Journal of Physical Chemistry C*, vol. 123, no 1, p. 185-194, 2019, https://doi.org/10.1021/acs.jpcc.8b10090.

14

Alginate Nanocomposite Membranes for Wastewater Treatment

Vandana Patyal, Prashant Mahajan, and Dipika Jaspal

Symbiosis Institute of Technology (SIT), Symbiosis International (Deemed University), (SIU), Pune, Maharashtra, India

14.1 Introduction of Alginate

14.1.1 Sources and Extraction Methods

Alginate is a naturally occurring polymer found in the bacterial capsules of certain bacteria (*Azotobacter* sp. and *Pseudomonas* sp.) and cell walls of brown algae commonly known as brown seaweeds [1]. Alginate has been largely extracted from brown algae by chemical companies globally. The main genera used for the extraction are Laminaria and Macrocystis [2].

There are multiple steps involved in the extrication of alginate from brown algae [3]. First, the algae are reacted with mineral acid for the formation of free alginic acids (Figure 14.1). Then the soluble sodium alginate is produced after treatment with sodium carbonate/hydroxide. The soluble sodium alginate is recovered by calcium chloride as calcium alginate fiber precipitates. It can also be recovered as an alginic acid gel posttreatment with mineral acid. Both the recovered forms are blended with sodium carbonate to produce sodium alginate. Alginate can also be recovered as potassium alginate and ammonium alginate by mixing suitable alkali with alginic acid gel. It can be extracted as the salt of sodium, calcium, potassium, and ammonium for biomedical and water treatment applications [4]. Recovery of alginate in the form of sodium alginate salt is the largest industry for commercial application [5].

Alginates can be applied as a matrix to transport or capture a wide range of particles which makes it suitable for use in a variety of industries [6]. Due to their unique rheological qualities, they are currently extensively employed in the textile, cosmetic, pharmaceutical, and food industries. In the biotechnology

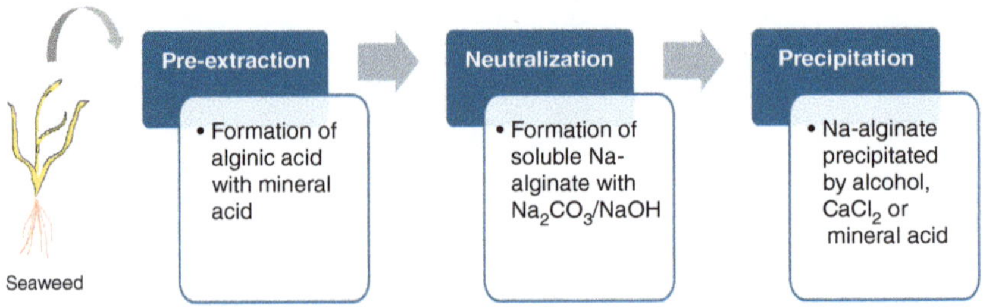

Figure 14.1 Alginate extraction from algae.

industry, their utilization is increasing due to their ability to retain water, gelling capability, and stabilizing properties [7]. Modified alginate materials/nanofibers are used for contaminant adsorption from wastewater. Research suggests their competency in the expulsion of dye, heavy metals, aromatic compounds, antibiotics, and nutrient recovery from wastewater [8–10].

14.1.2 Physicochemical Properties

Alginates consist of (1,4) linked β-D-mannuronic (M) and α-L-guluronic acids (G) residues. Alginates are block copolymers with irregular repeated blocks which can be homogeneous arrangements with MM or GG or heterogeneous arrangements with GM or MG blocks [11–13]. The arrangement of the blocks varies depending on algal source, as shown in Figure 14.2.

A very vital property of the alginate is its gel formation ability by selective binding with multivalent cations. For the formation of homogeneous alginate gel, it is important to control the cross-linking process of ions. The techniques generally used for gelatinating are diffusion and internal setting [11]. Alginate gelatinating is a complex process that is influenced by factors such as M/G ratio, calcium ion source, and method of preparation [14].

Alginates are susceptive to chain degradation in the contiguity of acids and bases but can also degrade at neutral pH in the proximity of reducing agents. It is very important that during the synthesis of alginate derivatives, the reaction conditions and the chemistries are carefully selected to attain targeted properties.

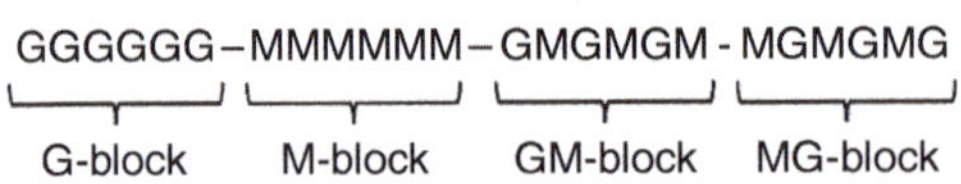

Figure 14.2 Representative block distribution in alginate structure.

14.2 Synthesis of Alginate Membranes

Alginate is an environmentally friendly biopolymer considered ideal for the synthesis of high-adsorption membranes for the treatment of waste effluent. Alginates are biodegradable, biocompatible, and economical which makes them an ideal choice for composite membranes [15, 16]. Alginates possess abundant hydroxyl and carboxyl functional groups which participate in chelation and ion exchange which leads to the adsorption of numerous pollutants from the wastewater [17]. Alginate alone cannot cope with the intricate wastewater environment. Gao et al. [8] reported poor stability, low mechanical strength, and low thermal endurance for sodium alginate matrix for pollutant remediation application. Numerous materials such as activated carbon, magnetic materials, nanoparticles, and graphene have been utilized with alginate for creating composite membranes to boost the performance and stability for real-time application in wastewater treatment [18–20].

The pristine alginate membrane has low stability and separation factor which can be improved by cross-linking, polymer blending, chemical modification, layer-by-layer assembly, and incorporating nanofillers in the composite membrane [21–23]. All these techniques help in improving the physicochemical properties and separation factor of the alginate membrane [24]. Incorporating inorganic filler in the alginate membrane increases its mechanical, chemical, and thermal stability. The addition of nanomaterials further amplifies the stability, permeability, sustainability, and selectivity of the composite membrane [25]. In recent times, blending inorganic nanomaterials in alginate membranes has sparked attention due to improved thermal and humid environment endurance [26].

The commonly used methods for the fabrication of alginate nanocomposite membranes are shown in Figure 14.3.

In the interfacial polymerization method, the polymerization takes place at the interface of two immiscible monomers. The distinctive formation of a thin layer on the substrate by this method has superior membrane flux [27]. The phase inversion method is a mixing procedure wherein the polymer solution is transformed into a porous solid due to the exchange between solvent and nonsolvent [28]. In the blending method, the nanoscale structures and the alginate polymer are directly mixed. There are two ways of blending: solution blending (appropriate solvent is used for dissemination) and melt blending (this involves melting and mixing at high temperatures) [29].

In the electrospinning method, a fibrous membrane is produced by the application of a high voltage between the solution of polymer droplets containing

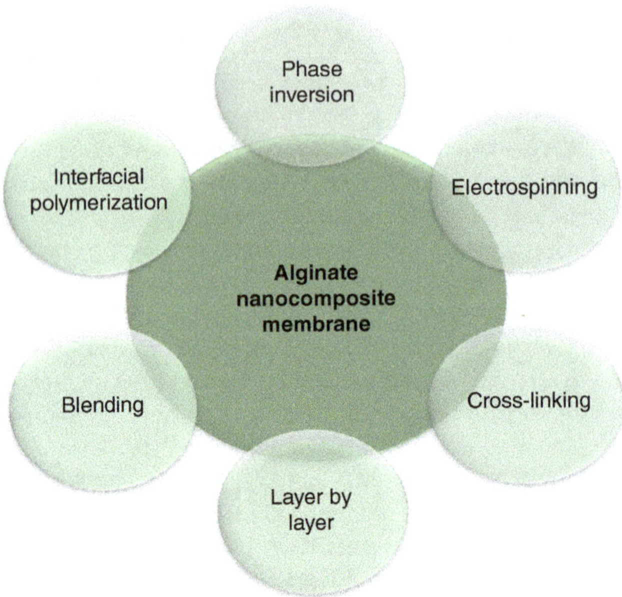

Figure 14.3 Preparation of nanocomposite membranes.

nanoscale entities and a grounded collector. The porous nanofiber membrane is formed at the collector which is the preferred choice for filtration application [30]. Alginate membrane can be modified by ionic cross-linking [31] or by organic cross-linkers [32]. Ionic cross-linking depends on the chelation between the carboxyl functional group on alginate and multivalent ions. In the second modification method, there is an interaction between the cross-linker's functional group and the alginate. Fabrication of composite membrane by layer-by-layer method requires the active and support layers to have good mechanical strength and improve the segregation performance [33].

The method adopted for the fabrication of a nanocomposite membrane depends on the nature of alginate and the desired physicochemical structure of the membrane which depends upon the specifications of the targeted pollutant.

14.3 Wastewater Treatment

The application of alginate nanocomposite membranes has been done for the removal of organic and inorganic substances from wastewater. Research suggests enhanced performance in the expulsion of dye, heavy metals, nutrients, and pharmaceutical compounds from the aqueous phase. They have also been successfully tested for oil/alcohol–water separation and desalination (Figure 14.4).

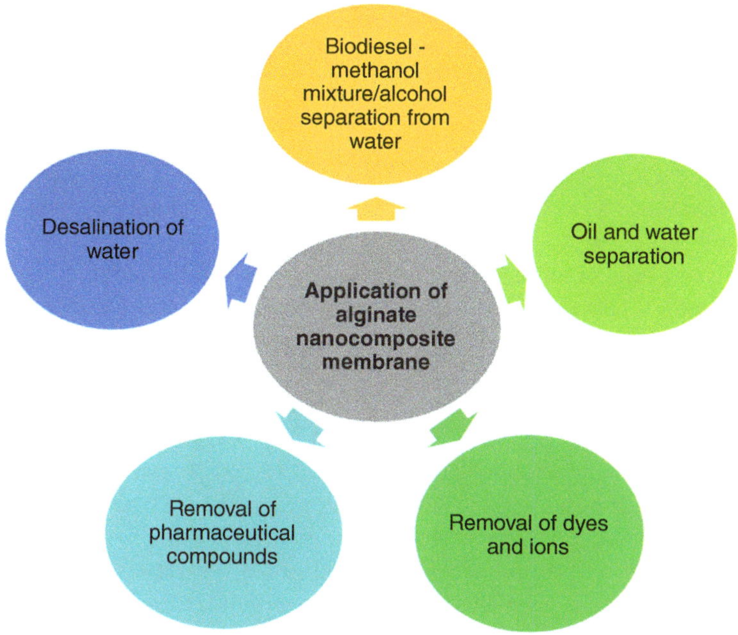

Figure 14.4 Application of alginate nanocomposite membrane in the treatment of wastewater.

14.3.1 Removal of Dyes and Ions

Kazemi et al. [34] employed a layer-by-layer approach to create FeO-doped WO_3 photocatalytic nanoparticles and then to build a polyethersulfone ultrafiltration membrane utilizing their created nanoparticles. Under visible light illumination, the new photocatalytic membranes removed large amounts of Cr(VI) ions [34].

Amiri et al. [35] created polyvinyl alcohol (PVA)-graphene oxide (GO)-sodium alginate (SA) (PVA-GO-SA) hydrogel (HG) blended polyethersulfone (PES) nanofiltration membranes. The hydrogel blended membranes rejected Lanasol blue 3R dye by more than 83% [35].

ZnO nanoparticles embedded in alginate (Alg)-based membranes were produced by Dodero et al. [36] using electrospinning in order to serve as biosorbent materials. UV-Vis spectrophotometric studies were done to assess membrane uptake and release performance using aqueous Congo Red (CR) and Methylene Blue (MB). According to the findings, electrospun membranes have the potential to serve as a novel, affordable, and diverse absorbent system for the administration of drugs and part of purification processes for pollutant abatement [36].

GO-filled SA membranes were created by Nigiz [18] for research on pervaporation desalination. All membranes achieved overall rejection values of more than

99.4% at all temperatures tested. Using a 2 wt.% GO-loaded NaAlg membrane at 60 °C resulted in a maximum flux of 8110 g m^{-2} h^{-1} and 99.41% rejection [18].

Sokhandan et al. [37] sought to reduce fouling in membrane bioreactors (MBR) by combining sodium alginate as a hydrophilic agent and ZnO as an antibacterial agent in the framework. The MBR had a flux of antibacterial hydrophilic polymeric membranes (AHPM) polyacrylonitrile membranes (PANM), and antibacterial polymeric membranes (APM) at 48, 37, and 51 l m^{-2} h, with chemical oxygen demand removal rates of 93.5%, 81%, and 96.7%, respectively [37].

Metecan et al. [38] combined the chelating capabilities of alginate and polyethyleneimine (PEI) with Fe^{3+} and Zn^{2+} ions to create loose NF membranes. The prepared membrane eliminated dyes and colored compounds in actual wastewater after 72 hours of continuous filtration [38].

Iqbal et al. [39] used the dissolving casting method to create a novel composite membrane comprising SA, polyvinyl pyrrolidine (PVP), and functionalized multi-walled carbon nanotubes (F-MWCNTs) with distinctive properties suitable for reverse osmosis (RO) desalination. The PVP/SA mix membrane has shown a penetration flux of 5.5 L m^{-2} h^{-1} and steeped water content [39].

Anokhina et al. [40] used a non-solvent-induced phase separation technique in conjunction with cation cross-linking to create nanofiltration membranes from NaAlg. It has been demonstrated that $AgNO_3$ or $AlCl_3$ must be precipitated and cross-linked in aqueous solutions simultaneously to prepare composite membranes with an extremely good selective recovery of medicinal chemicals from the organic phase [40].

Wang et al. [41] created a variety of thin-film composite (TFC) nanofiltration membranes using a contra-diffusion technique that involved the cross-linking of alginate and Cu^{2+}. Because of the strong attraction between alginate and Cu^{2+}, the composite membrane outperformed other metal ion/alginate composites in terms of performance. The Donnan effect caused the strongly negatively charged membrane to reject anionic dyes and divalent anions with high efficiency [41].

Amiri et al. [42] created a highly permeable membrane with an antifouling nature by surface coating sodium alginate (SA) and PVA, GO as a hydrophilic addition on polysulfone support, and glutaraldehyde (GA) as a cross-linker. The PVA-SA-GO membrane eliminated As(III) (88.4%), in cyclic continuous filtration [42].

A composite membrane with a SA matrix including titanium dioxide (TiO_2) nanotubes (TNT) was developed by Arla et al. in 2023. Furthermore, these composite membranes were used to adsorb an aqueous MB. Within 180 minutes at room temperature, a 50 mg membrane obtained an estimated degradation ratio of 85% for 10 mg L^{-1} MB concentration [43].

Cai et al. [44] created an innovative polysaccharide-based hydrogel membrane by incorporating micron-sized biochar (BC) and cellulose nanofiber (CNF) into SA hydrogel, with polyethylene glycol (PEG) serving as a pore-forming agent.

Under 0.35 MPa pressure, the flux reached 61.5 L/ (m^2 h), with outstanding anti-fouling properties due to hydrophilicity. In filtering tests, Cr(VI) and Cr(III) rejection rates of 50 ppm concentration were 91.4% and 96.8%, respectively [44].

In 2023, Spoială et al. developed a new adsorbent composite membrane of alginate using a direct casting approach for decontamination of lead. The caffeic acid (CA) and silver nanoparticles (Ag NPs) retained at low quantities on the composite membrane were sufficient to activate the antibacterial action of the membrane. The prepared composite membranes were appropriate for heavy metal ion elimination and antimicrobial treatment [45].

In another research by Wen et al. in 2023, a unique thin-film nanocomposite (TFN) forward osmosis FO membrane supported by an Alg HG interlayer scaffolded with Mxene was developed. The fertilizer-drawn FO method combined with sludge thickening resulted in a 2.4-fold dilution of a two molar potassium chloride solution after 12 hours of operation. Additionally, the MLSS sludge concentration increased from 2000 to 5183 ppm [46]. Table 14.1 gives the methods for the fabrication and various uses of alginate nanocomposite membranes.

14.3.2 Alcohol and Water Separation

Mesoporous molecular sieve (MCM-41)-filled SA and SA nanocomposite membranes were made by solution casting method and GA as cross-linker. Both membranes were examined for isopropanol (IPA) dehydration at 303 K. The 20% MCM-41-filled SA membrane had a flux extending between 110 and 555 g/ (m^2 h) [47].

Bhat and Aminabhavi [48] developed a unique hybrid composite membrane by adding sodium montmorillonite (Na^+MMT) clay particles in SA and cross-linking with GA. The hybrid composite membrane dehydrated tetrahydrofuran (THF), 1,4-dioxane, and IPA from aqueous solutions at 303 K [48].

Patil et al. [49] produced and studied SA matrix membranes loaded with aluminum-containing mesoporous silica (Al-MCM41) and cross-linking using GA. However, the cross-linked SA membrane when evaluated for 10 wt.% water content feed had flux and selectivity values of 27 g cm^{-2} h and 35 for water and ethanol mixture, whereas for water and IPA mixtures, the values were 67 g m^{-2} h and 653 [49].

Lokesh et al. [50] created a unique SA nanocomposite membrane that was loaded with polyaniline (PANI) and plated with TiO_2. The membrane was fabricated by using nanosized modified TiO_2 with polyaniline as filler in SA matrix and cross-linked with GA [50].

Patil et al. [51] employed SA-Al-MCM-41 for PV dehydration of aqueous solutions of THF and 1,4-dioxane at 303 K. It was noticed that flow and selectivity are directly proportional to the quantity of Al-MCM-41 particles added in the SA matrix [51].

Table 14.1 Methods for fabrication and uses of alginate nanocomposite membranes.

Sr. no.	Nanocomposite membrane used	Method of alginate NCM fabrication	Dyes/ions/ salts removed	% rejections	Permeation flux/ water permeability	Pressure	Reference
1.	PES/ $(CHI\text{-}ALG)_{3.5}$/ FeO@WO_3 membrane PES/ $(CHI\text{-}ALG)_{3.5}$ membrane	Thin film deposition	Cr(VI) (5 mg L^{-1})	99.2% 56.3%	—	1 bar	[34]
2.	PVA-GO-SA hydrogel blended PES NF membranes	Phase inversion method	Lanasol blue 3R (100 ppm)	>83%	115.7 (L m^{-2} h)	3 bar	[35]
3.	ZnO embedded alginate based membrane	Electrospinning	CR and MB (10 ppm)	77% 97%	—	—	[36]
4.	2% wt GO-SA 1% wt GO- SA	Cross linking	Desalination	99.41% 99.95%	8.11 kg m^{-2} h^{-1} 3.46 kg m^{-2} h^{-1}	Atmospheric pressure	[18]
5.	PAN-ZnO@SA	Blending	COD removal	81%	37 L/ (m^2 h)	0.5 bar	[37]
6.	PAI/PEI-Alg-Fe^{3+} PAI/PEI-Alg-Zn^{2+}	Nonsolvent-induced phase inversion technique	Methylene Blue 0.1 g L^{-1}	77.1% 67%	—	2 bar	[38]
7.	F-MWCNTs/ PVP/SA	Dissolving casting method	Salt	85.4%	5.5 L/ (m^2 h)	55.2 bar	[39]

8.	$AgNO_3$/SA or $AlCl_3$/SA membrane	Cross-linking	Pharmaceutical compounds	90–93% 99%	1.5–2.5 kg/(m^2·h·bar) 0.44–7.8 kg/(m^2·h bar)	—	[40]
9.	Cu^{2+}/alginate composite membrane	Contra-diffusion technique	Na_2SO_4 dyes	96.2% 99.5%	50.2 L m^{-2} h^{-1}	0.6 MPa	[41]
10.	PVA-SA-GO membrane	In situ poly-merization	As(III)	88.4%	15.8 L m^{-2} h^{-1} bar^{-1}	—	[42]
11.	TNTs/SA nanocomposite membrane	Immobilization	Methylene blue (10 ppm)	85%	—	—	[43]
12.	BC-CNG-SA-PEG membrane	In situ ionic cross-linking	Cr(III) and Cr(VI) (50 ppm)	96.8% and 91.4%,	61.5 L m^{-2} h^{-1}	0.35 MPa	[44]
13.	Alg/Ag NPs/CA Composite Membranes	Direct casting method	Lead	60.09 to 98.76%	—	—	[45]
14.	Alg HG @Mxene-TFN FO membrane	Thin-film deposition	1.0 M NaCl	>80%	45.6 LMH	—	[46]

Mali and Gokavi [52] employed the solution casting process to create nanocomposite membranes by introducing various amounts of Preyssler-type heteropolyacid nanoparticles into NaAlg, such as 6, 8, and 10 wt.%. The prepared membrane was examined for sorption and permeation features from aqueous IPA solutions at various concentrations [52].

Solution casting and GA cross-linking were used by Adoor et al. [53] to develop nanocomposite membranes by combining ammonium carbonate as a nanofiller in different compositions with phosphotungstic acid (PTA) in both modified and unmodified forms. The nanocomposite membrane outperformed pristine NaAlg membranes in terms of ethanol and IPA dehydration using the PV technique [53].

In order to enhance IPA dehydration by PV, Suhas et al. [54] developed graphene-integrated SA nanocomposite membranes. The membrane performance was optimal at the lowest graphene content, with a selectivity of 4623 and 3122 GPU value of permeation for a 10 wt.% water input mixture at 303 K [54].

Pristine GO (pGO) and reduced GO (rGO) nanosheets with distinct structures were synthesized and integrated into the SA matrix to create two types of hybrid membranes. PV dehydration was accomplished using an ethanol/water mixture to investigate the water permeative performance of hybrid membranes. The rGO exhibited superior performance, and at 1.6 wt.% content, the optimal performance was attained with a permeation flux of 1.699 kg/ (m^2 h) and a separation factor of 1566 was reported [55].

To construct composite membranes with increased PV performance, Suhas et al. [56] incorporated clay particles treated with para-toluene sulfonic acid (p-TSA) into the SA matrix. The modified clay-filled (10 wt.%) SA membranes demonstrated the greatest permeance of 3423 GPU with a selectivity of 5781 for water-IPA mixture (10 wt.%) at 303 K [56].

By incorporating g-C_3N_4 nanosheets (CNs) into a SA matrix, Cao et al. [57] created an incredibly hybrid membrane from a water-ethanol solution. The membrane integrated with 3 wt.% CNs had a separation factor of 1653 and a permeation flux of 2.469 kg m^{-2} h at 76 °C at 10 wt.% input water concentration [57].

A bacterial cellulose (BC) hydrogel was changed by Suratago et al. [58], by the diffusion of SA and the use of a $CaCl_2$ solution as a cross-linker. BC/alginate (BCA) membranes were designed, described, and employed for the PV of ethanol-water solutions [58].

In 2015, Uragami et al. created membranes from sodium deoxyribonucleate (DNA-Na) and SA cross-linked by dipping them in a CH_3OH solution of $MaCl_2$ or $CaCl_2$. The investigations showed that excessive immersion in cross-linker solution reduced the hydrophilicity of the membrane [59].

BCA nanocomposite film was employed by Suratago et al. [60] as a selective membrane for the PV-assisted water dissociation from biodiesel-methanol mixtures. The permeate contained 94.5% water by weight, and the BCA membrane fully rejected the methyl ester [60].

Zwitterionic poly (sulfobetaine methacrylate) graphene oxides (PSBMA@GO) nanosheets were produced by Zhao et al. [61] and integrated onto a SA membrane for the purpose of separating alcohol from water. The PSBMA@GO incorporated membranes have obtained optimal performance with a permeation flux of 2.140 kg/ (m^2 h) and a separation factor of 1370 for the water/alcohol mixture [61].

Wang et al. [62] used PEG—functionalized polyoctahedral oligomeric silsesquioxanes (PEG@POSS) nanoparticles in a SA matrix to create nanocomposite membranes. The best separation performance was achieved with a separation factor of 1077 and a permeation flux of 2.5 kg/ (m^2h) for dehydration of 10/90 wt.% water/ethanol input [62].

In order to study the impact of different concentrations of ZnO, Fe_3O_4, and Ag_2O particles on ethanol and water permeability, Dudek et al. [63] created alginate membranes loaded with these metal oxide (ZnO, Fe_3O_4, and Ag_2O) particles. The alginate membrane with 15% Ag_2O particles had the highest flux [63].

MoS_2 nanosheets were combined with sodium alginate (SA) in 2018 by Song et al. to separate water and ethanol. The quick transit of water molecules on the MoS_2 surface led to optimal ethanol dehydration performance at a MoS_2 content of 2 wt.%. At 350 K in a water/ethanol (10/90 wt.%) solution, the separation factor was 1229 and the permeation flux was 1.839 kg/ (m^2 h) [64].

By combining graphene oxide quantum dots (GOQDs) with a SA matrix, Wang et al. [65] created a novel nanocomposite membrane for ethanol dehydration. The nanocomposite membrane performed well in ethanol dehydration, with a total permeation flux of 2432 ± 58 g/ (m^2h), 60% greater than the virgin alginate membrane, and a separation factor of 1152 ± 48 [65].

A new dense matrix and supporting membrane based on SA modification with 5 weight percent fullerenol was developed by Dmitrenko et al. [66]. It was discovered that the developed supported cross-linked SA-5/PAN-$CaCl_2$ membrane exhibited high permeation abilities for the PV separation of the water-IPA mixture at 295 K [66].

Kulkarni et al. [67] created polystyrene-4-sulfonic acid co-maleic acid sodium salt (PSSAMA-Na) capped Ag-Np ingrained SA nanocomposite membranes to increase bioethanol PV dehydration. The values for the water permeation flux and total permeation flux were close to one another, indicating that the membranes could be effectively used for the removal of water from bioethanol [67].

Using a solution-casting technique, Mokhtarzadeh et al. [68] produced biodegradable nanocomposite membranes by including different amounts of GO as a nanocomponent into a 3:1 blend of SA/poly(ε-caprolactone). The flux values rose from 0.506 to 0.779 kg m^{-2} h when the water concentration increased from 20 to 80 wt.% [68].

SA-based membranes cross-linked with PSSAMA were produced by Premakshi et al. [69] using a solution technique. An equilibrium swelling experiment was carried out utilizing various compositions of water and isopropanol mixture. The membranes containing 40% TiO_2 (SA-TiO_2) had a flux of 0.1861 kg m^{-2}·h and a separation factor of 24,092 at 303 K with 10% water in the feed [69].

Dmitrenko et al. [70] created two types of membranes using SA and fullerene-arginine derivative SA/C60-Arg (5%) on a polyacrylonitrile (PAN) matrix for support. The cross-linked composite membrane had the best transportation properties for IPA dehydration [70].

Grzybek et al. [71] examined ethanol dehydration via PV using an SA membrane containing manganese dioxide (MnO_2) and a filler mixture containing MnO_2 on magnetite core MnO_2@Fe_3O_4 particles. The MnO_2@Fe_3O_4-Alg (7%) membrane showed the best performance with a flux of 1220 g/ (m^2 h) and a separation factor of 483 [71].

Development and manufacture of dual-layer hybrid membranes was done for the dehydration of ethanol by Pan et al. [72], taking advantage of electrostatic interactions. PEI with carboxyl-decorated carbon nanotubes (CNT-C) was spin-coated as one active layer over a porous hydrolyzed polyacrylonitrile (HPAN) base, followed by a layer of SA. The synthesized membrane was used for PV dehydration of an ethanol aqueous solution (90 wt.%), and the dual-layer hybrid membranes considerably improved segregation performance [72].

Unlu [73] created an SA and chitosan hybrid membrane that was embedded with zeolitic imidazolate framework-90 (ZIF-90) and used for butanol desiccation via PV. SA-ZIF-90 outperformed all other hybrid membranes. While the maximum separation factor and flux values of 1978 and 987 g m^{-2} h were reported with SA hybrid membrane [73].

A hybrid membrane was developed by Gallardo et al. [74], for the PV separation of aqueous IPA solution using two different montmorillonite (MMT) nanoclays in SA solution. The surfactant dialkyldimethyl ammonium chloride was used to transform Na-MMT into organo-montmorillonite (O-MMT) [74].

Ni (nickel) and Ag nanowires were added to the alginate membrane for improving the ability to separate a water/ethanol mixture [75]. The membrane containing 50 mg (5 mL) of nanowires had the highest separation coefficient. The separation coefficient and flux were 878 and 1100 g/ (m^2 h) for Ag, and 1415 and 1600 g/ (m^2h) for Ni nanowire membranes, respectively [75]. Table 14.2 gives the methods for the fabrication and various uses of alginate nanocomposite membranes.

Table 14.2 Methods for the fabrication and uses of alginate nanocomposite membranes in the dehydration of organic compounds.

Sr. no.	Nanocomposite membrane used	Method of alginate NCM fabrication	Compound dehydrated	Membrane flux	Selectivity/ separation factor	Reference
1.	20% MCM-41-loaded SA membrane	Solution casting and cross-linking	Isopropanol	0.106 kg cm^{-2} h	29,991	[47]
2.	SA–Na^+ MMT particles	Cross-linking	Isopropanol, 1,4-dioxane, and THF	0.050 kg·m^{-2}·h^{-1} 0.089 kg·m^{-2}·h^{-1} 0.148 kg·m^{-2}·h^{-1}	Infinity Infinity Infinity	[48]
3.	Al-MCM-41 incorporated SA membranes	Cross-linking	Isopropanol and ethanol	0.067 kg m^{-2} h 0.027 kg cm^{-2} h	653 35	[49]
4.	SA-PANI (TiO_2) membrane	Cross-linking	1,4-Dioxane	0.034 kg cm^{-2} h	12,848	[50]
5.	SA-Al-MCM-41 membranes	Cross-linking	1,4-Dioxane and THF	0.131 kg cm^{-2} h	Infinity	[51]
6.	Preyssler type heteropolyacid nanoparticles into SA	—	Isopropanol	0.059 kg cm^{-2} h	89,991	[52]
7.	SA-mPTA10	Solution casting and cross-linking	Isopropanol and ethanol	0.316 kg m^{-2} h 0.100 kg m^{-2} h	8991 298	[53]
8.	FGS-loaded SA	Solution-casting and solvent evaporation	Isopropanol	3122 GPU	4623	[54]

(Continued)

Table 14.2 (Continued)

Sr. no.	Nanocomposite membrane used	Method of alginate NCM fabrication	Compound dehydrated	Membrane flux	Selectivity/ separation factor	Reference
9.	rGO-SA membrane	Physical Blending	Ethanol	1699 g/ (m^2h)	1566	[55]
10.	SA and 10 wt.% p-TSA treated clay	Cross-linking	Isopropanol	3423 GPU	5781	[56]
11.	SA-CN_x	Physical blending	Ethanol	2469 g m^{-2} h	1633	[57]
12.	BCA membranes	Cross-linking	Ethanol	33 g m^{-2} h	429.9	[58]
13.	SA-DNA/Ca^{2+} NaAlg-DNA/Mg^{2+}	Cross-linking	Ethanol	0.01 kg m^{-2} h 0.01 kg m^{-2} h	5500 6500	[59]
14.	BCA membranes	Cross-linking	Biodiesel-methanol mixtures	148 g/ (m^2h)	332	[60]
15.	SA-PSBMA @GO(X)/PAN	Spin-coating method	Alcohol	2140 g/ (m^2 h)	1370	[61]
16.	SA-PEG@POSS(X)	Spin-coating method	Ethanol	2.5 kg/ (m^2 h)	1077	[62]
17.	Ag_2O loaded membrane	Cross-linking	Ethanol	8.49 kg/ (m^2 h)	108.15	[63]
18.	SA-MoS_2-X/PAN	Spin-coating method	Ethanol	1839 g/ (m^2 h)	1229	[64]
19.	SA-GOQDs	Spin-coating method	Ethanol	2432 ± 58 g m^{-2} h^{-1}	1152 ± 48	[65]
20.	SA-5/PAN $CaCl_2$ membrane	Cross-linking	Isopropanol	0.64–2.9 kg/ (m^2 h)	26–73,326	[66]

21.	PSSAMA-Na capped Ag-NPs SA membranes	Cross-linking	Bioethanol	13.40×10^{-2} kg/(m^2 h)	406	[67]
22.	SA/PCL/GO-6	Solution-casting technique	Isopropanol	0.779 kg m^{-2} h^{-1}	25	[68]
			Ethanol	0.757 kg m^{-2} h^{-1}	15	
			Methanol	0.569 kg m^{-2} h^{-1}	6	
23.	SA-TiO_2	Cross-linking	Isopropanol	18.61×10^{-2} kg m^{-2} h	24,092	[69]
24.	SA-C60-Arg5/PAN$CaCl_2$	Cross-linking	Isopropanol	0.507 kg/ (m^2h)	73,326	[70]
25.	MnO_2@Fe_3O_4-Alg membrane	Cross-linking	Ethanol	1.22 kg/ (m^2 h)	483	[71]
26.	SA/(PEI@CNT-CX)/HPAN	—	Ethanol	2825 g/ (m^2h)	2148	[72]
27.	SA-ZIF-90 membrane	Cross-linking	Butanol	0.987 kg m^{-2} h	1978	[73]
28.	SAO-MMT	—	Isopropanol	4764 g/ (m^2 h)		[74]
29.	Alg-Ag-NWs and	—	Ethanol	1.1 kg m^{-2} h^{-1}	878	[75]
	Alg-Ni-NWs membrane			1.6 kg m^{-2} h^{-1}	1415	

14.3.3 Oil and Water Separation

With the addition of GO nanosheets, Ehsan et al. [26] concentrated on creating hydrophilic nanocomposite membranes that could be employed for oil/water separation at a low cost and with no environmental impact. The findings show that adding 1.5 wt.% GO to the SA matrix enhanced oil antifouling performance with a flux recovery ratio of more than 88%, and removal efficiency by 93.26%, repeated over three filtration cycles [26].

14.4 Conclusion

To combat the challenges of water pollutants, it has become imperative to develop novel membranes. Alginate-based nanocomposite membranes are eco-friendly and have excellent adsorption performance and environmental stability. The study provides evidence that alginate, when combined with iron oxide, graphene oxide, zinc oxide, manganese dioxide, carbon nanotubes, biochar, cellulose, montmorillonite clay, silica, polyaniline, fullerenol, and nanoparticles/nanosheet as fillers and additives, resulted in promising membranes for treatment. The incorporation of appropriate filler and nanomaterial in the fabrication of the membrane depends on the type of pollutant targeted in the treatment. Alginate and the above-discussed materials performed synergistically in synthesized composite membranes for the elimination of organic and inorganic contaminants from the aqueous phase.

Despite the superior treatment efficiency of the nanocomposite membranes, most of the studies reported are at lab scale and pilot scale. High-resolution characterization techniques are required to understand the interaction between the alginate, organic/inorganic additive, and nanomaterials in the synthesis of membranes. This will help in the optimization of the desired characteristics of the composite membrane. More research is required on the long-term performance and stability of the fabricated composite membrane in real-time application.

References

1 Raus, R.A., Nawawi, W.M.F.W., and Nasaruddin, R.R. (2021). Alginate and alginate composites for biomedical applications. *Asian Journal of Pharmaceutical Sciences* 16 (3): 280–306.

2 Hay, I.D., Rehman, Z.U., Moradali, M.F. et al. (2013). Microbial alginate production, modification and its applications. *Microbial Biotechnology* 6 (6): 637–650.

3 Pawar, S.N. and Edgar, K.J. (2012). Alginate derivatization: a review of chemistry, properties and applications. *Biomaterials* 33 (11): 3279–3305.

4 Abka-Khajouei, R., Tounsi, L., Shahabi, N. et al. (2022). Structures, properties and applications of alginates. *Marine Drugs* 20 (6): 364.

5 Hallett, J.P. and Welton, T. (2011). Room-temperature ionic liquids: solvents for synthesis and catalysis. 2. *Chemical Reviews* 111 (5): 3508–3576.

6 Hirst, E.L., Jones, J.K.N., and Jones, W.O. (1939). Structure of alginic acid. *Nature* 143 (3629): 857–857.

7 Hecht, H. and Srebnik, S. (2016). Structural characterization of sodium alginate and calcium alginate. *Biomacromolecules* 17 (6): 2160–2167.

8 Gao, X., Guo, C., Hao, J. et al. (2020). Adsorption of heavy metal ions by sodium alginate based adsorbent-a review and new perspectives. *International Journal of Biological Macromolecules* 164: 4423–4434.

9 Guo, H., Qin, Q., Chang, J.S., and Lee, D.J. (2023). Modified alginate materials for wastewater treatment: application prospects. *Bio-resource Technology* 387: 129639.

10 Mao, W., Zhang, L., Zhang, Y. et al. (2022). Adsorption and photocatalysis removal of arsenite, arsenate, and hexavalent chromium in water by the carbonized composite of manganese-crosslinked sodium alginate. *Chemosphere* 292: 133391.

11 Draget, K.I. and Taylor, C. (2011). Chemical, physical and biological properties of alginates and their biomedical implications. *Food Hydrocolloids* 25 (2): 251–256.

12 Khajouei, R.A., Keramat, J., Hamdami, N. et al. (2018). Extraction and characterization of an alginate from the Iranian brown seaweed Nizimuddinia zanardini. *International Journal of Biological Macromolecules* 118: 1073–1081.

13 Ponce, N. and Stortz, C.A. (2020). A comprehensive and comparative analysis of the fucoidan compositional data across the Phaeophyceae. *Frontiers in Plant Science* 11: 556312.

14 Jiao, W., Chen, W., Mei, Y. et al. (2019). Effects of molecular weight and guluronic acid/mannuronic acid ratio on the rheological behavior and stabilizing property of sodium alginate. *Molecules* 24 (23): 4374.

15 Aburabie, J.H., Puspasari, T., and Peinemann, K.V. (2020). Alginate-based membranes: paving the way for green organic solvent nanofiltration. *Journal of Membrane Science* 596: 117615.

16 Shaari, N. and Kamarudin, S.K. (2020). Sodium alginate/alumina composite biomembrane preparation and performance in DMFC application. *Polymer Testing* 81: 106183.

17 Córdova, B.M., Jacinto, C.R., Alarcón, H. et al. (2018). Chemical modification of sodium alginate with thiosemicarbazide for the removal of Pb (II) and Cd (II) from aqueous solutions. *International Journal of Biological Macromolecules* 120: 2259–2270.

18 Nigiz, F.U. (2020). Graphene oxide-sodium alginate membrane for seawater desalination through pervaporation. *Desalination* 485: 114465.

19 Sharmin, N., Pang, C., Sone, I. et al. (2021). Synthesis of sodium alginate–silver nanocomposites using plasma activated water and cold atmospheric plasma treatment. *Nanomaterials* 11 (9): 2306.

20 Wang, B., Wan, Y., Zheng, Y. et al. (2019). Alginate-based composites for environmental applications: a critical review. *Critical Reviews in Environmental Science and Technology* 49 (4): 318–356.

21 Dudek, G., Krasowska, M., Turczyn, R. et al. (2019). Clustering analysis for pervaporation performance assessment of alginate hybrid membranes in dehydration of ethanol. *Chemical Engineering Research and Design* 144: 483–493.

22 Lecaros, R.L.G., Ho, S.Y., Tsai, H.A. et al. (2021). Ionically cross-linked sodium alginate and polyamidoamine dendrimers for ethanol/water separation through pervaporation. *Separation and Purification Technology* 275: 119125.

23 Xing, R., Pan, F., Zhao, J. et al. (2016). Enhancing the permeation selectivity of sodium alginate membrane by incorporating attapulgite nanorods for ethanol dehydration. *RSC Advances* 6 (17): 14381–14392.

24 Ehsan, M., Razzaq, H., Razzaque, S. et al. (2022). Recent advances in sodium alginate-based membranes for dehydration of aqueous ethanol through pervaporation. *Journal of Polymer Science* 60 (16): 2435–2453.

25 Liu, S., Li, Y., and Li, L. (2017). Enhanced stability and mechanical strength of sodium alginate composite films. *Carbohydrate Polymers* 160: 62–70.

26 Ehsan, M., Razzaq, H., Razzaque, S. et al. (2023). Engineering nanocomposite membranes of sodium alginate-graphene oxide for efficient separation of oil-water and antifouling performance. *Journal of Environmental Chemical Engineering* 11 (1): 109185.

27 Nasir, A., Masood, F., Yasin, T., and Hameed, A. (2019). Progress in polymeric nanocomposite membranes for wastewater treatment: preparation, properties and applications. *Journal of Industrial and Engineering Chemistry* 79: 29–40.

28 Aburabie, J., Villalobos, L.F., and Peinemann, K.V. (2017). Composite membrane formation by combination of reaction-induced and nonsolvent-induced phase separation. *Macromolecular Materials and Engineering* 302 (9): 1700131.

29 Tanahashi, M. (2010). Development of fabrication methods of filler/polymer nanocomposites: with focus on simple melt-compounding-based approach without surface modification of nanofillers. *Materials* 3 (3): 1593–1619.

30 Khajavi, R. and Abbasipour, M. (2017). Controlling nanofiber morphology by the electrospinning process. In: *Electrospun Nanofibers* (ed. M. Afshari), 109–123. Woodhead Publishing.

31 Lecaros, R.L.G., Ho, S.Y., Yang, H.L. et al. (2022). Acid-reinforced ionic cross-linking of sodium alginate/polyamidoamine dendrimer blended composite membranes for isopropanol dehydration through pervaporation. *Separation and Purification Technology* 288: 120660.

32 Rhimi, A., Zlaoui, K., Van der Bruggen, B. et al. (2021). Synthesis and characterization of crosslinked membranes based on sodium alginate/polyvinyl alcohol/graphene oxide for ultrafiltration applications. *Desalination and Water Treatment* 230: 204–218.

33 Li, S., Dai, J., Geng, X. et al. (2020). Highly selective sodium alginate mixed-matrix membrane incorporating multi-layered MXene for ethanol dehydration. *Separation and Purification Technology* 235: 116206.

34 Kazemi, M., Jahanshahi, M., and Peyravi, M. (2018). Chitosan-sodium alginate multilayer membrane developed by Fe_0@ WO_3 nanoparticles: photocatalytic removal of hexavalent chromium. *Carbohydrate Polymers* 198: 164–174.

35 Amiri, S., Asghari, A., Vatanpour, V., and Rajabi, M. (2020). Fabrication and characterization of a novel polyvinyl alcohol-graphene oxide-sodium alginate nanocomposite hydrogel blended PES nanofiltration membrane for improved water purification. *Separation and Purification Technology* 250: 117216.

36 Dodero, A., Vicini, S., Lova, P. et al. (2020). Nanocomposite alginate-based electrospun membranes as novel adsorbent systems. *International Journal of Biological Macromolecules* 165: 1939–1948.

37 Sokhandan, F., Homayoonfal, M., and Davar, F. (2020). Application of zinc oxide and sodium alginate for biofouling mitigation in a membrane bioreactor treating urban wastewater. *Biofouling* 36 (6): 660–678.

38 Metecan, A., Cihanoğlu, A., and Altinkaya, S.A. (2021). A positively charged loose nanofiltration membrane fabricated through complexing of alginate and polyethyleneimine with metal ions on the polyamideimide support for dye desalination. *Chemical Engineering Journal* 416: 128946.

39 Iqbal, S.S., Bahadar, A., Hossain, N. et al. (2021). Synergetic influence of F-MWCNTS on polyvinylpyrrolidone sodium alginate composite membrane for reverse osmosis. *Journal of Environmental Chemical Engineering* 9 (5): 106085.

40 Anokhina, T., Dmitrieva, E., and Volkov, A. (2022). Recovery of model pharmaceutical compounds from water and organic solutions with alginate-based composite membranes. *Membranes* 12 (2): 235.

41 Wang, Y.X., Zhu, C.Y., Huang, J. et al. (2022). Cu^{2+}/alginate nanofiltration membranes fabricated at the aqueous contra-diffusion "interface" for salt/dye rejection. *Desalination* 535: 115806.

42 Amiri, S., Vatanpour, V., and He, T. (2023). Antifouling thin-film nanocomposite NF membrane with polyvinyl alcohol-sodium alginate-graphene oxide nanocomposite hydrogel coated layer for As (III) removal. *Chemosphere* 322: 138159.

43 Arla, S.K., Godlaveeti, S.K., Sana, S.S. et al. (2023). The impact of titanium nanotubes (tnts) incorporated into a sodium alginate-based nanocomposite membrane on the adsorption of cationic dye from aqueous solutions. *ECS Journal of Solid State Science and Technology* 12 (12): 123001.

44 Cai, R., Chen, Y., Hu, J. et al. (2023). A self-supported sodium alginate composite hydrogel membrane and its performance in filtering heavy metal ions. *Carbohydrate Polymers* 300: 120278.

45 Spoială, A., Ilie, C.I., Dolete, G. et al. (2023). The development of alginate/Ag NPs/caffeic acid composite membranes as adsorbents for water purification. *Membranes* 13 (6): 591.

46 Wen, H., Xu, N., Soyekwo, F. et al. (2023). Towards enhanced performance of fertilizer-drawn forward osmosis process coupled with sludge thickening using a thin-film nanocomposite membrane interlayered with Mxene scaffolded alginate hydrogel. *Journal of Membrane Science* 685: 121899.

47 Bhat, S.D., Naidu, B.V.K., Shanbhag, G.V. et al. (2006). Mesoporous molecular sieve (MCM-41)-filled sodium alginate hybrid nanocomposite membranes for pervaporation separation of water–isopropanol mixtures. *Separation and Purification Technology* 49 (1): 56–63.

48 Bhat, S.D. and Aminabhavi, T.M. (2006). Novel sodium alginate–Na+ MMT hybrid composite membranes for pervaporation dehydration of isopropanol, 1, 4-dioxane and tetrahydrofuran. *Separation and Purification Technology* 51 (1): 85–94.

49 Patil, M.B., Veerapur, R.S., Patil, S.A. et al. (2007). Preparation and characterization of filled matrix membranes of sodium alginate incorporated with aluminum-containing mesoporous silica for pervaporation dehydration of alcohols. *Separation and Purification Technology* 54 (1): 34–43.

50 Lokesh, B.G., Rao, K.K., Reddy, K.M. et al. (2008). Novel nanocomposite membranes of sodium alginate filled with polyaniline-coated titanium dioxide for dehydration of 1, 4-dioxane/water mixtures. *Desalination* 233 (1-3): 166–172.

51 Patil, M.B., Veerapur, R.S., Bhat, S.D. et al. (2009). Hybrid composite membranes of sodium alginate for pervaporation dehydration of 1, 4-dioxane and tetrahydrofuran. *Desalination and Water Treatment* 3 (1-3): 11–20.

52 Mali, M.G. and Gokavi, G.S. (2012). Sorption and permeation studies for isopropanol+ water mixtures using alginate based highly water selective nanocomposite membranes. *Journal of Polymer Research* 19: 1–11.

53 Adoor, S.G., Rajineekanth, V., Nadagouda, M.N. et al. (2013). Exploration of nanocomposite membranes composed of phosphotungstic acid in sodium alginate for separation of aqueous–organic mixtures by pervaporation. *Separation and Purification Technology* 113: 64–74.

54 Suhas, D.P., Raghu, A.V., Jeong, H.M., and Aminabhavi, T.M. (2013). Graphene-loaded sodium alginate nanocomposite membranes with enhanced isopropanol dehydration performance via a pervaporation technique. *RSC Advances* 3 (38): 17120–17130.

55 Cao, K., Jiang, Z., Zhao, J. et al. (2014). Enhanced water permeation through sodium alginate membranes by incorporating graphene oxides. *Journal of Membrane Science* 469: 272–283.

56 Suhas, D.P., Aminabhavi, T.M., and Raghu, A.V. (2014). Para-Toluene sulfonic acid treated clay loaded sodium alginate membranes for enhanced pervaporative dehydration of isopropanol. *Applied Clay Science* 101: 419–429.

57 Cao, K., Jiang, Z., Zhang, X. et al. (2015). Highly water-selective hybrid membrane by incorporating g-C_3N_4 nanosheets into polymer matrix. *Journal of Membrane Science* 490: 72–83.

58 Suratago, T., Taokaew, S., Kanjanamosit, N. et al. (2015). Development of bacterial cellulose/alginate nanocomposite membrane for separation of ethanol–water mixtures. *Journal of Industrial and Engineering Chemistry* 32: 305–312.

59 Uragami, T., Banno, M., and Miyata, T. (2015). Dehydration of an ethanol/water azeotrope through alginate-DNA membranes cross-linked with metal ions by pervaporation. *Carbohydrate Polymers* 134: 38–45.

60 Suratago, T., Panitchakarn, P., Kerdlarpphon, P. et al. (2016). Bacterial cellulose-alginate membrane for dehydration of biodiesel-methanol mixtures. *Engineering Journal* 20 (5): 145–153.

61 Zhao, J., Zhu, Y., He, G. et al. (2016). Incorporating zwitterionic graphene oxides into sodium alginate membrane for efficient water/alcohol separation. *ACS Applied Materials & Interfaces* 8 (3): 2097–2103.

62 Wang, M., Xing, R., Wu, H. et al. (2017). Nanocomposite membranes based on alginate matrix and high loading of pegylated POSS for pervaporation dehydration. *Journal of Membrane Science* 538: 86–95.

63 Dudek, G., Gnus, M., Strzelewicz, A. et al. (2018). The influence of metal oxides on the separation properties of hybrid alginate membranes. *Separation Science and Technology* 53 (8): 1178–1190.

64 Song, Y., Jiang, Z., Gao, B. et al. (2018). Embedding hydrophobic MoS_2 nanosheets within hydrophilic sodium alginate membrane for enhanced ethanol dehydration. *Chemical Engineering Science* 185: 231–242.

65 Wang, M., Pan, F., Yang, L. et al. (2018). Graphene oxide quantum dots incorporated nanocomposite membranes with high water flux for pervaporative dehydration. *Journal of Membrane Science* 563: 903–913.

66 Dmitrenko, M., Liamin, V., Kuzminova, A. et al. (2020). Novel mixed matrix sodium alginate–fullerenol membranes: development, characterization, and study in pervaporation dehydration of isopropanol. *Polymers* 12 (4): 864.

67 Kulkarni, A.S., Sajjan, A.M., Ashwini, M. et al. (2020). Novel fabrication of PSSAMA_Na capped silver nanoparticle embedded sodium alginate membranes for pervaporative dehydration of bioethanol. *RSC Advances* 10 (38): 22645–22655.

68 Mokhtarzadeh, S., Hakimpour, F., Sarvari, R. et al. (2020). Nanocomposite membranes based on sodium alginate/poly (ε-caprolactone)/graphene oxide for methanol, ethanol and isopropanol dehydration via pervaporation. *Polymer Bulletin* 77: 3367–3387.

69 Premakshi, H.G., Kariduraganavar, M.Y., and Mitchell, G.R. (2020). Crosslinked nanocomposite sodium alginate-based membranes with titanium dioxide for the dehydration of isopropanol by pervaporation. *Molecules* 25 (6): 1298.

70 Dmitrenko, M., Liamin, V., Lahderanta, E. et al. (2021). Mixed matrix membranes based on sodium alginate modified by fullerene derivatives with L-amino acids for pervaporation isopropanol dehydration. *Journal of Materials Science* 56: 7765–7787.

71 Grzybek, P., Turczyn, R., and Dudek, G. (2021). Mixed manganese dioxide on magnetite core MnO_2@ Fe_3O_4 as a filler in a high-performance magnetic alginate membrane. *Materials* 14 (24): 7667.

72 Pan, F., Cao, C., Liu, G., and Jiang, Z. (2021). Enhanced water-selective performance of dual-layer hybrid membranes by incorporating carbon nanotubes. *Chemical Engineering Science: X* 11: 100102.

73 Unlu, D. (2021). Chitosan/sodium alginate hybrid membranes modified by zeolitic imidazolate framework-90 for pervaporative dehydration of butanol. *Iranian Polymer Journal* 30 (12): 1239–1249.

74 Gallardo, M.R., Duena, A.N., Ang, M.B.M.Y. et al. (2023). Improved pervaporation dehydration performance of alginate composite membranes by embedding organo-montmorillonite. *Separation and Purification Technology* 314: 123619.

75 Knyra, D., Dzido, G., Chrobak, A. et al. (2024). Novel use for metal nanowires–synthesis, characterization and application of nickel and silver nanowires in the process of ethanol dehydration via pervaporation. *Separation and Purification Technology* 333: 125894.

15

Photocatalytic Photothermal Membranes for Microbial Wastewater Remediation

Laila Noureen[1], Nadeem Muhammad[2,3], Noureddine El Messaoudi[4], Bushra Iqbal[5], and Muhammad Tayyab[6]

[1] *MTA-SZTE Lendület Biocolloids Research Group, Department of Physical Chemistry and Materials Science, Interdisciplinary Excellence Centre, University of Szeged, Szeged H-6720, Hungary*
[2] *School of Life Science, Wuchang University of Technology, Wuhan, 430223, China*
[3] *Department of Chemical Engineering, University of Louisville, Louisville, Kentucky, USA*
[4] *Laboratory of Applied Chemistry and Environment, Faculty of Sciences, Ibn Zohr University, Agadir 80000, Morocco*
[5] *Institute of Chemistry, University of Okara, Okara-56300, Pakistan*
[6] *Institute of Materials Research, Tsinghua Shenzhen International Graduate School, Tsinghua University, Shenzhen, Guangdong 518055, PR China*

15.1 Introduction

Microbial wastewater refers to wastewater that contains microorganisms, including bacteria, viruses, fungi, and protozoa, either as contaminants or as intentional agents for treatment processes [1, 2]. These microorganisms can originate from various sources, including domestic sewage, industrial effluents, agricultural runoff, and stormwater. In wastewater, microorganisms can pose significant challenges to public health and the environment if not properly treated [3, 4]. Pathogenic bacteria and viruses present in microbial wastewater can cause waterborne diseases, posing risks to human health when untreated wastewater enters drinking water sources or recreational water bodies [5]. Moreover, microbial contaminants can contribute to environmental pollution and ecosystem degradation, impacting aquatic life and biodiversity [6]. However, not all microorganisms in wastewater are harmful. Some bacteria and other microorganisms play beneficial roles in wastewater treatment processes, such as biological nutrient removal and biodegradation of organic pollutants [7]. These microorganisms can be harnessed in treatment systems, such as activated sludge processes, biofilters, and constructed wetlands, to remove contaminants and improve water quality [8].

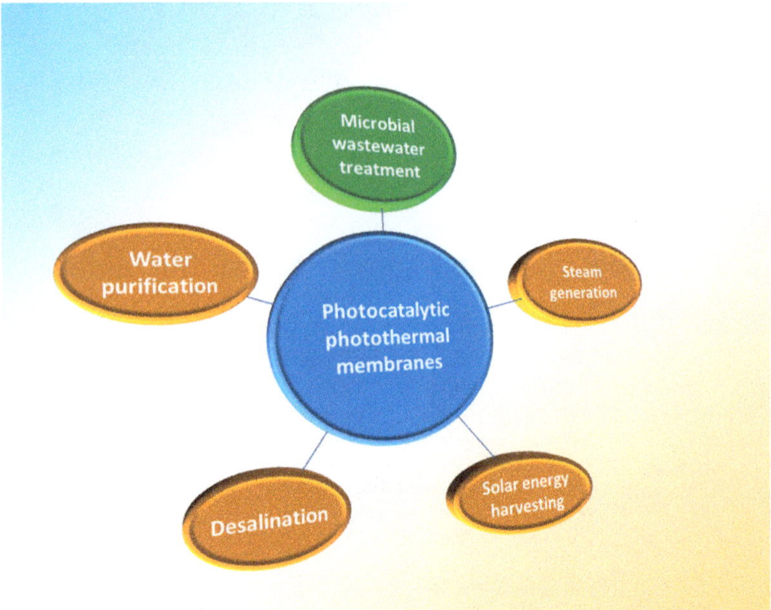

Figure 15.1 Graphical abstract of photocatalytic photothermal membranes.

Overall, the management of microbial wastewater involves a combination of physical, chemical, and biological treatment processes aimed at reducing or eliminating harmful microorganisms and organic pollutants [9]. Treatment technologies may include disinfection methods (e.g., chlorination, UV irradiation), biological treatment (e.g., activated sludge, trickling filters), and advanced treatment processes (e.g., membrane filtration, ozonation) to ensure the safe disposal or reuse of wastewater while protecting public health and the environment [10]. Among these, photocatalytic photothermal membranes have emerged as promising candidates for microbial wastewater remediation, offering the potential for simultaneous removal of organic pollutants, pathogens, and microbial contaminants [11, 12] (Figure 15.1).

Photocatalytic photothermal membranes are innovative materials that integrate the principles of photocatalysis with membrane technology to achieve simultaneous pollutant degradation and filtration in water treatment processes [13]. Photocatalytic photothermal membranes float at the air-water interface, absorb solar light, and thus enhance photocatalytic performance. These membranes combine the properties of photocatalytic materials, such as titanium dioxide (TiO_2), zinc oxide (ZnO), etc., with the selective purification capabilities of photothermal membranes, resulting in multifunctional structures with enhanced performance for microbial wastewater remediation [13, 14].

The photocatalytic component embedded within the membrane matrix catalyzes the degradation of organic pollutants and microbial contaminants when activated by ultraviolet (UV) or visible light irradiation [15, 16]. This enables the breakdown of organic compounds into harmless by-products, contributing to the overall purification of water [17, 18]. The membrane component of the composite facilitates the physical separation of suspended solids, microorganisms, and other particulate matter from the water matrix via enhanced photocatalytic properties due to thermal effect [14, 19]. The membrane acts as a selective barrier, allowing the passage of water molecules while retaining contaminants, thereby achieving purification and clarification of the water. By combining photocatalysis and with photothermal effect in a single material, these composite membranes offer synergistic benefits that enhance overall treatment performance [14]. The photocatalytic process can help prevent fouling and biofilm formation on the membrane surface, thereby extending membrane lifespan and reducing maintenance requirements [20]. Photocatalytic composite membranes can be tailored to target specific contaminants or address particular water quality challenges. Different types of photocatalytic materials, membrane polymers, and fabrication techniques can be employed to optimize membrane performance for various applications, including drinking water treatment, wastewater reuse, and environmental remediation [21–23].

Applications of photocatalytic photothermal membranes span a wide range of sectors, including municipal water treatment, industrial wastewater remediation, and decentralized water purification systems [24]. These membranes hold promise for addressing emerging contaminants, such as pharmaceuticals, personal care products, and microplastics, while offering sustainable and cost-effective solutions for water management challenges [25, 26]. Overall, photocatalytic photothermal membranes represent a promising avenue for advancing water treatment technologies, leveraging the synergies between photocatalysis and membrane filtration to achieve efficient and sustainable removal of pollutants from water sources [27]. Continued research and development in this field holds the potential to further optimize membrane performance, expand application possibilities, and contribute to global efforts to ensure access to clean and safe water resources.

This chapter aims to provide a comprehensive overview of the performance evaluation of the most advanced water purification techniques, i.e., photocatalytic photothermal composite membranes for microbial wastewater treatment. Through a critical examination of recent advancements and case studies, it seeks to elucidate the potential applications, challenges, and future directions in this rapidly evolving field. By exploring the intersection of photocatalysis and membrane technology, this chapter aims to contribute to the development of innovative solutions for addressing the growing challenges of microbial wastewater remediation in a sustainable and environmentally responsible manner.

15.2 Types of Photocatalytic Photothermal Membranes

Recently, an abrupt exploration of the photocatalytic photothermal membranes has been observed due to the limitation of photocatalysis and photothermal materials. These membranes have advantages of scalability, cost-effectiveness, and ease of fabrication over photocatalysis and photothermal materials. In addition, the photocatalytic photothermal membranes have high solar absorption efficiency and thermal conductivity, which rise their widespread use for efficiently capturing and utilizing solar energy. These membranes find applications in diverse fields, including water purification, desalination, sensing, microbial wastewater treatment, steam generation, CO_2 reduction, and H_2 generation. The main working principle of these membranes is similar with photocatalysis. However, in photocatalytic photothermal membranes the absorbed solar energy can also participate in heat generation (temperature rising). Therefore, in photocatalytic photothermal membranes dual effect is utilized (heat and excited energy charges) to speed up the chemical reaction (required application).

Until now various types of photocatalytic photothermal membranes have been reported for variety of applications. However, by keeping in mind nonresearchers, we have divided photothermal membranes into two categories: (i) inorganic photothermal membranes and (ii) polymeric photothermal membranes. The detail of each category along with explains from literature are mentioned below.

15.2.1 Inorganic Photothermal Membranes

Inorganic photothermal membranes typically consist of inorganic compounds with photothermal properties, such as metal chalcogenide, metal oxides, or nanoparticles, which can absorb efficient sunlight and convert it into heat due to their unique electronic and optical properties. These membranes are specially engineered to have high solar absorption efficiency and thermal conductivity, allowing them to capture and utilize solar energy. The absorbed solar energy can then be used to drive chemical reactions or processes, such as the degradation of microbe or organic pollutants in wastewater or the separation of salt from seawater through desalination. Inorganic photothermal membranes offer several advantages, including sustainability, as they rely on renewable solar energy, and versatility, as they can be tailored for specific applications by adjusting their composition and structure. Moreover, they can operate under ambient conditions without the need of additional energy sources, making them cost-effective and environmentally friendly solutions for various environmental and industrial challenges.

For instance, Yang et al. first reported the ultrathin 2D porous inorganic photothermal membrane of MoS_2 nanosheets combined with single-walled nanotube (SWNT), as shown in Figure 15.2a–c [28]. The optimized SWNT-MoS_2 (120 nm ultrathin film) demonstrates an efficient absorption in the entire solar spectrum (almost over 82%). Moreover, due to the effect of localized heat confinement, SWNT-MoS_2 films floating on water generated sharp temperature gradients. This thin and porous structure effectively facilitated the fast escape of water vapor, resulting in 91.5% evaporation efficiency. In another example, Jia et al. synthesized dual inorganic photothermal and photodynamic membrane of WSe_2 nanosheet via chemical exploration with bovine serum albumin (BSA) for biomedical application (cancer therapy), as shown in Figure 15.2d [29]. A comparison of dual-mode photothermal and photodynamic therapy showed that it greatly improved the therapeutic efficacy against cancer cells over single-mode treatment. These examples indicate the emerging potential of inorganic photothermal membranes in various applications.

15.2.2 Polymeric Photothermal Membranes

Polymeric photothermal membranes are a class of materials that harness the photothermal effect within a polymer matrix. These membranes typically consist of polymers doped or coated with photothermal agents, such as carbon-based nanomaterials (e.g., carbon nanotubes, graphene) or metal nanoparticles (e.g., silver, gold). The integration of these photothermal agents into the polymer matrix enables the membrane to absorb light energy and convert it into heat efficiently. For example, Wang et al. synthesized two-dimensional (2D) nanosandwich polymeric photothermal membrane of graphitic carbon nitride (g-C_3N_4) loaded with dense Ag nanoparticles which convert into Ag_2O nanolayers [31].

Amazingly, the Ag_2O@Ag/g-C_3N_4 nanosandwiches with improved broadband absorption and thermal conversion acquired both increased photo-triggered catalytic degradation and pronounced photothermal utilization for boosting degradation efficiency. In another example, Yang et al. explored a polymeric photothermal membrane of porous SWNT loaded with gold nanorod (AuNR), as shown in Figure 15.2e [31]. By the optimized SWNT/AuNR, they achieved nearly 94% water evaporation efficiency because of the synergic effect of enhanced photothermal performance, high solar spectrum absorption, interconnected porous structure, excellent mechanical performance, and thermal insulating feature. Both polymeric and inorganic photothermal membranes have their own advantages and disadvantages. Therefore, by keeping eyes on the emerging exploration of photocatalytic photothermal membranes, we have also presented the comparison of the inorganic and polymeric photothermal membranes in Figure 15.2.

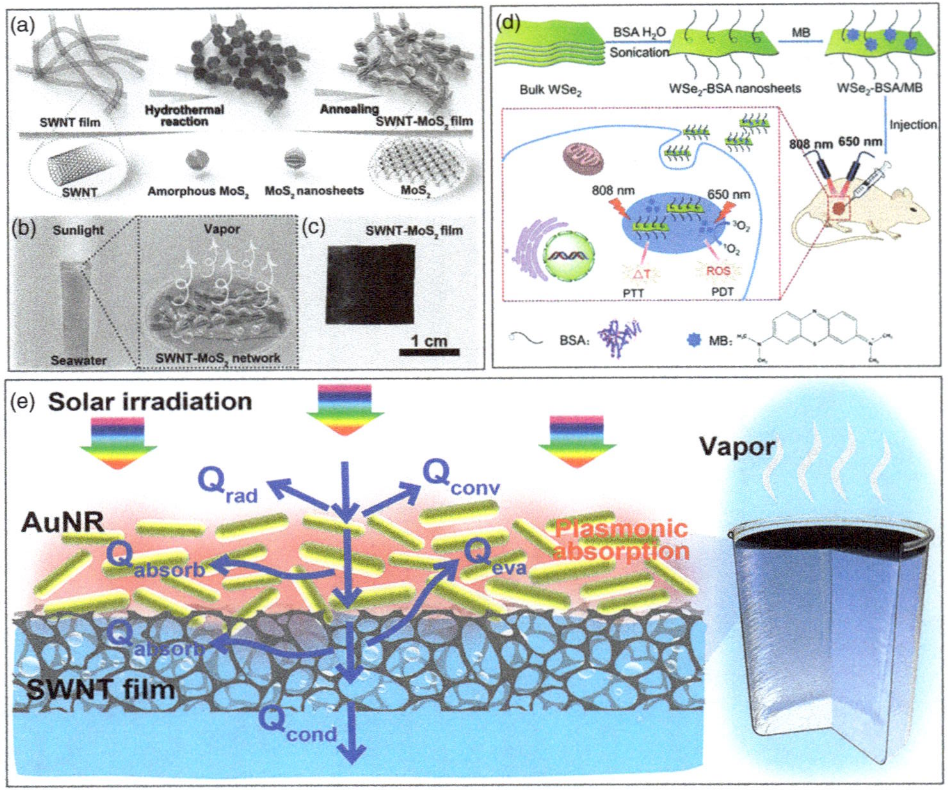

Figure 15.2 Photocatalytic photothermal membranes. (a–c) Inorganic photothermal membrane of MoS_2 nanosheets combined with single-walled nanotube (SWNT). *Source:* Reproduced with the permission of Ref. [28]. (d) WSe_2 nanosheet integrated with bovine serum albumin (BSA) for biomedical applications. *Source:* reproduced with the Ref. [29]. (e) Porous SWNT loaded with gold nanorod (AuNR). *Source:* Reproduced with the permission of Ref. [31].

15.3 Factors Effecting Efficiency of Photocatalytic Photothermal Membranes

15.3.1 Arrangement of Photocatalytic Photothermal Membranes

A crucial factor that can significantly influence the photocatalytic photothermal properties of these materials is getting the maximum utilization of solar light. This is accompanied by the superior solar light absorption property of the photothermal materials. When solar light strikes the surface of the photothermal material, it is both reflected as well as absorbed. The reflected light is scattered in the surroundings without being adequately utilized. Diminishing or recycling this reflected light is a practical approach to enhance the interfacial heating ability of photothermal materials. Lately, studies have focused on the development of techniques to track and recycle the reflected light by tweaking the design of photocatalytic photothermal materials. For instance, 2D graphdiyne was deposited on the surface of CuO nanowires in a multidimensional architecture, which achieved 91% interfacial heating efficiency in a single sun irradiation (Figure 15.3a(I)) [32]. With this method, the reflected solar light was absorbed and reused in multiple reflections, improving the utilization of solar energy by 8%. Another factor that can substantially amplify the interfacial heating property of the photothermal materials is the careful selection of the substrate. Optimization of the surface morphology of the photothermal materials can further augment their light-harvesting ability. For example, a material with a bioinspired 3D photothermal cone was reported to have superior light absorption ability. This photothermal cone was constructed by coating polypyrrole on a polyvinylidene fluoride (PVDF) membrane in a simple clipping manner (Figure 15.3a(II)) [33]. Due to the rational architecture of this cone, it was able to achieve exceptional light absorption by reducing the heat loss in reflection. The curved inner surface of the cone can trigger multiple reflections leading to an increase in the surface area for light absorption. In general, photothermal materials have a large top and a narrow bottom, which reduces the contact between the photothermal material and bulk water and minimizes heat loss during bulk water heating. This factor also paves a way to the enhanced properties of photothermal materials.

15.3.2 Structural Arrangements of the Photothermal Membranes

Another factor that has an impact on the efficiency of solar-driven steam generation is the deposition of salt on the surface of the photothermal material during the process of solar desalination. To prevent this accumulation, optimization of the device structure is necessary. Chen et al. conducted research in this regard and constructed an evaporator with the extraordinary ability of heat localization at the surface of the photothermal material attributed to its distinctive structural arrangement (Figure 15.3b) [34]. The evaporator has black fabric for the absorption of

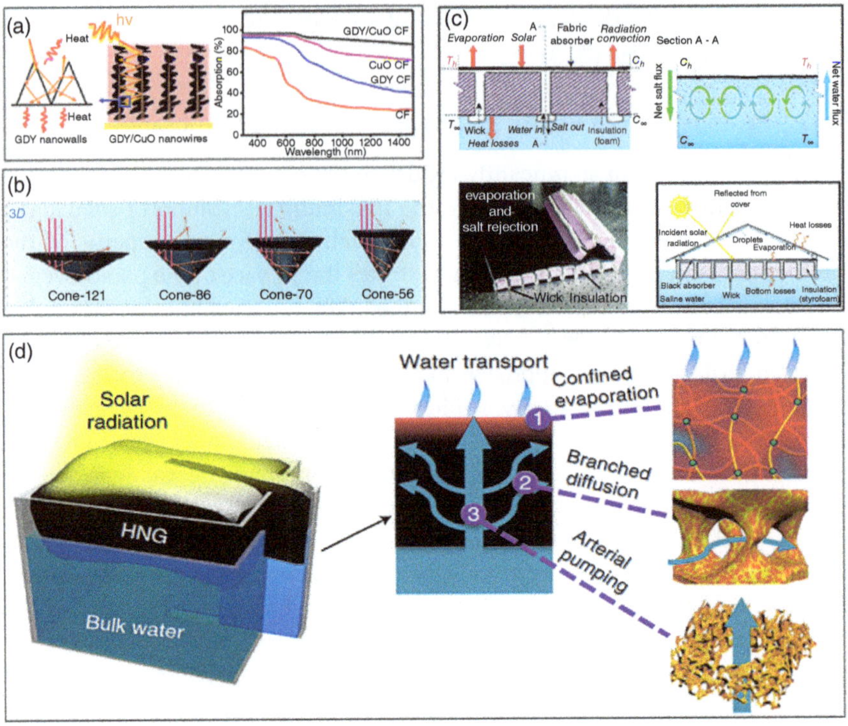

Figure 15.3 Strategies to improve photocatalytic photothermal performance. (a, b) Multidirectional architecture of the photocatalytic photothermal membrane. *Source:* Reproduced with the permission from Ref. [32, 33] / American Chemical Society. (c) Salt rejecting photothermal membrane. *Source:* Reproduced with the permission of [34] / Royal Society of Chemistry. (d) Hydrogel based photothermal membrane. *Source:* Reproduced with the permission of [37] / Springer Nature.

solar light, a wick of white fabric for the water supply, and polystyrene foam for the insulation. The permeable and hydrophilic characteristic of the white cellulose fabric can efficiently draw water for increased evaporation and proficient negation of the salt back into the water. The logical design of microchannels facilitates the salt exchange from higher to lower salt concentration via the diffusion process. Therefore, this mechanism is capable of sustaining a superior evaporation rate for longer periods (100 hours) even at a higher salt concentration level (20%).

15.3.3 Gels-based Photothermal Membranes

The interfacial heating mechanisms with insulating materials mentioned previously were able to considerably enhance solar-driven steam evaporation. However, heat losses in single-layer photothermal materials that are directly in contact with the water surface are still inevitable. So, improving solar-driven evaporation efficiency of such materials remains a challenge. Therefore, to minimize the heat required for evaporation, lowering the water content at the evaporative surface is greatly desired [36]. This can be accomplished by altering the surface morphology of the photothermal materials. Numerous characteristics of the photothermal materials such as water content optimization, required heat energy regulation for evaporation, robust light absorption efficiency, and adequate surface area for evaporation can be amplified by altering the surface structure of these materials. Fundamentally, the control of surface structure of photothermal materials endows water content regulation and heat energy confinement at the evaporating surface. In this regard, materials with rough and spongy top surface are proposed that are able to hold minor quantity of water while floating on the surface of water. Moreover, a substantial decline in the energy requirement for evaporation (enthalpy) is also observed which can be attributed to the porous structure of these materials. Following this strategy, researchers were able to create some hydrophilic materials having spongy and rugged surface with excellent steam evaporation efficiency. For instance, hydrogels are materials with rough and porous morphology due to specific cross-linked networks. These materials possess outstanding light absorption capability and are able to hold lower quantity of water at the evaporative surface [37]. When these materials were incorporated in interfacial heating systems, the heat losses in bulk water heating were drastically declined due to the regulated water quantity at the evaporative surface. Furthermore, pores of these materials demonstrate elevated osmotic pressure, which plays an integral role in minimization of energy required for the evaporation. In this regard, Yu et al. reported a polymeric hydrogel with proficient evaporation efficiency [35]. This PVA/PPy hydrogel possessed internal pores in a 3D network, provided ample supply of water to the evaporation surface, and damped heat energy losses in bulk water heating process. Another similar hydrogel comprised of PVA, and possessing sharp, arched upper surface aimed to optimize the water content at the evaporative surface was also reported. The surface area of this hydrogel expanded due to its greater surface roughness, and it was able to reach an evaporation rate of about 45% higher in comparison to a smooth surface hydrogel. Therefore, owing to their unique morphology, hydrogels present an effective platform to amplify the efficiency of solar-driven evaporation by optimizing the water content and controlling the heat at the evaporative surface.

15.3.4 Morphology of the Photocatalytic Photothermal Membranes

In addition to the previously mentioned materials with rough and porous structures, some more unique materials with integrated 3D structures possessing the ability of enhancing the solar-driven steam evaporation have also been reported. These materials demonstrate a superior evaporation rate than that of 1D and 2D porous materials [38]. This is attributed to the internal gaps and even higher surface area to supply adequate water to the evaporation surface. The sunlight gets reflected multiple times when it falls onto these 3D structures. These reflections enhance the light absorbing ability of these materials. Such structures also provide a bigger evaporation interface. Wang et al. reported such 3D material with above-mentioned features [33].

They demonstrated a photothermal cone with enhanced solar-driven steam evaporation. These materials were able to attain minimized reflection and heat loss. Water treatment by photothermal membranes is showing promising results in providing clean and safe water for consumption. However, one major challenge faced by this technology is membrane fouling. Membrane fouling refers to the accumulation of unwanted substances on the surface or inside the membrane, leading to a decrease in the permeation flux of water and a decline in the quality of the product water. This problem is omnipresent, and nearly all water treatment processes that use membrane technology face this issue due to the various types of impurities present in the water [28–31].

15.4 Applications of Photocatalytic Photothermal Membranes

Membrane fouling can occur due to several reasons such as pore blocking, cake formation, organic adsorption, inorganic precipitation, and biological fouling. These reasons lead to a temporary or permanent reduction in flux. Organic and biological pollutants are particularly challenging to remove, leading to membrane pollution and affecting their operation [39]. The pollutants responsible for membrane fouling can be divided into organic, inorganic, and biological. Inorganic impurities can be controlled by optimizing the operational conditions during water pretreatment. However, it is challenging to remove organic and biological pollutants through pretreatment, and they eventually accumulate on the membrane surface and lead to fouling [40]. Biological fouling is responsible for more than 45% of membrane fouling and is particularly challenging to stop [41, 42]. Several strategies have been developed to prevent membrane fouling, depending on the type of pollutant. For instance, physical cleaning, chemical cleaning, backwashing, and air scouring are some of the methods used to prevent inorganic and organic fouling. However, these strategies can introduce hazardous contaminants and increase operational costs [43, 44].

To prepare different types of antifouling membranes, various methods have been explored, including surface coating, grafting, bioadhesion, physical blending, and segregation. Surface coating and grafting are the most common methods used to prepare different types of antifouling membranes. Surface coating involves the deposition of a thin layer of material on the membrane surface, while grafting involves the attachment of a polymer or other material to the membrane surface [44–46].

Membrane fouling is a challenging issue faced by water treatment technology. The antifouling of oceans mainly focuses on organic and biological pollutants, which are challenging to remove through pretreatment. Although several strategies have been developed to prevent membrane fouling, most of them add additional operational costs and can introduce hazardous contaminants. Therefore, there is a need for continuous research to develop effective and cost-efficient methods for preventing membrane fouling.

15.4.1 Antibacterial Properties

Nanoparticles of silver (AgNPs) have been extensively utilized to modify various membrane materials due to their excellent antibacterial properties. These silver-modified membranes have been found to exhibit exceptional antibiofouling and antibacterial properties, which makes them highly effective in preventing the growth of harmful bacteria. The reason behind this effectiveness is the release of silver ions, which can destroy cell membranes and proteins, providing excellent germicidal properties. Zhang et al. conducted research on polyethersulfone (PES) membranes coupled with silver-modified graphitic carbon nitride ($Ag/g\text{-}C_3N_4$) and found that they exhibited outstanding antibacterial characteristics. They reported that silver-based nanocomposites can release silver ions into the surrounding environment, resulting in excellent germicidal properties. The researchers concluded that these silver-based nanocomposites can effectively destroy bacteria and other harmful microorganisms. The use of silver-based nanocomposites in membrane materials can significantly enhance their antibacterial properties. The release of silver ions can effectively destroy harmful bacteria and prevent biofouling. With the development of new nanocomposites, these materials have the potential to be used in various applications to improve safety and hygiene [47].

A study conducted by Jiang et al. focused on developing a photothermal membrane that could prevent bacterial growth [48]. They used a composite ultrafiltration membrane made from reduced graphene oxide and bacterial nanocellulose, which had exceptional antibiofouling properties (Figure 15.4a) [48]. To create this membrane, reduced graphene oxide sheets were added to the film solution during the growth of bacterial nanocellulose. This resulted in a membrane that had an excellent capacity to absorb light and could generate heat when exposed to light. In germicidal testing, the membrane was covered with *Escherichia coli* and then exposed to light, causing its temperature to rise above 70 °C. This temperature increase killed all the bacteria in

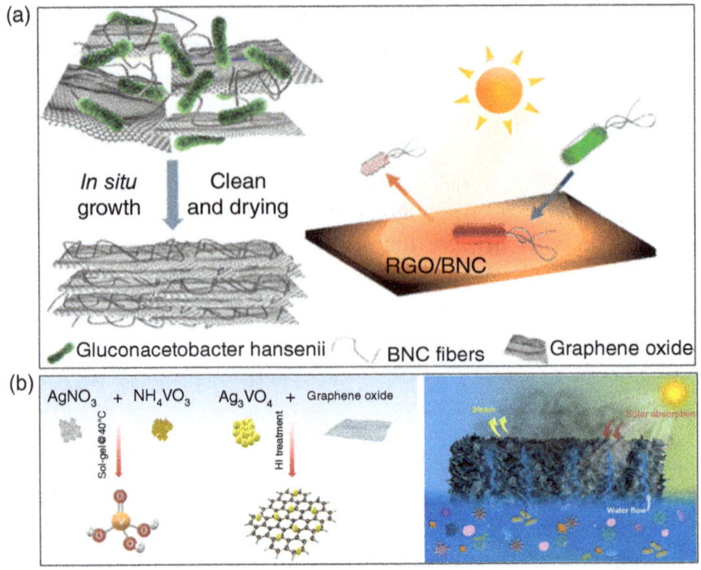

Figure 15.4 Microbial wastewater treatment. (a) rGO-BNC antibiofouling membranes. *Source:* [48] / with permission of American Chemical Society. (b) Antibacterial, antibiofouling aerogels. *Source:* [49] / with permission of American Chemical Society.

under 3 minutes, making the membrane highly effective in preventing biofouling. The reduced graphene oxide sheets in the composite membrane also helped to prevent further bacterial growth [48].

Another study by Zha et al. focused on creating a flexible MXene/cellulose membrane that could resist bioadhesion. To create this membrane, they dip-coated the surface of a cellulose film with MXene nanoflakes. The resulting membrane had excellent bioadhesion resistance and could absorb light efficiently. To test the effectiveness of the MXene/cellulose membrane, *Escherichia coli* and *Staphylococcus aureus* were used. The results of the testing showed that MXene particles effectively avoided the growth of biofilm with 99.9% efficiency. This makes the MXene/cellulose membrane highly effective in preventing biofouling and bacterial growth, making it a promising solution for various applications such as water filtration [50].

Li et al. conducted an in-depth study on photothermal membranes with antibacterial properties, using a vacuum filtration method for solar-driven water evaporation. The membrane was made of a composite material consisting of carbon nanotubes, polyamine, and mixed cellulose ester. This composite material was chosen after careful evaluation of individual materials and their properties, with the aim of creating a membrane with excellent photothermal properties and antibacterial capabilities [51]. The membrane's excellent photothermal properties were attributed to its high efficiency in absorbing light. In another report, Noureen et al. reported multifunctional coated textiles that exhibited microbial wastewater treatment via solar driven water evaporation photodecomposition of organic and disinfection of the toxic microbes present in raw water [52]. This photothermal membrane was coated with Ag_3PO_4-rGO nanocomposites, which showed excellent photocatalytic and antibacterial properties due to the presence of Ag-based nanoparticles. The membrane's surface was found to heat up quickly and efficiently when exposed to light, making it an ideal material for solar-driven water evaporation. In addition to its photothermal properties, polyamine, an antibacterial agent, was incorporated into the membrane to provide antibacterial properties. The study revealed that the membrane's surface was free of bacterial growth, and its antimicrobial stability was retained even after eight testing cycles.

15.4.2 Antibiofouling Control of Photothermal Membranes

Biofouling is a type of fouling that occurs when live microorganisms adhere to the surface of a membrane, creating a biofilm. Biofouling can lead to reduced efficiency and performance of the membrane, making it a significant issue in various industries. To address this problem, there has been an increasing interest in the development of antibacterial membranes that can prevent or inhibit biofouling. These membranes have unique surfaces that can kill bacteria or prevent their attachment. The use of these membranes has shown promising results in mitigating biofouling and increasing the longevity and efficiency of the membrane. Mixed matrix photocatalytic membranes have been found to possess significant antibacterial properties. In order to study these properties, experiments were conducted with UV or Vis light illuminations. The antibacterial effect of these membranes is attributed to the reactive oxygen species produced by the photocatalyst or direct irradiation of bacteria under UV light. The literature suggests that the antibacterial effect of hydroxyl radical attack and lipid peroxidation reaction is the main mechanism of antibacterial efficiency of photocatalytic membranes [53].

TiO_2 membranes are commonly used in wastewater treatment, coupled with UV light irradiation to enhance the antibiofouling of the photocatalytic membranes. For instance, Kwak et al. constructed a TiO_2-coated hybrid polyamide TFC membrane that showed decreased *E. coli* numbers when exposed to UV light. Similarly, Kim et al. successfully fabricated a photocatalytic membrane with TiO_2 self-assembly that was able to destroy microorganisms from the membrane surface [54, 55]. Vatanpour et al. reported that the bactericidal property of TiO_2 is dependent on the crystallinity degree of the nanocomposite [56]. In their experiments, they found that TiO_2 composed of anatase-type had the exceptional capability of blocking membrane pores, which decreased membrane application.

They also observed that the antibacterial property of the photothermal membrane relied on the location of the catalyst in the membrane structure. The surface of the photocatalytic membrane exhibited superior antibacterial properties because of its higher exposure to UV light. The crystallinity degree of the nanocomposite and the location of the catalyst in the membrane structure are vital factors to consider when designing photocatalytic membranes for antibacterial applications.

Another approach to combat the biofouling of thermal membranes involves incorporating gold nanostars, GO nanosheets, and hydrophilic poly (ethylene glycol) into polyamide, which is a commonly used Reverse Osmosis (RO) membrane material. This combination results in a high-performance membrane that exhibits excellent resistance to mineral scaling and organic fouling. The gold nanostars are particularly effective due to their high photothermal conversion efficiency. When the surface of the membrane is exposed to 808 nm laser illumination, the gold nanostars convert the light into heat, allowing the membrane to reach a locally high temperature. This high temperature is powerful enough to kill almost all the bacteria that have attached themselves to the membrane's surface. In addition to the membrane's antibiofouling properties, the hydrophilic poly(ethylene glycol) enhances the membrane's surface wettability, which helps prevent biofilm formation [57]. The GO nanosheets also contribute to the membrane's performance by acting as a physical barrier that prevents the penetration of organic matter and other impurities. More recently, Noureen et al. reported a solar evaporator, which possesses super photothermal properties including photodegradation, desalination, and antibiofouling activities. This solar evaporator was prepared by the incorporation of Ag_3VO_4 nanoparticles within rGO matrix and was developed into an aerogel (Figure 15.4b) [49].

Overall, this innovative membrane technology holds great promise for addressing the persistent problem of biofouling in thermal membranes. All these reports evidence that the photocatalytic photothermal membranes have opened an all-in-one strategy for wastewater purification and energy utilization.

15.5 Conclusion and Outlook

Photocatalytic photothermal membranes are providing an emerging and promising technology for freshwater production and many other water-and energy-related applications. Although many successful demonstrations have been achieved, there is still a huge gap between the practical application and the fundamental materials and devices, which are summarized as follows:

The scaling and interweaving of the photothermal materials in their real applications require further considerations. Since these membranes usually requires long-term contact with seawater, lake water, or other complicated water sources, the stability of the photothermal membranes and structural performances in these complicated situations must be taken into careful consideration. In addition, the possible leakage of materials during the application and their impacts on the environment should be also considered. Meanwhile, the cost of the materials involved in the preparation of the photothermal membranes and evaporators should be deceased for practical applications.

Rational designs of the photocatalytic photothermal membranes are another strategy that may be useful in improving the potentials of these microbial wastewater treatment systems. The structural design of the membrane, energy enhancement (e.g., utilization of the latent heat generated by water condensation), and pure water collection are also important in microbial wastewater treatment and distillation applications. This requires researchers from different disciplines to work together, including material science, mechanical engineering, physical science, and interfacial science.

References

1 Kaur, R., Yadav, B., and Tyagi, R.D. (2020). Microbiology of hospital wastewater. In: *Current Developments in Biotechnology and Bioengineering* (ed. R.D. Tyagi et al.), 103–148. Elsevier.

2 Saini, S., Tewari, S., Dwivedi, J., and Sharma, V. (2023). Biofilm-mediated wastewater treatment: a comprehensive review. *Materials Advances* 4 (6): 1415–1443.

3 Selvakumar, A. and Borst, M. (2006). Variation of microorganism concentrations in urban stormwater runoff with land use and seasons. *Journal of Water and Health* 4 (1): 109–124.

4 Khalid, S., Shahid, M., Natasha et al. (2018). A review of environmental contamination and health risk assessment of wastewater use for crop irrigation with a focus on low and high-income countries. *International Journal of Environmental Research and Public Health* 15: https://doi.org/10.3390/ijerph15050895.

5 Haldar, K., Kujawa-Roeleveld, K., Hofstra, N. et al. (2022). Microbial contamination in surface water and potential health risks for peri-urban farmers of the Bengal delta. *International Journal of Hygiene and Environmental Health* 244: 114002.

6 Mishra, R.K., Mentha, S.S., Misra, Y., and Dwivedi, N. (2023). Emerging pollutants of severe environmental concern in water and wastewater: a comprehensive review on current developments and future research. *Water-Energy Nexus* 6: 74–95.

7 Swapnil, P., Singh, L.A., Mandal, C. et al. (2023). Functional characterization of microbes and their association with unwanted substance for wastewater treatment processes. *Journal of Water Process Engineering* 54: 103983.

8 de Campos, S.X. and Soto, M. (2024). The use of constructed wetlands to treat effluents for water reuse. *Environments* 11 (2): 35.

9 Mannina, G., Gulhan, H., and Ni, B.-J. (2022). Water reuse from wastewater treatment: the transition towards circular economy in the water sector. *Bioresource Technology* 363: 127951.

10 Bairagi, S. and Ali, S.W. (2020). Conventional and advanced technologies for wastewater treatment. In: *Environmental Nanotechnology for Water Purification*, 33–56.

11 Silva, J.A. (2023). Wastewater treatment and reuse for sustainable water resources management: a systematic literature review. *Sustainability* 15 (14): 10940.

12 Al-Nuaim, M.A., Alwasiti, A.A., and Shnain, Z.Y. (2023). The photocatalytic process in the treatment of polluted water. *Chemical Papers* 77 (2): 677–701.

13 Martins, P.M., Santos, B., Salazar, H. et al. (2022). Multifunctional hybrid membranes for photocatalytic and adsorptive removal of water contaminants of emerging concern. *Chemosphere* 293: 133548.

14 Zhang, J., Wu, H., Shi, L. et al. (2024). Photocatalysis coupling with membrane technology for sustainable and continuous purification of wastewater. *Separation and Purification Technology* 329: 125225.

15 Mousa, S.A., Abdallah, H., and Khairy, S.A. (2023). Low-cost photocatalytic membrane modified with green heterojunction TiO_2/ZnO nanoparticles prepared from waste. *Scientific Reports* 13 (1): 22150.

16 Li, N., Lu, X., He, M. et al. (2021). Catalytic membrane-based oxidation-filtration systems for organic wastewater purification: a review. *Journal of Hazardous Materials* 414: 125478.

17 El Messaoudi, N., Ciğeroğlu, Z., Şenol, Z.M. et al. (2024). Chapter fourteen—green synthesis of nanoparticles for remediation organic pollutants in wastewater by adsorption. In: *Advances in Chemical Pollution, Environmental Management and Protection* (ed. A. Kumar, M. Bilal, and L.F.R. Ferreira), 305–345. Elsevier.

18 Ciğeroğlu, Z., el Messaoudi, N., Şenol, Z.M. et al. (2024). Clay-based nanomaterials and their adsorptive removal efficiency for dyes and antibiotics: a review. *Materials Today Sustainability* 26: 100735.

19 Asif, M.B. and Zhang, Z. (2021). Ceramic membrane technology for water and wastewater treatment: a critical review of performance, full-scale applications, membrane fouling and prospects. *Chemical Engineering Journal* 418: 129481.

20 Jeong, E., Byun, J., Bayarkhuu, B., and Hong, S.W. (2021). Hydrophilic photocatalytic membrane via grafting conjugated polyelectrolyte for visible-light-driven biofouling control. *Applied Catalysis B: Environmental* 282: 119587.

21 Abdel-Khalek, A.A., Mahmoud, S.A., and Zaki, A.H. (2018). Visible light assisted photocatalytic degradation of crystal violet, bromophenol blue and eosin Y dyes using AgBr-ZnO nanocomposite. *Environmental Nanotechnology, Monitoring & Management* 9: 164–173.

22 Arslan, D.Ş., Ertap, H., Şenol, Z.M. et al. (2024). Preparation of polyacrylamide titanium dioxide hybrid nanocomposite by direct polymerization and its applicability in removing crystal violet from aqueous solution. *Journal of Polymers and the Environment* 32 (2): 573–587.

23 Kundu, S. and Karak, N. (2022). Polymeric photocatalytic membrane: an emerging solution for environmental remediation. *Chemical Engineering Journal* 438: 135575.

24 Mishra, S. and Sundaram, B. (2023). A review of the photocatalysis process used for wastewater treatment. *Materials Today: Proceedings.*

25 Ouda, M., Pandey, R.P., Banat, F., and Hasan, S.W. (2023). Advancing water treatment sustainability: investigating electrified $Ti_3C_2T_x$ composite membranes for minimizing microplastic fouling. *Chemosphere* 343: 140224.

26 Ramírez-Malule, H., Quiñones-Murillo, D.H., and Manotas-Duque, D. (2020). Emerging contaminants as global environmental hazards. A bibliometric analysis. *Emerging Contaminants* 6: 179–193.

27 Yu, R., Chen, W., Zhang, J. et al. (2024). Catalytic membranes for water treatment: perspectives and challenges. *Journal of Hazardous Materials Advances* 13: 100414.

28 Yang, X., Yang, Y., Fu, L. et al. (2018). An ultrathin flexible 2D membrane based on single-walled nanotube–MoS2 hybrid film for high-performance solar steam generation. *Advanced Functional Materials* 28 (3): 1704505.

29 Xiaodan, J., Jing, B., Zhifang, M., and Xiue, J. (2017). BSA-exfoliated WSe2 nanosheets as a photoregulated carrier for synergistic photodynamic/photothermal therapy. *Journal of Materials Chemistry B* 5: 269–278.

30 Lee, S., Cho, J., and Elimelech, M. (2004). Influence of colloidal fouling and feed water recovery on salt rejection of RO and NF membranes. *Desalination* 160 (1): 1–12.

31 Yanbing, Y., Xiangdong, Y., Linna, F. et al. (2018). Two-dimensional flexible bilayer janus membrane for advanced photothermal water desalination. *ACS Energy Letters* 3 (1): 1165–1171.

32 Gao, X., Ren, H., Zhou, J. et al. (2017). Synthesis of hierarchical graphdiyne-based architecture for efficient solar steam generation. *Chemistry of Materials* 29 (14): 5777–5781.

33 Wang, Y., Wang, C., Song, X. et al. (2018). Improved light-harvesting and thermal management for efficient solar-driven water evaporation using 3D photothermal cones. *Journal of Materials Chemistry A* 6 (21): 9874–9881.

34 Ni, G., Zandavi, S.H., Javid, S.M. et al. (2018). A salt-rejecting floating solar still for low-cost desalination. *Energy & Environmental Science* 11 (6): 1510–1519.

35 Zhao, F., Zhou, X., Shi, Y. et al. (2018). Highly efficient solar vapour generation via hierarchically nanostructured gels. *Nature Nanotechnology* 13 (6): 489–495.

36 An, N., Chen, Y., and Li, Q. (2023). Hydrogel-based solar-driven interfacial evaporation: current progress and future challenges. *Green Energy and Resources* 1 (2): 100011.

37 Noureen, L., Xie, Z., Hussain, M. et al. (2021). $BiVO_4$ and reduced graphene oxide composite hydrogels for solar-driven steam generation and decontamination of polluted water. *Solar Energy Materials and Solar Cells* 222: 110952.

38 Sun, P., Zhang, W., Zada, I. et al. (2020). 3d-structured carbonized sunflower heads for improved energy efficiency in solar steam generation. *ACS Applied Materials & Interfaces* 12 (2): 2171–2179.

39 Zhang, R., Liu, Y., He, M. et al. (2016). Antifouling membranes for sustainable water purification: strategies and mechanisms. *Chemical Society Reviews* 45 (21): 5888–5924.

40 Pan, Z., Mao, Y., Ai, X. et al. (2023). Photothermal membranes for water treatment and anti-fouling performances: a review. *Journal of Cleaner Production* 412: 137335.

41 Flemming, H.C., Schaule, G., Griebe, T. et al. (1997). Biofouling—the Achilles heel of membrane processes. *Desalination* 113 (2): 215–225.

42 Baker, J.S. and Dudley, L.Y. (1998). Biofouling in membrane systems—a review. *Desalination* 118 (1): 81–89.

43 Alkhatib, A., Ayari, M.A., and Hawari, A.H. (2021). Fouling mitigation strategies for different foulants in membrane distillation. *Chemical Engineering and Processing Process Intensification* 167: 108517.

44 Yang, R., Jang, H., Stocker, R., and Gleason, K.K. (2014). Synergistic prevention of biofouling in seawater desalination by zwitterionic surfaces and low-level chlorination. *Advanced Materials* 26 (11): 1711–1718.

45 Yang, R., Xu, J., Ozaydin-Ince, G. et al. (2011). Surface-tethered zwitterionic ultrathin antifouling coatings on reverse osmosis membranes by initiated chemical vapor deposition. *Chemistry of Materials* 23 (5): 1263–1272.

46 Marine, J., Myers, C.P., Picquet, G.A. et al. (2018). Reduction of bacterial attachment on hydroxyapatite surfaces: using hydrophobicity and chemical functionality to enhance surface retention and prevent attachment. *Colloids and Surfaces B: Biointerfaces* 167: 531–537.

47 Zhang, M., Liu, Z., Gao, Y., and Shu, L. (2017). Ag modified g-C_3N_4 composite entrapped PES UF membrane with visible-light-driven photocatalytic antifouling performance. *RSC Advances* 7 (68): 42919–42928.

48 Jiang, Q., Ghim, D., Cao, S. et al. (2019). Photothermally active reduced graphene oxide/bacterial nanocellulose composites as biofouling-resistant ultrafiltration membranes. *Environmental Science & Technology* 53 (1): 412–421.

49 Noureen, L., Wang, Q., Ismail, P.M. et al. (2024). Multifunctional aerogel with antibiofouling properties for efficient solar steam generation and seawater desalination. *Nano Today* 54: 102130.

50 Zha, X.-J., Zhao, X., Pu, J.H. et al. (2019). Flexible anti-biofouling MXene/cellulose fibrous membrane for sustainable solar-driven water purification. *ACS Applied Materials & Interfaces* 11 (40): 36589–36597.

51 Li, Y., Cui, X., Zhao, M. et al. (2019). Facile preparation of a robust porous photothermal membrane with antibacterial activity for efficient solar-driven interfacial water evaporation. *Journal of Materials Chemistry A* 7 (2): 704–710.

52 Noureen, L., Xie, Z., Gao, Y. et al. (2020). Multifunctional Ag_3PO_4-rGO-coated textiles for clean water production by solar-driven evaporation, photocatalysis, and disinfection. *ACS Applied Materials & Interfaces* 12 (5): 6343–6350.

53 Nasrollahi, N., Ghalamchi, L., Vatanpour, V., and Khataee, A. (2021). Photocatalytic-membrane technology: a critical review for membrane fouling mitigation. *Journal of Industrial and Engineering Chemistry* 93: 101–116.

54 Kwak, S.-Y., Kim, S.H., and Kim, S.S. (2001). Hybrid organic/inorganic reverse osmosis (RO) membrane for bactericidal anti-fouling. 1. Preparation and characterization of TiO_2 nanoparticle self-assembled aromatic polyamide thin-film-composite (TFC) membrane. *Environmental Science & Technology* 35 (11): 2388–2394.

55 Chae, H.-R., Lee, J., Lee, C.H. et al. (2015). Graphene oxide-embedded thin-film composite reverse osmosis membrane with high flux, anti-biofouling, and chlorine resistance. *Journal of Membrane Science* 483: 128–135.

56 Vatanpour, V., Madaeni, S.S., Khataee, A.R. et al. (2012). TiO_2 embedded mixed matrix PES nanocomposite membranes: influence of different sizes and types of nanoparticles on antifouling and performance. *Desalination* 292: 19–29.

57 Jun, Y.S., Wu, X., Ghim, D. et al. (2019). Photothermal membrane water treatment for two worlds. *Accounts of Chemical Research* 52 (5): 1215–1225.

16

Nanocellulose Biocomposite Membranes for Wastewater Treatment

Subhrata Dutta[1], *Sathish Kumar Palaniappan*[2], *Sanjay Mavinkere Rangappa*[2], and *Suchart Siengchin*[2]

[1]*Department of Chemistry, Faculty of Paramedical Science, Assam Down Town University, Guwahati, Assam, India*
[2]*Natural Composites Research Group Lab, Department of Materials and Production Engineering, The Sirindhorn International Thai-German Graduate School of Engineering, King Mongkut's University of Technology North Bangkok, Bangsue, Thailand*

16.1 Introduction

There is severe water and soil contamination because of industrialization, population growth, and fast urbanization. Pollutants including saturated salts, organic compounds, heavy metals, oil emulsions, colors, and even microbes are mostly found in wastewater. Modern society faces a major environmental problem in the form of treating wastewater to eliminate pollutants and toxins [1]. This requires creative solutions that are both environmentally friendly and effective. In recent years, there has been growing interest in developing more sustainable and eco-friendly approaches to wastewater treatment, leveraging the unique properties of nanomaterials and biopolymers. Cellulose, the most abundant biopolymer on Earth, has long captivated researchers due to its remarkable properties such as renewability, biodegradability, and biocompatibility [2, 3]. The potential of cellulose at the nanoscale has been revealed by recent advances in nanotechnology, opening the door to the creation of materials based on nanocellulose that have never been possible [4]. Nanocellulose, a renewable and biodegradable material derived from plant sources, has emerged as a promising candidate for the development of advanced membrane technologies for wastewater treatment. The unique structure of nanocellulose (Figure 16.1) offers a versatile platform for the design of membranes with tunable pore sizes, surface chemistries, and hydrophilic/hydrophobic properties, making it an ideal material for the fabrication of high-performance membranes [5]. The inherent biocompatibility of nanocellulose further promotes biofouling resistance,

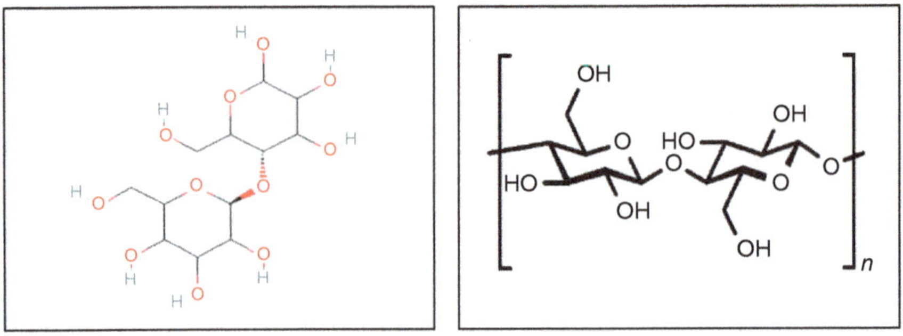

Figure 16.1 The chemical structures of cellulose.

ensuring prolonged membrane lifespan and reduced maintenance costs. They can effectively remove a wide range of contaminants, including organic matter, heavy metals, dyes, and microplastics, through various mechanisms such as size exclusion, adsorption, and catalytic degradation [6]. Moreover, the inherent biodegradability and eco-friendliness of nanocellulose reduce the environmental impact associated with membrane disposal and contribute to a circular economy approach [7]. The combination of nanocellulose with other materials, such as polymers, ceramics, or metals, enables nanocellulose to form biocomposite membranes with enhanced properties specially designed for specific wastewater treatment applications [8].

This chapter highlights the importance of sustainable wastewater treatment methods, introduces nanocellulose as a promising material for membrane development, and emphasizes the advantages of nanocellulose biocomposite membranes in addressing wastewater treatment challenges. Throughout this chapter, the latest advancements in nanocellulose-based membrane technologies were explored, ranging from fundamental principles to practical applications in wastewater treatment. By elucidating the underlying mechanisms governing membrane performance and highlighting key research findings, insights that will catalyze further innovations in this burgeoning field were provided, ultimately contributing to the sustainable management of water resources and protection of environmental ecosystems.

16.2 Characteristics and Classification of Nanocellulose

Nanocellulose, derived from different renewable sources like wood pulp or agricultural leftovers, makes it an attractive material for various amazing mechanical qualities, some of those characteristics are mentioned as follows:

1) Nanocellulose has an exceptionally high strength-to-weight ratio, making it one of the strongest materials [9]. It is stronger than steel but much lighter,

Figure 16.2 SEM images of three types of nanocellulose, i.e., (a) CNF [11], (b) CNC [12], and (c) BNC [13].

making it attractive for applications where strength and lightweight properties are desirable.

2) Nanocellulose has a high specific surface area due to its nanoscale dimensions, which can be beneficial for applications such as adsorption, catalysis, etc.
3) Transparency: Nanocellulose can be produced in a transparent form, making it suitable for applications in optics, displays, and packaging materials.
4) Being derived from natural sources, nanocellulose is generally biocompatible and biodegradable, which makes it an attractive material for biomedical applications, such as wound dressings, tissue engineering, and drug delivery [10].
5) Nanocellulose exhibits good thermal stability, which makes it suitable for applications that require high-temperature processing or operation.
6) Nanocellulose has excellent barrier properties against gases, moisture, and oils, making it suitable for food packaging and protective coatings.
7) Nanocellulose is derived from renewable and sustainable sources, such as plants and bacteria, making it an environmentally friendly material.
8) Nanocellulose can be chemically modified or combined with other materials to utilize its properties for specific applications, like improving hydrophobicity, electrical conductivity, or biological activity.

These unique characteristics of nanocellulose have led to extensive research and development efforts in various fields, including materials science, electronics, biomedical engineering, and energy storage, among others. Based on their dimensions, functions, preparation methods employed, and the source of production, nanocellulose can be classified into different types (Figure 16.2) that are mentioned below.

16.2.1 Cellulose Nanofiber (CNF)

CNF can be synthesized from different sources like plants, animals, and bacteria [14]. They are mainly extracted from by-products of rice, corn, barley, banana, wheat, sugarcane, etc. [15]. These flexible fibers consist of alternating crystalline and amorphous domains with a high aspect ratio, having diameter ranging from

5 to 30 nm and length of several micrometers [16]. The mechanical treatments and the unprocessed natural sources have the greatest influence on the dimensions, crystallinity, and morphology of the resulting fibrillated cellulose. Compared to the other methods, like enzymatic, mechanical, and chemical treatments, the mechanical treatment is considered as the most popular technique for extracting CNF [17]. Different kinds of mechanical treatments like high-pressure homogenization, steam explosion, micro grinding, micro fluidization, high-intensity ultrasonication, and cryo-crushing are used for nanofibrillation [18]. Increased shearing forces caused by the mechanical treatment improve defibrillation by the cellulose bundles.

16.2.2 Cellulose Nanocrystal (CNC)

The methods of acid hydrolysis, including enzymatic treatments and the use of mineral acids like hydrochloric, sulfuric, or phosphoric acid, are widely known as isolating CNC [19]. These are the crystalline cellulose that forms with a crystallinity ranging from 54% to 88% and 100% cellulose content [20]. It has a high aspect ratio, high specific strength, stiffness, and higher surface area. These rod-like, whisker-shaped particles that range in length from 100 to 600 nm and width from 2 to 70 nm are rich in Iβ crystal structure [21]. The crystalline fraction of cellulose can range from 50% to 90% depending on its source, and its elastic moduli have been found to be between ~105 and 168 GPa [22]. Tunicates, which have a diameter of 5–50 nm and a length of 100–300 nm, are the source of the longest form of CNC [23]. The addition of acid leads the charge on the CNC surface during the acid hydrolysis process that helps in the removal of the amorphous sections in the cellulose fibers and increases the crystalline content [24]. At certain concentrations, CNC has a shear-thinning characteristic and behaves like liquid crystals, forming asymmetrical nematic phases [25]. The presence of negatively charged sulfate groups surrounding the cellulose chains gives the sulfuric acid-produced nanocellulose its great colloidal stability [26]. These surface groups have repulsive nature that stops the cellulose chains from further agglomeration [27].

16.2.3 Bacterial Nanocellulose (BNC)

Other than the two methods, here a bottom-up technique is used to form this kind of cellulose where the glucose molecules are metabolized first. Conditions during cultivation, such as oxygen delivery and pH fluctuations, affect the mechanical properties, crystallinity, and form of the resulting BNC [28]. Bacteria species like Acetobacter, Azotobacter, Pseudomonas, and Sarcina Ventriculi can make BNC, which starts off as floccus, grows into fiber, and interweaves to form pellicles [29].

Biodegradable, BNC is often produced as fibers with a diameter of 10–100 nm and a crystallinity range of 75–96% [30]. Numerous methods have been employed to determine the mechanical properties of the single fiber; Young's moduli, which surpass CNF and CNC, range from 78 to 114 GPa [31].

16.3 Membrane Fabrication Techniques

Nanocellulose, a promising nanomaterial derived from renewable sources like wood pulp or agricultural waste, can be fabricated through various methods, each offering unique advantages. Depending on the degradability and compatibility, different kinds of methods, like phase inversion, solvent casting, in situ polymerization, extrusion, and melt intercalation, are some common membrane fabrication techniques used with different sources for producing nanocellulose [32]. Table 16.1 summarizes the common fabrication techniques.

When using the in situ polymerization approach, an initiator is present when the monomer and nanofiller are combined. The monomer is subsequently polymerized. The most common method for creating composites is solvent casting [35], which is also the easiest to utilize. In this case, the polymer is dissolved in

Table 16.1 Different fabrication techniques used in different sources for producing nanocellulose [33, 34].

Source	Membrane fabrication technique
Wood	1. Mechanical treatment (grinding, high-pressure homogenization, cryocrushing) 2. Chemical treatment (acid hydrolysis, TEMPO oxidation) 3. Electrospinning
Plants (e.g., cotton, hemp, flax)	1. Mechanical treatment (grinding, high-pressure homogenization) 2. Chemical treatment (acid hydrolysis, TEMPO oxidation) 3. Electrospinning
Algae	1. Mechanical treatment (high-pressure homogenization) 2. Chemical treatment (acid hydrolysis, TEMPO oxidation) 3. Solvent casting
Bacteria (e.g., *Gluconacetobacter xylinus*)	1. Biosynthesis (static or agitated culture) 2. Solvent casting 3. Electrospinning
Tunicate (e.g., sea squirts)	1. Mechanical treatment (grinding, high-pressure homogenization) 2. Chemical treatment (acid hydrolysis, TEMPO oxidation) 3. Solvent casting

an appropriate solvent, and once it has done so, the filler is added to the solution and the mixture is cast in a casting chamber. It is possible to carefully remove the polymer composite from the casting chamber once the solvent has completely evaporated. Cast polymer composite films have weak strength, stiffness, and crack development as downsides [36]. To ensure adequate dispersion, the thermoplastic was melted over the Tm and combined with nanofillers during the melt intercalation process. Nanocomposites were later created by gradually cooling the solution. The benefits of melt intercalation include the absence of a chemical reaction and the need for a solvent. Compounding plastic and cellulose fiber is a relatively new process known as extrusion. The materials so created can then be molded using techniques like injection, compression, and so on. Some other chemical methodologies have also been developed to alter the hydrophobicity of nanocellulose materials and boost the efficiency of the isolation process.

16.3.1 Ionic Charge Transfer to Nanocellulosic Surfaces

Nanocellulose, with its abundant hydroxyl groups on the surface, readily interacts with ions present in solution. This interaction leads to the adsorption of ions onto the nanocellulosic surface, resulting in changes in surface charge density and chemical reactivity. Such ionic interactions enable the functionalization of nanocellulose for various applications, including wastewater treatment, drug delivery, and sensor development [37]. Additionally, the ionic charge transfer process can facilitate the binding of nanoparticles or polymers onto nanocellulosic surfaces, further expanding their versatility in nanocomposite fabrication. Using this method, the C6 alcohol functions of the glucose unit are converted to carboxylic acid [38]. Anionic carboxylate is created when cellulose is oxidized at position six using $NaOCl_2$/NaOCl/TEMPO (2,2,6,6-tetramethylpiperidine-1-oxyl) as a catalyst. This high charge level results in improved water dispersibility [39].

16.3.2 Cellulose Graft Polymerization

It involves introducing many functional groups to the polymer in order to modify its chemical and physical properties [40]. The effective properties of two polymeric units can be merged into a single physical unit using the graft copolymerization technique [41]. Grafting a polymeric branch to the cellulose material allows for the graft copolymerization of cellulose, thereby inducing certain properties without sacrificing the substrate's inherent characteristics. The graft copolymerization process is based on three major approaches: "grafting-to," "grafting-from," and "grafting-through" [42]. The polymer's reactive terminal end group attached to the hydroxyl group of the cellulose and polymers, like polystyrene, polypropylene, etc., are joined to the cellulose in case of "grafting-to" technique. In "grafting-from" method, cellulose is first functionalized with an initiator and then monomers are polymerized directly from the surface. By using this method instead of the "grafting-to" approach, higher polymer densities can be achieved. Using monomers that contain polymerizable vinyl, cellulose or its derivatives are functionalized by the "grafting-through" process. Subsequently, the comonomer is mixed with the functionalized cellulose and the polymerization process is started. A variety of methods, which can be categorized as (i) free radical polymerization, (ii) ionic and ring-opening polymerization, and (iii) live radical polymerization, are typically used to graft monomers to cellulose and its derivatives [43].

16.3.3 Production of Hydrophobic Surface on the Nanocellulosic Materials

Cellulosic alcohols are acetylated, which causes plasticization of lignocellulosic fibers [44]. Acetylation is frequently used to increase the hydrophobicity of cellulosic fibers. Dry acetic acid and acid anhydride are added slowly next to sulfuric acid. Two primary acetylation processes have been developed, where in the fibrous method, toluene is added to the reaction solution to make the acetylated cellulose insoluble and this preserves the native structure while achieving a higher level of acetylation [45]. On the other hand, the other one, i.e., homogeneous method, helps to control all the reaction conditions to make the reactants present in the same phase, usually liquid. It leads to higher yields and selectivity.

16.4 Applications in Wastewater Treatment

Nanocellulose biocomposite membranes special qualities and ability to solve some of the problems related to membrane technologies have led to their prospective uses in wastewater treatment. The ongoing research and advancements in material design and membrane fabrication techniques continue to improve their performance and potential for large-scale applications. The general applications are shown in Figure 16.3, where nanocellulose biocomposite membrane is used.

The following are some important details about the wastewater treatment where the nanocellulose biocomposite membranes are applied.

a) ***Improved membrane performance:***
   Biocomposite membranes are a relatively new development in wastewater treatment, created by adding nanocellulose, which is abundant and renewable.

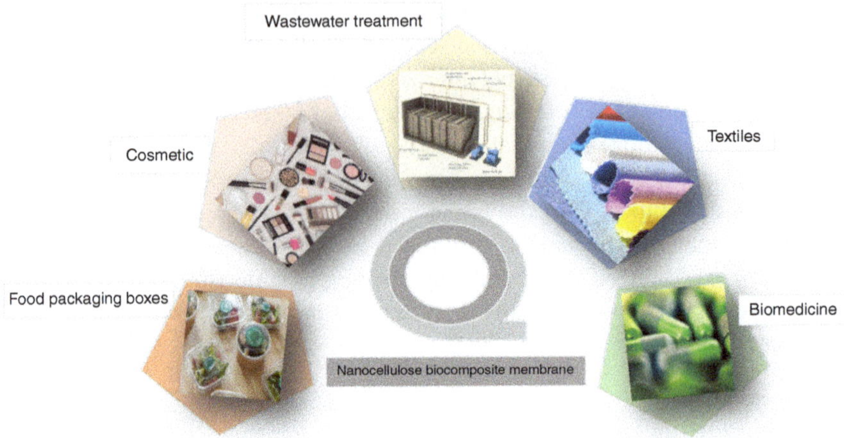

Figure 16.3 The different applied forms of nanocellulose biocomposite membrane.

The addition of nanocellulose improves the membranes mechanical strength, thermal stability, and fouling resistance, which results in better performance and longer operational lifetimes [46].

b) ***Antifouling properties:***
Fouling is a significant problem in membrane-based wastewater treatment since it lowers membrane efficiency and raises operating expenses. However, nanocellulose is hydrophilic and inhibits foulants from adhering to the membrane surface; nanocellulose biocomposite membranes have shown strong antifouling characteristics [47].

c) ***Selective separation:***
Heavy metals, dyes, and organic pollutants are just a few examples of the contaminants that can be specifically separated from wastewater using nanocellulose biocomposite membranes. By altering the nanocellulose composition and manufacturing parameters, the membranes' pore size and surface chemistry may be managed [4].

d) ***Biodegradability and environmental compatibility:***
Nanocellulose biocomposite membranes are more sustainable and environmentally friendly than traditional synthetic membranes since they come from renewable and biodegradable sources [5]. This feature helps to promote a circular economy by lowering the environmental impact of disposing of membranes.

e) ***Potential for low-cost production:***
The overall cost of producing the membranes can be decreased by obtaining nanocellulose from abundant and inexpensive sources, such as forestry

and agricultural waste [48]. Because of this, nanocellulose biocomposite membranes provide both industrialized and developing countries a viable alternative for treating wastewater.

Applications of nanocellulose in wastewater treatment can be divided into two main categories.

16.4.1 Removal of Organic Contaminants

i) Adsorption: Nanocellulose has a high surface area and can be functionalized with various groups, making it an effective adsorbent for removing organic pollutants like dyes, pharmaceuticals, and pesticides from wastewater [6].
ii) Membrane filtration: Nanocellulose membranes can be used for microfiltration, ultrafiltration, and nanofiltration processes to remove organic compounds, bacteria, and viruses from wastewater [49].
iii) Flocculation and coagulation: Nanocellulose can be used as a coagulant or flocculant aid to remove organic matter, colloidal particles, and suspended solids from wastewater [50].
iv) Photocatalysis: Nanocellulose can be functionalized with photocatalytic materials like titanium dioxide (TiO_2) or zinc oxide (ZnO) to degrade organic pollutants under UV or visible light irradiation [49].

16.4.2 Removal of Inorganic Contaminants

i) Adsorption: Nanocellulose can be chemically modified or functionalized with various groups to enhance its adsorption capacity for removing inorganic contaminants like heavy metals, radionuclides, and fluoride from wastewater [6].
ii) Ion exchange: Nanocellulose can be functionalized with ion exchange groups (e.g., carboxyl, sulfonate, or quaternary ammonium groups) to selectively remove ionic species like metal ions, nitrates, and phosphates from wastewater [51].
iii) Membrane filtration: Nanocellulose membranes can be used for nanofiltration and reverse osmosis processes to remove dissolved inorganic contaminants, such as salts and heavy metals, from wastewater [49].
iv) Flocculation and coagulation: Nanocellulose can be used as a coagulant or flocculant aid to remove colloidal and suspended inorganic particles from wastewater [50].
v) Catalysis: Nanocellulose can be functionalized with catalytic nanoparticles or enzymes to facilitate the degradation or transformation of inorganic contaminants in wastewater [49].

The specific application of nanocellulose in wastewater treatment depends on the type and concentration of contaminants present, as well as the desired treatment goals and process requirements.

16.5 Environmental Impact

Nanocellulose pollutants biocomposite membranes offer several environmental advantages and reduced environmental impact compared to conventional synthetic membranes used in wastewater treatment. As nanocellulose is derived from renewable sources like wood pulp, bacteria, agricultural wastes, etc., it reduces the dependency on nonrenewable sources and thus is less harmful toward the environment. With the help of this feature, the problem of plastic pollution will decrease as waste builds up in landfills and marine ecosystems. When compared to more traditional membrane materials like plastics or metals, the manufacture of nanocellulose membranes frequently uses less energy [46]. The overall environmental impact and carbon footprint are reduced because of this decreased energy demand. Water purification is one of the main uses for nanocellulose biocomposite membranes. Without the addition of dangerous chemicals, these membranes can efficiently filter impurities, diseases, and pollutants out of water. Human health and ecosystems are supported by cleaner water. Since nanocellulose is produced by photosynthesis in plants, it is naturally carbon-neutral [41]. Biocomposite membranes that use nanocellulose effectively trap carbon from the environment, contributing to attempts to mitigate climate change. Nanocellulose biocomposite membranes often show minimal levels of toxicity, in contrast to certain synthetic materials [42]. This property is effective for the environment and human health, particularly in situations where these membranes come into close contact with soil or water. Although the nanocellulose biocomposite membranes can break down naturally, attempts are being undertaken to find ways to recycle them. By increasing the materials lifespan and lowering the demand for raw material extraction, recycling would further lessen its negative effects on the environment [43]. To properly evaluate the environmental impact of these membranes, it is necessary to consider every stage of their lifecycle, including production, use, and disposal. Moreover, factors including the source of raw materials, transportation, methods of manufacturing, and end-of-life management techniques might affect how sustainable they are overall. In terms of human health, nanocellulose biocomposite membranes present promising advantages and some potential concerns. On the positive side, their biocompatibility makes them suitable for various biomedical applications, ranging from wound dressings to drug delivery systems [52]. Additionally, their renewable and biodegradable nature reduces the reliance on nonrenewable resources and mitigates environmental pollution. However, there are lingering questions regarding the safety of nanocellulose particles, particularly concerning their long-term effects on human health and potential risks associated with occupational exposure during production and handling. Further research is going on to assess any adverse health impacts to ensure the responsible use of nanocellulose-based materials.

16.6 Conclusions

Nanocellulose biocomposite membranes represent a promising and environmentally sustainable solution for wastewater treatment applications. Derived from renewable sources, biodegradable and often requiring less energy to produce, these membranes contribute to reducing reliance on nonrenewable resources, mitigating pollution, and combating climate change. Their low toxicity and potential for lower greenhouse gas emissions and reduced chemical requirements contribute to mitigating environmental impacts associated with conventional synthetic membranes. Additionally, the improved performance and extended operational lifetimes of nanocellulose biocomposite membranes can lead to significant water and energy savings, further enhancing their environmental credentials. While ongoing research is necessary to optimize their design and production processes, these membranes align with the principles of a circular economy and sustainable development, making them a promising technology for addressing the growing challenges of wastewater treatment while minimizing environmental burdens.

References

1. Rashid, R., Shafiq, I., Akhter, P. et al. (2021). A state-of-the-art review on wastewater treatment techniques: the effectiveness of adsorption method. *Environmental Science and Pollution Research* 28 (8): 9050–9066.
2. Bhaladhare, S. and Das, D. (2022). Cellulose: a fascinating biopolymer for hydrogel synthesis. *Journal of Materials Chemistry B* 10 (12): 1923–1945.
3. Popa, L., Ghica, M.V., Tudoroiu, E.-E. et al. (2022). Bacterial cellulose—a remarkable polymer as a source for biomaterials tailoring. *Materials* 15 (3): 1054.
4. Chandel, N., Jain, K., Jain, A. et al. (2023). The versatile world of cellulose-based materials in healthcare: from production to applications. *Industrial Crops and Products* 201: 116929.
5. Jaffar, S.S., Saallah, S., Misson, M. et al. (2022). Recent development and environmental applications of nanocellulose-based membranes. *Membranes* 12: 3.
6. Abdelhamid, H.N. and Mathew, A.P. (2021). Cellulose-based materials for water remediation: adsorption, catalysis, and antifouling. *Frontiers in Chemical Engineering* 3.
7. Majumder, S., Moharana, S., and Kim, K.H. (2024). Cellulose nano-papers: a comprehensive review of their synthesis methods, applications, and influence on the circular economy. *Journal of Cleaner Production* 451: 142045.

8 Thomas, B., Raj, M.C., B, A.K. et al. (2018). Nanocellulose, a versatile green platform: from biosources to materials and their applications. *Chemical Reviews* 118 (24): 11575–11625.

9 Rajinipriya, M., Nagalakshmaiah, M., Robert, M., and Elkoun, S. (2018). Importance of agricultural and industrial waste in the field of nanocellulose and recent industrial developments of wood based nanocellulose: a review. *ACS Sustainable Chemistry & Engineering* 6 (3): 2807–2828.

10 Curvello, R., Raghuwanshi, V.S., and Garnier, G. (2019). Engineering nanocellulose hydrogels for biomedical applications. *Advances in Colloid and Interface Science* 267: 47–61.

11 Perdoch, W., Cao, Z., Florczak, P. et al. (2022). Influence of nanocellulose structure on paper reinforcement. *Molecules* 27: 4696.

12 Fornari, A., Rossi, M., Rocco, D., and Mattiello, L. (2022). A review of applications of nanocellulose to preserve and protect cultural heritage wood, paintings, and historical papers. *Applied Sciences* 12: 12846.

13 Ilyas, R.A., Sapuan, S.M., Sanyang, M.L. et al. (2018f). Nanocrystalline cellulose reinforced starch-based nanocomposite: a review. *Current Analytical Chemistry* 14: 203–225.

14 Blanco, A., Monte, M.C., Campano, C. et al. (2018). Chapter 5—nanocellulose for industrial use: cellulose nanofibers (CNF), cellulose nanocrystals (CNC), and bacterial cellulose (BC). In: *Handbook of Nanomaterials for Industrial Applications*, Micro and Nano Technologies (ed. C.M. Hussain), 74–126. Elsevier.

15 Ighalo, J.O., Adeniyi, A.G., and Adelodun, A.A. (2021). Recent advances on the adsorption of herbicides and pesticides from polluted waters: performance evaluation via physical attributes. *Journal of Industrial and Engineering Chemistry* 93: 117–137.

16 Afshari, M., Sikkema, D.J., Lee, K., and Bogle, M. (2008). High performance fibers based on rigid and flexible polymers. *Polymer Reviews* 48 (2): 230–274.

17 Nechyporchuk, O., Belgacem, M.N., and Bras, J. (2016). Production of cellulose nanofibrils: a review of recent advances. *Industrial Crops and Products* 93: 2–25.

18 Tanpichai, S., Mekcham, S., Kongwittaya, C. et al. (2022). Extraction of nanofibrillated cellulose from water hyacinth using a high speed homogenizer. *Journal of Natural Fibers* 19 (13): 5676–5696.

19 Tang, Y., Shen, X., Zhang, J. et al. (2015). Extraction of cellulose nano-crystals from old corrugated container fiber using phosphoric acid and enzymatic hydrolysis followed by sonication. *Carbohydrate Polymers* 125: 360–366.

20 Bansal, P., Hall, M., Realff, M.J. et al. (2010). Multivariate statistical analysis of X-ray data from cellulose: a new method to determine degree of crystallinity and predict hydrolysis rates. *Bioresource Technology* 101 (12): 4461–4471.

21 Yusuf, J., Sapuan, S.M., Ansari, M.A. et al. (2024). Exploring nanocellulose frontiers: a comprehensive review of its extraction, properties, and pioneering applications in the automotive and biomedical industries. *International Journal of Biological Macromolecules* 255: 128121.

22 Benítez, A.J. and Walther, A. (2017). Cellulose nanofibril nanopapers and bioinspired nanocomposites: a review to understand the mechanical property space. *Journal of Materials Chemistry A* 5 (31): 16003–16024.

23 Cao, L., Fu, X., Xu, C. et al. (2017). High-performance natural rubber nanocomposites with marine biomass (tunicate cellulose). *Cellulose* 24 (7): 2849–2860.

24 Pereira, P.H.F., Ornaghi Júnior, H.L., Coutinho, L.V. et al. (2020). Obtaining cellulose nanocrystals from pineapple crown fibers by free-chlorite hydrolysis with sulfuric acid: physical, chemical and structural characterization. *Cellulose* 27 (10): 5745–5756.

25 Abbasi, M.A. (2022). Chiral liquid crystalline properties of cellulose nanocrystals: fundamentals and applications. *ACS Omega* 7 (35): 30673–30699.

26 Zhu, J.Y. and Agarwal, U.P. (2022). Nanocellulose: native state, production, and characterization. In: *Emerging Nanotechnologies in Nanocellulose*, (ed. Liangbing Hu, Feng Jiang, Chaoji Chen), 1–39. Cham: Springer International Publishing.

27 Koshani, R., Tavakolian, M., and van de Ven, T.G.M. (2020). Cellulose-based dispersants and flocculants. *Journal of Materials Chemistry B* 8 (46): 10502–10526.

28 Chen, G., Wu, G., Alriksson, B. et al. (2018). Scale-up of production of bacterial nanocellulose using submerged cultivation. *Journal of Chemical Technology & Biotechnology* 93 (12): 3418–3427.

29 Mohamad Misenan, M.S.B., Ahmad Farabi, M.S., Nor Akhlisah, Z., and Azlisham, N.A.F. (2024). 16—preparation, characterization and application of nanocellulose from tunicate for electronic applications. In: *Polymer Composites Derived from Animal Sources*, Woodhead Publishing in Materials (ed. S.M. Sapuan, C.H. Azhari, and N.M. Nurazzi), 295–319. Woodhead Publishing.

30 Doagou-Rad, S., Islam, A., and Merca, T.D. (2020). An application-oriented roadmap to select polymeric nanocomposites for advanced applications: a review. *Polymer Composites* 41 (4): 1153–1189.

31 Borbolla-Jiménez, F.V., Peña-Corona, S.I., Farah, S.J. et al. (2023). Films for wound healing fabricated using a solvent casting technique. *Pharmaceutics* 15 (7): 1914.

32 Huang, C., Liu, J., Zhao, L. et al. (2023). Advances in atomic oxygen resistant polyimide composite films. *Composites Part A: Applied Science and Manufacturing* 168: 107459.

33 Jonoobi, M., Oladi, R., Davoudpour, Y. et al. (2015). Different preparation methods and properties of nanostructured cellulose from various natural resources and residues: a review. *Cellulose* 22: 935–969.

34 Kargarzadeh, H., Ioelovich, M., Ahmad, I. et al. (2017). Methods for extraction of nanocellulose from various sources. In: *Handbook of Nanocellulose and Cellulose Nanocomposites*, (ed. Hanieh Kargarzadeh, Ishak Ahmad, Sabu Thomas, Alain Dufresne), vol. 1, 1–49. Wiley.

35 Norrrahim, M.N., Knight, V.F., Nurazzi, N.M. et al. (2022). The frontiers of functionalized nanocellulose-based composites and their application as chemical sensors. *Polymers* 14 (20): 4461.

36 Reddy, M.V., Nandan Reddy, V.U., and Chang, Y.-C. (2022). Integration of anaerobic digestion and chain elongation technologies for biogas and carboxylic acids production from cheese whey. *Journal of Cleaner Production* 364: 132670. https://doi.org/10.1016/j.jclepro.2022.132670.

37 Omran, B. and Baek, K.-H. (2022). Graphene-derived antibacterial nanocomposites for water disinfection: current and future perspectives. *Environmental Pollution* 298: 118836. https://doi.org/10.1016/j.envpol.2022.118836.

38 Zhang, J., Qi, Y., Shen, Y., and Li, H. (2022, https://doi.org/10.5755/j02.ms.25485). Research progress on chemical modification and application of cellulose: a review. *Materials Science* 28: 60–67.

39 Wu, Y.-Y., Figueira, F.L., Edeleva, M. et al. (2022). Cost-efficient modeling of distributed molar mass and topological variations in graft copolymer synthesis by upgrading the method of moments. *AIChE Journal* 68 (4): e17559. https://doi.org/10.1002/aic.17559.

40 Purohit, P., Bhatt, A., Mittal, R.K. et al. (2023). Polymer grafting and its chemical reactions. *Frontiers in Bioengineering and Biotechnology* 10: 1044927.

41 Hasan, A., Rabbi, M.S., and Maruf Billah, M. (2022). Making the lignocellulosic fibers chemically compatible for composite: a comprehensive review. *Cleaner Materials* 4: 100078. https://doi.org/10.1016/j.clema.2022.100078.

42 Souza Da Rosa, T., Trianoski, R., Michaud, F. et al. (2022). Efficiency of different acetylation methods applied to cellulose fibers waste from pulp and paper mill sludge. *Journal of Natural Fibers* 19 (1): 185–198.

43 Norfarhana, A.S., Ilyas, R.A., and Ngadi, N. (2022). A review of nanocellulose adsorptive membrane as multifunctional wastewater treatment. *Carbohydrate Polymers* 291: 119563. https://doi.org/10.1016/j.carbpol.2022.119563.

44 Cherian, R.M., Tharayil, A., Varghese, R.T. et al. (2022). A review on the emerging applications of nano-cellulose as advanced coatings. *Carbohydrate Polymers* 282: 119123. https://doi.org/10.1016/j.carbpol.2022.119123.

45 Mateo, S., Peinado, S., Morillas-Gutiérrez, F. et al. (2021). Nanocellulose from agricultural wastes: products and applications—a review. *Processes* 9: https://doi.org/10.3390/pr9091594.

46 Karade, S.R. and Jami, T. (2025). Application of vegetal concrete for carbon-neutral built environment. In: *Handbook of Climate Change Mitigation and Adaptation*, (ed. Maximilian Lackner, Baharak Sajjadi, Wei-Yin Chen), 2647–2688. Cham: Springer Nature Switzerland.

47 Khalid, M.Y., Al Rashid, A., Arif, Z.U. et al. (2021). Recent advances in nanocellulose-based different biomaterials: types, properties, and emerging applications. *Journal of Materials Research and Technology* 14: 2601–2623. https://doi.org/10.1016/j.jmrt.2021.07.128.

48 Chen, Q., Lai, X., Chen, J. et al. (2023). Comparative environmental impacts of different hydrometallurgical recycling and remanufacturing technologies of lithium-ion batteries considering multi-recycling-approach and temporal-geographical scenarios in China. *Separation and Purification Technology* 324: 124642. https://doi.org/10.1016/j.seppur.2023.124642.

49 Dhakshinamoorthy, A., Li, Z., and Garcia, H. (2018). Catalysis and photocatalysis by metal organic frameworks. *Chemical Society Reviews* 47 (22): 8134–8172.

50 Prakash, N.B., Sockan, V., and Jayakaran, P. (2014). Waste water treatment by coagulation and flocculation. *International Journal of Engineering Science and Innovative Technology* 3 (2): 479–484.

51 Anirudhan, T.S. and Rejeena, S.R. (2012). Adsorption and hydrolytic activity of trypsin on a carboxylate-functionalized cation exchanger prepared from nanocellulose. *Journal of Colloid and Interface Science* 381 (1): 125–136.

52 Gumrah Dumanli, A. (2017). Nanocellulose and its composites for biomedical applications. *Current Medicinal Chemistry* 24 (5): 512–528.

17

Wastewater Treatment Plants Based on Biocomposite Materials

Abdeladim Oulguidoum[1]*, Rachid Bouhfid*[2]*, and Nadia Zari*[1]

[1]*Moroccan Foundation for Advanced Science, Innovation and Research (MAScIR), Mohammed VI Polytechnic University, Lot 660-Hay Moulay Rachid, 43150 Ben Guerir, Morocco*
[2]*International Water Research Institute (IWRI), Mohammed VI Polytechnic University, Lot 660-Hay Moulay Rachid, Ben Guerir, 43150, Morocco*

17.1 Introduction

In the new millennium, industrial development has resulted in various pollutants being discharged into water bodies, leading to a decline in water quality [1]. This degradation threatens multiple sectors, including agriculture, commerce, domestic use, and industry [2, 3]. These pollutants can adversely affect human and animal health and contribute to issues such as eutrophication and the inhibition of photosynthesis [4]. Water pollution is defined as the physical, chemical, biological, or bacteriological alteration of water's natural properties due to human activity. The most common contaminants found in freshwater sources include dyes [5], antibiotics, pharmaceuticals [6, 7], heavy metals [8, 9], metalloids, and microplastics [10, 11]. The urgent need to purify water contaminated by pollutants is becoming increasingly important due to their negative impact on ecosystems. This issue is vital for industry, public health, and environmental protection, which is why effective wastewater treatment is necessary before discharging it into the environment [12]. Wastewater treatment plants serve as the first line of defense in the fight against water pollution and in efforts to restore water quality [13]. These systems are specifically designed to remove a certain quantity of pollutants from contaminated water. The treatment process involves various stages, including physical, chemical, and biological techniques [14]. For instance, biological processes such as activated sludge utilize bacteria and other microorganisms to eliminate pollutants through assimilation. This method is especially effective in reducing the organic pollutant load in wastewater [15]. Additionally, physical and chemical methods include sedimentation, filtration, flotation, coagulation, flocculation, advanced oxidation processes, and adsorption, all of which specifically target several pollutants [16, 17]. The selection of technologies for water treatment is influenced by the characteristics of the water being addressed and the specific requirements of each treatment facility. Each plant customizes its processes to ensure that water quality meets environmental standards. These technologies are essential for effectively tackling the challenges posed by water pollution and for environment protection. There has been a growing interest from both industry and academia in researching and developing efficient and affordable wastewater treatment methods [18].

Multidisciplinary research teams comprising engineers, chemists, physicists, ecologists, biologists, materials, and environmental scientists collaborate to create innovative materials and systems allowing pollutant separation or transformation. As a result, numerous alternative materials have been recently developed, including membrane filters [19], adsorbents [20], and catalytic surfaces [21]. With the aim of purifying water, commonly used materials include clays, zeolites [22], activated carbon [23], biochar [24], graphene, carbon nanotubes [25], etc. Although good removal rates have been recorded with these materials, the high cost associated with their synthesis and use represents a major limitation to their application. To address this challenge, several research initiatives focused on the use of biocomposite materials as effective alternatives for wastewater treatment. Biocomposites have attracted great interest as environmentally friendly materials because of their low price, wide availability, ecological characteristics, and exceptional ability to remove toxic pollutants from water. They are an advanced class of materials that combine the benefits of reinforcements and matrices to achieve improved performance [26]. The rational development of these functional materials requires a comprehensive understanding of their fundamental properties.

This chapter aims to examine the current state of the water crisis and provide an updated summary of recent advances in wastewater treatment technologies. The focus is on the different types of wastewater treatment plants, as well as their role and effectiveness in addressing water pollutants. An overview of biocomposites, including their composition and applications in wastewater treatment plants, is then presented. Finally, the chapter concludes with a discussion of the challenges associated with the use of biocomposites for wastewater treatment.

17.2 Global Water Crisis and Wastewater Management

17.2.1 Global Water Crisis

Since water is necessary for the existence of many species on our planet, it plays a crucial role in nature; all living things on earth, such as microorganisms, plants,

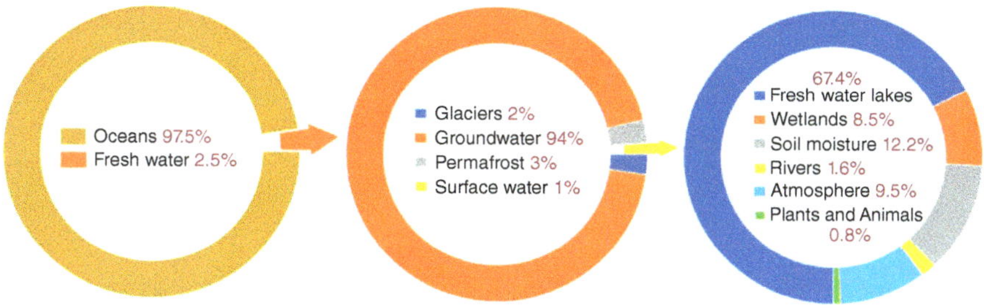

Figure 17.1 Water sources distributed around the world [29].

animals, people, and even our brains, are made primarily of water. In addition, water is used in many areas, including domestic consumption, industrial and energy production, irrigation, and purification [27]. Although more than 70% of the Earth's surface is covered with water, 97.5% of saltwater is found in oceans and marginal seas, leaving only 2.5% of water as freshwater, which is primarily stored in ice caps, groundwater, lakes, and rivers. People, animals, and plants depend on this fresh water for their needs [28]. Figure 17.1 shows the distribution of water resources worldwide. Owing to population growth and increasing water demands, groundwater reserves, an important source of fresh water, are being exploited at a rate that exceeds their renewal capacity. Surface waters, accounting for just 1% of the world's freshwater, are highly susceptible to pollution from both human activities and natural events. Much of this accessible freshwater is unfit for consumption and must be treated before use [30, 31].

The total water mass in the hydrosphere does not vary over time. During its cycle, water changes state, but its total quantity remains unchanged for billions of years. However, water scarcity is increasing worldwide because of its uneven distribution [32]. With a global population of 8 billion people today, the availability of clean water has become a major problem. According to the World Health Organization, approximately 11% of the world's population remains without access to basic drinking water. In 2015, one-third of the world's population still lacked adequate sanitation, and nearly 263 million people per world spent more than 30 minutes per trip trying to obtain drinking water from an improved source [33]. In developing countries, a lack of access to drinking water is a major challenge that urgently needs to be addressed.

Owing to population growth and economic development, traditional freshwater supplies are being overexploited in many regions of the world, leading to water shortages. It is estimated that 33 countries will experience extremely high levels of water stress by 2040, indicating increased demand for the world's water resources (Figure 17.2). Currently, 95% of people impacted by water scarcity reside in developing countries [35], with the majority situated in Asia and North Africa.

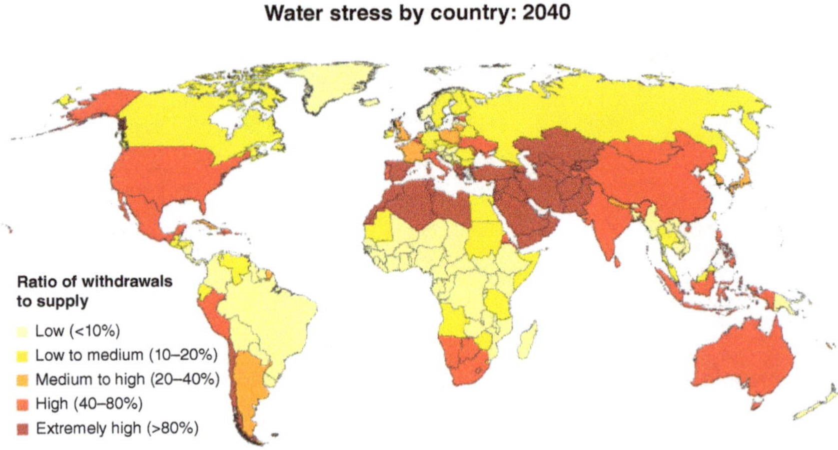

Figure 17.2 Anticipated worldwide water stress by 2040 [34].

It is estimated that 22–34% of the world's population will experience water scarcity at least once a year, affecting approximately 57% of the population in many countries [36].

Almost 47% of the world's population, approximately 3.6 billion people, live in regions with a water shortage for at least 1 month each year [37]. In addition, pollution deteriorates the quality of water regardless of its source or type, which has serious consequences for human health and causes numerous diseases, such as dysentery, cholera, typhoid fever, hepatitis, dracunculiasis, and poliomyelitis. Each year, more than 500,000 people, mainly children under 5 years of age, develop diarrheal diseases [29, 38].

The agricultural, municipal, and industrial sectors are likely to suffer from the consequences of clean water shortages, as they are sources of pollutants that negatively impact water quality. Current and future generations will pay a heavy price for the conflict between limited resources and economic development. The water consumption and wastewater production of the primary water-using sectors are depicted in Figure 17.3. As shown in Figure 17.3, the amount of wastewater consumed is greater than the amount of water consumed in each of the three main water use areas, meaning that the withdrawal is very high compared with the actual demand [39].

Water pollution from industrial, agricultural, and municipal wastes exacerbates the situation and increases pressure on available freshwater resources. It is becoming a serious challenge in many parts of the world, as it negatively impacts both the quality and quantity of water needed for human consumption and the preservation of ecosystems. According to the United Nations World Water Development Report (2017), 56% of the freshwater extracted for municipal, industrial,

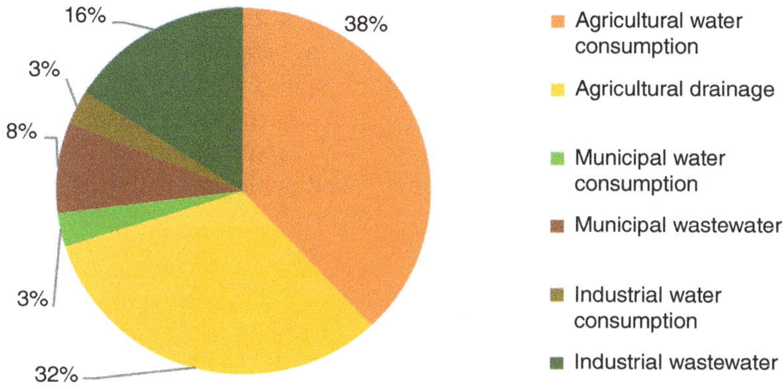

Figure 17.3 Worldwide water consumption and wastewater by sector [39].

and agricultural purposes is discharged as wastewater, 80% of which is released into the environment without prior treatment. In some developing countries, this percentage can reach 95% [29]. Owing to population growth and increasing water demands, groundwater, a major source of fresh water, is being developed faster than it can be replenished. The urgent need to treat polluted water is becoming increasingly critical because of its detrimental impact on ecosystems. This issue is of vital importance for industry, public health, and environmental protection.

17.2.2 Characteristics, Sources, and Composition of Wastewater

Wastewater is any water that flows into sewers whose natural properties have been altered by domestic, industrial, agricultural, and other uses. This also includes rainwater that flows into these pipes [40, 41]. By definition, domestic wastewater includes household wastewater such as toilet, laundry, and kitchen water. Consequently, domestic wastewater contains mineral substances such as chlorides, phosphates, sulfates, etc., as well as organic substances composed of ternary compounds such as sugars and fats, consisting of carbon, oxygen, and hydrogen, as well as nitrogen and, in some cases, other elements such as sulfur, phosphorus, and iron [41, 42].

The term "industrial wastewater" refers to any wastewater produced when water is used for purposes other than domestic use. This definition includes discharges from factories, craft or commercial establishments, laundries, restaurants, medical analysis laboratories, etc. Industrial wastewater comes in a wide variety of forms. Some of these compounds are toxic to aquatic flora and fauna or to humans. It is characterized by highly variable flow rates and compositions, depending on the type of industry. In addition to organic, nitrogenous, or phosphorous substances, it may also contain organic matter and fats (food processing industries, rendering), metallic salts (surface treatment, metallurgy, electroplating, steel industry), various mineral materials (chemical industries, tanning, quarries, mining), hydrocarbons (refineries), hot water (cooling circuits of thermal power plants), and radioactive materials (nuclear power plants, radioactive waste treatment) [43, 44]. Urban wastewater includes domestic sewage and runoff from roofs, streets, and public spaces, which carry various mineral and organic wastes, such as soil, plant debris, and fine particles. It also contains micropollutants such as hydrocarbons, pesticides, detergents, and rubber debris from vehicles. These contaminants complicate the treatment of urban wastewater, making its management challenging [45]. Generally, pollutants from industrial, domestic, or other sectors can have long-lasting, bioaccumulative, and synergistic impacts on ecosystems, food production, and human health and well-being. Consequently, wastewater must be treated through various processes to ensure its purification. To achieve this goal, a wastewater management system is put in place to gather and treat this water before it is released into the environment.

17.2.3 Typical Wastewater Treatment Processes

Industrialization and population growth have increased in recent years, resulting in increased wastewater generation. This trend poses significant threats to ecosystems and the environment, and implementing effective wastewater treatment techniques is essential. These techniques are crucial for either reusing treated water or safely returning it to the environment [46]. Wastewater treatment involves a range of methods aimed at removing pollutants and contaminants before discharge. The primary objective is to purify water by reducing toxins and pollutants to levels below the maximum allowable limits, thereby safeguarding the environment and public health [47]. In recent years, various techniques have been developed to find economical and effective methods for treating wastewater containing organic, inorganic, biological, and other pollutants. Wastewater treatment plants typically consist of preliminary, primary, secondary, and tertiary treatments, which incorporate physical, chemical, and biological processes (Figure 17.4).

Figure 17.4 illustrates the pathways through which emerging contaminants (ECs), such as pharmaceuticals, personal care products, and industrial chemicals, enter wastewater treatment plants (WWTPs) from various sources, such as hospitals, agriculture, and domestic use. It highlights the stages of treatment—primary, secondary, and tertiary—emphasizing the dominant mechanisms (sorption, biodegradation, and oxidation) used to remove these contaminants before discharging treated wastewater into the environment.

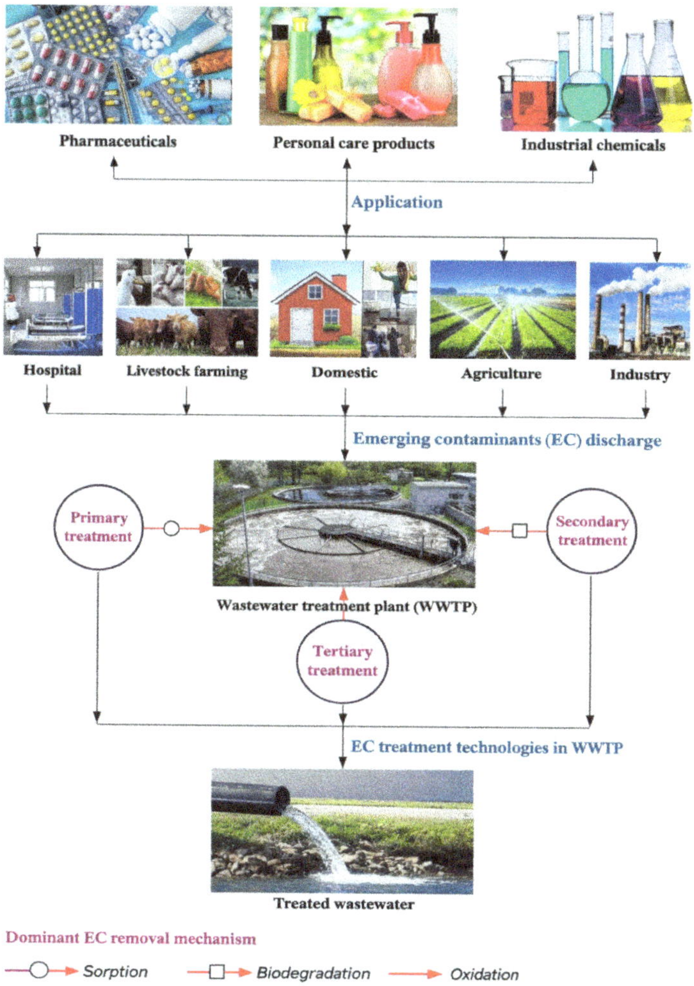

Figure 17.4 Emerging contaminant (EC) pathways and treatment technologies in wastewater treatment plants (WWTPs) [48].

The first step in the wastewater treatment process, known as preliminary treatment, attempts to eliminate large and/or heavy particles, such as clogs and damage to pumps and pipelines, to avoid issues later in the process. In general, there are two stages: sieving and gravel removal. Screening eliminates large amounts of floating waste, such as plastic, paper, and rags. Following the sieving process, larger particles such as gravel, sand, coffee grounds, broken bones, and corn kernels are eliminated by sinking into gravel channels [29].

Once the initial treatment is complete, the effluent undergoes primary treatment (also called primary sedimentation), which can be purely mechanical or enhanced by the addition of chemicals (flocculants and coagulants). This step generally includes a sedimentation procedure to segregate the suspended solids. The effluent is stored in tanks for a predetermined amount of time, which allows the suspended particles to sink to the bottom or build up as a layer of foam on top. The sludge that gathered as a solid residue at the bottom of the tank was then removed, and the foam layer was skimmed off. The oxygen demand for biochemistry (i.e., after primary treatment, the amount of oxygen needed by microorganisms to break down organic matter) decreases by approximately 40%. During this phase of treatment, 80–90% of the suspended particles and 55% of the fecal coliforms are removed [49]. The primary treatment method can effectively remove a variety of pollutants, such as heavy metals and organic matter (including nitrogen and organic phosphorus); however, it is not effective at removing colloidal or dissolved materials.

Secondary treatment systems, which are used after primary treatment, consist of both biological treatment to remove organic and mineral contaminants (nitrogen), which are sources of nutrients for microorganisms, and chemical treatment to improve the quality of wastewater (phosphorus removal) by adding ferric chloride and sludge flocculation by adding polymers. The principle of this treatment is to bring the organic substances contained in the wastewater into contact with a population of bacteria. The bacteria then assimilate the organic matter for development. These systems allow the intensification and localization of the processes of transformation and degradation of organic matter on reduced surfaces, as they occur in a natural environment [50]. In cases where specific wastewater constituents require additional removal following secondary treatment, tertiary treatment is implemented. Physical methods achieve separation by utilizing natural forces such as electromagnetic and gravity forces. Screens, filters, and membranes are examples of physical barriers that are frequently used to filter water and wastewater. Additional physicochemical methods for treating water include adsorption, flotation, sedimentation, and flocculation/coagulation. Chemical reagents and reactions are typically used in chemical processes to extract the desired substances from wastewater. Chemical processes for disinfecting wastewater include chemical precipitation, chemical oxidation, chemical neutralization, and ion exchange [16, 17]. Figure 17.5 provides an illustration of the wastewater treatment methods used.

Figure 17.6 shows a schematic diagram of a sewage water treatment plant, highlighting the three main treatment stages: primary, secondary, and tertiary. Following the primary treatment, the water undergoes secondary treatment, which is then succeeded by tertiary treatment. The tertiary stage focuses on advanced purification methods that incorporate innovative materials. For example, biocomposites are utilized as adsorbents, catalysts,

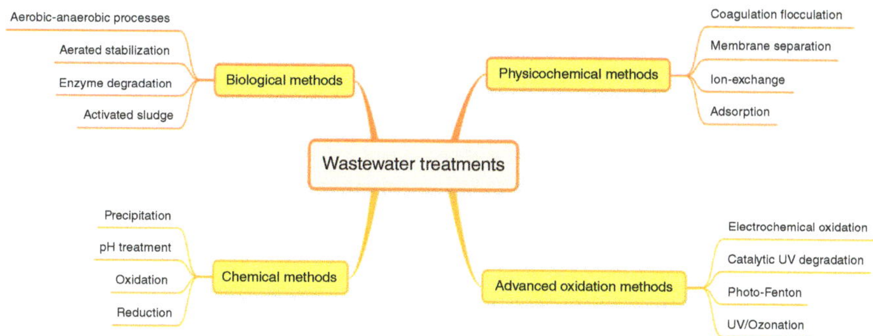

Figure 17.5 Different techniques for treating wastewater.

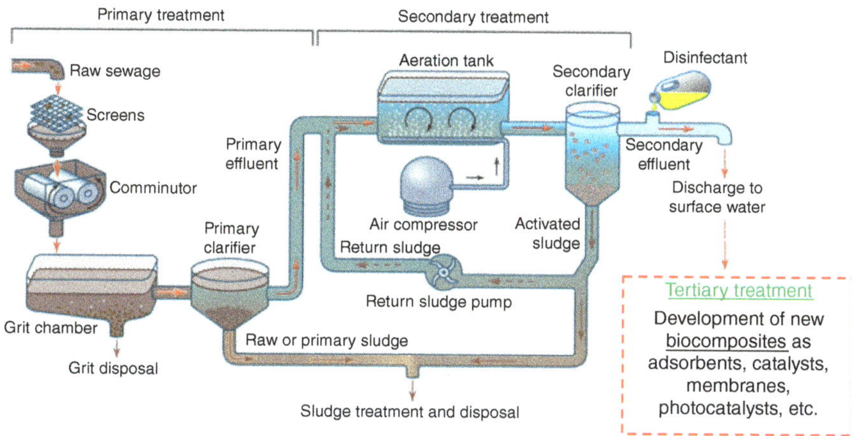

Figure 17.6 Primary and secondary treatment of sewage using the activated sludge process and biocomposite in tertiary treatment.

membranes, and photocatalysts, significantly enhancing the removal of residual pollutants from prior treatments and improving the overall efficiency of the water remediation process.

17.2.4 Types of Wastewater Treatment Plants

Wastewater treatment plants play a critical role in wastewater management by reducing pollutants before they enter the environment. It is essential to understand the type of discharge, the various pollutants carried (physical, chemical, and microbiological), their contents, and the treatment techniques employed to treat wastewater and choose the best treatment processes. Understanding these species is critical for selecting the appropriate treatment process to ensure that treated wastewater can be safely discharged back into the environment or reused. Different types of wastewater treatment plants exist, each of which uses different treatment methods to meet specific needs. Wastewater treatment plants are classified into several main types, such as hospital wastewater treatment plants (H-WWTPs), pharmaceutical wastewater treatment plants (P-WWTPs), livestock wastewater treatment plants (L-WWTPs), municipal wastewater treatment plants (M-WWTPs), and commercial wastewater treatment plants (C-WWTPs), etc. Bina et al. [51] conducted a study on the determination of organic pollutants in different types of wastewater and their removal rates in several wastewater treatment plants in Iran. The results are presented in Table 17.1, which describe the efficiency of different kinds of wastewater treatment plants in removing specific pollutants, particularly 4-nonylphenol (4-NP) and 4-tert-octylphenol (4-t-OP), which are common organic pollutants found in wastewater.

Over the past year, advances in nanotechnology have led to the development of biomaterials used as adsorbents, membranes, catalysts, etc., to remove contaminants from wastewater and restore the quality of water resources. In this context, we present biocomposites, focusing on their composition, properties, and applications in wastewater treatment.

17.3 Biocomposite Materials in the Management of Wastewater

17.3.1 Definition and Composition of the Biocomposites

Compared with their individual components, biocomposites represent a broad category of materials in which two or more substances are combined to form a material with superior properties compared to those of its individual components [26], these materials are also known as "eco-composites" or "agro-composites" and align closely with the growing concepts of green and sustainable chemistry. Although there is currently no formal definition of biocomposite, it is commonly agreed that at least one of the components must be derived from natural resources that have undergone mechanical and/or chemical modification [52]. A biocomposite is a composite material made of natural reinforcement and a polymer matrix sourced from either synthetic or natural origins. Owing to their natural origins, biocomposites offer desirable economic and environmental benefits for various industries [53].

In terms of reinforcement, three main types of fibers have been used. The three major types are chemical fibers (both synthetic and artificial) sourced from fossil resources or biomass, natural fibers, and inorganic fibers obtained through fusion or pyrolysis [54].

Table 17.1 Efficiency of different types of wastewater treatment plants in removing organic pollutants.

Type of wastewater treatment	Description	Efficiency
Municipal wastewater treatment plants (M-WWTPs)	M-WWTPs treat domestic wastewater from households, industries, and sometimes livestock. They often use conventional activated sludge (CAS) processes and may also include aerated lagoon (AL) systems.	The results showed that M-WWTPs are effective in removing organic matter and nutrients, with estrogen and alkylphenol removal rates ranging from 79.93% to 91.42% depending on the method used.
Livestock wastewater treatment plants (L-WWTPs)	L-WWTPs are specifically designed for wastewater management from livestock farms and often use activated sludge processes with wetlands. Due to the intensive use of detergents and surfactants on farms, they contain the highest concentrations of organic pollutants such as 4-nonylphenol (4-NP).	L-WWTPs exhibit high removal rates, up to 93.17% for 4-NP.
Hospital wastewater treatment plants (H-WWTPs)	H-WWTPs are located in large hospitals and primarily use conventional activated sludge (CAS) processes to treat hospital wastewater. H-WWTPs are responsible for managing large quantities of medications and personal care products.	The results show that wastewater treatment plants (WWTPs) have effective removal rates for estrogens and alkylphenols as well as for 4-NP.
Commercial wastewater treatment plants (C-WWTPs)	C-WWTPs treat wastewater from commercial complexes and use moving bed biofilm reactors (MBBRs)	C-WWTPs are particularly effective at removing contaminants, with removal rates of 91.85% for 4-NP and 94.40% for 4-t-OP.
Rural wastewater treatment plants (R-WWTPs)	R-WWTPs, which typically use aerated lagoon systems (AL), are designed to treat wastewater from rural areas.	Although R-WWTPs are less efficient than other types of plants, they play a crucial role in wastewater treatment in remote areas, with removal rates of 66.40% for 4-NP and 71.56% for 4-t-OP.

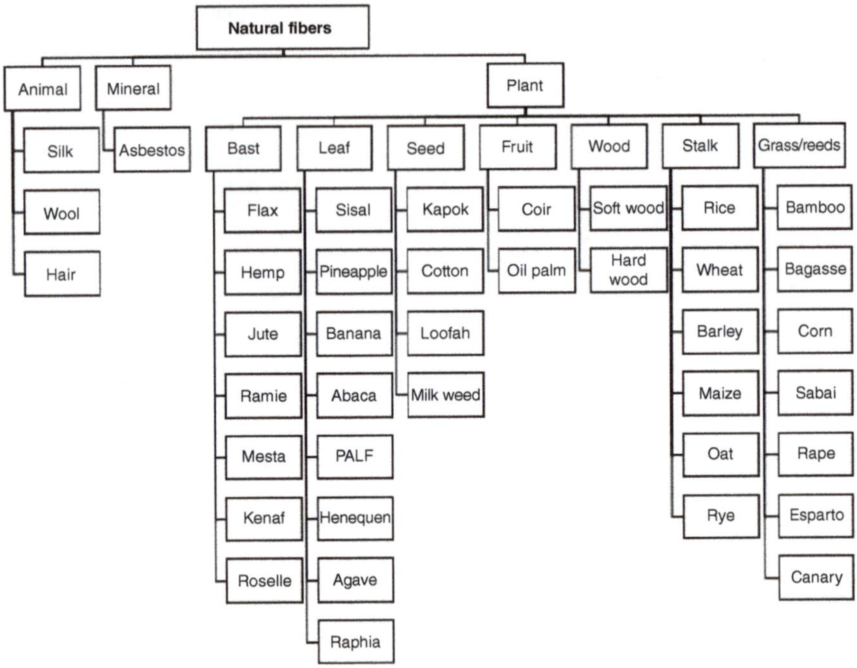

Figure 17.7 Classification of natural fibers used as reinforcements in biocomposites [56].

Environmental and economic pressures drive markets toward the use of natural fibers, which are abundant, renewable, and have minimal environmental impact. These fibers present an eco-friendly alternative for various applications, including technical components. Owing to their inherent properties, natural fibers not only reduce the carbon footprint and enhance recyclability but also impart valuable characteristics, such as improved vibration resistance and weight reduction, to composites. These fibers can be further classified into three types: animal fibers, plant fibers, and vegetable fibers [26, 55, 56]. Figure 17.7 shows the classification of natural fibers used as reinforcements in biocomposites.

Animal and vegetable fibers appear to be the most prevalent, with the majority being used in the textile and clothing industries [57]. Other vegetable fibers, particularly those derived from plants with rigid stems, are used in more technical applications where they reinforce polymer matrices in various structural applications. A variety of natural fibers, including jute, linen, hemp, sisal, hardwoods, softwoods, silk, wool, and others, are employed as reinforcements in polymer matrices to enhance the mechanical properties of the composite. The main components of natural fibers are cellulose, hemicellulose, and lignin. Figure 17.8 shows the typical chemical structures of these elements.

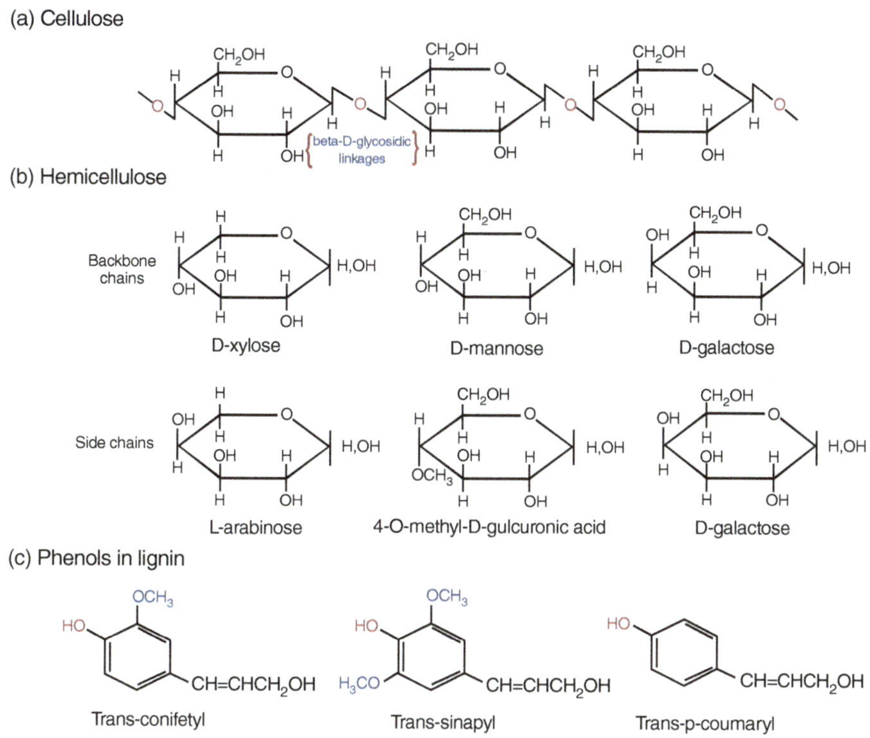

Figure 17.8 Chemical composition of the main components of natural fibers: (a) Cellulose, (b) Hemicellulose, (c) Phenols in lignin [26].

Cellulose, hemicellulose, and lignin are classified among the natural fibers of various biological structures. They also contain extractives, proteins, and certain inorganic compounds in small proportions [58, 59]. Cellulose is a linear macromolecule that can have an amorphous or crystalline structure, whereas hemicelluloses have a branched structure. The degree of polymerization of hemicelluloses generally ranges between 200 and 300, and their structure depends on the plant species [60]. Cellulose and hemicelluloses form the group of holocelluloses. On the other hand, cellulose macromolecules have good mechanical strength and a hydrophilic nature, whereas lignin consists of a complex three-dimensional network that provides rigidity, hydrophobicity, and good resistance to decomposition [61]. Additionally, waxes, tannins, minerals, and pectins are present. In the following table, we present the proportions of various components of natural fibers found in the literature, along with their mechanical properties.

As Table 17.2 shows, the contribution of each component varies significantly, which is characteristic of natural fibers. These variations notably affect the

Table 17.2 Composition and mechanical properties of various natural fibers [26].

Fiber	Chemical composition						Mechanical properties
	Cellulose (%)	Hemi-cellulose (%)	Lignin (%)	Pectin (%)	Wax (%)	Minerals (%)	a = Density (g/cm^3), b = Tensile strength (MPa), c = E-modulus (GPa), d = Elongation at break (%)
Jute	51–84	12–24	5–14	0.2–4.5	0.4–0.8	0.17–0.7	**a** = 1.3–1.45, **b** = 393–77, **c** = 10–30, **d** = 1.2–1.8
Bamboo	26–43	15–30	21–31	—	—	—	**a** = 0.6–1.1, **b** = 140–441, **c** = 11–17.
Cotton	90	6.3	—	—	0.7	1.1	**a** = 1.5–1.6, **b** = 287–800, **c** = 5.5–12.6, **d** = 7–8
Hemp	57–92	6–22	2.8–13	0.8–2.5	0.7–0.8	0.7–3	**a** = 1.48, **b** = 690, **c** = 30–70, **d** = 1.6–4
Banana	60–65	6–19	5–10	3–5	—	—	—
Kenaf	36–72	20–21	9–19	2	—	—	**b** = 930, **c** = 22–60, **d** = 1.6
Ramie	68–85	3–17	0.5–1	1.9–2.1	0.3	0.3	**a** = 1.5, **b** = 400–938, **c** = 44–128, **d** = 1.2–3.8
Sisal	43–88	10–15	4–14	0.5–10	0.2–2	—	**a** = 1.5, **b** = 468–640, **c** = 9.4–22, **d** = 3–7
Abaca	56–68	109–25	5–13	0.5–1	0.2–3	—	**a** = 1.5, **b** = 400, **c** = 12, **d** = 3–10
Flax	60–81	10–21	1–5	0.9–3.8	1.3–3	1–1.5	**a** = 1.5, **b** = 345–1100, **c** = 27.6, **d** = 0.2–3.2

overall properties of the fibers. In particular, the cellulose content plays a crucial role in the mechanical properties of fibers, as it imparts rigidity [62]. For example, the cellulose content in kenaf fibers can vary from one to two, with values reported in the literature ranging from 36% to 72%. This observation also applies to sisal, where the cellulose content can vary between 43% and 88%, with an average value of approximately 66%. Likewise, jute fibers vary widely, with an average cellulose content of 65% but ranging from 51% to 84%. These fibers exhibit excellent mechanical properties, making them particularly suitable as reinforcements in biocomposites for environmental applications. For example, they can be used as effective membranes [63] or absorbents [64] for wastewater treatment, contributing to sustainable and environmentally friendly solutions.

In terms of matrices, numerous studies are currently being conducted on biomass-derived matrices. However, these matrices still exhibit limited mechanical and thermal properties. Therefore, they are suitable for applications where the priority is environmental considerations (renewable resources, health, biodegradability, or recyclability) rather than technical performance. The matrix plays a crucial role in ensuring the overall durability of the material and significantly influences its appearance, shape, and surface resistance to environmental factors. They are also responsible for the effective adhesion of fibers, facilitating the transfer of load between them, and can be classified into thermoplastics and thermosets, depending on their processing methods and the types of bonds they form. Figure 17.9 illustrates the polymers used as matrices in composites [55].

Biobased thermoplastics made from renewable raw materials are becoming increasingly popular as alternatives to petroleum-based thermoplastics [65]. In this section, several of the most important biobased thermoplastics are presented. The earliest biobased matrices were produced from cellulose, which is well known in the textile and photographic film industries. The production processes for cellulose acetate and viscose (or regenerated cellulose) enable the creation of artificial polymers from wood or cotton cellulose [66]. Additionally, thermoplastic matrices such as polylactic acid and polyhydroxyalkanoate are available on the market [67]. These matrices are linear aliphatic polyesters produced by bacteria, primarily through the fermentation of sugars derived from corn or beets [68]. We also find a class of polyhydroxybutyrates, polymers that are both derived from biomass and biodegradable [69]. Various forms of polyamides exist, particularly those that can be synthesized entirely or partially from castor oil and polyethylene. These polyamides are biobased but not biodegradable [70]. Another type of polymer is an elastomer, which can be prepared from biobased polyols derived from cereals or algae [71]. Additionally, a new generation of composites has been developed from wood fibers and a biobased elastomeric matrix of the poly(ether-b-amide) type, demonstrating remarkable results in terms of thermal stability and mechanical properties [72].

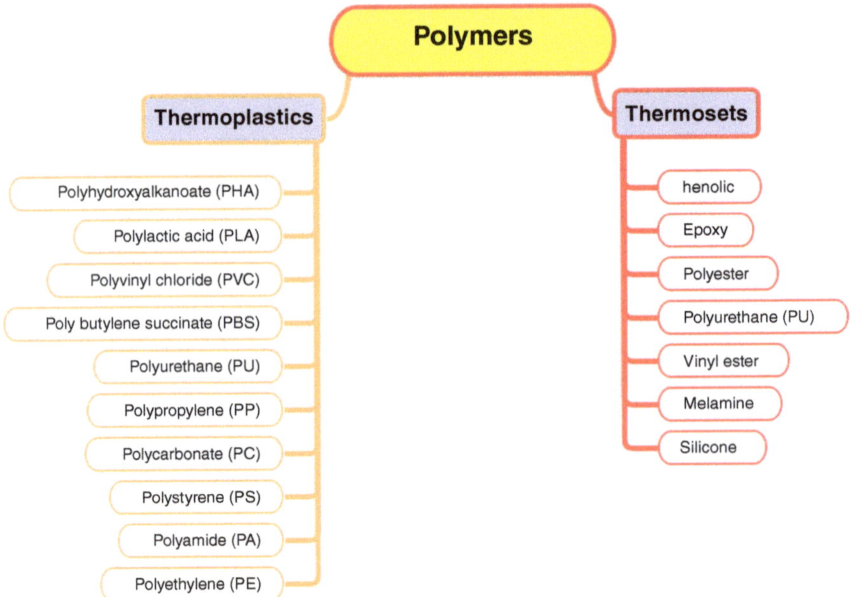

Figure 17.9 Principal polymers utilized in composites as a matrix.

The raw material for biobased thermosets often comes in the form of oil or other liquids extracted from various plants or animals. For example, thermosetting resins can be derived from fish oil, corn oil, soybean oil, linseed oil, and cashew nut shells. In the field of thermosetting matrices, few resins are derived entirely from biomass. However, cardanol resin, synthesized from cashew nut bark, is a notable example that has enabled the manufacture of composites with sufficiently robust mechanical characteristics for commercial applications, notably in the automotive and furniture industries [73]. Epoxy resins, one of the most functional thermosetting polymers, are generally synthesized by reacting polyols, polyphenols, or other active hydrogen compounds with epichlorohydrin [74]. Similarly, phenolic resins, which are valued for their mechanical strength, dimensional stability, and resistance to heat, fire, and solvents, continue to attract industrial and commercial interest even a century after their introduction [75]. A great deal of research is underway to replace these petroleum-based resins with biosourced alternatives via liquefaction and pyrolysis processes. On the other hand, polyester resins, widely used in the marine and automotive industries, are the most economical resin systems for engineering applications. Produced by a condensation reaction between a glycol and an unsaturated dibasic acid, these resins offer a good balance between mechanical, thermal, electrical, and chemical properties while being easy to process and relatively inexpensive [76].

17.3.2 Biocomposites for Water and Wastewater Treatment

The use of biocomposites as agents for wastewater treatment offers numerous advantages, primarily due to their remarkable surface properties, low toxicity, and high chemical stability. Additionally, their production from natural resources makes them an environmentally friendly and economically viable solution for water purification. Numerous laboratory studies have been conducted to assess the ability of biocomposites to retain various pollutants, including heavy metals, dyes, pharmaceuticals, and pesticides. This research demonstrated that biocomposites can achieve significant efficiencies in the removal of both organic and inorganic pollutants. Table 17.3 provides an overview of the applications of biocomposites in water treatment.

Despite extensive research on the use of biocomposites as adsorbents, membranes, catalysts, and photocatalysts for water treatment, most of this work remains confined to the laboratory scale. The large-scale application of biocomposites, particularly in wastewater treatment plants, is still limited and presents significant challenges. In this section, we highlight studies on the application of biocomposites in real water treatment plants, emphasizing their effectiveness. Montoya et al. investigated the performance of calcium-based biocomposites prepared from banana peels and eggshells (BPESs) for the removal and recovery of phosphate (P) from domestic wastewater. The biocomposites were produced by pyrolyzing mixtures of banana peel (BP) and eggshell (ES) at 700 °C with different mass percentages. The results indicated that the BPES biocomposite with a 1:2 ratio of banana peel:eggshell had the highest efficiency in terms of P adsorption, achieving a maximum adsorption capacity of $214 \pm 5\ \text{mg g}^{-1}$. The retention of P on the prepared biocomposite led to the precipitation of calcium phosphates, notably apatite ($Ca_5(PO_4)_3OH$), a key component in fertilizer production. The biocomposite demonstrated dual efficacy: it could remove up to 90% of phosphate in tertiary wastewater treatment applications, and the adsorbed P could be easily recovered as a slow-release fertilizer, reducing the risk of nutrient leaching into the environment. This makes the prepared biocomposite a promising candidate for use in wastewater treatment plants, offering a lower production cost than commercial adsorbents such as activated carbon. As a result, it presents an environmentally friendly, cost-effective alternative that aligns with the principles of the circular economy. Figure 17.10 illustrates the role of the calcium-based biocomposite, prepared from BPESs, as an environmentally friendly material [87].

The same research team developed a low-cost biocomposite by calcining a mixture of orange peel (OP) and eggshell (ES) at 700 °C. This biocomposite exhibited a high phosphorus adsorption capacity, with a maximum adsorption capacity of $315.25\ \text{mg g}^{-1}$. Additionally, the biocomposite demonstrated strong potential for

Table 17.3 Few examples of biocomposites used in water treatment.

Biocomposite	Remediation approach	Target pollutant	Removal efficiency	Reference
Dialdehyde nanocellulose -carboxymethyl chitosan, incorporating nanobentonite	Adsorption	Bromophenol blue, direct blue 6, organic solvents, and oils.	29.842 $g\,g^{-1}$ (bromophenol blue), 20.927 $g\,g^{-1}$ (direct blue 6), and absorb up to 50 times its weight in oils and organic solvents.	[77]
Magnetic alginate-coated chitosan core@shell beads	Adsorption	Ciprofloxacin (CPX) antibiotic, Cu^{2+}, Cd^{2+}, Co^{2+}, Ni^{2+}, Pb^{2+}, Zn^{2+}, and Hg^{2+}.	Ciprofloxacin: removal efficiency of 86%. Heavy metals: Removal efficiencies ranging from 74.89% to 99.86%.	[78]
Polyhydroxy-butyrate -Cellulose Acetate	Adsorption	Methylene blue	89% elimination of methylene blue	[79]
polyaniline-@oak acorn	Adsorption	Orange G dye	Removal efficiency of 99.46%.	[80]
Biochar-alginate beads	Adsorption	Methylene blue	Removal efficiency of 96.4 %	[81]
Silver nanoparticle-decorated cellulose beads	Catalysis	4-nitrophenol	100% conversion of 4-nitrophenol	[82]
Cellulose beads supported $CoFe_2O_4$	Catalysis	Rhodamine B. Orange G. Ofloxacin and phenol.	Rhodamine B: 98.51%. Orange G: Approximately 99.01%. Ofloxacin: Approximately 94.65% Phenol: 61.39%.	[78]
Nanochitosan/ carboxymethyl cellulose/TiO_2	Photo-catalytic degradation	Crystal violet dye	95% elimination of Crystal violet under visible light irradiation	[83]
Alginate/ bentonite impregnated TiO_2 beads	Adsorption and photo-catalysis	Methylene blue (MB)	98% discoloration of methylene blue	[84]

Table 17.3 (Continued)

Biocomposite	Remediation approach	Target pollutant	Removal efficiency	Reference
Polyvinylidene Fluoride–Quaternized Hemicellulose.	Membrane	Simulated wastewater including surfactants, oil, and saline conditions.	The Biocomposite membrane demonstrated a 99.99% salt rejection rate and reduced oil adhesion by 97% compared to pristine PVDF after 50 hours of continuous operation.	[85]
Chitosane/montmorillonite	Membrane	Cu^{2+}	86% removal efficiency of Cu(II) ions.	[86]
Polysaccharide-based materials such as chitosan molecules, including chitosan acetic acid solution, carboxymethyl chitosan, and sodium alginate.	Flocculation and sedimentation	Turbidity, ammonia nitrogen (NH_4^+-N), total nitrogen (TN), chemical oxygen demand (COD), total phosphate (TP), and harmful marine vibrio.	Turbidity: 66.1–78.6%, ammonia nitrogen (NH_4^+-N): 44.2–50.9%, total nitrogen (TN): 49.8–53.2%, chemical oxygen demand (COD): 64.3–69.6%, total phosphate (TP): 47.5–51.2%, harmful marine vibrio: 81.3–93.8%.	[85]

phosphorus recovery from domestic wastewater, achieving a recovery rate of 98.6% in the form of apatite. This material could also be used as an alternative fertilizer, contributing to the implementation of the 3R principle (reduce, reuse, recycle) within a circular economy. This work highlights the potential of agricultural waste as a valuable resource for developing biocomposite materials that are effective in wastewater treatment and nutrient recovery while also providing an environmentally friendly solution for waste management [88].

Araujo et al. [89] studied the use of an agar-graphene oxide hydrogel biocomposite (Agar-GO) for the treatment of textile wastewater in a pilot-scale adsorption plant. This biocomposite, synthesized from 80% agar and 20% graphene oxide, features a porous three-dimensional structure that provides excellent mechanical stability and enhanced adsorption properties due to the presence of active functional groups. Adsorption tests were conducted in a

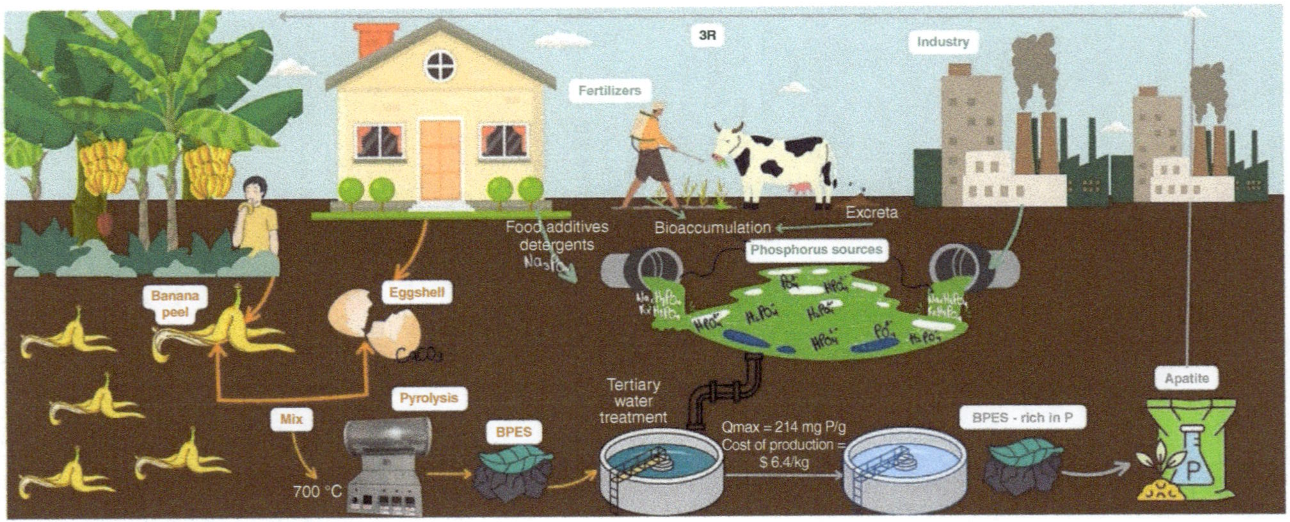

Figure 17.10 Diagram of the incorporation of biocomposite (BPESs) in environmental remediation [87].

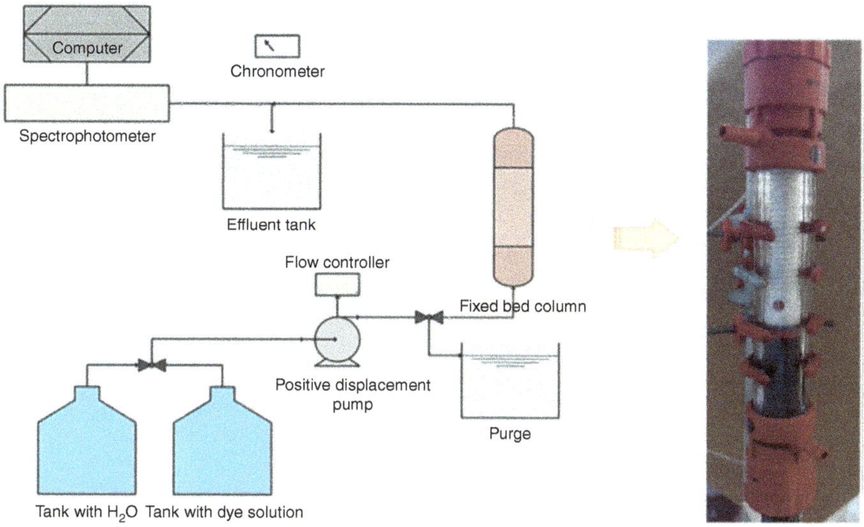

Figure 17.11 Process diagram illustrating the experimental setup for fixed-bed adsorption tests (pilot-scale adsorption plant) [89, 90].

fixed-bed adsorption column made of borosilicate glass. The column, with an internal diameter of 1.6 cm and a height of 7 cm, was filled with 6 g of wet Agar-GO biocomposite, equivalent to 0.13 g on a dry basis. This configuration enabled the evaluation of the material's effectiveness in removing dyes from solutions and synthetic wastewater. Figure 17.11 illustrates the application of the agar-graphene oxide biocomposite in a pilot-scale adsorption plant.

The results indicate that the adsorption capacities achieved in the fixed bed were 226.46 mg g^{-1} for Nile blue A, 79.51 mg g^{-1} for methylene blue, 58.25 mg g^{-1} for Malachite green, and 38.11 mg g^{-1} for basic fuchsin. The biocomposite demonstrated the ability to treat over 6 L of wastewater while maintaining a high regeneration capacity, highlighting its potential for industrial applications. Furthermore, the analyses revealed selective dye adsorption, which could facilitate the recovery and reuse of dyes in industrial processes.

The same research group utilized the biocomposite (A-GO) for the continuous treatment of water contaminated with chloroquine diphosphate (CQ) and the cationic dye Safranin-O (SO). Fixed-bed column adsorption tests demonstrated adsorption capacities of 66.15 mg g^{-1} for CQ and 101.52 mg g^{-1} for SO at a flow rate of 2.0 mL min^{-1}. These capacities decreased slightly when the flow rate was increased to 4.5 mL min^{-1}, attributed to a reduction in contact time. Regeneration tests indicated that the biocomposite (Agar-GO) can be effectively reused without significant loss of adsorption capacity after adsorption/desorption cycles using a 0.30 M NaOH solution [90].

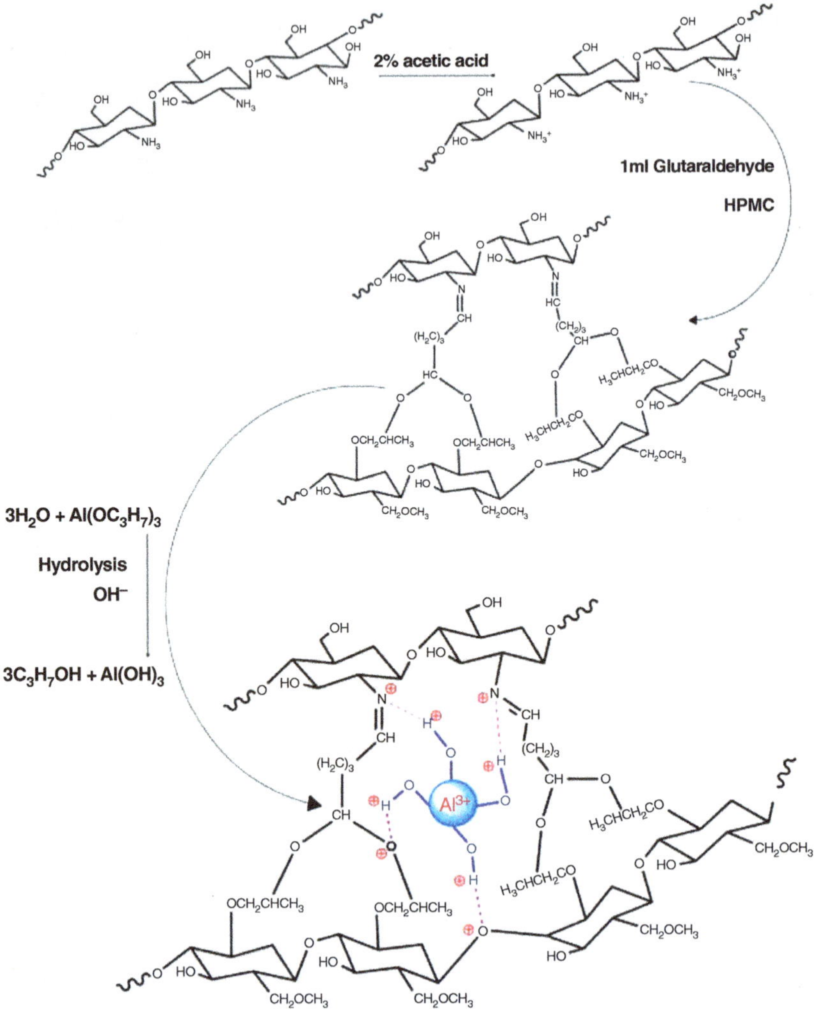

Figure 17.12 Proposed mechanism for the preparation of the CgHA biocomposite [91].

Another study focused on a novel biocomposite film, designated CgHA, which is composed of alumina, cross-linked chitosan, and hydroxypropyl methylcellulose (HPMC). Figure 17.12 illustrates the proposed mechanism for the preparation of the CgHA biocomposite.

This biocomposite was developed to treat real groundwater samples collected from a treatment plant, addressing the need for environmentally friendly materials capable of effectively removing two major contaminants: fluoride ions and azo dyes. The prepared biocomposite demonstrated enhanced thermal stability and a surface structure favorable for pollutant removal. Adsorption studies revealed that CgHA has a superior fluoride removal capacity and exceptional decolorization efficiency in textile wastewater. Furthermore, CgHA exhibited excellent regeneration capacity, maintaining significant efficacy after multiple cycles of use. This work highlights the potential of the CgHA biocomposite as a sustainable and effective solution for water purification [91].

Biocomposites have proven to be an ecological and effective solution for treating water contaminated with pharmaceuticals, dyes, and other pollutants. They exhibit remarkable potential for regeneration and reuse. However, despite these advantages, their application remains largely confined to the laboratory scale, with their implementation in wastewater treatment plants still being limited.

17.4 Conclusion, Challenges, and Future Recommendations

This chapter highlights the crucial importance of developing advanced, sustainable water treatment technologies, particularly in response to the increasing pollution of water resources caused by industrial, agricultural, and domestic activities. In this context, biocomposites are emerging as promising materials owing to their eco-friendliness, affordability, and notable effectiveness in removing pollutants. Laboratory research has proven their ability to eliminate various pollutants, such as heavy metals, drugs, dyes, and many others, suggesting that biocomposites are an economical and sustainable alternative to traditional materials such as activated carbon and clays, etc. Nevertheless, although biocomposites offer many benefits, their use in water treatment remains limited to the laboratory. On an industrial scale, there are many challenges, such as the variety of wastewater and the complex mix of pollutants it contains. It is difficult to completely purify wastewater with a single type of biocomposite, as it contains a variety of organic and inorganic contaminants, as well as solid particles and microorganisms. Another major challenge is the cost of raw materials and production methods. Specific techniques for extracting and processing biopolymers to create biocomposites in different forms (beads, membranes, gels, hydrogels, etc.) are costly and energy intensive. In addition, the use of analytical-grade chemicals, which are frequently employed in the laboratory, is not always possible on a large scale, resulting in increased expenditures. A major challenge also lies in the high energy consumption of the mechanical disintegration techniques used to increase biocomposite production.

The diversity of biopolymer characteristics, such as crystallinity, molecular weight, and number of functional groups, also leads to difficulties in the manufacture and efficiency of biocomposites. For example, the degree of deacetylation influences the number of functional groups present in chitin, while the characteristics of tannic acid differ depending on the source of the material. Biocomposite performance is impacted by these variations and must be taken into consideration when selecting raw materials and extraction techniques. Although many studies have demonstrated that biocomposites are nontoxic or slightly toxic, some studies suggest that high concentrations can have adverse consequences on cell viability and proliferation. For example, prolonged exposure to nanocellulose can lead to cytotoxic and inflammatory effects on human and animal lung cells. It is therefore essential to deepen our understanding of biocomposite degradation mechanisms, characterize their potential environmental consequences, and develop strategies for their recovery, stability, and efficient reuse of adsorbed contaminants. Biocomposites represent a promising opportunity for more environmentally friendly water treatment solutions, in line with global sustainability objectives. They have immense potential for large-scale use, particularly in the context of circular economy models where waste can be reused as a treatment agent. Future studies should focus on improving the performance of biocomposites, increasing their large-scale production, and assessing their long-term stability and regeneration capabilities in real-world water treatment plants.

References

1 Rathi, B.S., Kumar, P.S., and Vo, D.V.N. (2021). Critical review on hazardous pollutants in water environment: occurrence, monitoring, fate, removal technologies and risk assessment. *Science of the Total Environment* 797: 149134.

2 Kinuthia, G.K., Ngure, V., Beti, D. et al. (2020). Levels of heavy metals in wastewater and soil samples from open drainage channels in Nairobi, Kenya: community health implication. *Scientific Reports* 10: 1–13.

3 Al-Gheethi, A.A., Azhar, Q.M., Senthil Kumar, P. et al. (2022). Sustainable approaches for removing Rhodamine B dye using agricultural waste adsorbents: a review. *Chemosphere* 287: 132080.

4 Al-Tohamy, R., Ali, S.S., Li, F. et al. (2022). A critical review on the treatment of dye-containing wastewater: ecotoxicological and health concerns of textile dyes and possible remediation approaches for environmental safety. *Ecotoxicology and Environmental Safety* 231: 113160.

5 Oulguidoum, A., Fatine, S., Abbadi, M. et al. (2025). Development of stable and porous Ag_2O-loaded hydroxyapatite nanocomposites for enriched photocatalytic degradation of Rhodamine B under visible light. *Environmental Science and Pollution Research* 32: 2267–2279.

6 Labrag, J., El Bekkali, C., Oulguidoum, A. et al. (2021). Porous and bifunctional ZnO-hydroxyapatite nanostructure for photocatalytic degradation of

paracetamol and methylene blue in water. *Iranian Journal of Catalysis* 11: 389–395.

7 El Bekkali, C., Labrag, J., Oulguidoum, A. et al. (2022). Porous ZnO/hydroxyapatite nanomaterials with effective photocatalytic and antibacterial activities for the degradation of antibiotics. *Nanotechnology For Environmental Engineering* 7: 333–341.

8 Oulguidoum, A., Bouyarmane, H., Laghzizil, A. et al. (2019). Development of sulfonate-functionalized hydroxyapatite nanoparticles for cadmium removal from aqueous solutions. *Colloid and Interface Science Communications* 30: 100178.

9 Oulguidoum, A., Bouiahya, K., Bouyarmane, H. et al. (2021). Mesoporous nanocrystalline sulfonated hydroxyapatites enhance heavy metal removal and antimicrobial activity. *Separation and Purification Technology* 255: 117777.

10 Carnevale Miino, M., Galafassi, S., Zullo, R. et al. (2024). Microplastics removal in wastewater treatment plants: a review of the different approaches to limit their release in the environment. *Science of the Total Environment* 930: 172675.

11 Angulo-Bejarano, P.I., Puente-Rivera, J., and Cruz-Ortega, R. (2021). Metal and metalloid toxicity in plants: an overview on molecular aspects. *Plants* 10: 1–28.

12 Akpor, O.B. and Muchie, M. (2011). Environmental and public health implications of wastewater quality. *African Journal of Biotechnology* 10: 2379–2387.

13 Suschka, J. (2006). Integration of waste-water treatment and sludge handling on small scale, for the conservation and restoration of water and land quality. *Ecohydrology and Hydrobiology* 6: 189–195.

14 Crini, G. and Lichtfouse, E. (2019). Advantages and disadvantages of techniques used for wastewater treatment. *Environmental Chemistry Letters* 17: 145–155.

15 Sarvajith, M. and Nancharaiah, Y.V. (2020). Granulation of the autochthonous planktonic bacterial community of seawater for saline wastewater treatment. *Environmental Science: Water Research & Technology* 6: 1902–1916.

16 Peng, H. and Guo, J. (2020). Removal of chromium from wastewater by membrane filtration, chemical precipitation, ion exchange, adsorption electrocoagulation, electrochemical reduction, electrodialysis, electrodeionization, photocatalysis and nanotechnology: a review. *Environmental Chemistry Letters* 18: 2055–2068.

17 Malaeb, L. and Ayoub, G.M. (2011). Reverse osmosis technology for water treatment: state of the art review. *Desalination* 267: 1–8.

18 Hernández-Chover, V., Castellet-Viciano, L., Fuentes, R., and Hernández-Sancho, F. (2023). Circular economy and efficiency to ensure the sustainability in the wastewater treatment plants. *Journal of Cleaner Production* 384: 135563.

19 Lustenberger, S. and Castro-Muñoz, R. (2022). Advanced biomaterials and alternatives tailored as membranes for water treatment and the latest

innovative European water remediation projects: a review. *Case Studies in Chemical and Environmental Engineering* 5: 100205.

20 Vakili, M., Rafatullah, M., Salamatinia, B. et al. (2014). Application of chitosan and its derivatives as adsorbents for dye removal from water and wastewater : a review. *Carbohydrate Polymers* 113: 115–130.

21 El Allaoui, B., Benzeid, H., Zari, N. et al. (2024). Cellulose beads supported $CoFe_2O_4$: a novel heterogeneous catalyst for efficient rhodamine B degradation via advanced oxidation processes. *International Journal of Biological Macromolecules* 259: 128893.

22 Kithinji Kinoti, I., Ogunah, J., Muturia M'Thiruaine, C., and Marangu, J.M. (2022). Adsorption of heavy metals in contaminated water using zeolite derived from agro-wastes and clays: a review. *Journal of Chemistry* 2022: 1–25.

23 Ali, M.M.S., Imam, D.M., and El-Nadi, Y.A. (2021). Vanadium(V) removal and recovery by adsorption onto modified activated carbon derived from natural hydroxyapatite. *Journal of the Iranian Chemical Society* 18: 2771–2784.

24 Dong, L., Li, S., Jin, Y. et al. (2021). Applied Surface Science Enhanced adsorption of Eu (III) from wastewater using Solidago canadensis -derived biochar functionalized by Ca/Al-LDH and hydroxyapatite. *Applied Surface Science* 567: 150794.

25 RR, R., Julius, A., Suman, T.Y., and Samrot, A.V. (2024). Advancements in nanocomposites for wastewater treatment addressing emerging pollutants and contaminants. *Nanotechnology For Environmental Engineering* 9: 99–110.

26 Bahrami, M., Abenojar, J., and Martínez, M.Á. (2020). Recent progress in hybrid biocomposites: mechanical properties, water absorption, and flame retardancy. *Materials* 13.

27 Flörke, M., Kynast, E., Bärlund, I. et al. (2013). Domestic and industrial water uses of the past 60 years as a mirror of socio-economic development: a global simulation study. *Global Environmental Change* 23: 144–156.

28 Mishra, R.K. (2023). Fresh water availability and it's global challenge. *J Mar Sci Res* 2: 01–03.

29 Khodakarami, M. and Bagheri, M. (2021). Recent advances in synthesis and application of polymer nanocomposites for water and wastewater treatment. *Journal of Cleaner Production* 296: 126404.

30 Petersen, L., Heynen, M., and Pellicciotti, F. (2021). *Water Availability: A Hydrological View*, 9–21. Springer.

31 Petersen, L., Heynen, M. and Pellicciotti, F. (2016). Freshwater resources: past, present, future. In: *International Encyclopedia of Geography: People, the Earth, Environment and Technology* (eds. D. Richardson, N. Castree, M.F. Goodchild, A. Kobayashi, W. Liu and R.A. Marston), 1–11.

32 Mancosu, N., Snyder, R.L., Kyriakakis, G., and Spano, D. (2015). Water scarcity and future challenges for food production. *Water (Switzerland)* 7: 975–992.

33 UNICEF, WHO (2015). *Progress on Sanitation and Drinking Water: 2015 Update and MDG Assessment*. World Heal Organ.

34 Luo, T., Young, R., and Reig, P. (2015). Aqueduct projected water stress country rankings. *Technical Note* 16: 1–6.

35 Ryder, G. and Al, E. (2017). The United Nations world water development report, 2017. Wastewater Untapped Resour .

36 Boretti, A. and Rosa, L. (2019). Reassessing the projections of the world water development report. *npj Clean Water* 2: 15.

37 Koncagül, E., Tran, M., Connor, R. et al. (2018). *The United Nations World Water Development Report 2018: Nature-Based Solutions for Water; Facts and Figures*. UNESCO.

38 World Health Organization (2016). *Protecting Surface Water for Health. Identifying, Assessing and Managing Drinking-water Quality Risks in Surface-Water Catchments*. World Health Organization.

39 Kuski, L., Maia, E., Moura, P. et al. (2020). ScienceDirect development of a decentralized monitoring system of domestic water consumption. *Energy Reports* 6: 856–861.

40 Ahmad, H.R., Aziz, T., Zia-ur-Rehman, M. et al. (2016). Sources and composition of waste water: threats to plants and soil health. In: *Soil Science: Agricultural and Environmental Prospectives* (eds. K. Hakeem, J. Akhtar, and M. Sabir), 349–370. Springer, Cham.

41 Rajasekhar, B., Nambi, I.M., and Govindarajan, S.K. (2018). Human health risk assessment of ground water contaminated with petroleum PAHs using Monte Carlo simulations: a case study of an Indian metropolitan city. *Journal of Environmental Management* 205: 183–191.

42 Najy, M., Talbi, F.Z., Ech-Chafay, H. et al. (2020). Exploitation of the domestic wastewater treatment plant by activated sludge in the airport area of the city ben slimane (Morocco). *The International Archives of the Photogrammetry, Remote Sensing and Spatial Information Sciences* 44: 303–308.

43 Prabu, Sakthivel Lakshmana, et al. "Effective Waste Water Treatment and its Management." Toxicity and Waste Management Using Bioremediation, edited by Ashok K. Rathoure and Vinod K. Dhatwalia, IGI Global Scientific Publishing, 2016, pp. 312–334.

44 Haq I, Kalamdhad AS. Advanced and Ecofriendly Technologies for the Treatment of Industrial Wastewater to Constrain Environmental Pollution BT - Innovations in Environmental Biotechnology. In: Arora S, Kumar A, Ogita S, Yau Y-Y, editors., Singapore: Springer Nature Singapore; 2022, p. 561–73.

45 Alnazza Alhamad, A., Ben Amor, I., Zeghoud, S. et al. (2024). Biobased nanomaterials in wastewater treatment. In: *Biobased Nanomaterials: Applications in Biomedicine, Food Industry, Agriculture, and Environmental Sustainability*, 451–473. Singapore: Springer Nature Singapore.

46 Ahmadi, F., Zinatizadeh, A.A., Asadi, A. et al. (2020). Simultaneous carbon and nutrients removal and PHA production in a novel single air lift bioreactor treating an industrial wastewater. *Environmental Technology and Innovation* 18: 100776.

47 Aboelfetoh, E.F., Elabedien, M.E.Z., and Ebeid, E.-Z.M. (2021). Effective treatment of industrial wastewater applying SBA-15 mesoporous silica modified with graphene oxide and hematite nanoparticles. *Journal of Environmental Chemical Engineering* 9: 104817.

48 Rout, P.R., Zhang, T.C., Bhunia, P., and Surampalli, R.Y. (2021). Treatment technologies for emerging contaminants in wastewater treatment plants: a review. *Science of the Total Environment* 753: 141990.

49 Kalfa, A., Shapira, B., Shopin, A. et al. (2020). Capacitive deionization for wastewater treatment: opportunities and challenges. *Chemosphere* 241: 125003.

50 Chen, G.-H., van Loosdrecht, M.C.M., Ekama, G.A., and Brdjanovic, D. (2020). *Biological Wastewater Treatment: Principles, Modeling and Design*. IWA Publishing.

51 Bina, B., Mohammadi, F., Amin, M.M. et al. (2018). Determination of 4-nonylphenol and 4-tert-octylphenol compounds in various types of wastewater and their removal rates in different treatment processes in nine wastewater treatment plants of Iran. *Chinese Journal of Chemical Engineering* 26: 183–190.

52 Guna, V., Ilangovan, M., Ananthaprasad, M.G., and Reddy, N. (2018). Hybrid biocomposites. *Polymer Composites* 39: E30–E54.

53 Ghasemi, S., Sibi, M.P., Ulven, C.A. et al. (2020). A preliminary environmental assessment of epoxidized sucrose soyate (ESS)-based biocomposite. *Molecules* 25: 2797.

54 Bernava, A., Manins, M., and Kukle, S. (2012). Natural fibers woven structures for composites reinforcements. *Journal of Biobased Materials and Bioenergy* 6: 449–455.

55 Andrew, J.J. and Dhakal, H.N. (2022). Sustainable biobased composites for advanced applications: recent trends and future opportunities—a critical review. *Composites Part C: Open Access* 7: 100220.

56 Nunes, F.M., Morais, J.A.R., Machado, Ê.L. et al. (2023). Classifications and applications of biocomposites: a bibliometric review and analysis. *Concilium* 23: 438–465.

57 Tridico SR, Natural animal textile fibres: Structure, characteristics and identification, in Identification of textile fibers, edited by MM Houck. Woodhead Publishing, 2009. p. 27–67.

58 Panthapulakkal, S., Zereshkian, A., and Sain, M. (2006). Preparation and characterization of wheat straw fibers for reinforcing application in injection molded thermoplastic composites. *Bioresource Technology* 97: 265–272.

59 Rowell, R.M. and Rowell, J. (1996). *Paper and Composites from Agro-based Resources*. CRC Press.

60 Fengel, D. and Wegener, G. (2011). *Wood: Chemistry, Ultrastructure, Reactions*. Walter de Gruyter.

61 Meshitsuka, G. and Isogai, A. (2017). Chemical structures of cellulose, hemicelluloses, and lignin. In: *Chemical Modification of Lignocellulosic Materials* (ed. David N.-S. Hon), 11–33. Routledge.

62 Königsberger, M., Lukacevic, M., and Füssl, J. (2023). Multiscale micromechanics modeling of plant fibers: upscaling of stiffness and elastic limits from cellulose nanofibrils to technical fibers. *Materials and Structures* 56: 13.

63 Oprea, M. and Voicu, S.I. (2020). Recent advances in applications of cellulose derivatives-based composite membranes with hydroxyapatite. *Materials (Basel)* 13: 2481.

64 Gupta, V.K., Carrott, P.J.M., Singh, R. et al. (2016). Cellulose: a review as natural, modified and activated carbon adsorbent. *Bioresource Technology* 216: 1066–1076.

65 Hofmann, M., Garrido, M., Machado, M. et al. (2022). Development of high-performance partially biobased thermoset polyester using renewable building blocks from isosorbide, 1, 3-propanediol, and fumaric acid. *Journal of Applied Polymer Science* 139: e53029.

66 QI, Haisong. Platforms for functionalization of cellulose. In: Novel Functional Materials Based on Cellulose, edited by Haisong Qi, Cham: Springer International Publishing, 2016. p. 11–23.

67 Roy, I. and Philip, S.E. (2008). *Polyhydroxyalkanoates: A New Generation of Biotechnologically Produced Biodegradable Polymers*. Nova Science Publishers.

68 Vink, E.T.H., Rabago, K.R., Glassner, D.A., and Gruber, P.R. (2003). Applications of life cycle assessment to NatureWorks™ polylactide (PLA) production. *Polymer Degradation and Stability* 80: 403–419.

69 Freier, T., Kunze, C., Nischan, C. et al. (2002). In vitro and in vivo degradation studies for development of a biodegradable patch based on poly (3-hydroxybutyrate). *Biomaterials* 23: 2649–2657.

70 Khedr, M.S.F. (2023). Bio-based polyamide. *Physical Sciences Reviews* 8: 827–847.

71 Głowińska, E., Gotkiewicz, O., and Kosmela, P. (2023). Sustainable strategy for algae biomass waste management via development of novel bio-based thermoplastic polyurethane elastomers composites. *Molecules* 28: 436.

72 Sliwa, F., Marin, G., Charrier, F., and Malet, F. (2012). A new generation of wood polymer composite with improved thermal stability. *Polymer Degradation and Stability* 97: 496–503.

73 Maffezzoli, A., Calò, E., Zurlo, S. et al. (2004). Cardanol based matrix biocomposites reinforced with natural fibres. *Composites Science and Technology* 64: 839–845.

74 Baroncini, E.A., Kumar Yadav, S., Palmese, G.R., and Stanzione, J.F. III, (2016). Recent advances in bio-based epoxy resins and bio-based epoxy curing agents. *Journal of Applied Polymer Science* 133: 133.

75 Pilato, L. (2010). *Phenolic Resins: A Century of Progress*, vol. 11. Springer.

76 Mehdipour-Ataei, S. and Aram, E. (2023). High-end applications of unsaturated polyester composites. In: *Applications of Unsaturated Polyester Resins* (eds. S. Thomas and C. J. Chirayil), 421–439. Elsevier.

77 Sharma, V., Shahnaz, T., Subbiah, S., and Narayanasamy, S. (2020). New Insights into the remediation of water pollutants using nanobentonite incorporated nanocellulose chitosan based aerogel. *Journal of Polymers and the Environment* 28: 2008–2019.

78 Chkirida, S., El Mernissi, N., Zari, N. et al. (2024). In-situ magnetic alginate coated chitosan core@shell beads with excellent performance in simulated and real wastewater treatment: behavior, mechanisms, and new perspectives. *International Journal of Biological Macromolecules* 260: 129389.

79 Villabona-Ortíz, Á., Ortega-Toro, R., and Pedroza-Hernández, J. (2024). Biocomposite based on polyhydroxybutyrate and cellulose acetate for the adsorption of methylene blue. *Journal Of Composites Science* 8: 234.

80 Doughmi, O., Salihi, M., Hsini, A. et al. (2024). Development and investigation of a new polyaniline@oak acorn biocomposite as a potential adsorbent for efficiently removing orange G dye from aqueous solution. *Materials Science and Engineering B* 307: 117493.

81 Bahsaine, K., Benzeid, H., Zari, N. et al. (2024). Biochar-alginate beads derived from argan nutshells for effective methylene blue removal: a sustainable approach to wastewater treatment. *International Journal of Biological Macromolecules* 282: 136853.

82 El Allaoui, B., Chakhtouna, H., Ouhssain, A. et al. (2024). Silver nanoparticle-decorated cellulose beads: eco-friendly catalysts for efficient 4-nitrophenol reduction and antibacterial performance. *International Journal of Biological Macromolecules* 273: 133078.

83 Sugashini, S., Gomathi, T., Devi, R.A. et al. (2022). Nanochitosan/carboxymethyl cellulose/TiO_2 biocomposite for visible-light-induced photocatalytic degradation of crystal violet dye. *Environmental Research* 204: 112047.

84 Chkirida, S., Zari, N., Achour, R. et al. (2021). Highly synergic adsorption/photocatalytic efficiency of alginate/bentonite impregnated TiO_2 beads for wastewater treatment. *Journal of Photochemistry and Photobiology A: Chemistry* 412: 113215.

85 Liu, R., Li, Q., Gan, L. et al. (2024). Robust, versatile, and hemicellulose-derived biocomposite Janus membrane for saline wastewater desalination. *Journal of Membrane Science* 697: 122566.

86 Vedula, S.S. and Yadav, G.D. (2022). Superior efficacy of biocomposite membranes of chitosan with montmorillonite and kaolin vs pure chitosan for removal of Cu(II) from wastewater. *Journal of Chemical Sciences* 134: 134.

87 Ospina-Montoya, V., Pérez, S., Muñoz-Saldaña, J. et al. (2024). Performance of novel Ca-biocomposites produced from banana peel and eggshell for highly efficient removal and recovery of phosphate from domestic wastewater. *Journal of Environmental Management* 352: 120029.

88 Pérez, S., Giraldo, S., Forgionny, A. et al. (2024). Eco-friendly reuse of agricultural wastes to produce biocomposites with high potential in water treatment and fertilizers. *Biomass Conversion and Biorefinery* 14: 8537–8547.

89 de Araujo, C.M.B., Ghislandi, M.G., Rios, A.G. et al. (2022). Wastewater treatment using recyclable agar-graphene oxide biocomposite hydrogel in batch and fixed-bed adsorption column: bench experiments and modeling for the selective removal of organics. *Colloids and Surfaces A: Physicochemical and Engineering Aspects* 639: 128357.

90 Bezerra de Araujo, C.M., Wernke, G., Ghislandi, M.G. et al. (2023). Continuous removal of pharmaceutical drug chloroquine and Safranin-O dye from water using agar-graphene oxide hydrogel: selective adsorption in batch and fixed-bed experiments. *Environmental Research* 216: 114425.

91 Barik, B., Nayak, P.S., Achary, L.S.K. et al. (2019). Synthesis of alumina-based cross-linked chitosan-HPMC biocomposite film: an efficient and user-friendly adsorbent for multipurpose water purification. *New Journal of Chemistry* 44: 322–337.

18

Biocomposites for Wastewater Remediation: Present Trends and Challenges for the Future

Kenza Bahsaine[1,2], Hanane Benzeid[2], Nadia Zari[1], and Rachid Bouhfid[3]

[1] *Moroccan Foundation for Advanced Science, Innovation and Research (MAScIR), Mohammed VI Polytechnic University, Lot 660-Hay Moulay Rachid, 43150 Ben Guerir, Morocco*
[2] *Laboratory of Analytical Chemistry, Faculty of Medicine and Pharmacy, Mohammed V University, Rabat 10100, Morocco*
[3] *International Water Research Institute (IWRI), Mohammed VI Polytechnic University, Lot 660 Hay Moulay Rachid, Ben Guerir 43150, Morocco*

18.1 Introduction

Global demand for water is increasing, and it is predicted that by 2030, more than 160% of the world's current water supply will be required to meet water needs [1].

Providing an adequate supply of uncontaminated water to households and businesses is a major obstacle [2]. Similarly, the discharge of untreated water containing pollutants such as pesticides, pigments, heavy metals, pathogens, pharmaceuticals, etc., has led to a sharp decline in water quality [3]. The discharge of wastewater into aquatic ecosystems can cause significant degradation [4], adversely affect biodiversity, and lead to the transmission of waterborne diseases such as typhoid, cholera, and dysentery [5]. The prevalence of harmful contaminants in drinking water has attracted the interest of the research community because of its adverse effects not only on human well-being but also on our planet. The limited availability of clean water underlines the importance of using treated wastewater for applications such as irrigation, aquaculture, and domestic use [6]. Conventional wastewater treatment techniques such as filtration, oxidation, adsorption, sedimentation, ion exchange, coagulation-flocculation, membrane separation, and reverse osmosis are widely used [7], but are unable to achieve the level of purification required to meet accurate and cost-effective discharge standards. Innovative wastewater treatment methods include electro-precipitation, photocatalysis, advanced oxidation processes, UV or visible light photolysis, and activated carbon adsorption. However, these approaches are both costly and energy intensive [8]

The inherent limitations of composite materials have led to significant progress in the development of biocomposites. Biopolymers are considered abundant and inexpensive natural resources that offer numerous advantages such as nontoxicity, cost effectiveness compared to synthetic polymers, and biocompatibility. In addition, biopolymers are composed of renewable materials, which accelerates their biodegradation [9]. The incorporation of functional groups with reactive properties such as carbonyl, amide, carboxyl, and hydroxyl into the structure of biopolymers makes them highly suitable for wastewater treatment. Given the inadequate characteristics of these green polymers, particularly their limited physical, chemical, and mechanical resistance, low degradation temperature, and significant gas and water absorption, the incorporation of reinforcements into the biopolymer matrix is proposed.

Recently, the term "biocomposite" has become more widely used, particularly in the field of wastewater treatment. This emerging class of well-structured bio-hybrid materials consists of reinforcements combined with a biopolymer. These compounds offer improved physical, structural, and functional properties thanks to the integration of the inherent properties of the biopolymers and the advantages offered by the additional reinforcements. This intelligent combination makes it possible to reduce or eliminate certain risks associated with nanoparticles, while developing new versatile and lightweight biocomposites [10, 11].

18.2 Fundamentals of Biocomposites

The trend toward replacing petroleum-based polymers with biopolymers has extended to the field of composite materials. Over the last decade, increasing attention has been paid to modifying biopolymers with natural fillers to develop new materials known as biocomposites. These have been developed to overcome the shortcomings of polymeric materials, in particular low mechanical strength and poor long-term durability. A biocomposite is a substance made by combining two or more materials, at least one of which is of natural origin. This combination results in a new material whose performance is superior to that of the individual materials. The resulting materials consist of a matrix and a reinforcement. The matrix acts as a cohesive agent, consolidating the reinforcing components and providing mechanical strength. The categorization of the polymer matrix used in biocomposites can be divided into two distinct groups: a nonbiodegradable petroleum-based matrix and a biodegradable polymer-based matrix. The latter are produced by reinforcing the biopolymers with appropriate fillers such as metals, metal oxides, and natural fibers [12]. Biopolymers are often produced by polymerization of monomers, solvent extraction, or fermentation. Polymers can be reinforced by the introduction of fillers by molding, extrusion, solvent casting, electrospinning, or intercalation [13]. Compared with traditional composites, biocomposites have a great ability to minimize energy consumption and CO_2 emissions during processing. In addition, biocomposites offer advantages such as durability, light weight, high strength and rigidity, high thermal and acoustic insulation capacity, and cost-effectiveness [14].

18.2.1 Biopolymers

Green technologies and sustainable development are increasingly popular these days. The use of biopolymers has attracted particular attention in the field of water treatment due to their unique properties and wide applicability. These substances are often nontoxic, compatible with living organisms, and able to decompose naturally, reducing the risk of further pollution. In addition, their adaptability means they can be modified and functionalized to improve their effectiveness and focus on specific pollutants. Biopolymers are made up of long chains of repeating monomer units, such as amino acids, monosaccharide units, and nucleic acids. They have several advantages, including large surface area, wide accessibility, high sorption efficiency, and reasonable price. Figure 18.1 illustrates some of the most common biopolymers.

18.2.1.1 Cellulose

The most common and widely used biopolymer is cellulose, which is characterized by its large surface area, mechanical strength, and chemical surface. Cellulose is known as a linear polymer formed by the condensation of 1,4-glycosidic linkages with D-glucose (Figure 18.2). In 1839, the French chemist Anselme Payen was the first to isolate cellulose from plant debris. Cellulose can be extracted from a variety of sources, including wood, plants, algae, tunicates, and bacteria. Each monomer in this biopolymer has three hydroxyl groups, which are responsible for forming hydrogen bonds that can affect the crystallinity and physical properties of the cellulose material. Multiple cellulose chains align in parallel and self-assemble to form microfibrils, which are composed of crystalline segments characterized by well-ordered cellulose chains and amorphous regions with a lower degree of order due to van der Waals forces and intermolecular hydrogen bonding. Cellulose is therefore characterized as a semi-crystalline material whose degree of crystallinity is influenced by its origin, the extraction method used, and any subsequent treatment [15].

Cellulose has unique properties compared to other polysaccharides due to its structural composition. It is associated with many desirable physical properties, including a high degree of crystallinity (89%) and polymerization (14,400) and a

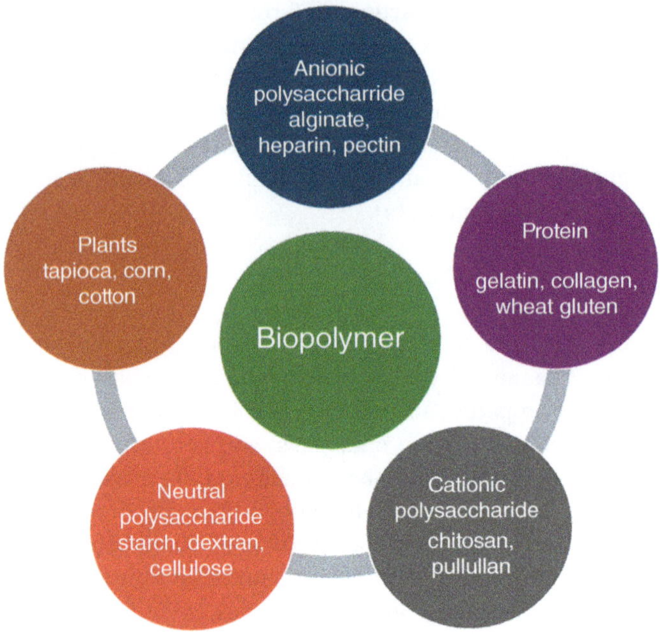

Figure 18.1 Most common biopolymers.

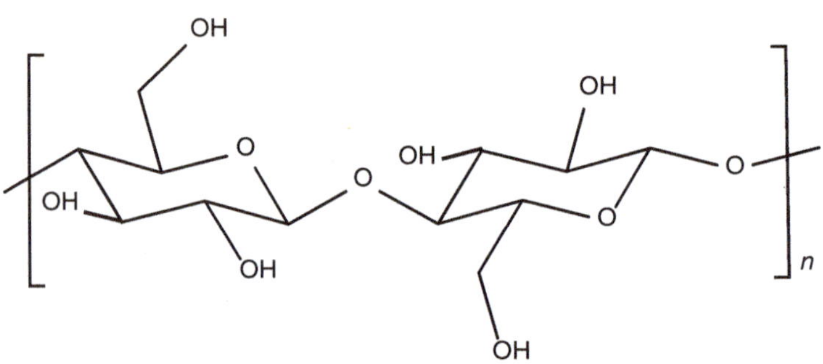

Figure 18.2 Chemical structure of cellulose.

specific surface area (37 m^2 g^{-1}). A cellulose fibril can reach a Young's modulus of the order of 114 GPa [16]. The length of the polymer chain, its dimensions, its degree of crystallinity, and its thermal resistance can also vary depending on the extraction method and its source (bacteria or plants) [17]. Cellulose is virtually insoluble in water and other organic solvents due to the strong hydrogen bonding between its molecules. However, the mechanical properties and adsorption capacity of unmodified cellulose are limited and require physical or chemical modification.

The main advantages of this material are its low cost, renewable nature, and biodegradability. However, it has two major disadvantages compared to synthetic polymers: high hygroscopicity due to the hydroxyl groups and limited processability due to its rapid degradation [18]. In terms of applications, cellulose is a highly adaptable natural polymer with a wide range of properties, and cellulose-based materials have been investigated for their potential applications in diverse fields such as biomedicine, energy, packaging, and environment. Due to its intrinsic hydrophilic properties, cellulose has been used as a hydrophilic coating on membranes to improve flow and prevent fouling. In addition, cellulose-based functional materials have high adsorption capacities, making them suitable for water treatment applications.

In addition, research has shown that this biopolymer presents a high capacity to adsorb pollutants after chemical modification of its surface. This modification specifically targets molecules with basic groups, particularly those with high concentrations of nitrogen, sulfur, and oxygen [19]. Cellulosic materials can be processed mechanically or chemically to create cellulose nanofibers (CNFs), nanowhiskers (CNWs), and nanocrystals (CNCs), having a great ability to purify water [20].

18.2.1.2 Chitosan

Chitosan, a compound derived from the deacetylation of chitin found abundantly in shells of crustaceans, is a linear polysaccharide made of linked N-acetylglucosamine units. Through a multistep process, chitin's acetyl groups are removed, yielding a material called chitosan with over half of its acetyl groups cleaved away (Figure 18.3). Solubility in dilute acids below ($pH < 6$) is a distinguishing trait of chitosan given its positive charge under acidic conditions [21].

Amino group molecules have distinctive characteristics, as their protonation increases the solubility of NH^{3+} and facilitates the intermolecular development of nanostructures with biological activities as effluent coagulants [22]. Amino groups are very reactive in nucleophilic addition to aldehydes, which is then followed by dehydration to form imine bonds, commonly known as Schiff bases. Through electrostatic attraction and coagulation, chitosan's large surface area and positive charge enable it to effectively remove organic contaminants, heavy metals, and dyes from water. Chitosan-based composites effectively remove organic contaminants, heavy metals, and dyes from water [23].

18.2.1.3 Alginate

Alginate, a polysaccharide derived from kelp, is composed of α-l-guluronic acid and β-D-mannuronic acid residues joined linearly through 1,4-glycosidic linkages. The carboxyl moieties within the G units impart an overall negative charge at $pH = 7$, typically balanced by sodium cations. Alginate gelation arises through cross-linking of alginate polymers with divalent cations, commonly

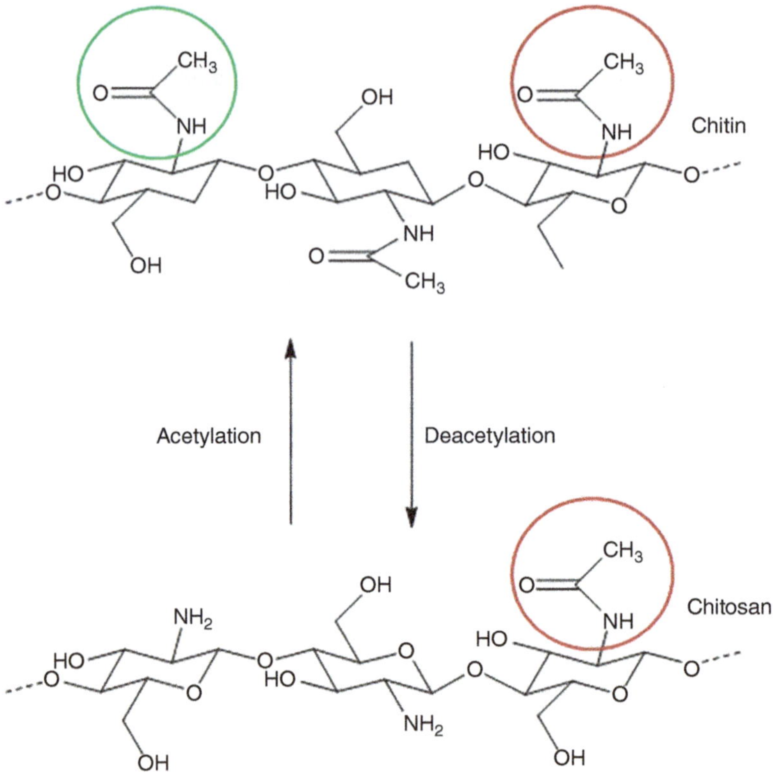

Figure 18.3 Chemical structure of both chitin and chitosan.

calcium, adhering to a mechanism known as the "egg-box" model of gelation. In mammalian tissues, alginate hydrogels have seen considerable investigation as synthetic extracellular matrices for tissue engineering applications due to their favorable mechanical properties and mild gelation process. However, their success depends upon achieving tunable gel structures through control of block distributions and precise stoichiometry of divalent cross-linkers [24]. Alginate is a hydrophilic and anionic polymer that was first isolated from bacteria and brown algae like *Ascophyllum nodosum*, *Durvillaea antarctica*, *Laminaria digitata*, *Laminaria hyperborea*, *Macrocystis pyrifera*, *Ecklonia maxima*, *Laminaria japonica*, *Sargassum* spp., and *Lessonia nigrescens* [25, 26]. It is mostly produced by brown algae varieties. These polymers can be degraded into monosaccharides and oligosaccharides by enzymes like alginate lyases. The diverse range of functional characteristics, including biocompatibility, gelation ease, and biodegradability, are the primary factors contributing to the extensive applications of alginate (Figure 18.4). It is frequently employed in water treatment for many purposes,

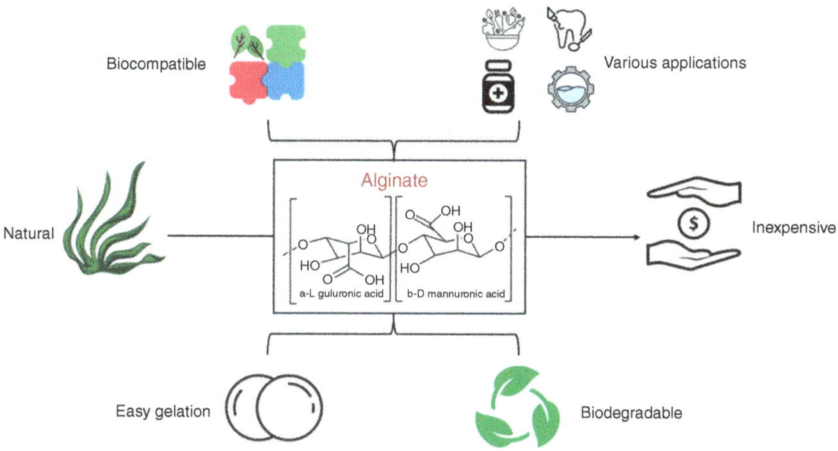

Figure 18.4 Alginate properties.

including encapsulating, immobilizing enzymes or microbes, and eliminating metal ions. Alginate beads and membranes have demonstrated potential in metals and dyes removal using ion exchange and adsorption processes. However, alginate gel has disadvantages such as high stiffness and brittleness, as well as poor elasticity and mechanical properties. Composite materials made up of organic and inorganic components are being developed to improve the mechanical and thermal stability, in addition to the swelling properties, of pure alginate gel [27].

18.2.1.4 Starch

Starch, the second most abundant polysaccharide in nature and a low-cost substance, is composed of glucopyranose units organized in a homopolymer $(C_6H_{10}O_5)_n$. Amylopectin, a complex polymer with several branches, constitutes the primary constituent of most starches and significantly influences the properties of starch. Whereas, amylose, a linear polymer, accounts for 25% of the starch [28]. Each of these polymers acquires distinct characteristics as a consequence of this conformational alteration. For example, the crystalline region of the particles is created by amylopectin's short branching at the 1,6-glycosidic connections [29]. Amylopectin is the biggest naturally occurring molecule, with an average degree of polymerization of about 2 million units, whereas amylose molecules have a degree of polymerization ranging from 840 to 22,000 glucose units. Amylopectin has a molecular weight that is approximately one thousand times greater than that of amylose [30]. Starch is a naturally occurring biopolymer that is abundant, renewable, biocompatible, and biodegradable. It serves as a secure carbohydrate in multiple plant species and is frequently encountered in different parts of plants including stalks, roots, and crop seeds. The most common

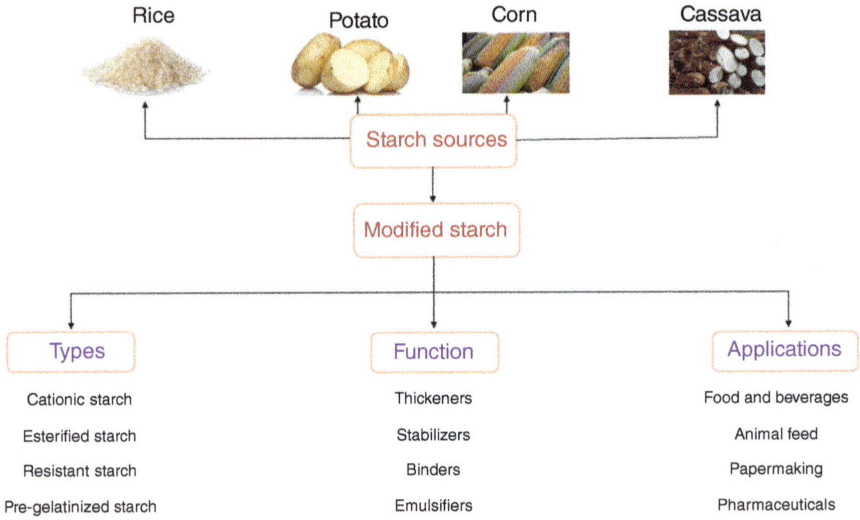

Figure 18.5 Raw materials, types, uses, and market applications of starch.

sources of starch include cassava, wheat, rice, maize or corn, and potatoes, among others. Native starches undergo modifications to adjust their responsiveness to pH and temperature fluctuations, which are essential variables in adsorption. Incorporating particular functional groups, such as amino, carbonyl, carboxyl, ester, etc., enhances the ability of starch to attract heavy metal ions or dyes. Additionally, these carbohydrates are biodegradable to a significant extent. Throughout the past 10–15 years, a variety of modified starches, including starch phosphate, starch xanthate, starch sulfate, starch carbamate, and carboxyl methyl starch, have been engaged as effective adsorbents for heavy metal treatment [31]. The raw materials, different forms of altered starch, their intended uses, and the applications that are currently available on the market are all depicted in Figure 18.5.

18.2.1.5 Xanthan Gum

Xanthan gum (XG) possesses an inherent anionic structure consisting of a recurring pentasaccharide unit comprised of d-glucose, d-mannose, and d-glucuronic acid in a molar ratio of 2:2:1. The XG backbone mostly consists of a linear β-(1–4)-d-glucose structure, which bears resemblance to the cellulose chain. However, the side backbone is mainly composed of d-glucuronic acid and terminal β-d-mannose connected by β-(1-4) in a structure that is attached to d-mannose by α-(1-2) (Figure 18.6) [32]. Xanthan gum, having a chemical formula of $C_{35}H_{49}O_{29}$ and a molecular weight of 933.75 g mol^{-1}, is one of the most extensively utilized microbial polysaccharides in commercial applications. Xanthomonas

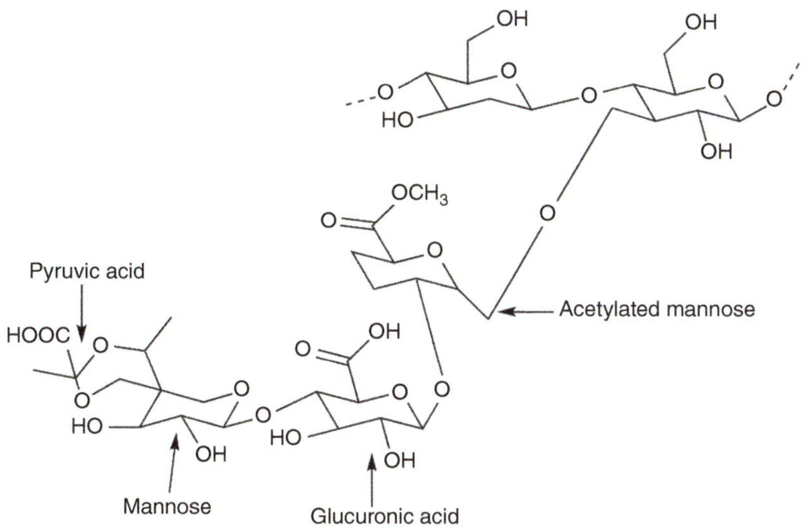

Figure 18.6 Xanthan gum structure.

campestris primarily produces an extracellular heteropolysaccharide by the aerobic fermentation of corn [33]. Gums are hydrophilic polysaccharides that have limited solubility in organic solvents. Furthermore, they possess the qualities of being nontoxic, lacking odor, and flavor. Due to their intermolecular interactions and high molecular weights, they are unable to form a transparent solution with ease. Xanthan gum is considered a vital natural and industrial biopolymer due to its remarkable characteristics, including nontoxicity, cell development capability, and nonsensitizing features. The presence of several functional groups, such as hydroxyl and carboxyl, on its structure allows for effective adsorption of various water contaminants through a chelation mechanism. This biopolymer has the ability to combine with other polymers or biopolymers in order to create composite sorbents that have been modified for certain functions [34].

18.2.1.6 Keratin

Protein-derived biomaterials have garnered significant interest from researchers in the last few years considering their remarkable versatility in a wide range of products. Keratin is a good protein alternative since it is both biodegradable and biocompatible as well as an abundant indigestible protein present in hair, horns, wool, nails, and feathers (Figure 18.7). It consists of several amino acids, each with a distinct side chain with a distinctive chemical structure, bonding capacity, charge, and reactivity [35]. By deploying various chemical treatments, such chains can serve a variety of purposes in the removal of pollutants from wastewater. These amino acid residues are connected to one another through

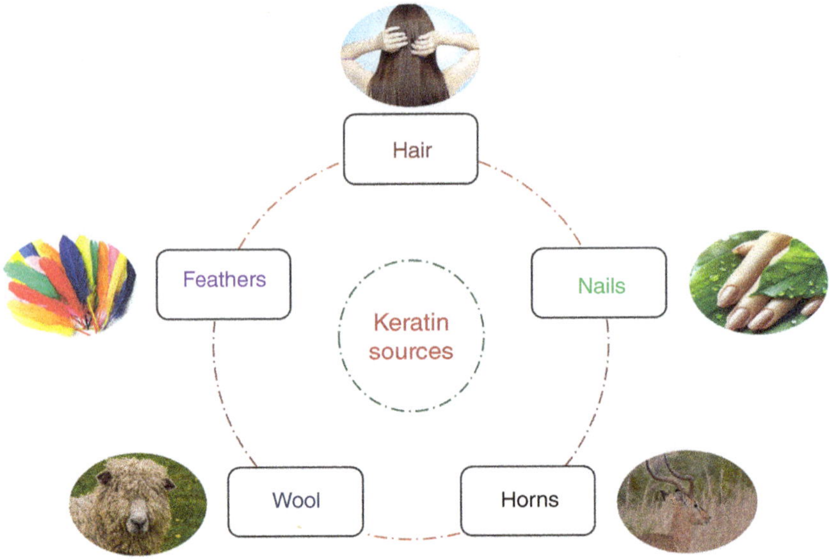

Figure 18.7 Keratin sources.

peptide bonds, resulting in a structure that resembles a ladder. This structure exhibits a dense packing of α-helix and β-sheet structures [36]. Strong disulfide (—S—S—) interactions between the cysteine residues in keratin allow thiol —SH groups to cross-link the entire matrix. Chemical treatment can disrupt the cross-linked disulfide bonds, thereby revealing unbound functional groups that have a substantial capacity for adsorbing trace elements [37].

The property of biodegradability is significant when considering biopolymers, as it pertains specifically to their ability to break down naturally over time. It should be noted that this property is not directly related to the specific raw materials utilized in the production process. A biodegradable polymer can be derived from fossil fuels. Biodegradation is an inherent biological chemical reaction that occurs during composting, resulting in the release of carbon dioxide, water, inorganic particles, and biomass. For a substance to be classified as biodegradable, it must undergo degradation at a comparable rate to that of certified compostable materials, while also ensuring the absence of any potentially harmful residue [12]. If a plastic material could undergo decomposition, it is important to assess whether it adheres to the predetermined standards. In cases where the plastic is biodegradable but fails to meet the established criteria, it should be categorized as noncompostable. Biopolymers are categorized into three distinct categories according to their synthesis technique and origin and can be obtained from many natural sources including carbohydrates and proteins. These natural substances, also known as monomers, such as lactic acid, serve as the fundamental building blocks for the synthesis of biopolymers. Moreover, it is worth noting that various other biopolymers, including polyhydroxyalkanoate (PHA), can be derived from microorganisms [13]. Commercially available biopolymers can be classified into several categories. These include polyhydroxyalkanoates (PHAs/PHBs), polylactic acid (PLA), polyamides, polyols, bio-PET, butyl rubber, and cellulose acetate. These latter are commonly included in a wide range of applications.

Unlike natural biopolymers, which are synthesized within living organisms, synthetic biopolymers are constructed through chemical synthesis reactions such as follows:

-*Poly(lactic acid) (PLA)* is widely recognized as one of the most promising polymers in use today and has earned the distinction of being referred to as the "polymer of the 21st century" [38]. The synthesized compound in question is unique in that it possesses the combined characteristics of biocompatibility, biodegradability, and being derived from biological sources. PLA represents a category of aliphatic biobased polyesters generated from lactic acid, or 2-hydroxypropionic acid. This compound is predominantly derived from various animal and plant sources, including cellulose, starch, corn, fish, and kitchen waste [39]. PLA demonstrates favorable durability, biocompatibility, biodegradability, and well-developed compostability. These characteristics make it a propitious substitute for conventional petroleum-based polymers employed in the production of membranes. Additionally, PLA offers the advantage of being recyclable through composting methods [40].

-*Polyethylene furanoate (PEF)* is a polyester derived entirely from renewable sources, making it a 100% biobased material. It is synthesized through the polymerization of 2,5-furandicarboxylic acid (FDCA) and monoethylene glycol (MEG). FDCA, a monomer obtained through bio fermentation from sugars, can serve as a substitute for oil-based terephthalic acid (TA) in the production of polyethylene terephthalate (PET), resulting in the creation of PEF [41]. The production of PEF can be accomplished by exploitation of either first-generation bio-feedstock, which includes sugars derived from corn or wheat, or second-generation feedstock, which includes waste, wood, wheat straw, corn stover, or bagasse, provided that these feedstocks remain economically feasible. PEF exhibits notable advantages in terms of its thermal and barrier properties. Notably, PEF demonstrates a higher glass transition temperature, indicating enhanced thermal stability compared to other materials. Additionally, PEF possesses a lower melting point, which can be advantageous in various applications. In terms of barrier properties, PEF exhibits a higher resistance to the permeation of oxygen, carbon dioxide, and water, making it an attractive choice for packaging and other protective applications [42].

-Polyhydroxyalkanoates (PHAs) meet all the requirements outlined in the definition of a natural polymer or a biopolymer. PHA biopolymers, a class of bio polyesters, occur naturally and are manufactured by bacteria using renewable carbon sources. PHAs are biodegradable as they may be converted by natural processes into carbon dioxide (CO_2) and water, along with around 10% of organic fertilizer or humus. This biodegradation process is like what occurs with cellulose or cotton. PHAs are biocompatible, causing no harm to any living organisms. As a result, they can be safely employed in medical applications within the body, such as scaffolds, stents, or meshes. The PHAs will ultimately undergo complete absorption by the body, leaving no residual evidence [43]. PHAs are biopolymers produced by various microbes by sugar fermentation under nutrient-deprived circumstances. These microbial polyesters possess thermoplastic characteristics comparable to conventional plastics and can be manufactured by a diverse spectrum of microorganisms. Among the class of biopolymers known as PHAs, several forms—including poly(3-hydroxybutyrate) and polyhydroxybutyrate-co-valerate—are already commercially viable. PHAs are entirely derived from natural sources and can be broken down by natural processes. The characteristics of these polyesters can be adjusted extensively by altering their chemical compositions. The synthesis of PHA by microorganisms involves multiple stages, during which the characteristics of the product can be modified by manipulating the substrate and the circumstances of fermentation [44]. Table 18.1 gives a concise overview of biobased polymers.

18.2.2 Natural Fillers

The term “filler” refers to a solid particulate substance that exhibits irregularities in both its shape and size. Fillers serve a dual purpose in composites: reducing costs and improving performance. The utilization of fillers in combination with a polymer matrix serves to enhance specific properties in targeted applications. Conversely, fillers can be integrated with other reinforcing materials and the polymer matrix to enhance specific desirable characteristics that cannot be sufficiently attained. Common names for fillers include functional fillers and extending fillers that can serve primarily to augment the overall volume of a substance, while functional fillers are employed to enhance the material’s specific properties [54]. However, it has been observed that certain extending fillers, when subjected to a reduction in particle size or surface treatment, exhibit the ability to function as functional fillers as well. Fillers are commonly employed in various applications to enhance dimensional stability, mechanical properties, thermal stability, and other desirable characteristics. Nevertheless, the alterations in the characteristics of the composites are influenced by multiple parameters, including the composition, surface, size, concentration, shape, and dispersion of the filler within the matrix.

Table 18.1 A brief description and classification of biobased polymers.

Class	Chemical group	Example	References
Polymers derived from biomass	Polysaccharides Proteins and lipids	• Pectin (chitosan/ chitin, gums); • Carbohydrates (potatoes, wheat, maize); • Cellulosic and ligno-cellulosic materials (straws, wood); • Whey, casein, collagen/gelatin; • Zein, soya, gluten.	[45–47]
Microbial fermentation-derived polymers	Polyhydroxy-alkanoates (PHAs)	• Poly (hydroxybutyrate) (PHB) • Poly (hydroxybutyrate-cohydroxyvalerate) (PHBV)	[48–50]
Polymers are chemically produced from agro-resource monomers	Poly (hydroxyacid)	• Poly(lactic acid) (PLA) • Polyglycolic acid (PGA)	[51–53]

Fillers can be categorized according to their chemical composition, form, and size (Figure 18.8). It should be emphasized that these fillers are usually inflexible and do not mix well with the polymer matrix. Their impact on chemical composition, form, and size can differ depending on the organic and inorganic components employed as fillers.

18.2.2.1 Inorganic Fillers

Industrial applications have historically relied heavily on the use of inorganic fillers. When incorporated into composite laminates, the inclusion of inorganic fillers can contribute to a significant proportion, ranging from 40% to 65% by weight. There exists a multitude of inorganic fillers that have the potential to be utilized in conjunction with polymer matrices or composites. Research has been undertaken to investigate the reinforcement mechanisms of semi-crystalline polymers via the incorporation of inorganic fillers. The following are some of the often-employed inorganic fillers:

- Calcium carbonate ($CaCO_3$): Calcium carbonate is a highly prevalent inert mineral filler that is obtained from limestone or marble. In the past, $CaCO_3$ was

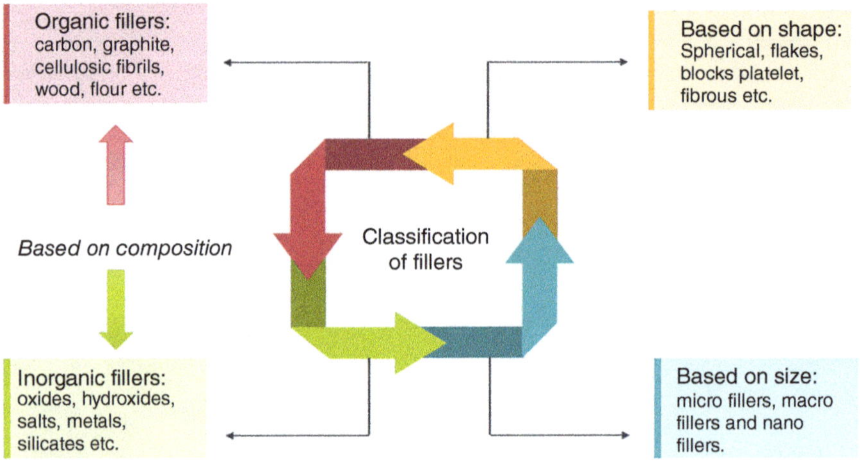

Figure 18.8 Fillers classification.

employed as a filler in composites to reduce costs. However, the subsequent usage of coupling agents in combination with $CaCO_3$ allowed for its utilization as a filler that enhances toughness [55].

- Silicon dioxide (SiO_2): It is a highly intricate and widely abundant natural chemical consisting of silicon and oxygen. It is obtained through the process of mining and purifying the mineral [56].
- Kaolin: Also known as hydrated aluminosilicate, it is a type of industrial filler that is frequently utilized in the plastic and rubber industries. It possesses a structure similar to that of a plate with a high aspect ratio. Therefore, obtaining improved properties depends critically on the size and dispersion of the particles. Its reinforcing function can provide the materials more strength and rigidity [57].
- Montmorillonite (MMT): It is a mineral filler belonging to the phyllosilicate family. It is known to be formed through the precipitation of microscopic crystals from aqueous solutions. They possess a soft texture and are commonly employed in several applications, including their use as absorbents for heavy metals, fillers for both organic coatings and composites [58].

18.2.2.2 Organic Fillers

Despite not being as widely used as mineral or inorganic fillers, natural or organic fillers are attracting increasing interest due to their distinct advantages over conventional fillers [59]. Their notable benefits encompass cost-effectiveness, reduced density, high resilience to abrasion, minimal energy use, biodegradability, and renewability. Material scientists have recognized that the utilization of natural fillers as reinforcing agents in polymer composites is both unavoidable and a very efficient method for producing environmentally friendly composites without compromising their performance [60]. The growing fascination with organic fillers might be ascribed to their renewable nature, ease of separation, ability to sequester carbon dioxide, and nonabrasiveness to equipment [61]. Below, we will discuss some of the organic fillers that are commonly used in various applications:

- Wood flour (WF) is obtained from a variety of wood by-products, such as wood shavings, sawdust, chips, and other leftovers generated by wood processing companies. WF is of importance due to its ability to maintain its shape and improve the stiffness of materials. A significant limitation of WF is its inadequate adherence with the polymer matrix and lower breakdown temperatures [62].
- Lignocellulosic fillers encompass wood, natural fibers, and residues from many plant species. These fillers are readily accessible, cost-effective, and sustainable. Lignocellulosic fibers encompass a diverse range of materials, including cellulose, cotton, flax, sisal, kenaf, jute, hemp, bamboo, banana, pineapple, coir, ramie, and starch, among others. In addition, many agricultural by-products, such as powdered wasted tea leaves, powdered spent coffee beans, powdered eggshells, powdered banana peels, powdered tamarind nuts, and powdered turmeric, are also utilized as fillers [63].
- Carbon-based nanofillers, including single-walled carbon nanotubes, multi-walled carbon nanotubes, carbon nanofibers, graphene, and carbon black, are widely utilized in polymer matrix composites for their exceptional structural and functional characteristics, making them highly suitable for engineering uses [64].

A schematic depicting the production of biocomposite materials with organic/inorganic fillers is shown in Figure 18.9. Additionally, scanning electron microscopy (SEM) micrographs of some fillers are provided as examples.

18.2.3 Preparation Methods for Biocomposites

The specific difficulties and demands presented by the aqueous environment must be carefully considered when developing biocomposites for wastewater treatment applications. These materials need to possess exceptional mechanical, thermal, and chemical features and be specifically engineered to endure the challenging conditions commonly found in wastewater treatment systems. Although biopolymers possess favorable biological and biodegradable characteristics, they exhibit deficiencies in certain mechanical properties, including limited chemical resistance, reduced processing capacity, and limited storage duration. The aforementioned issues can be effectively addressed through the utilization

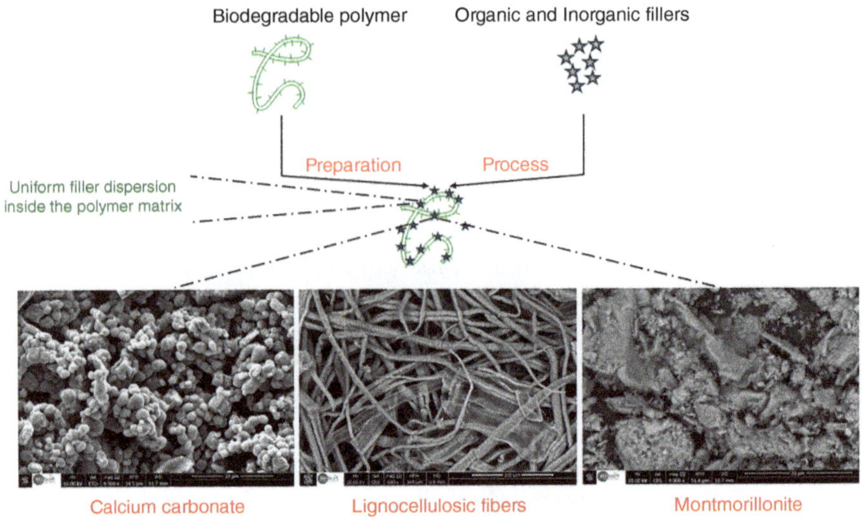

Figure 18.9 Biocomposite materials production with organic/inorganic fillers as well as some SEM micrographs of some fillers.

of biopolymer composites, wherein biopolymers are combined with fillers of varying sizes, whether they be of natural or synthetic origin. The formulation methods chosen for the production of polymeric composites must be straightforward and enable the modification of the surface properties of the mixture according to the particular pollutant being addressed, while simultaneously guaranteeing the chemical resilience of the composite. However, selecting the appropriate preparation procedure is crucial in developing polymer composites with acceptable characteristics [65]. Some of the most frequently employed techniques are depicted in Figure 18.10.

18.2.3.1 Casting

Casting is undeniably the most straightforward method used in the production of polymeric films, particularly membranes [66]. The process known as phase inversion involves the mechanical distribution of a forming solution in order to create polymeric films. Typically, this method produces thicker films in the micrometer range. These films depend on the porosity to attain the necessary characteristics for water treatment, such as ultrafiltration, catalysis, or adsorption [67]. The incorporation of additives and active materials into the forming fluid can enhance the active characteristics, porosity, and surface area of the membrane. Alternatively, the establishment of a porous structure may be realized through the selection of a specific polymer and the regulation of conditions during the film fabrication process. The occurrence of phase inversion, defined by

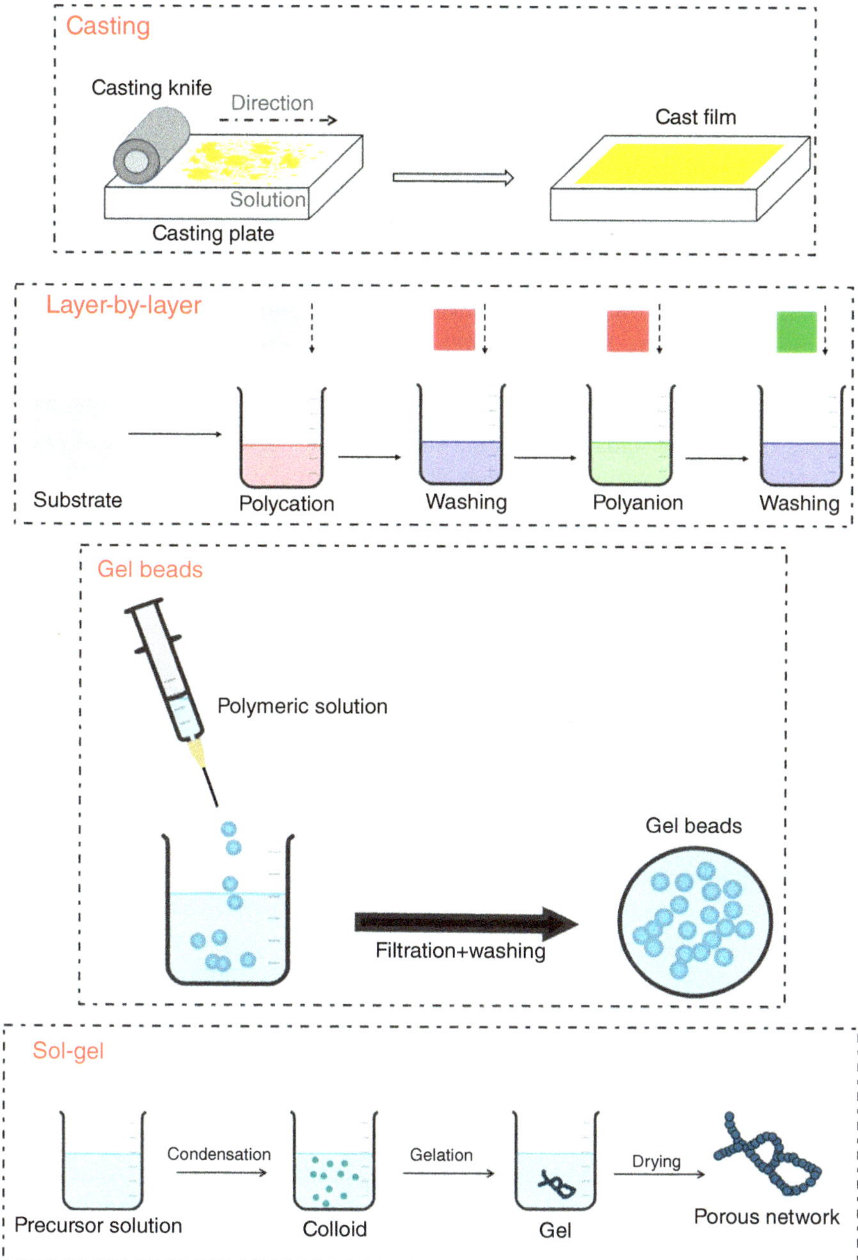

Figure 18.10 An illustration of the procedures to develop polymeric composites used to remove contaminants from wastewater.

the segregation of components within the developing solution, can be facilitated via two primary approaches: lowering the temperature of the solution or adding an antisolvent to encourage the precipitation of the polymeric phase. The films are subsequently washed to remove the additive or templating phase, resulting in a porous structure alongside a large surface area.

18.2.3.2 Layer-by-Layer

In general, the term "Layer-by-Layer" (LBL) refers to a group of processes that involve depositing thin layers of nanoscale materials onto a substrate in a sequential manner. This can be done using various methods such as spin-coating, dip-coating, or vapor deposition [68]. Within the realm of composites for water treatment, the process of self-assembly of polymeric layers, particularly polyelectrolytes, emerges as a highly effective and advantageous method for creating catalytic designs with a large surface area. This approach involves the sequential and alternating deposition of polyelectrolytes with opposite charges on a selected substrate. The result is the formation of thin films, which can range in thickness from a few nanometers (20–40 nm) to a few micrometers, based on the overall number of layers. Second-order electrostatic interactions hold the charged polymers together after they are deposited via adsorption. The composition, salinity, and pH of the solution affect the structure of the resultant film. These factors have the ability to alter the configuration of the polymer chains. As soon as the chains are mainly ionized, electric repulsion causes them to adopt a linear arrangement. Conversely, if the ionization rate is poor, the chains become tangled.

18.2.3.3 Gel Beads

Producing polymeric beads is a straightforward method for immobilizing and reusing active water-treatment substances. The beads typically have a size on the order of millimeters, which allows for their easy recovery from the treated water without the need for any intricate processes. The utilization of this technique is frequently observed in the immobilization of metallic nanoparticles for the purpose of microbiological treatment, as well as in the application of clay particles as carriers in the process of adsorption. Biopolymers are commonly utilized for the production of water-treatment beads because of their lack of toxicity, ability to break down naturally, capacity to allow substances to pass through, and affinity for water. During a standard technique, a highly concentrated solution of polymer is injected into a mixture of antisolvent, which includes metallic nanoparticles and/or clay particles, using a syringe. Upon introduction of the active component, the polymer undergoes rapid precipitation, resulting in the formation of spherical beads. This process effectively entraps the active component within the polymer structure [69].

18.2.3.4 Sol-Gel

The sol-gel method is a commonly employed technique for the synthesis of inorganic nanostructures with diverse sizes and properties. It is particularly adopted for the production of nanoporous inorganic films. The growth of crystalline chains of oxides is achieved by using metal precursors, typically in the form of organic compounds, through the process of acid or basic hydrolysis. While sol-gel films often demonstrate significant surface activity, they lack crucial characteristics necessary for their use as a recovered component in water treatment, such as mechanical flexibility and adhesion. Polymeric composites have been developed to tackle these challenges. One common approach is to incorporate sol-gel nanoparticles into thin films using the aforementioned techniques. However, it is worth noting that there is often a reduction inefficiency associated with this process [70].

18.3 Application of Biocomposites in Wastewater Treatment

The World Health Organization (WHO) has anticipated that almost 50% of the world's population will reside in areas facing water scarcity due to factors such as population expansion, climate and geographical changes, and poor water management practices. Water toxicity causes the spread of numerous dangerous and deadly water-borne illnesses, resulting in about 48,500 diarrheal fatalities annually [71]. Wastewater sources can be categorized according to the location at which they are discharged. The four main categories encompass domestic sources (which include commercial institutions), industrial sources (notably the pharmaceutical and petrochemical sectors), infiltration, and stormwater management. The 2017 edition of the United Nations World Water Development Report indicates that approximately 56% of freshwater utilized for urban, industrial, and agricultural activities is subsequently discharged as wastewater. Furthermore, it reveals that 80% of the worldwide sewage is released into the environment without undergoing adequate treatment. The quality of water is of utmost importance for the well-being of humans, the success of the economy, and the sustainability of environmental processes in the long run. The availability of water depends not only on the quantity of water sources but also on the water's quality.

To facilitate the implementation and utilization of advanced pollution prevention and wastewater treatment innovations, it is imperative to establish a comprehensive legislative and regulatory framework. This framework will provide the necessary support and guidance for management and investment decisions in this field. The advancement of cutting-edge wastewater treatment technologies enhances the economy in the long run through the promotion of

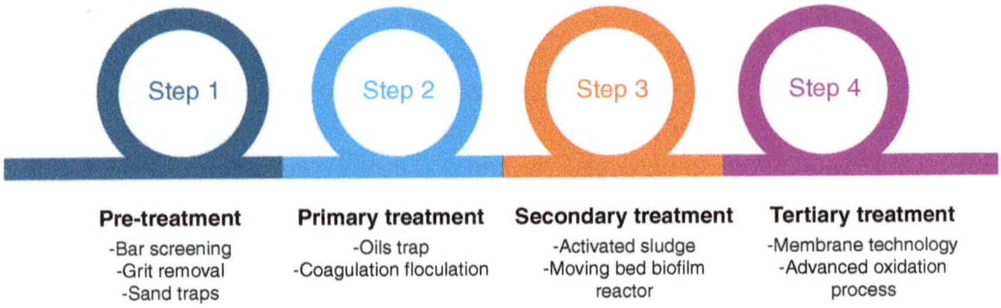

Figure 18.11 Schematic representation of the wastewater treatment process.

water reuse, reduction of water resource depletion, and generation of energy and other secondary products [72]. An adequate policy framework and cost-effective technology are necessary for the efficient management of wastewater treatment systems and the prevention of water contamination. The selection of an acceptable and cost-effective technology is contingent upon the volume of wastewater, the character of pollution, and the intended application of clean water. In general, the typical procedure for treating wastewater consists of three basic steps: primary, secondary, and tertiary treatments. These treatments involve the use of physical, biological, and chemical approaches, as shown in Figure 18.11. However, the existing monitoring systems, along with the significant operational costs and energy consumption associated with conventional wastewater treatment processes, have prompted numerous researchers to explore alternative methods and materials for effectively decontaminating wastewater while minimizing environmental impact and energy consumption.

The essential criteria for water purification are suitable materials with desirable characteristics, including a significant adsorption capacity, the potential for reusability, a low cost, and the capability to modify its porosity. By harnessing the capacity to enhance various attributes such as mechanical strength, surface area, porosity, hydrophobicity, and dispersibility, biobased materials exhibit promising potential as a viable option for water purification. These materials are derived from a wide range of polymeric substances. In recent years, water sources have experienced the accumulation of various pollutants such as dyes, pharmaceutical wastes, heavy metals, and other substances (Figure 18.12) that is connected to the rapid growth of industrialization [73]. Heavy metal ions belong to the category of inorganic substances, while colors, hydrocarbons, insecticides, and pharmaceuticals are classified as organic compounds. Organic compounds, which are essential components in numerous industries, are produced through a wide range of processes. The entry of contaminants into sewage systems and subsequent pollution thereof is a well-documented phenomenon. Organic compounds, including pesticides, have been found to exhibit significant toxic effects on various organisms. Due to the adverse effects caused by these substances, it is imperative to remove them from water sources [74].

Organic pollutants

- Dyes (malachite green, basic yellow, methylene blue, methyl violet, Congo red, etc.)
- Drugs, pesticides, and insecticides
- Hydrocarbon compounds (phenol, benzene, toluene, aromatic compounds, etc.)

Inorganic pollutants

- Sulphates, nitrates, phosphates
- Heavy metals (Pb, Cd, Cr, As, etc.)

Figure 18.12 Different pollutants in water.

Recent years have seen a notable increase in the attention given to biomaterials-based adsorbents, mostly because of their high efficacy and ecologically benign properties in the domain of wastewater treatment. There is a significant interest in the advancement of adsorbents that are both more efficient and cost-effective, particularly those derived from natural materials. Agricultural by-products such as rice and wheat husks, peanut hulls, sugarcane bagasse, and cellulosic biomass have proven to be efficient adsorbents. Nevertheless, biosorbents must undergo several modifications as their native state does not meet the required performance standards. The adsorption capacity of these chemicals can be enhanced by forming composites with additional natural or synthetic components. It has been reported in the literature that composite adsorbents demonstrate significantly greater adsorption activity when compared to individual adsorbents [75]. There has been a significant focus recently on the development of biobased composite materials to meet the specific needs of various industries, including biobased packaging, biomedical, pharmaceutical, textiles, paper, and more. This research is driven by the increasing global concerns regarding the environmental impact of our reliance on nonrenewable petroleum-based resources, which are not sustainable in the long term [76]. The production, application, and disposal of conventional synthetically generated polymers or composite constructions are being investigated more closely due to the growing ecological awareness. The green biocomposites are composed of biopolymers that are both biocompatible and biodegradable. These biopolymers are derived from renewable natural sources through the action of microorganisms. Furthermore, these biocomposites are known for their biodegradability, biocompatibility, and nontoxicity. Thus, it is highly desired to manufacture green composites employing one or more distinct biopolymers. The need for sustainability is altering the dynamics of the materials field and presenting fresh possibilities.

Several factors affect the selection of a polymer composite for treating a particular pollutant. These factors include the kind and concentration of the contaminant in the wastewater, the capacity of the composite to remove the pollutant, and the efficiency-to-cost ratio. Additionally, the composite's capacity for regeneration and its ability to be exploited multiple times without altering or compromising its physical–chemical characteristics is crucial to consider.

For such reasons, biocomposites possess the necessary attributes to be effectively utilized as a means of water decontamination, particularly in environments with high pressure. As a result of their compact nature, exceptional mechanical characteristics, and ability to regenerate, in addition to their substantial surface area and capacity for adsorption, these materials possess notable attributes. The following section highlights significant and recent progress of biocomposites for wastewater remediation, with an emphasize on heavy metals, dyes, and pharmaceuticals removal.

18.3.1 Removal of Dyes

The excessive discharge of industrial wastewaters into water bodies has a detrimental impact on water resources. Water-soluble dyes are recognized as one of the main types of pollutants. Dye is a term that refers to organic substances that are highly stained or vividly colored. The textile industry is the primary consumer of dyestuffs, which accounts for around 70% of their usage [77]. The textile industry represents a major contributor to global pollution levels, which consumes around 77.5×10^9 tons of freshwater on an annual basis [78]. They are chromatic chemical compounds, widely employed in the textile area for the purpose of coloring fibers. The elimination of such pollutants from textile wastewater that has been discharged is a highly challenging task due to their strong capacity to dissolve in water, resistance to degradation by light, and vibrant colors even at low levels of concentration. Furthermore, it should be noted that dyes possess a high level of toxicity, posing a significant risk to both microbial populations and human beings. In addition to their harmful effects, dyes also exhibit resistance to aerobic digestion and lack the ability to undergo biological degradation [79]. Given their broad versatility, high adsorption capacities, and ease of operation, biocomposites have been extensively researched in the scientific community as a potential material for dye removal.

Dyes are commonly used as model molecules to evaluate the adsorption and pollutant removal effectiveness of various materials. Consequently, numerous studies have been reported on the utilization of biocomposites for dye removal. Cui et al. developed a method for creating a Cu_2O/CuO composite within cellulose-based carbon nanosheets utilizing a freeze-drying-calcination technique. This method was specifically designed for the photocatalytic elimination of methylene blue [80]. Photocatalytic tests show that the adsorbent efficiently degrades methylene blue, as predicted by the first-order kinetics equation. The improved reaction conditions resulted in an ideal degradation rate of 99.09% in 60 minutes. As part of their study, Mokhtar et al. have chosen to explore the synthesis of biopolymer-sodium silicate mineral biocomposites, which presents a unique and intriguing approach. The focus of this study is on the development of

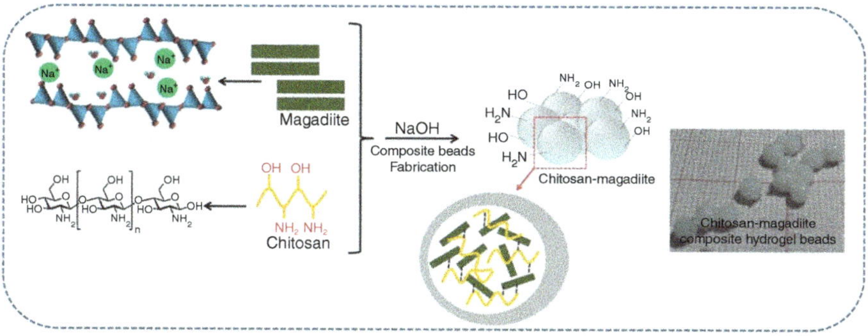

Figure 18.13 Schematic representation of preparation procedure of chitosan-magadiite composite hydrogel beads.

chitosan-magadiite composite beads as shown in Figure 18.13 for the purpose of enhancing the removal efficiency of Congo red (CR) and methylene blue (MB) dyes [81]. The biopolymer and silicate mineral have synergistically combined, resulting in an increased adsorption capacity for both dyes, which were found equal to 135.77 (CR) and 45.25 $mg\,g^{-1}$ (MB). The adsorption investigation indicated that the adsorption of MB dye followed the pseudo-second-order kinetic and Langmuir isotherm models, while the removal mechanism suggests that both hydrogen bonding and electrostatic forces play a role in the process. The process of enhancing water-soluble biopolymers with clay presents a compelling approach for optimizing their functionality as multifunctional adsorbents for contaminants in aqueous environments.

Kausar et al. have enhanced cellulose by including clay, resulting in an environmentally acceptable and efficient biocomposite for Drimarine Yellow HF-3GL removal from wastewater [82]. It was reported that the composites have a coarse and permeable surface texture, which becomes smooth once dye is absorbed (Figure 18.14). The cellulose/clay biocomposites exhibited high efficacy in the treatment of dyes, achieving a removal rate in the range of 88.64–89.95% for Drimarine Yellow HF-3GL dye under optimal conditions (pH = 2, 60 minutes, $T = 30\,°C$). In the same vein, a clay/cellulose biocomposite was developed by combining chitosan extracted from crab shells with clay to remove an anionic dye (Congo red) from synthetic wastewater [83]. The polymer's effectiveness is boosted by grafting it onto the clay surface, as evidenced by the approximately 92% dye removal accomplished under optimal conditions (20 mg dosage, pH = 2, 30-minute reaction period, and 20 $mg\,L^{-1}$ dye concentration).

Chkirida et al. formulated a novel biocomposite aimed at the degradation of cationic dyes, which comprises an alginate biopolymer reinforced with bentonite clay and is surface impregnated with TiO_2 [84]. The distinctive aspect of this study lies in the development of an integrated adsorbent/photocatalyst, as well as

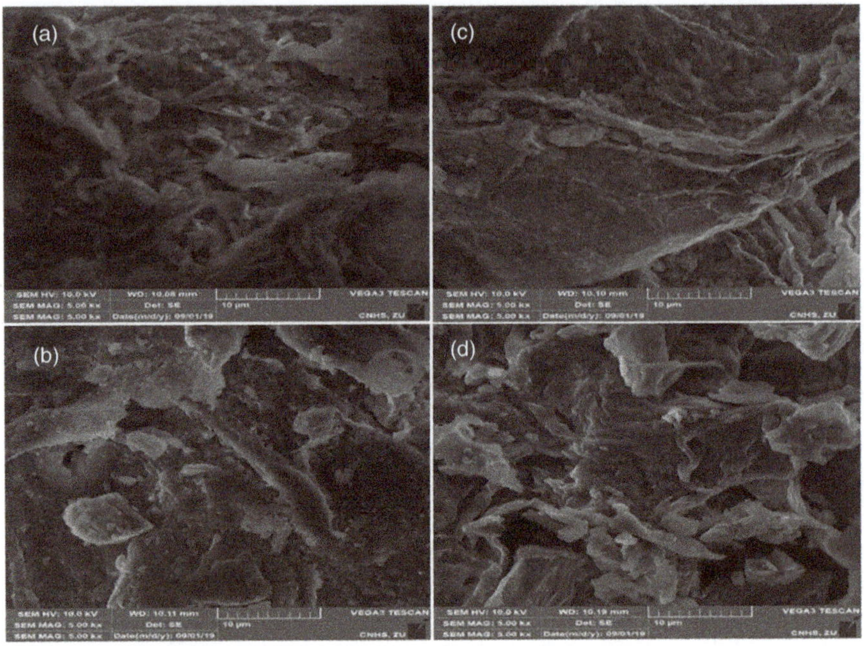

Figure 18.14 SEM analysis of unloaded cellulose/clay composite (a, b) and dye loaded composite (c, d).

the novel approach of fixing TiO_2 onto the surface of bentonite/alginate using a nonthermal impregnation technique to obtain a well-designed bead in $CaCl_2$ solution (Figure 18.15). This approach enables direct interaction between the photocatalyst and the irradiation source, therefore preventing any obstruction of the irradiation light's propagation onto the photocatalyst. The remarkable synergistic adsorption-photocatalytic efficacy of this bionanocomposite resulted in a discoloration rate of around 98% and an outstanding COD abatement of 93%. The analysis verified the degradation of dyes' chromophore groups, not just their adsorption.

In another intriguing approach, M. Thamer et al. aimed to synthesize a novel biocomposite hydrogel by incorporating activated carbon derived from pomegranate peels (PPAC) into tragacanth gum (TG) [85]. The hydrogel formed from the combination of PPAC and TG exhibited a porous architecture, with a surface charge that became negative when the pH exceeded 4.9. Furthermore, this hydrogel demonstrated excellent stability when exposed to water. The adsorption efficiency of the hydrogel was evaluated in terms of its ability to remove crystal violet (CV) dye. The equilibrium data was analyzed using the Sips isotherm model, which was found to be the most suitable model for describing the

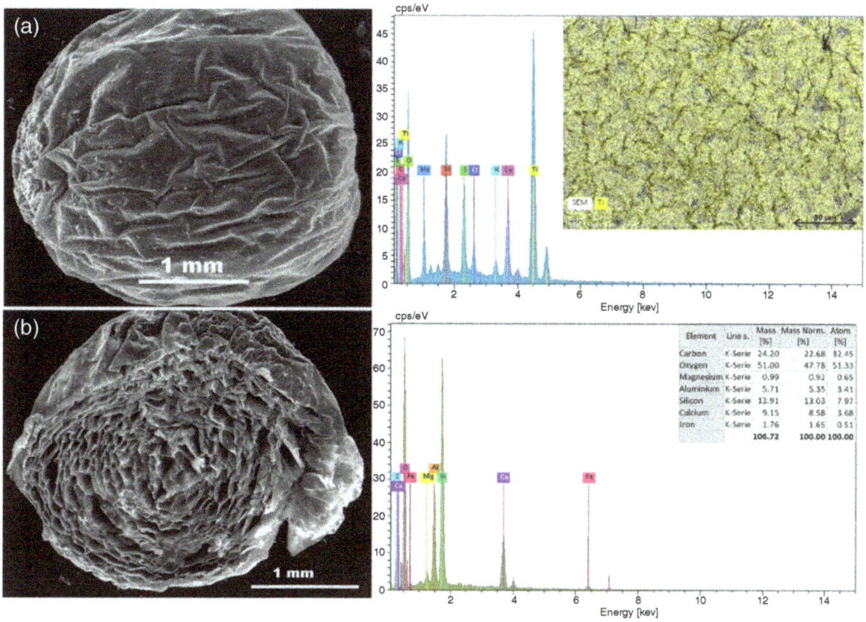

Figure 18.15 (a) SEM-EDX micrographs illustrating the surface bead of TiO_2-Bnt-Alg along with its Titanium mapping. (b) SEM-EDX micrographs depicting the cross section of the biocomposite bead.

adsorption process. The maximum adsorption capacities at temperatures of 25, 30, and 35 °C were found to be 455.61, 470.86, and 477.37 mg g^{-1}, respectively. Kinetic investigations confirmed the pseudo-second-order model, while thermodynamic study provided evidence that the adsorption process was both spontaneous and endothermic. Furthermore, the hydrogel exhibited remarkable reusability, retaining its effectiveness even after five applications. Ultimately, PPAC/TG demonstrated significant promise as a sustainable adsorbent for eliminating cationic dyes in water treatment solutions.

Table 18.2 summarizes some biocomposite materials for the removal of dyes from water-based solutions.

18.3.2 Removal of Heavy Metals

The presence of heavy metals in the environment, even at minimal concentrations, is recognized as a considerable risk to various forms of life, including fauna, flora, and humans. This risk is largely due to the tendency of these metals to bioaccumulate within biological systems [94]. Although sources of heavy metals vary, the chemical industrial sectors are mostly responsible for their

Table 18.2 Different biocomposites for dyes removal.

Biocomposites	Dyes	Qmax ($mg\,g^{-1}$)	Reference
AC–bentonite–alginate beads	Methylene blue	756.97	[86]
SA/poly(N-vinyl-2-pyrrolidone) beads	Reactive red 120	116.82	[87]
Cellulose nanoparticles/graphene oxide aerogels	Methylene blue	111.2	[88]
Alginate–montmorillonite composite beads	Methylene blue	181.8	[89]
Carboxymethyl cellulose/graphene oxide aerogel	Rhodamine B	161.29	[90]
XG/acrylic acid/acrylamide superporous hydrogel	Methyl violet	287.54	[91]
CS/TPP bead	BB7 BRO	232 259	[92]
XG/TMC hydrogel	Crystal violet	555.6	[93]

pollution and output. Conventional methods of removing heavy metals consider biological, physical, and chemical approaches. Coagulation/flocculation and costly membrane filtering procedures have proven to be effective in heavy metals remediation. However, they contribute to the creation of sizable particles and dense sludge. Alternative approaches like electrochemical treatment and oxidations are efficient and convenient for removing some metals. However, these methods require a significant amount of energy. Furthermore, scientific evidence has demonstrated that biological treatments are ineffective [95]. The use of biocomposite created a new, affordable, and environmentally friendly path for handling heavy metal contamination. Their extensive specific surface area and the prevalence of surface functional groups render them highly suitable for metal binding through adsorption, while ensuring the absence of any detrimental by-products.

By grafting (1,4,7,10-tetraazacyclododecane-N,N′,N,N′-tetraacetic acid) onto a recycled shrimp shell and sludge platform, Xu et al. explored a simple approach to produce inexpensive and very effective adsorbents (DOTA@Sludge@Chitosan) (Figure 18.16). The mechanical and adsorption characteristics were established and examined for the elimination of Cr^{6+} and Pb^{2+} ions from water [96].

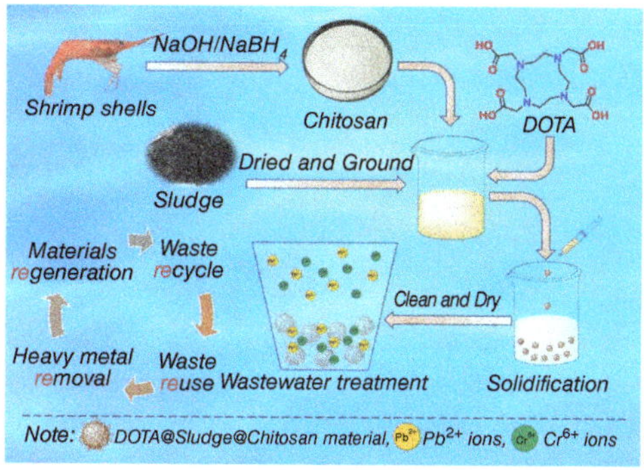

Figure 18.16 Development of a DOTA@Sludge@Chitosan composite for chromium and lead removal.

The DOTA@Sludge@Chitosan composite has excellent adsorption capacity for Pb^{2+} and Cr^{6+} because of its high pore volume ($0.069\,cm^3\,g^{-1}$), surface area ($52.009\,m^2\,g^{-1}$), and both hydroxyl and amino moieties. At 20 °C, the composite had an adsorption capacity of $329.4\,mg\,g^{-1}$ for Pb^{2+} and up to $273.3\,mg\,g^{-1}$ for Cr^{6+}. Adsorbent regeneration was investigated for six adsorption–desorption cycles. Due to its long-term stability, the DOTA@Sludge@Chitosan adsorbent preserved good adsorption capacity, resulting in an attractive heavy metals adsorbent for industrial effluent.

Zhang and colleagues synthesized in situ a poly(acrylic acid) grafted chitosan and biochar composite (PAA/CTS/BC) with exceptional efficiency and selectivity in capturing heavy metals. The biocomposite had high maximum adsorption capacities of Cu ($111.11\,mg\,g^{-1}$), Zn ($114.94\,mg\,g^{-1}$), Ni ($99.01\,mg\,g^{-1}$), Pb ($476.19\,mg\,g^{-1}$), Cd ($370.37\,mg\,g^{-1}$), and Mn ($138.89\,mg\,g^{-1}$), respectively. In addition, heavy metal removal approached 90% within 1 hour, showing outstanding adsorption kinetics. The pH dependence of HMs adsorption on PAA/CTS/BC was found to be rather mild within the investigated pH range of 2–7. This implies that the adsorption of HMs on PAA/CTS/BC may be attributed on only a partial basis to ion exchange facilitated by electrostatic attraction. Finally, Cr^{3+} adsorbed on PAA/CTS/BC exhibited nearly total resistance to desorption using a $0.01\,mol\,L^{-1}$ Na_2EDTA solution, whereas the other heavy metals (HMs) could be substantially recovered (>80%). Thoroughly conscious of the precarious situation, Isawi has synthesized polyvinyl alcohol/sodium alginate (PVA/SA) beads by combining PVA with SA, together with glutaraldehyde as a cross-linking agent. The zeolite was then integrated, leading to the formation of Zeo/PVA/SA beads as illustrated in SEM micrographs (Figure 18.17) designed for the elimination of certain heavy metals from wastewater [97]. The experimental data on adsorption kinetics and isotherms are most accurately described by the pseudo-first order and Langmuir models. At a pH of 6.0, the biocomposite beads exhibited maximum removal efficiencies of 99.5%, 99.2%, 98.8%, 97.2%, 95.6%, 93.1%, 92.4%, and 74.5% for

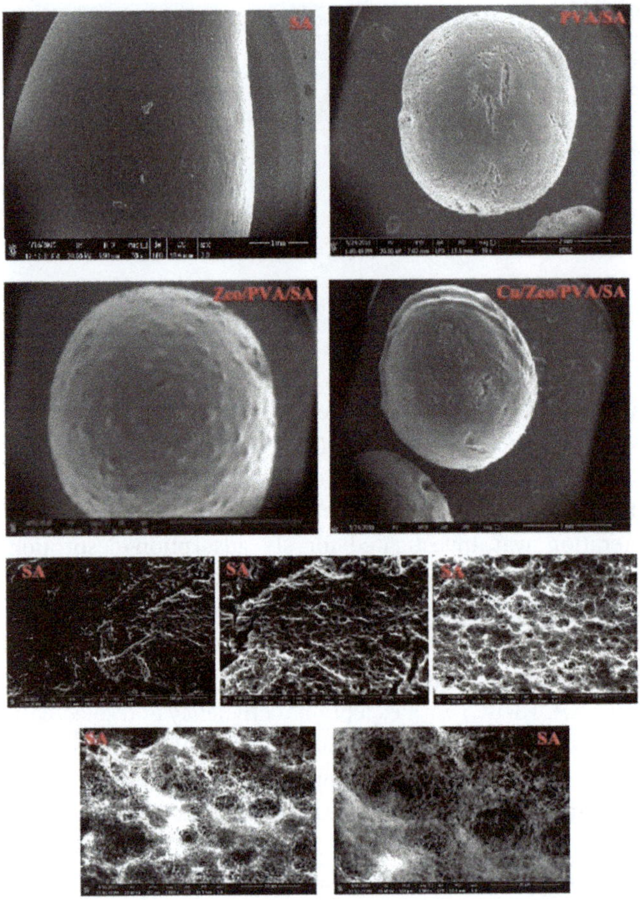

Figure 18.17 SEM images at varying magnifications depicting the surfaces and cross sections of sodium alginate (SA), polyvinyl alcohol/sodium alginate (PVA/SA), and zeolite/polyvinyl alcohol/sodium alginate (Zeo/PVA/SA).

Pb(II), Cd(II), Sr(II), Cu(II), Zn(II), Ni(II), Mn(II), and Li(II) correspondingly. The conducted experiments on reusability indicate that the Zeo/PVA/SA NC modified beads exhibited a significant decrease in sorption capacity following 10 successive cycles. The results obtained suggest that the synthesized Zeo/PVA/SA NC beads are regarded as safe, easy to use, and environmentally friendly for the purpose of wastewater purification.

Always with chitosan, Wang et al. provided a detailed account of the synthesis process employed to create a composite material known as magnetic xanthate-modified polyvinyl alcohol and chitosan (XMPC) [98]. The main aim of this study was to establish a highly efficient technique for the extraction and recovery of heavy metal ions from water-based solutions. The process of synthesizing XMPC hydrogel beads entailed the integration of polyvinyl alcohol, chitosan, and magnetic $Fe_3O_4@SiO_2$ particles. The resultant biocomposite was further analyzed using characterization methods to ascertain its specific features. Furthermore, the adsorption efficiency of the XMPC in eliminating heavy metal ions was examined. Following approximately 120 minutes contact time at 303 K, the XMPC achieved adsorption equilibrium. The observed removal rates for Pd(II), Cu(II), and Cd(II) ions were quantified as 67, 100, and 307 $mg\,g^{-1}$, respectively. The XMPC demonstrated a spontaneous and endothermic process for the adsorption of metal ions, which was precisely characterized using the quasi-secondary kinetic model and the Langmuir adsorption isotherm. Furthermore, the magnetic hydrogel particles demonstrated a metal ion removal rate that surpassed 50% following five cycles of application.

Modified 3D porous graphene (PG) materials have been experimentally shown to be a very suitable choice for water remediation. In this sense, in their study, Zhou et al. employed a hydrothermal technique in an aqueous solution to produce a 3D composite consisting of graphene, lignin, and sodium alginate (3D PG/L/SA). The primary objective of their research was to investigate the potential of this material for scavenging Cd(II) and Pb(II) ions [99]. Figure 18.18 illustrates a scheme depicting the manufacture process of the composite material. Through the utilization of diverse characterization techniques, it was demonstrated that the porous structure exhibited a high level of integration within the three-dimensional macroscopic structure composed of graphene. The obtained composite displayed clearly improved adsorption capacity for

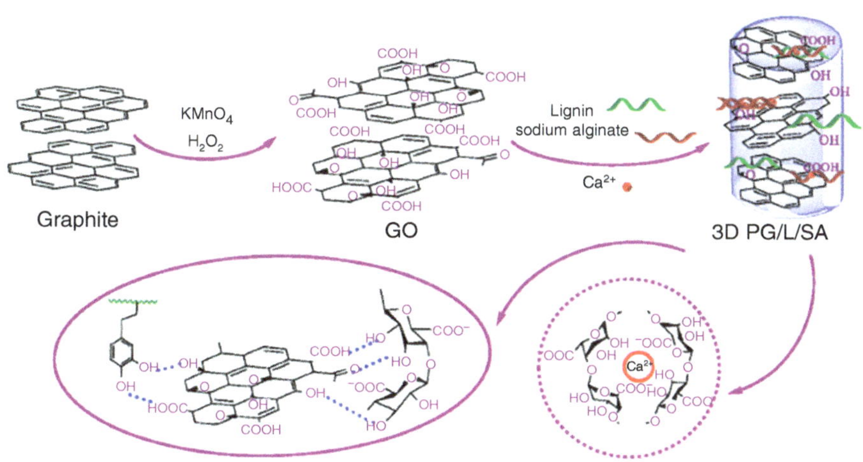

Figure 18.18 Illustration depicting the fabrication process and the underlying formation mechanism of 3D PG/L/SA.

Pb(II) and Cd(II) remediation owing to the biopolymer assistance and the inclusion of cross-linker. The graphene/lignin/sodium alginate composite exhibited substantially improved adsorption characteristics compared to graphene. The optimal adsorption capacity for Cd(II) and Pb(II) was determined to be 79.88 and 225.24 mg g^{-1}, respectively. The observed adsorption kinetics and isotherm processes of two metal ions demonstrated a significant alignment with the pseudo-second-order kinetic model and the Langmuir isotherm equation, respectively. This indicates that the chemisorption reactions serve as the rate-limiting step, and that the adsorption process is uniform, occurring in the form of a monolayer.

Apart from these, Salih et al. devised a method to create a reusable adsorbent by applying a layer of chitosan onto vermiculite. This coated vermiculite was then shaped into spherical particles using the commonly employed drop-wise technique [100]. This approach aimed to improve both abundance and accessibility of reactive moieties on the adsorbent for metal ions entrapment. The researchers subsequently conducted a thorough investigation into the selective adsorption mechanism of this adsorbent. The Fourier transform infrared (FTIR) spectra of the synthesized composite display characteristic bands commonly associated with both pristine chitosan and vermiculite, indicating the presence of beneficial interactions between these materials (Figure 18.19). By virtue of its porous morphology and enlarged surface area, the composite exhibited enhanced

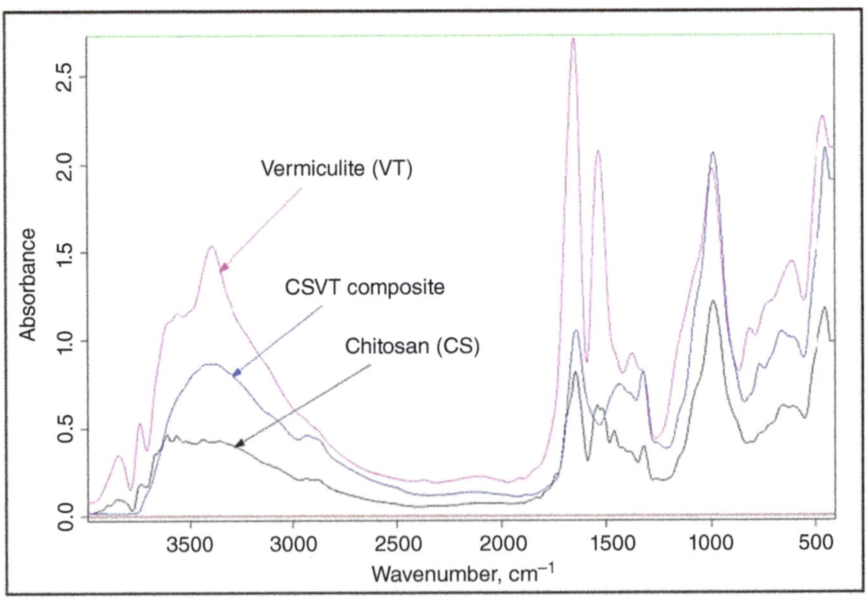

Figure 18.19 FTIR spectra of vermiculite, chitosan as well as the composite (CSVT).

Table 18.3 Biocomposites for heavy metals remediation.

Biocomposites	Target metal	Qmax ($mg\,g^{-1}$)	References
AM/AO/AEBI-CS	Cu(II)	190.70	[101]
CS@activated carbon composites	Cd(II)	84.750	[102]
CS/Bentonite	Cr(VI)	133.85	[103]
Alginate/PEI	Cu(II)	322.6	[104]
Chitosan- carbon nanotubes	Pb(II)	83.20	[105]
PANI@OA	Cr(VI)	249.09	[106]
Oystershell biocomposite	Pb(II) Cd(II)	78.98 97.54	[107]
CNC@Bt	Cr(VI)	1.85	[108]
Magnetic alginate-bentonite@chitosan	Cu^{2+}, Cd^{2+}, Co^{2+}, Ni^{2+}, Pb^{2+}, Zn^{2+}, Hg^{2+}	12–20	[109]

adsorption capacity in comparison to pure chitosan. The Langmuir adsorption isotherms performed in batch studies revealed that the maximal adsorption capacities of Cu(II) and Cd(II) were 116.22 and 147.64 $mg\,g^{-1}$, respectively. Moreover, the regenerated chitosan-vermiculite (CSVT) composite demonstrated a consistent capacity for the adsorption of Cu(II) and Cd(II) metal ions, exhibiting no significant reduction in performance after undergoing five successive cycles of adsorption and desorption. Following the completion of the fifth cycle, a notable reduction in the adsorption capacities for Cu(II) and Cd(II) is observed, decreasing from 107.78 and 133.22 $mg\,g^{-1}$, respectively, to 88.33 and 110.67 $mg\,g^{-1}$.

Table 18.3 presents a compilation of various biocomposites along with their respective applications in the extraction of heavy metals.

18.3.3 Removal of Pharmaceuticals

The presence of pharmaceuticals in aquatic environments has garnered attention as a significant concern over the last two decades, attributed to their notable persistence, limited biodegradability, and adverse effects on both aquatic organisms and the health of humans and animals [110]. These substances are classified into several classes, including antibiotics, anti-inflammatory medicines, blood-lipid regulators, and steroidal hormones [111]. Principal sources of these contaminants in water are the pharmaceutical industry, personal hygiene products, hospital waste, and the excretion of animals or humans through urine or feces. Pharmaceuticals have been found in a number of aquatic environmental matrices, including drinking water, surface water (rivers, lakes), ground water, and wastewater treatment facility effluents, in a number of different nations. This widespread occurrence of pharmaceutical compounds in aquatic environments poses a significant environmental concern [112]. It is necessary to establish protocols and technologies to fully eliminate pharmaceuticals from water before their release into the ecosystem. Earlier methods for removing pharmaceuticals from water have involved chemical, physical, and biological approaches. These methods are distinct from one another in terms of their efficiency, practicability, and cost, and they each have their own set of benefits and drawbacks [20]. Since many conventional choices fall short in providing sufficient performance, the development of affordable and effective biosorbents is absolutely vital. This has prompted the application of techniques like chemical modification in order to raise their efficacy. Meanwhile, specific materials were employed as adsorbents for decontaminating wastewater owing to their exceptional effectiveness, a significant obstacle is their lack of biodegradability, making natural adsorbents more desirable than synthetic ones. Biocomposites that are cost-effective, plentiful, highly efficient, environmentally benign, biocompatible, and capable of being reused are extensively exploited.

In this context, the research study carried by Li et al. examined the adsorption of tetracycline and cefotaxime using a magnetic $NiFe_2O_4$–COF-chitosan-terephthalaldehyde nanocomposite film (NCCT) that combines both chitosan and covalent organic framework (COF) components, as illustrated in Figure 18.20. The findings of their investigation revealed that the synergistic interaction

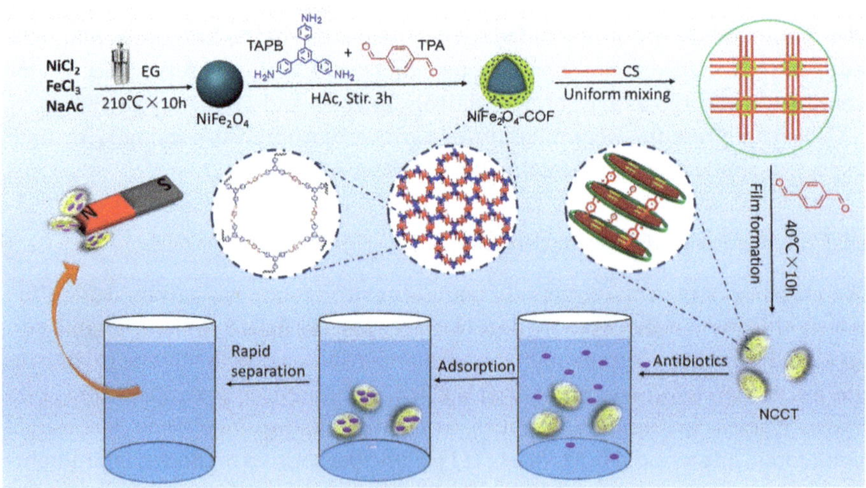

Figure 18.20 Schematic representation of the preparation process for $NiFe_2O_4$/COF/CS/TPA films (NCCT).

between the hydroxyl and amino groups of chitosan and the phenyl group of COF and terephthalaldehyde greatly enhanced the adsorption of antibiotics onto NCCT [113]. The chitosan has a crucial function, as its chemical cross-linking with terephthalaldehyde limits the reaction of NCCT to acid. The sample achieved a highest adsorption capacity equal to 388.52 mg g^{-1} for tetracycline (TC) at pH 8 and 309.26 mg g^{-1} for cefotaxime (CTX) at medium of pH 4.

In another successful attempt, efficient antibiotic remediation from wastewater was achieved by Mirizadeh et al. through developing two environmentally acceptable biosorbents, derived from magnetite, chitosan, and either Chlorella vulgaris (MCC) or Arthrospira platensis (MCA) [114]. Characterized by SEM, FTIR, and thermogravimetric analysis (TGA), both biosorbents showed excellent adsorption capacities for tetracycline (834.0 mg g^{-1} for MCA, 831.1 mg g^{-1} for MCC), ciprofloxacin (394.9 mg g^{-1} for MCA, 374.2 mg g^{-1} for MCC), and amoxicillin (150.8 mg g^{-1} for MCA, 140.2 mg g^{-1} for MCC). The equilibrium data was better described by the Langmuir isotherm, indicating a monolayer adsorption process. Furthermore, the pseudo-second-order kinetic model reliably characterized the adsorption process, indicating a chemisorption mechanism. The observed phenomena of adsorption in the experimental setting were determined to be spontaneous, suggesting that it occurred without any external interference. Furthermore, it was established that the process was endothermic, indicating that it absorbed thermal energy from the surroundings while undergoing adsorption. The biosorbents demonstrated the potential for multiple reuses, with a negligible decrease in capacity observed. Specifically, the capacity loss was found to be $\leq$7% for TC, 4% for ciprofloxacin (CIP), and 9% for amoxicillin (AMX). In their study, Rusu et al. focused on enhancing the efficiency of the process and assessing the adsorption capacity in a batch system. The researchers implemented a biosorbent formulated from the residual biomass of *Saccharomyces pastorianus*, which was immobilized in calcium alginate. The aim was to assess the performance of this biosorbent in eliminating the antibacterial medication, ethacridine lactate (EL), from water-based solutions (Figure 18.21) [115]. The biosorption of EL was optimized through the implementation of a response surface methodology utilizing a Box–Behnken design to improve removal efficiency by systematically manipulating three key variables: initial EL concentration, contact time, and agitation speed. The experimental results demonstrated that the optimum parameter values for achieving the highest removal (89.49%) and biosorption capacity (26.04 mg g^{-1}) were observed at an EL concentration of 59.73 mg L^{-1}, a contact time of 94.26 minutes, and an agitation speed of 297.57 rpm. Moreover, an investigation of kinetics revealed that the biosorption process was most accurately characterized by both Avrami and pseudo-first-order models. The findings emphasized the crucial significance of agitation speed in maximizing the efficiency of biosorption.

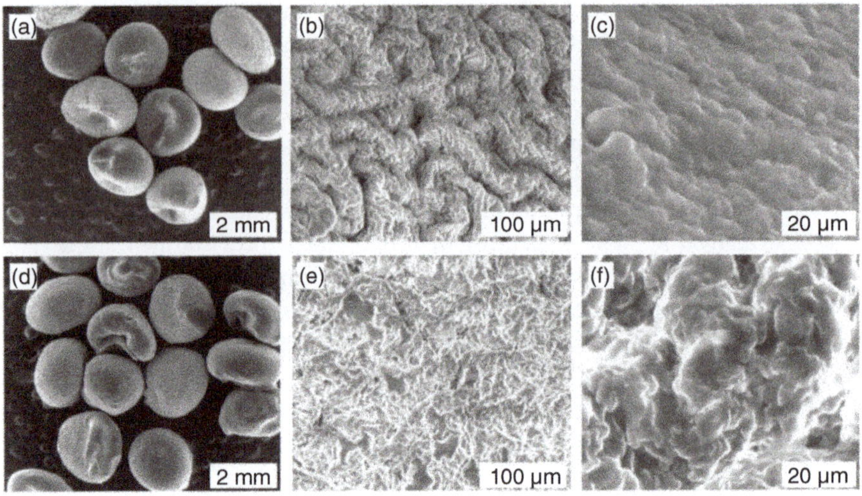

Figure 18.21 SEM images of the *Saccharomyces pastorianus* residual biomass (SPRBA) 5% composite material (*Saccharomyces pastorianus* residual biomass immobilized in calcium alginate with 5% dry mass) (a–c) before and (d–f) after EL biosorption.

An innovative method was suggested by Araujo et al. to effectively eliminate both chloroquine and safranin-O from water. A novel hydrogel named agar-graphene oxide (A-GO) was presented, characterized by its simplicity in synthesis and biodegradability. The present hydrogel is synthesized by a straightforward one-step jellification procedure, therefore offering a practical and effective approach for the ongoing and ecologically sustainable elimination of harmful pollutants from water [116]. The A-GO biocomposite was subjected to thorough characterization, revealing its remarkable property of being exceptionally lightweight. Furthermore, it exhibited a distinctive macrostructure characterized by an airy and porous 3D architecture. During the process of selective batch adsorption, it was noticed a drop in the adsorption capacities by approximately 10 $mg\,g^{-1}$ owing to the presence of multiple contaminants. As a result, the adsorption capacity for Safranin-O was measured to be 41 $mg\,g^{-1}$, while for Chloroquine it was found to be 31 $mg\,g^{-1}$. The results of the fixed-bed column experiments revealed significantly higher adsorption capacities for Chloroquine, exceeding 63 $mg\,g^{-1}$, and for Safranin-O, surpassing 100 $mg\,g^{-1}$. These findings indicate the excellent regenerative capacity of the biocomposite used in the study.

Table 18.4 provides an overview of biocomposites and their efficiency toward pharmaceuticals removal.

Table 18.4 Pharmaceuticals removal using different biocomposites.

Biocomposites	Pharmaceuticals	Qe (mg g^{-1})	Removal (%)	Desorption (%)	Reference
Chitosan-based magnetic composite	Diclofenac sodium	164	>70	Ethanol and water, >60	[117]
	Tetracycline hydrochloride	40.20	>55		
Phragmites-Australis	Ciprofloxacin	17.30	76.66	—	[118]
Magnetic waste tyre activated carbon-chitosan	Methylparaben	85.90	100	Methanol, >90	[119]
	Propylparaben	90	100		
Alginate/carbon-based films	Diclofenac	29.9	—	Acetic acid, 66.7	[120]
PVA-copper alginate gel beads	Tetracycline	231.431	97.8	—	[121]
Chitosan/Zr-MOF (UiO-66) composite foams	Ketoprofen	209.7	—	Methanol, 104.9	[122]
Ball-Milled Silk Fibroin Films	Carbamazepine	281	53	—	[123]
Chitosan-alginate-bentonite composites	Tetracycline	—	97.7	—	[124]
Chitosan-carbon black waste composite beads	Amoxicillin	12	>70	—	[125]
	Tetracycline	39	>60	—	

18.4 Conclusion and Future Prospects

The inherent resistance of synthetic plastic trash to degradation within the ecological system has prompted a surge in the investigation of more environmentally sustainable materials. Furthermore, there has been development of ecologically sustainable solutions, such as biocomposites. The recyclability, lightweight profile, and cost-effectiveness of biocomposites make them a compelling field of study for researchers [126]. Biocomposite components, including biobased polymers and fillers, are derived from renewable natural sources and represent a potential substitute for nonrenewable plastics derived from oil [127]. Although biocomposites have been commercially accessible for a considerable period, their adoption has been restrained, and the scientific literature continues to emphasize their promise rather than their practical uses. Notwithstanding the superior environmental benefits in comparison to synthetic alternatives, there are concerns regarding the compatibility of natural fibers with conventional applications. These concerns primarily stem from their hydrophilic nature, low thermal resistance, flammability, and variability across different types of fibers.

Effective implementation of biocomposites requires focused development and research efforts to tackle key challenges like the creation of novel modified polymers, blending, reinforcing, additives, cross-linking, and other related aspects. Considering that biocomposites are intended to serve as a replacement for materials that are nonrecyclable, highly polluting, and outdated, it is of the utmost importance that they preserve their environmental protection properties. Although biocomposites show significant potential for using in wastewater treatment, it is important to acknowledge their specific constraints. An inherent obstacle lies in their suitability for extensive industrial implementations. The pursuit of scalability in synthesis techniques as well as their incorporation into current wastewater management systems is a challenging endeavor. The use of biocomposites in wastewater decontamination poses both considerable obstacles and attractive opportunities. This section goes into the complexity of their use, as well as future prospects to improve their effectiveness in treating wastewater.

18.4.1 Challenges of Biocomposites in Wastewater Remediation

a-Reusability and Regeneration: A notable obstacle is in guaranteeing the durability and recyclability of biocomposites. Although several scientific investigations have demonstrated that biopolymeric composites can retain their efficacy during several cycles, other studies suggest that their performance may deteriorate over time, especially when exposed to different environmental conditions [128]. Ensuring constant performance is essential for practical applications.

18.4.2 Cost-effectiveness

Production of biocomposites typically requires costly raw ingredients and intricate synthesis procedures. As an example, although biocomposites demonstrate excellent adsorption properties, the expenses associated with their production can be too exorbitant for widespread use. The development of economically efficient manufacturing techniques and the acquisition of inexpensive raw materials are crucial in order to establish biocomposites as a feasible substitute for conventional wastewater treatment technologies [129].

18.4.3 Stability and Durability

Maintaining the stability and lifetime of biocomposites under wastewater treatment operations presents one of the main difficulties. Particularly in demanding chemical conditions, many biocomposites, especially those based on natural polymers like chitosan or polysaccharides, can break down over time. Particularly with regard to heavy metals and organic contaminants, this degradation can reduce adsorption capacity and general efficacy in removing pollutants [130]. It is imperative to establish regulations for the life cycle of biocomposites, despite their inherent ability to completely decompose under natural conditions and their environmentally friendly nature. When materials possess a limited life cycle, they are unable to be utilized in their natural environment, whereas materials with an extended life cycle exhibit a lack of ability to undergo decomposition. Therefore, it is imperative to ensure that the durability of green biocomposites is appropriately tailored to align with the life cycle of the particular situations in which they are employed.

18.4.4 Pollutant Specificity and Selectivity

Biocomposites frequently demonstrate diverse efficacy in selectively eliminating various categories of contaminants. The efficacy of a biocomposite can be influenced by variables like pollutants' composition (heavy metals, dyes, or pharmaceuticals) and the unique properties of the biocomposite material. Optimizing biocomposites to selectively target particular contaminants while preserving their overall efficacy continues to be a considerable obstacle [129].

18.4.5 Conformity with Environmental and Regulatory Requirements

Biocomposites' environmental effects have to be carefully evaluated to guarantee they do not add any more contaminants into the wastewater system. The rules controlling the use of biocomposites in wastewater treatment are still developing, so following such guidelines could provide difficulties for field practitioners [131].

18.4.6 Perspectives on the Future of Biocomposites in Wastewater Remediation

18.4.6.1 Environmental Sustainability and Circular Economy

Biocomposites, especially those produced from waste materials, are highly compatible with environmental sustainability and the circular economy. Taking advantage of agricultural by-products or industrial waste as raw materials for biocomposites manufacturing can effectively minimize waste generation and mitigate the environmental impact of wastewater treatment procedures. In addition to lowering pollutants, this method also enhances resource efficiency [132].

18.4.6.2 Combination with Current Technologies

The integration of biocomposites with traditional wastewater treatment methods, including activated sludge processes or membrane filtration systems, may significantly enhance the overall efficiency of treatment. This hybrid methodology can use the advantages of both biocomposite materials and conventional techniques, resulting in enhanced rates of pollution removal and decreased operational expenses.

18.4.6.3 Research and Development

Further investigation is imperative in order to effectively tackle the existing constraints of biocomposites in the field of wastewater remediation. Examining the intricate dynamics between biocomposites and diverse pollutants, alongside refining their synthesis and application methodologies, will be of paramount importance in propelling the progress of this domain. Collaboration among academia and industries is crucial in promoting the advancement of biocomposites in wastewater treatment.

Although the challenges related to biocomposites in wastewater remediation are considerable, the prospects for their future application appear to be promising. Through the implementation of innovative research methodologies, the utilization of sustainable practices, and the seamless integration with preexisting technologies, biocomposites have the potential to assume a pivotal role in augmenting the efficiency of wastewater treatment processes, thereby rendering significant support regarding the achievement of environmental sustainability goals.

References

1 Azhoni, A., Jude, S., and Holman, I. (2018). Adapting to climate change by water management organisations: enablers and barriers. *Journal of Hydrology* 559: 736–748. https://doi.org/10.1016/j.jhydrol.2018.02.047.

2 Ricart, S. and Rico, A.M. (2019). Assessing technical and social driving factors of water reuse in agriculture: a review on risks, regulation and the yuck factor. *Agricultural Water Management* 217 (March): 426–439. https://doi.org/10.1016/j.agwat.2019.03.017.

3 Aruna, N.B., Sharma, A.K., and Kumar, S. (2021). A review on modified sugarcane bagasse biosorbent for removal of dyes. *Chemosphere* 268: 129309. https://doi.org/10.1016/j.chemosphere.2020.129309.

4 Wei, S., Chen, T., Hou, H., and Xu, Y. (2023). Recent advances in electrochemical sterilization. *Journal of Electroanalytical Chemistry* 937: 117419. https://doi.org/10.1016/j.jelechem.2023.117419.

5 Sun, S., Liu, H., Zhang, J. et al. (2023). Application of a novel coagulant in reservoir water treatment in Qingdao. *Desalination and Water Treatment* 284: 49–60. https://doi.org/10.5004/dwt.2023.29209.

6 Almuktar, S.A.A.N., Abed, S.N., and Scholz, M. (2018). Wetlands for wastewater treatment and subsequent recycling of treated effluent: a review. *Environmental Science and Pollution Research* 25 (24): 23595–23623. https://doi.org/10.1007/s11356-018-2629-3.

7 Yadav, A., Bagotia, N., Sharma, A.K., and Kumar, S. (2021). Advances in decontamination of wastewater using biomass-basedcomposites: a critical review. *Science of the Total Environment* 784: 147108. https://doi.org/10.1016/j.scitotenv.2021.147108.

8 Pan, X., Gu, Z., Chen, W., and Li, Q. (2021). Preparation of biochar and biochar composites and their application in a Fenton-like process for wastewater decontamination: a review. *Science of the Total Environment* 754: 142104. https://doi.org/10.1016/j.scitotenv.2020.142104.

9 Bahsaine, K., Benzeid, H., el Allaoui, B. et al. (2024). Porous polyvinyl fluoride coated cellulose beads for efficient removal of Cd(II) from phosphoric acid. *International Journal of Biological Macromolecules* 254: 127867. https://doi.org/10.1016/j.ijbiomac.2023.127867.

10 Shchipunov, Y. (2012). Bionanocomposites: green sustainable materials for the near future. *Pure and Applied Chemistry* 84 (12): 2579–2607. https://doi.org/10.1351/PAC-CON-12-05-04.

11 Darder, M., Aranda, P., Ferrer, M.L. et al. (2011). Progress in bionanocomposite and bioinspired foams. *Advanced Materials* 23 (44): 5262–5267. https://doi.org/10.1002/adma.201101617.

12 Brizga, J., Hubacek, K., and Feng, K. (2020). The unintended side effects of bioplastics: carbon, land, and water footprints. *One Earth* 3 (1): 45–53. https://doi.org/10.1016/j.oneear.2020.06.016.

13 Reichert, C.L., Bugnicourt, E., Coltelli, M.B. et al. (2020). Bio-based packaging: materials, modifications, industrial applications and sustainability. *Polymers (Basel)* 12 (7): https://doi.org/10.3390/polym12071558.

14 Ferrari, F., Striani, R., Fico, D. et al. (2022). An overview on wood waste valorization as biopolymers and biocomposites: definition, classification, production, properties and applications. *Polymers (Basel)* 14 (24): 5519. https://doi.org/10.3390/polym14245519.

15 AI-Jawhari, I.F.H. (2021). Nanocellulose for Sustainable Future Applications. In: Kharissova, O.V., Torres-Martínez, L.M., Kharisov, B.I. (eds) *Handbook of Nanomaterials and Nanocomposites for Energy and Environmental Applications*. Springer, Cham. https://doi.org/10.1007/978-3-030-36268-3_16.

16 Peng, B., Yao, Z., Wang, X. et al. (2020). Cellulose-based materials in wastewater treatment of petroleum industry. *Green Energy & Environment* 5 (1): 37–49. https://doi.org/10.1016/j.gee.2019.09.003.

17 Romero-Montero, A., Valencia-Bermúdez, J.L., Rosas-Meléndez, S.A. et al. (2023). Biopolymeric fibrous aerogels: the sustainable alternative for water remediation. *Polymers (Basel)* 15 (2): https://doi.org/10.3390/polym15020262.

18 Balart, R., Garcia-Garcia, D., Fombuena, V. et al. (2021). Biopolymers from natural resources. *Polymers (Basel)* 13 (15): 1–9. https://doi.org/10.3390/polym13152532.

19 Shaghaleh, H., Xu, X., and Wang, S. (2018). Current progress in production of biopolymeric materials based on cellulose, cellulose nanofibers, and cellulose derivatives. *RSC Advances* 8 (2): 825–842. https://doi.org/10.1039/c7ra11157f.

20 Wang, D. (2019). A critical review of cellulose-based nanomaterials for water purification in industrial processes. *Cellulose* 26 (2): 687–701. https://doi.org/10.1007/s10570-018-2143-2.

21 Takeshita, S., Zhao, S., Malfait, W.J., and Koebel, M.M. (2021). Chemistry of chitosan aerogels: three-dimensional pore control for tailored applications. *Angewandte Chemie, International Edition* 60 (18): 9828–9851. https://doi.org/10.1002/anie.202003053.

22 Fan, S., Chen, J., Fan, C. et al. (2021). Fabrication of a CO_2-responsive chitosan aerogel as an effective adsorbent for the adsorption and desorption of heavy metal ions. *Journal of Hazardous Materials* 416: 126225. https://doi.org/10.1016/j.jhazmat.2021.126225.

23 Kolya, H. and Kang, C.W. (2023). Next-generation water treatment: exploring the potential of biopolymer-based nanocomposites in adsorption and membrane filtration. *Polymers (Basel)* 15 (16): https://doi.org/10.3390/polym15163421.

24 Rehm, B.H.A. (2009). *Alginates:Biology and Applications*. Berlin, Heidelberg: Springer Berlin Heidelberg.

25 Kaidi, S. et al. (2022). Isolation and structural characterization of alginates from the kelp species *Laminaria ochroleuca* and *Saccorhiza polyschides* from the Atlantic Coast of Morocco. *Colloids Interfaces* 6 (4): 1–12. https://doi.org/10.3390/colloids6040051.

26 El Atouani, S., Bentiss, F., Reani, A. et al. (2016). The invasive brown seaweed Sargassum muticum as new resource for alginate in Morocco: spectroscopic and rheological characterization. *Phycological Research* 64 (3): 185–193. https://doi.org/10.1111/pre.12135.

27 Thakur, S., Pandey, S., and Arotiba, O.A. (2016). Development of a sodium alginate-based organic/inorganic superabsorbent composite hydrogel for adsorption of methylene blue. *Carbohydrate Polymers* 153: 34–46. https://doi.org/10.1016/j.carbpol.2016.06.104.

28 Oladoja, N.A., Unuabonah, E.I., Amuda, O.S., and Kolawole, O.M. (2017). Mechanistic insight into the coagulation efficiency of polysaccharide-based coagulants. In: *Polysaccharides as a Green and Sustainable Resources for Water and Wastewater Treatment* (ed. N.A. Oladoja, E.I. Unuabonah, O.S. Amuda, and O.M. Kolawole), 13–35. Springer International Publishing https://doi.org/10.1007/978-3-319-56599-6_2.

29 Apriyanto, A., Compart, J., and Fettke, J. (2022). A review of starch, a unique biopolymer—structure, metabolism and in planta modifications. *Plant Science* 318 (December 2021): 111223. https://doi.org/10.1016/j.plantsci.2022.111223.

30 Gupta, A.D., Rawat, K.P., Bhadauria, V., and Singh, H. (2021). Recent trends in the application of modified starch in the adsorption of heavy metals from water: a review. *Carbohydrate Polymers* 269: 117763. https://doi.org/10.1016/j.carbpol.2021.117763.

31 Xiao, Z., Wang, L., Lv, C. et al. (2020). Preparation and characterization of pH-responsive Pickering emulsion stabilized by grafted carboxymethyl starch nanoparticles. *International Journal of Biological Macromolecules* 143: 401–412. https://doi.org/10.1016/j.ijbiomac.2019.10.261.

32 Kumar, A., Rao, K.M., and Han, S.S. (2017). Application of xanthan gum as polysaccharide in tissue engineering: a review. *Carbohydrate Polymers* 180 (August): 128–144. https://doi.org/10.1016/j.carbpol.2017.10.009.

33 Qiu, H., Yan, J., Lan, G. et al. (2016). Removal of Cu2+ from wastewater by modified xanthan gum (XG) with ethylenediamine (EDA). *RSC Advances* 6 (86): 83226–83233. https://doi.org/10.1039/c6ra11423g.

34 Hassanzadeh-Afruzi, F., Ranjbar, G., Salehi, M.M. et al. (2023). Thiacalix[4]arene-functionalized magnetic xanthan gum (TC4As-XG@Fe_3O_4) as a hydrogel adsorbent for removal of dye and pesticide from water medium. *Separation and Purification Technology* 306: 122700. https://doi.org/10.1016/j.seppur.2022.122700.

35 Elgarahy, A.M., Eloffy, M.G., Guibal, E. et al. (2023). Use of biopolymers in wastewater treatment: a brief review of current trends and prospects. *Chinese Journal of Chemical Engineering* 64: 292–320. https://doi.org/10.1016/j.cjche.2023.05.018.

36 Ullah, A. and Wu, J. (2013). Feather fiber-based thermoplastics: effects of different plasticizers on material properties. *Macromolecular Materials and Engineering* 298 (2): 153–162. https://doi.org/10.1002/mame.201200010.

37 Khademian, E., Salehi, E., Sanaeepur, H. et al. (2020). A systematic review on carbohydrate biopolymers for adsorptive remediation of copper ions from aqueous environments-part A: classification and modification strategies. *Science of the Total Environment* 738: 139829. https://doi.org/10.1016/j.scitotenv.2020.139829.

38 Balla, E., Daniilidis, V., Karlioti, G. et al. (2021). Poly(lactic acid): a versatile biobased polymer for the future with multifunctional properties.applications. *Polymers (Basel)* 13 (11): 1822.

39 Jin, F.L., Hu, R.R., and Park, S.J. (2018). Improvement of thermal behaviors of biodegradable poly(lactic acid) polymer: a review. *Composites. Part B, Engineering* 164 (August): 287–296. https://doi.org/10.1016/j.compositesb.2018.10.078.

40 Shen, P., Tu, K., Yang, C.Y. et al. (2014). Preparation of anti-fouling poly(lactic acid)(PLA) hollow fiber membranes via non-solvent induced phase separation. *Advances in Materials Research* 884-885: 112–116. https://doi.org/10.4028/www.scientific.net/AMR.884-885.112.

41 Eerhart, A.J.J.E., Faaij, A.P.C., and Patel, M.K. (2012). Replacing fossil based PET with biobased PEF; process analysis, energy and GHG balance. *Energy & Environmental Science* 5 (4): 6407–6422. https://doi.org/10.1039/c2ee02480b.

42 Louw, J., Farzad, S., and Görgens, J.F. (2023). Polyethylene furanoate: technoeconomic analysis of biobased production. *Biofuels, Bioproducts and Biorefining* 17 (1): 135–152. https://doi.org/10.1002/bbb.2430.

43 Rodriguez-Contreras, A. (2019). Recent advances in the use of polyhydroyalkanoates in biomedicine. *Bioengineering* 6 (3): 1–14. https://doi.org/10.3390/bioengineering6030082.

44 Chen, G.Q. (2009). A microbial polyhydroxyalkanoates (PHA) based bio- and materials industry. *Chemical Society Reviews* 38 (8): 2434–2446. https://doi.org/10.1039/b812677c.

45 Gontard, N., Sonesson, U., Birkved, M. et al. (2018). A research challenge vision regarding management of agricultural waste in a circular bio-based economy. *Critical Reviews in Environmental Science and Technology* 48 (6): 614–654. https://doi.org/10.1080/10643389.2018.1471957.

46 Ghanbarzadeh B, Almasi H. Biodegradable polymers. In: Chamy R, editor. *Biodegradation - Life of Science. IntechOpen*; 2013. p. 380.

47 Raza, S., Zhang, J., Ali, I. et al. (2020). Recent trends in the development of biomass-based polymers from renewable resources and their environmental applications. *Journal of the Taiwan Institute of Chemical Engineers* 115: 293–303. https://doi.org/10.1016/j.jtice.2020.10.013.

48 Shah, T.V. and Vasava, D.V. (2019). A glimpse of biodegradable polymers and their biomedical applications. *E-Polymers* 19: 385–410.

49 Karande, R.D., Abitha, V.K., Rane, A.V., and Mishra, R.K. (2016). Preparation of polylactide fromsynthesized lactic acid and effect ofreaction parameters on conversion. *Journal of Materials Science and Engineering with Advanced Technology* 12 (1–2): 1–37. https://doi.org/10.18642/jmseat.

50 Hernández-García, E., Vargas, M., González-Martínez, C., and Chiralt, A. (2021). Biodegradable antimicrobial films for food packaging: effect of antimicrobials on degradation. *Foods* 10 (6): https://doi.org/10.3390/foods10061256.

51 Tan, B.H., Muiruri, J.K., Li, Z., and He, C. (2016). Recent progress in using stereocomplexation for enhancement of thermal and mechanical property of polylactide. *ACS Sustainable Chemistry & Engineering* 4 (10): 5370–5391. https://doi.org/10.1021/acssuschemeng.6b01713.

52 Rosen, T., Goldberg, I., Venditto, V., and Kol, M. (2016). Tailor-made stereoblock copolymers of poly(lactic acid) by a truly living polymerization catalyst. *Journal of the American Chemical Society* 138 (37): 12041–12044. https://doi.org/10.1021/jacs.6b07287.

53 Nakajima, H., Dijkstra, P., and Loos, K. (2017). The recent developments in biobased polymers toward general and engineering applications: polymers that are upgraded from biodegradable polymers, analogous to petroleum-derived polymers, and newly developed. *Polymers (Basel)* 9 (10): 1–26. https://doi.org/10.3390/polym9100523.

54 Plackett, D., Andersen, T.L., Pedersen, W.B., and Nielsen, L. (2003). Biodegradable composites based on L-polylactide and jute fibres. *Composites Science and Technology* 63 (9): 1287–1296. https://doi.org/10.1016/S0266-3538(03)00100-3.

55 Demjén, Z., Pukánszky, B., and Nagy, J. (1998). Evaluation of interfacial interaction in polypropylene/surface treated $CaCO_3$ composites. *Composites. Part A, Applied Science and Manufacturing* 29 (3): 323–329. https://doi.org/10.1016/S1359-835X(97)00032-8.

56 Shivamurthy, B. and Prabhuswamyc, M.S. (2009). Influence of SiO fillers on sliding wear resistance and mechanical properties of compression moulded glass epoxy composites. *Journal of Minerals and Materials Characterization and Engineering* 08 (07): 513–530. https://doi.org/10.4236/jmmce.2009.87045.

57 Fellahi, S., Chikhi, N., and Bakar, M. (2001). Modification of epoxy resin with kaolin as a toughening agent. *Journal of Applied Polymer Science* 82 (4): 790–798. https://doi.org/10.1002/app.1908.

58 Atta, A.M., El-Saeed, A.M., Al-Lohedan, H.A., and Wahby, M. (2017). Effect of montmorillonite nanogel composite fillers on the protection performance of epoxy coatings on steel pipelines. *Molecules* 22 (6): https://doi.org/10.3390/molecules22060905.

59 Kumar, T.S.M., Rajini, N., Huafeng, T. et al. (2019). Improved mechanical and thermal properties of spent coffee bean particulate reinforced poly(propylene carbonate) composites. *Particulate Science and Technology* 37 (5): 639–646. https://doi.org/10.1080/02726351.2017.1420116.

60 Majeed, K., Jawaid, M., Hassan, A. et al. (2013). Potential materials for food packaging from nanoclay/natural fibres filled hybrid composites. *Materials and Design* 46: 391–410. https://doi.org/10.1016/j.matdes.2012.10.044.

61 Ku, H., Wang, H., Pattarachaiyakoop, N., and Trada, M. (2011). A review on the tensile properties of natural fiber reinforced polymer composites. *Composites. Part B, Engineering* 42 (4): 856–873. https://doi.org/10.1016/j.compositesb.2011.01.010.

62 Matuana, N.M.S. (2015). 20 - The use of wood fibers as reinforcements in composites. In: *Biofiber Reinforcements in Composite Materials* (ed. O. Faruk and M. Sain), 648–688. Woodhead Publishing ISBN 9781782421221, https://doi.org/10.1533/9781782421276.5.648.

63 Sanjay, M.R., Madhu, P., Jawaid, M. et al. (2018). Characterization and properties of natural fiber polymer composites: a comprehensive review. *Journal of Cleaner Production* 172: 566–581. https://doi.org/10.1016/j.jclepro.2017.10.101.

64 Mittal, G., Dhand, V., Rhee, K.Y. et al. (2015). A review on carbon nanotubes and graphene as fillers in reinforced polymer nanocomposites. *Journal of Industrial and Engineering Chemistry* 21: 11–25. https://doi.org/10.1016/j.jiec.2014.03.022.

65 Pandey, N., Shukla, S.K., and Singh, N.B. (2017). Water purification by polymer nanocomposites: an overview. *Nanocomposites* 3 (2): 47–66. https://doi.org/10.1080/20550324.2017.1329983.

66 Karami, P., Khorshidi, B., McGregor, M. et al. (2020, https://doi.org/10.1016/j.jclepro.2019.119447). Thermally stable thin film composite polymeric membranes for water treatment: a review. *Journal of Cleaner Production* 250: 119447.

67 Tan, X.M. and Rodrigue, D. (2019). A review on porous polymeric membrane preparation. Part I: production techniques with polysulfone and poly (vinylidene fluoride). *Polymers (Basel)* 11 (8): 1160.

68 Nunes, B.N., Paula, L.F., Costa, Í.A. et al. (2017). Layer-by-layer assembled photocatalysts for environmental remediation and solar energy conversion. *Journal of Photochemistry and Photobiology C Photochemistry Reviews* 32: 1–20. https://doi.org/10.1016/j.jphotochemrev.2017.05.002.

69 Raota, C.S., Cerbaro, A.F., Salvador, M. et al. (2019). Green synthesis of silver nanoparticles using an extract of Ives cultivar (Vitis labrusca) pomace: characterization and application in wastewater disinfection. *Journal of Environmental Chemical Engineering* 7 (5): 103383. https://doi.org/10.1016/j.jece.2019.103383.

70 Vebber, M.C., da Silva Crespo, J., and Giovanela, M. (2019). Self-assembled thin films of PAA/PAH/TiO_2 for the photooxidation of ibuprofen. Part I: optimization of photoactivity using design of experiments and surface response methodology. *Chemical Engineering Journal* 360: 1447–1458. https://doi.org/10.1016/j.cej.2018.10.189.

71 Subash, A., Naebe, M., Wang, X., and Kandasubramanian, B. (2023). Biopolymer—a sustainable and efficacious material system for effluent removal. *Journal of Hazardous Materials* 443: 130168. https://doi.org/10.1016/j.jhazmat.2022.130168.

72 Khodakarami, M. and Bagheri, M. (2021). Recent advances in synthesis and application of polymer nanocomposites for water and wastewater treatment. *Journal of Cleaner Production* 296: 126404. https://doi.org/10.1016/j.jclepro.2021.126404.

73 Li, R., Wu, Y., Lou, X. et al. (2023). Porous biochar materials for sustainable water treatment: synthesis, modification, and application. *Water (Switzerland)* 15 (3): https://doi.org/10.3390/w15030395.

74 Akpor, O., Otohinoyi, D.A., Olaolu, T.D., and Aderiye, B.I. (2014). Pollutants in wastewater effluents: impacts and remediation processes. *International Journal of Environmental Research and Earth Science* 3 (3): 050–059.

75 Mohd-Salleh, S.N.A., Mohd-Zin, N.S., and Othman, N. (2019). A review of wastewater treatment using natural material and its potential as aid and composite coagulant. *Sains Malaysiana* 48 (1): 155–164. https://doi.org/10.17576/jsm-2019-4801-18.

76 Iqbal, H.M.N., Kyazze, G., and Keshavarz, T. (2013). Advances in the valorization of lignocellulosic materials by biotechnology: an overview. *BioResources* 8 (2): 3157–3176. https://doi.org/10.15376/biores.8.2.3157-3176.

77 Said, B., Souad M', R., and Ahmed, E.H. (2020). A review on classifications, recent synthesis and applications of textile dyes. *Inorganic Chemistry Communications* 3 (March): 107891.

78 López-Rodríguez, D., Micó-Vicent, B., Jordán-Núñez, J. et al. (2023). Printing with hydrotalcite hybrid nanoclays and dispersed, reactive, direct dyes. *Dyes and Pigments* 210 (July 2022): https://doi.org/10.1016/j.dyepig.2022.111022.

79 Chkirida, S., Zari, N., Bouhfid, R., and Qaiss, A. (2021). Insight into the bionanocomposite applications on wastewater decontamination : review. *Journal of Water Process Engineering* 43: 102198. https://doi.org/10.1016/j.jwpe.2021.102198.

80 Cui, R., Jin, D., Jiao, G. et al. (2023). Cuprous oxide/copper oxide interpenetrated into ordered mesoporous cellulose-based carbon aerogels for efficient photocatalytic degradation of methylene blue. *Frontiers of Chemical Science and Engineering* 17 (7): 918–929. https://doi.org/10.1007/s11705-023-2305-0.

81 Mokhtar, A., Abdelkrim, S., Djelad, A. et al. (2020). Adsorption behavior of cationic and anionic dyes on magadiite-chitosan composite beads. *Carbohydrate Polymers* 229 (September 2019): 115399. https://doi.org/10.1016/j.carbpol.2019.115399.

82 Kausar, A., Shahzad, R., Iqbal, J. et al. (2020). Development of new organic-inorganic, hybrid bionanocomposite from cellulose and clay for enhanced removal of Drimarine Yellow HF-3GL dye. *International Journal of Biological Macromolecules* 149: 1059–1071. https://doi.org/10.1016/j.ijbiomac.2020.02.012.

83 Bellaj, M., Naboulsi, A., Aziz, K. et al. (2024). Bio-based composite from chitosan waste and clay for effective removal of Congo red dye from contaminated water: experimental studies and theoretical insights. *Environmental Research* 255: 119089. https://doi.org/10.1016/j.envres.2024.119089.

84 Chkirida, S., Zari, N., Achour, R. et al. (2021). Highly synergic adsorption/photocatalytic efficiency of Alginate/Bentonite impregnated TiO_2 beads for wastewater treatment. *Journal of Photochemistry and Photobiology A: Chemistry* 412: 113215. https://doi.org/10.1016/j.jphotochem.2021.113215.

85 Thamer, B.M., Al-aizari, F.A., and Abdo, H.S. (2023). Activated carbon-incorporated tragacanth gum hydrogel biocomposite: a promising adsorbent for crystal violet dye removal from aqueous solutions. *Gels* 9 (12): https://doi.org/10.3390/gels9120959.

86 Benhouria, A., Islam, M.A., Zaghouane-Boudiaf, H. et al. (2015). Calcium alginate-bentonite-activated carbon composite beads as highly effective adsorbent for methylene blue. *Chemical Engineering Journal* 270: 621–630. https://doi.org/10.1016/j.cej.2015.02.030.

87 İnal, M. and Erduran, N. (2015). Removal of various anionic dyes using sodium alginate/poly(N-vinyl-2-pyrrolidone) blend hydrogel beads. *Polymer Bulletin* 72 (7): 1735–1752. https://doi.org/10.1007/s00289-015-1367-7.

88 Wang, Z., Song, L., Wang, Y. et al. (2021). Construction of a hybrid graphene oxide/nanofibrillated cellulose aerogel used for the efficient removal of methylene blue and tetracycline. *Journal of Physics and Chemistry of Solids* 150: 109839. https://doi.org/10.1016/j.jpcs.2020.109839.

89 Uyar, G., Kaygusuz, H., and Erim, F.B. (2016). Methylene blue removal by alginate–clay quasi-cryogel beads. *Reactive and Functional Polymers* 106: 1–7. https://doi.org/10.1016/j.reactfunctpolym.2016.07.001.

90 Xiang, C., Wang, C., Guo, R. et al. (2019). Synthesis of carboxymethyl cellulose-reduced graphene oxide aerogel for efficient removal of organic liquids and dyes. *Journal of Materials Science* 54 (2): 1872–1883. https://doi.org/10.1007/s10853-018-2900-5.

91 Mittal, H., Babu, R., and Alhassan, S.M. (2020). Utilization of gum xanthan based superporous hydrogels for the effective removal of methyl violet from aqueous solution. *International Journal of Biological Macromolecules* 143: 413–423. https://doi.org/10.1016/j.ijbiomac.2019.11.008.

92 Morais da Silva, P.M., Camparotto, N.G., Figueiredo Neves, T. et al. (2020, https://doi.org/10.1016/j.scp.2020.100348). Effective removal of basic dye onto sustainable chitosan beads: batch and fixed-bed column adsorption, beads stability and mechanism. *Sustainable Chemistry and Pharmacy* 18 (November): 100348.

93 Abu Elella, M.H., ElHafeez, E.A., Goda, E.S. et al. (2019). Smart bactericidal filter containing biodegradable polymers for crystal violet dye adsorption. *Cellulose* 26 (17): 9179–9206. https://doi.org/10.1007/s10570-019-02698-1.

94 Bahsaine, K., Mekhzoum, M.E.M., Benzeid, H. et al. (2023, https://doi.org/10.1007/s11356-023-27241-5). Chromium (III) adsorption from the phosphoric acid medium using DETA grafted Merrifield resin. *Environmental Science and Pollution Research* 30: 67720–67729.

95 Ariffin, N., Abdullah, M.M., Zainol, M.R. et al. (2017). Review on adsorption of heavy metal in wastewater by using geopolymer. *MATEC Web of Conferences* 97 (01023): 1–8.

96 Xu, K., He, T., Li, L. et al. (2024). DOTA functionalized adsorbent DOTA@Sludge@Chitosan derived from recycled shrimp shells and sludge and its application for lead and chromium removal from water. *International Journal of Biological Macromolecules* 255 (November 2023): 128263. https://doi.org/10.1016/j.ijbiomac.2023.128263.

97 Isawi, H. (2020). Using Zeolite/Polyvinyl alcohol/sodium alginate nanocomposite beads for removal of some heavy metals from wastewater. *Arabian Journal of Chemistry* 13 (6): 5691–5716. https://doi.org/10.1016/j.arabjc.2020.04.009.

98 Wang, S., Liu, Y., Yang, A. et al. (2022). Xanthate-modified magnetic Fe_3O_4@SiO_2-based polyvinyl alcohol/chitosan composite material for efficient removal of heavy metal ions from water. *Polymers (Basel)* 14 (6): https://doi.org/10.3390/polym14061107.

99 Zhou, F., Feng, X., Yu, J., and Jiang, X. (2018). High performance of 3D porous graphene/lignin/sodium alginate composite for adsorption of Cd(II) and Pb(II). *Environmental Science and Pollution Research* 25 (16): 15651–15661. https://doi.org/10.1007/s11356-018-1733-8.

100 Salih, S.S., Shihab, M.A., Mohammed, H.N. et al. (2024). Chitosan-vermiculite composite adsorbent : preparation, characterization, and competitive adsorption of Cu(II) and Cd(II) ions. *Journal of Water Process Engineering* 59: 105044.

101 He, Y., Gou, S., Zhou, L. et al. (2021). Amidoxime-functionalized polyacrylamide-modified chitosan containing imidazoline groups for effective removal of Cu^{2+} and Ni^{2+}. *Carbohydrate Polymers* 252 (July 2020): 117160. https://doi.org/10.1016/j.carbpol.2020.117160.

102 Saheed, I.O., Yusof, E.N.M., Da Oh, W. et al. (2023). Fabrication of chitosan@activated carbon composites in EmimAc for Cd(II) adsorption from aqueous solution: experimental, optimization and DFT study. *International*

Journal of Biological Macromolecules 242 (P2): 124798. https://doi.org/10.1016/j.ijbiomac.2023.124798.

103 Jia, J., Liu, Y., and Sun, S. (2021, https://doi.org/10.1155/2021/6681486). Preparation and characterization of chitosan/bentonite composites for Cr (VI) removal from aqueous solutions. *Adsorption Science and Technology* 2021.

104 Godiya, C.B., Liang, M., Sayed, S.M. et al. (2019). Novel alginate/polyethyleneimine hydrogel adsorbent for cascaded removal and utilization of Cu^{2+} and Pb^{2+} ions. *Journal of Environmental Management* 232 (August 2018): 829–841. https://doi.org/10.1016/j.jenvman.2018.11.131.

105 Wang, H., Shang, H., Sun, X. et al. (2020). Preparation of thermo-sensitive surface ion-imprinted polymers based on multi-walled carbon nanotube composites for selective adsorption of lead(II) ion. *Colloids and Surfaces A: Physicochemical and Engineering Aspects* 585 (August 2019): 124139. https://doi.org/10.1016/j.colsurfa.2019.124139.

106 Doughmi, O., Ennajih, Z., el-bardai, R. et al. (2024). Creating a new polyaniline@oak acorn biocomposite to remove chromium ions from wastewater effectively. *Journal of Environmental Chemical Engineering* 12 (5): 113878. https://doi.org/10.1016/j.jece.2024.113878.

107 Adeleke, A.O., Royahu, C.O., Afolabi, A.A. et al. (2024). A novel oyster shell biocomposite for the efficient adsorptive removal of cadmium and lead from aqueous solution: synthesis, process optimization, modelling and mechanism studies. *PLoS One* 19 (2 February): 1–29. https://doi.org/10.1371/journal.pone.0294286.

108 Deng, Z., Wu, Z., Wu, Q. et al. (2024). Cellulose nanocrystals intercalated clay biocomposite for rapid Cr(VI) removal. *Environmental Science and Pollution Research* 31 (20): 29719–29729. https://doi.org/10.1007/s11356-024-33066-7.

109 Chkirida, S., El Mernissi, N., Zari, N. et al. (2024). In-situ magnetic alginate coated chitosan core@shell beads with excellent performance in simulated and real wastewater treatment: behavior, mechanisms, and new perspectives. *International Journal of Biological Macromolecules* 260 (P2): 129389. https://doi.org/10.1016/j.ijbiomac.2024.129389.

110 Taoufik, N., Boumya, W., Janani, F.Z. et al. (2020). Removal of emerging pharmaceutical pollutants: a systematic mapping study review. *Journal of Environmental Chemical Engineering* 8 (5): https://doi.org/10.1016/j.jece.2020.104251.

111 Ahmed, M.J. and Hameed, B.H. (2017). Removal of emerging pharmaceutical contaminants by adsorption in a fixed-bed column: a review. *Ecotoxicology and Environmental Safety* 149 (December): 257–266. https://doi.org/10.1016/j.ecoenv.2017.12.012.

112 Khan, H.K., Rehman, M.Y.A., and Malik, R.N. (2020). Fate and toxicity of pharmaceuticals in water environment: an insight on their occurrence in South Asia. *Journal of Environmental Management* 271: 111030. https://doi.org/10.1016/j.jenvman.2020.111030.

113 Li, Z., Liu, Y., Zou, S. et al. (2020). Removal and adsorption mechanism of tetracycline and cefotaxime contaminants in water by $NiFe_2O_4$-COF-chitosan-terephthalaldehyde nanocomposites film. *Chemical Engineering Journal* 382 (July 2019): 123008. https://doi.org/10.1016/j.cej.2019.123008.

114 Mirizadeh, S., Solisio, C., Converti, A., and Casazza, A.A. (2024). Efficient removal of tetracycline, ciprofloxacin, and amoxicillin by novel magnetic chitosan/microalgae biocomposites. *Separation and Purification Technology* 329 (July 2023): 125115. https://doi.org/10.1016/j.seppur.2023.125115.

115 Rusu, L., Grigoraş, C.G., Simion, A.I. et al. (2022). Investigation into biosorption of pharmaceuticals from aqueous solutions by biocomposite material based on microbial biomass and natural polymer: process variables optimization and kinetic studies. *Polymers (Basel)* 14 (16): https://doi.org/10.3390/polym14163388.

116 de Araujo, C.M.B., Wernke, G., Ghislandi, M.G. et al. (2023). Continuous removal of pharmaceutical drug chloroquine and safranin-O dye from water using agar-graphene oxide hydrogel: selective adsorption in batch and fixed-bed experiments. *Environmental Research* 216 (March 2022): https://doi.org/10.1016/j.envres.2022.114425.

117 Zhang, S., Dong, Y., Yang, Z. et al. (2016). Adsorption of pharmaceuticals on chitosan-based magnetic composite particles with core-brush topology. *Chemical Engineering Journal* 304: 325–334. https://doi.org/10.1016/j.cej.2016.06.087.

118 Karoui, S., Ben Arfi, R., Mougin, K. et al. (2020). Synthesis of novel biocomposite powder for simultaneous removal of hazardous ciprofloxacin and methylene blue: central composite design, kinetic and isotherm studies using Brouers-Sotolongo family models. *Journal of Hazardous Materials* 387 (October 2019): 121675. https://doi.org/10.1016/j.jhazmat.2019.121675.

119 Mashile, G.P., Mpupa, A., Nqombolo, A. et al. (2020). Recyclable magnetic waste tyre activated carbon-chitosan composite as an effective adsorbent rapid and simultaneous removal of methylparaben and propylparaben from aqueous solution and wastewater. *Journal of Water Process Engineering* 33 (October 2019): 101011. https://doi.org/10.1016/j.jwpe.2019.101011.

120 Shamsudin, M.S., Azha, S.F., Sellaoui, L. et al. (2022). Performance and interactions of diclofenac adsorption using alginate/carbon-based films: experimental investigation and statistical physics modelling. *Chemical Engineering Journal* 428 (August 2021): 131929. https://doi.org/10.1016/j.cej.2021.131929.

121 Liao, Q., Rong, H., Zhao, M. et al. (2022). Strong adsorption properties and mechanism of action with regard to tetracycline adsorption of double-network polyvinyl alcohol-copper alginate gel beads. *Journal of Hazardous Materials* 422 (July 2021): 126863. https://doi.org/10.1016/j.jhazmat.2021.126863.

122 Chen, J., Ouyang, J., Chen, W. et al. (2022). Fabrication and adsorption mechanism of chitosan/Zr-MOF (UiO-66) composite foams for efficient removal of ketoprofen from aqueous solution. *Chemical Engineering Journal* 431 (P2): 134045. https://doi.org/10.1016/j.cej.2021.134045.

123 Kgomo, H., Dube, S., and Nindi, M.M. (2022). Evaluating the performance of ball-milled silk fibroin films for simultaneous adsorption of eight pharmaceuticals from water. *International Journal of Environmental Research and Public Health* 19 (22): https://doi.org/10.3390/ijerph192214922.

124 Nunes Filho, F.G., Silva Filho, E.C., Osajima, J.A. et al. (2023, https://doi.org/10.1016/j.clay.2023.106952). Adsorption of tetracycline using chitosan–alginate–bentonite composites. *Applied Clay Science* 239 (January): 106952.

125 Yaqubi, O., Tai, M.H., Mitra, D. et al. (2021). Adsorptive removal of tetracycline and amoxicillin from aqueous solution by leached carbon black waste and chitosan-carbon composite beads. *Journal of Environmental Chemical Engineering* 9 (1): 104988. https://doi.org/10.1016/j.jece.2020.104988.

126 Kumar, A., Jyske, T., and Möttönen, V. (2020). Properties of injection molded biocomposites reinforced withwood particles of short-rotation aspen and willow. *Polymers (Basel)* 12 (2): https://doi.org/10.3390/polym12020257.

127 Das, O., Neisiany, R.E., Capezza, A.J. et al. (2020, https://doi.org/10.1016/j.scitotenv.2020.139611). The need for fully bio-based facemasks to counter coronavirus outbreaks: a perspective. *Science of the Total Environment* 736: 139611.

128 Mayet, A.M., Hijji, M., Saleh, E.A. et al. (2023). The role of biocomposites and nanocomposites in eliminating organic contaminants from effluents. *Water (Switzerland)* 15 (17): https://doi.org/10.3390/w15173093.

129 Annu, M., Mittal, S.T., and Shin, D.K. (2024). Biopolymeric nanocomposites for wastewater remediation: an overview on recent progress and challenges. *Polymers (Basel)* 16 (2): https://doi.org/10.3390/polym16020294.

130 Fakhri, V., Jafari, A., Layaei Vahed, F. et al. (2023). Polysaccharides as eco-friendly bio-adsorbents for wastewater remediation: current state and future perspective. *Journal of Water Process Engineering* 54: 103980. https://doi.org/10.1016/j.jwpe.2023.103980.

131 Issaka, E., Fapohunda, F.O., Amu-Darko, J.N.O. et al. (2022). Biochar-based composites for remediation of polluted wastewater and soil environments: challenges and prospects. *Chemosphere* 297: 134163. https://doi.org/10.1016/j.chemosphere.2022.134163.

132 Shanmugam, V., Mensah, R.A., Försth, M. et al. (2020). Circular economy in biocomposite development: state-of-the-art, challenges and emerging trends. *Composites Part C: Open Access* 5 (December): 100138. https://doi.org/10.1016/j.jcomc.2021.100138.

Index

h

i

n

Printed in the USA/Agawam, MA
June 24, 2026

906827.008